Springer Tracts in Civil Engineering

Springer Tracts in Civil Engineering (STCE) publishes the latest developments in Civil Engineering - quickly, informally and in top quality. The series scope includes monographs, professional books, graduate textbooks and edited volumes, as well as outstanding PhD theses. Its goal is to cover all the main branches of civil engineering, both theoretical and applied, including:

- Construction and Structural Mechanics
- Building Materials
- Concrete, Steel and Timber Structures
- Geotechnical Engineering
- Earthquake Engineering
- Coastal Engineering; Ocean and Offshore Engineering
- Hydraulics, Hydrology and Water Resources Engineering
- Environmental Engineering and Sustainability
- Structural Health and Monitoring
- Surveying and Geographical Information Systems
- Heating, Ventilation and Air Conditioning (HVAC)
- Transportation and Traffic
- Risk Analysis
- Safety and Security

Indexed by Scopus

To submit a proposal or request further information, please contact:
Pierpaolo Riva at Pierpaolo.Riva@springer.com (Europe and Americas) Wayne Hu at wayne.hu@springer.com (China)

More information about this series at http://www.springer.com/series/15088

Sašo Medved

Building Physics

Heat, Ventilation, Moisture, Light, Sound, Fire, and Urban Microclimate

Springer

Sašo Medved
Laboratory for Sustainable Technologies in Buildings
University of Ljubljana
Ljubljana, Slovenia

ISSN 2366-259X ISSN 2366-2603 (electronic)
Springer Tracts in Civil Engineering
ISBN 978-3-030-74389-5 ISBN 978-3-030-74390-1 (eBook)
https://doi.org/10.1007/978-3-030-74390-1

This Springer imprint is published by the registered company Springer Nature Switzerland AG
The registered company address is: Gewerbestrasse 11, 6330 Cham, Switzerland

Preface

Building physics is a research and applied scientific discipline that deals with the evaluation of physical phenomena related to heat transfer and energy use in buildings, water and water vapour transfer in building structures, daylighting and electric lighting of buildings, sound transmission in building structures and protection against noise, the occurrence and spread of fires in buildings and the thermal response of cities. Therefore, with the physical processes that affect the quality of the living and working comfort in indoor and outdoor environment, as well as impact of buildings on local and global environment, adequate knowledge in the field of building physics has become one of the basic engineering skills of building designers and urban planners.

In terms of content, the monograph *Building Physics: Heat, Ventilation, Moistening, Light, Sound, Fire, and Urban Microclimate* is related to a memorandum prepared by the Group of Professors of Building Physics at European Universities, initiated and led by Prof. Karl Gertis from the University of Stuttgart. The book presents the theoretical background of the considered physical processes, numerical and computational evaluation methods and numerous computational study examples. Each of the chapter guides the reader from basic physical descriptions to more complex scientific approaches. The book also contains numerous examples of experimental analyses performed by the staff of the Laboratory for Sustainable Technologies in Buildings at the Faculty of Mechanical Engineering, University of Ljubljana. In 2021, we are celebrating the 10th anniversary of the founding of the laboratory, and this book seems like an appropriate "birthday present".

Ljubljana, Slovenia — Prof. Sašo Medved

Acknowledgements

The author is grateful for the support and contribution of collaborators in research and application projects and involved master's and doctoral students. I would like to thank my colleagues Mr. Ciril Arkar, Mr. Boštjan Černe, Mr. Rok Fink, Mr. Tomaž Šuklje, Mr. Boris Vidrih and, especially, Mrs. Suzana Domjan without whose contribution this book would not have been written.

The book also includes the results of the research founded by the Slovenian Research Agency and by several industrial companies. I would like to thank them for their financial support.

I would like to express my gratitude to Prof. Ružica Nikolić from the University of Žilina and Prof. Karel Kabele from Czech Technical University in Prague, who reviewed the book, for their opinions and suggestions.

Finally, I would like to thank my family for unwaveringly encouraging and supporting.

Prof. Sašo Medved

Reviews

The book "Building Physics—Heat, Ventilation, Moisture, Light, Sound, Fire, Urban Microclimate" by Sašo Medved is a scientific monograph, which is a product of long-time research efforts (theoretical and experimental) by the author and his team from the Faculty of Mechanical Engineering, University of Ljubljana. This book is unique in that it manages to combine a scientific approach with examples of practical problems that we encounter in everyday life. The text is divided into seven chapters, titles of which are given in the subtitle of the book. The matter is presented in a logical order, the theoretical explanations are accompanied with numerous "case studies", actually examples and solved problems, which masterfully complement the theory and complete any explanation of the topic in question. All these are enriched with numerous adequate illustrations in the form of diagrams, pictures and photographs, combined with tables and explanations. What must be emphasized is that the book also gives the historical prospective of development of the building physics problems, for each topic is covered with examples of its progress in time. This book is also a modern guide both for scientists who can find in it the theoretical background of physics problems in civil engineering and for construction engineers who can use its numerous examples for solving practical problems. This is valid for the usually considered "classical" physics problems of heating and ventilation, of light and sound spreading and fire safety, as well as for the more recently studied problems of urban development, microclimate of growing cities, climate changes and increasing air pollution. Thus, the book is highly recommended to all types of readers, even to those who are starting to study these problems.

Professor Ružica Nikolić
University of Žilina

Submitted monograph, Building Physics—Heat, Ventilation, Moisture, Light, Sound, Fire, Urban Microclimate by Sašo Medved, summarizes the latest findings from building physics in the field of heat transfer, air flow, moisture transport, light and sound propagation, fire and environmental phenomena in urban areas. I consider the topic to be topical in connection with the intensive pressure to energy efficiency and increase the requirements for the indoor environmental quality of buildings.

The publication is divided into individual chapters, which contain an explanation of the basic theory of individual phenomena, methods of measuring and obtaining information and the application of the theory to solve current problems of construction research and practice. I appreciate the clearly structured elaboration of individual topics as well as practical examples, which clearly declare the use of knowledge summarized in the book for solving engineering problems. At the same time, I find it very useful to have a holistic and balanced view of the author on the individual areas addressed, which will allow the reader to understand the interrelationships and links that must be taken into account when seeking solutions to problems associated with construction and operation of buildings. The publication also has an overlap in the area of the quality of the internal environment of buildings, which must be evaluated not only by objectively measured physical quantities but also from the user's point of view.

The publication is prepared at an excellent language level, with correctly cited sources, carefully processed mathematical expressions with explanations of variables and precisely processed graphic images. The designation of variables used is consistent and complies with international standards. The publication shows the author's didactic ability to explain clearly even complex phenomena and reflects his excellent pedagogical skills. The size and number of pages of the publication are commensurate with the content.

Based on the above, I can state that the presented publication is an original and unique work and will be beneficial for all those interested in this issue. Its use will be not only for university students in the field of construction, architecture and technical equipment of buildings, but also for graduates in practice who get to the problems of building physics in the construction process and the general professional public. I am not aware that a publication of similar scope and content is available on the market.

I unequivocally recommend the monograph to be published and, if it meets all the conditions, to be supported by a grant project.

Professor Karel Kabele
Czech Technical University

Contents

Chapter 1
Heat Transfer in Buildings Structures and Thermal Comfort in Buildings

Abstract Nowadays we spend most of our time in buildings. The building designers are therefore responsible for creating a pleasant, productive and healthy indoor environment. Adequate thermal comfort plays a very important role in achieving this goal. Thermal comfort is also close related to the energy use. In the European Union, buildings are responsible for more than one-third of final energy consumption. As consequence the requirements for the energy efficiency of buildings becoming increasingly stringent, in particular due to global climate change mitigation targets. The thermal properties of building envelope have a decisive role in ensuring these objectives. This chapter describe the mechanisms of heat transfer in building structures in steady and dynamic outdoor and indoor environment conditions and the methods for assessment of thermal properties of opaque and transparent building structures and building envelope. The experimental and numerical assessment methods are explained. It also explains the principles and guidelines of the design and the evaluation methods for the in-situ assessment of indoor thermal comfort.

Learning objectives In this chapter you will learn about:

- thermodynamic laws and heat-transfer mechanisms;
- the specifics of the heat transfer in building structures;
- the numerical methods for determination of the thermal transmittances of opaque and transparent building structures and thermal bridges;
- the thermal response and heat accumulation in building structures for the case of transient heat transfer;
- how heat losses of the building envelope is determine;
- in-situ evaluation methods of evaluation of building envelope thermal properties;
- the indicators and criteria for indoor thermal comfort;

S. Medved, *Building Physics*, Springer Tracts in Civil Engineering,
https://doi.org/10.1007/978-3-030-74390-1_1

1.1 Symbols for Quantities and Units

A, A_{env}	Area, area of external surface of the building thermal envelope (env) (m^2)
A_g, A_f, A_w	Area of glazing (g), of frame (f), of window (w) (m^2)
A_{roof}, A_f	Area of roof building structure (roof), of building structure in contact with the soil (f) (m^2)
a	Acceleration (m^2/s)
a, a_{soil}	Thermal diffusivity, of soil (m^2/s)
α, α_s, α_{IR}	Absorptivity, of solar irradiation (s), of long wave irradiation (IR) (1)
α	Hellmann's exponent (1)
B'	Characteristic dimension of structure in contact with ground (m)
b	Thermal effusivity ($kJ/m^2Ks^{0.5}$), weighting factor of transmission heat losses (1)
β	Volumetric thermal expansion coefficient (1/K)
C	Specific thermal capacity (kJ/m^2K), constant (1)
c_p, $c_{p,soil}$	Specific heat at constant pressure, of the soil (J/kgK)
c_o	Speed of light (m/s)
χ	Point (thermal bridge) thermal transmittance (W/K)
D_i, D_e, D_{ti}	Inner pipe diameter (i) (m), outside pipe diameter (e), length of edge thermal insulation (ti) (m)
DD	Annual degree-day (Kday/an)
d, d_{ti}, d_a, d_f, d_w, d_n, d_p	Thickness, of thermal insulation layer (ti), of air gap (a), total equivalent thickness of building structure in contact with soil (f), of wall (w), of edge thermal insulation (n), periodic heat-penetration depth (p) (m)
Δn, n	Time lag of minimum yearly outdoor air temperature, running number of the day over the year (day)
$\Delta\theta_{SRT}$	Difference in surface radiant temperature of opposite surfaces (°C)
$\Delta\psi$, $\Delta\psi_f$, $\Delta\psi_{env}$	Thermal bridge factor of building structure (W/K), linear thermal transmittance of edge junction of building structure in contact with the soil (f) (W/mK), thermal bridge factor of building thermal envelope (env)(W/m^2K)
E	Energy (J)
ET, ET_o	Plant water evaporation rate, the reference plant water evaporation rate (o) (mm/m^2h)
ε, ε_{IR}, ε_{sky}	Emissivity, of long wave (infra-red) irradiation (IR), of the sky (1)
ε_v	Efficiency of ventilation of the building (1, %)

(continued)

(continued)

F	Force (N)
Fo	Fourier number (1)
F_{i-j}	Radiation exchange factor (view factor between surface i and j (1)
$F_{p,o}$, $F_{p,b}$, $F_{z,b}$	Dimensionless weighting factors (1)
F_i, F_e	Heat-accumulation factors of building structures (1)
f	Heat-flux damping factor of building structure (1)
f_o	Shape factor of building (m^{-1})
f_{RSi}	Inner-surface temperature factor (1)
φ_i, φ_e	Relative humidity of indoor air (i), relative humidity of outdoor air (e) (1, %),
$\Phi_{gr,\max}$ $\Phi_{gr,\max,m}$ $\Phi_{gr,f,\max}$, $\Phi_{gr,u,\max}$	Maximum annual heat flux through underground building structure, maximum monthly (gr, max, m) (W) Maximum heat flux through floor on the ground (gr, f, max), through underground basement (gr, u, max) (W)
$G_{glob,\beta}$	Global (total) solar irradiation on an arbitrary orientated surface (W/m^2)
Gr	Grashof dimensionless number (1)
g	Acceleration due to gravity (m^2/s), total solar energy transmittance of glazing (1)
H, H_a	Height, of air gap (m)
H	Enthalpy (J)
H_T, H_V, $H_{V,n}$, $H_{v,m}$	Transmission heat-transfer coefficient (T), ventilation heat-transfer coefficient (V), natural ventilation heat-transfer coefficient (V,n), mechanical ventilation heat-transfer coefficient (V,m) (W/K)
$H_{gr,f}$, $H_{gr,u,f}$, $H_{gr,u,w}$	Steady-state heat-transfer coefficient of floor on the ground (gf, f), of underground floor (gr, u, f), of underground wall (gr, u, w) (W/K)
H'_T, $H'_{T,\max}$	Thermal transmittance of building thermal envelope, maximum permissible thermal transmittance of building envelope (W/m^2K)
h	Specific enthalpy (J/kg), Planck's constant (Js)
h, h_c, h_r, $\overline{h}$	Surface heat-transfer coefficient, convective (c), radiative (r), average (W/m^2K)
h_{si}, h_{se}	Combined heat-transfer coefficient on internal surface (si), on external surface (se) (W/m^2K)
I_{el}	Electric current (A)
LAI	Leaf area index (m^2/m^2)
L^{2D}, L^{3D}	Specific heat flux that pass structure with linear thermal bridge, with point thermal bridge (W/K)
l	width of gap (m)

(continued)

(continued)

Ψ, Ψ_i, Ψ_e	Linear thermal transmittance (of linear thermal bridge), linear thermal transmittance determined by system of internal (i) and external (e) dimensions (W/mK)
Ψ_{spacer}	Linear thermal transmittance of glazing spacer (W/mK)
λ, λ_s, λ_{vis}, λ_{IR}	Wavelength, of solar irradiation (s), of light (vis), of infra-red radiation (IR) (nm, μm)
λ, λ_f, λ_{ti}, λ_{soil}	Thermal conductivity, of fluid (f), of thermal insulation layer (ti), of soil (W/mK),
λ', λ''	Substitutional thermal conductivity of non-homogeneous building structure (W/mK)
μ	Dynamic viscosity (Ns/m^2)
Nu	Nusselt dimensionless number (1)
ν	Frequency of electromagnetic wave (s^{-1}), kinematic viscosity (m^2/s)
P	Exposed perimeter of building structure in contact with soil (m)
Pr	Prandtl dimensionless number (1)
PMV	Predicted mean vote of thermal indoor comfort (1)
PPD	Predicted percentage of dissatisfied with thermal indoor comfort (%)
$PLOS$	Productivity loss index (%)
p, Δp, Δp_t	Air pressure, air-pressure difference, stagnation pressure difference (Pa)
R, R_{tot}	Thermal resistance, total thermal resistance (tot) (m^{2}K/W)
Ra, Re	Rayleigh dimensionless number, Reynolds dimensionless number (1)
R_λ, R_a	Conduction thermal resistance (λ) (m^{2}K/W), thermal resistance of closed air layer in building structure (a) (m^{2}K/W)
R'_p	Thermal resistance of insulation on a pipe (mK/W)
R_{si}, R_{se}	Combined convective and radiative surface thermal resistance on indoor (si), on outdoor (se) surface of building structure (m^{2}K/W)
$R_{c,si}$, $R_{c,se}$	Convective thermal resistance at inner surface of the pipe (c,si) (from pipe to air) (mK/W), thermal resistance at inner surface of the pipe (from fluid to pipe) (c,se) (mK/W)
R_T, $R_{\|\|}$	Thermal transmittance of non-homogeneous building structure, transverse (T), parallel (II) (m^{2}K/W)
r	Latent heat of water evaporation (J/kg)
ρ, ρ_i, ρ_e, ρ_{soil}	Density, of indoor air (i), of outdoor air (e), of soil (kg/m^3)
ρ, ρ_s, ρ_{IR}	Reflectivity, of solar irradiation (s), of longwave (infra-red) irradiation (IR) (1)
s	Distance (m)

(continued)

(continued)

σ	Stefan-Boltzmann constant (W/m^2K^4)
$\dot{Q}$	Heat flux (W)
Q	Heat (J)
q	Specific heat (J/m^2)
$\dot{q}, \dot{q_\lambda}, \dot{q_c}, \dot{q_r}, \dot{q_{ET}}$	Specific heat flux, conductive (λ), convective (c), radiant (r), evapotranspiraton (ET) (W/m^2)
$\dot{q_{r,\lambda}}, \dot{q_{r,b}}$	Spectral radiant (r, λ), black body radiant (r, b) (W/m^2)
$\dot{q_{r,o}}, \dot{q_{r,in}}, \dot{q_{1-2}}$	Specific outgoing radiant heat flux (r,o), received radiant heat flux (r,in), between surface 1 and surface 2 (W/m^2)
$\dot{q_o}, \dot{q_i}$	Specific heat flux at outer surface of building structure (o), at inner surface (i) (W/m^2)
$\dot{q}'_p$	Specific heat flux per unit of the pipe length (W/m)
$\theta, \overline{\theta}, \widehat{\theta}, \widetilde{\theta}$	Temperature, average temperature, amplitude of temperature, periodic temperature (°C)
$\theta_i, \theta_e, \theta_s, \theta_{si}, \theta_{se}\ \theta_o, \theta_v, \theta_{e,p}, \theta_{mrt}$	Temperature of indoor air (i), of outdoor air (e), of surface (s), of internal surface of building structure (si), of external surface of building structure (se), Temperature of fluid outside of the convective boundary layer (o), of vegetation blanket (v), design heating outdoor-air temperature (e, p), mean radiant temperature (mrt) (°C)
$\overline{\theta}_{e,an}, \widehat{\theta}_{e,an}$	Average annual outdoor air temperature, average annual amplitude of outdoor-air temperature (°C)
$\theta_{e,eq}, \theta_{sol-air}$	Equivalent outdoor air temperature, "sol–air" temperature of outdoor air (°C)
$\theta_{op}, \theta_{PET}$	Operative temperature (op), physiological equivalent temperature (PET) (°C)
$T, T_i, T_e, T_{sky}, T_{green}$	Absolute temperature, absolute temperature of indoor (i), outdoor (e) air, absolute temperature of the sky, absolute temperature of the greenery of a green roof (K)
Tu	Turbulence (%)
$t, t_p, \Delta t$	Time, length of time period (p), time lag (s)
τ, τ_s, τ_{vis}	Transmittivity, of solar irradiation (s), of light (vis) (1)
U, U_w, U_g, U_f $U_{gr,f}, U_{gr,w}, U_{gr,fc}, U_{gr,u,f}, U_{gr,u,w}$	Thermal transmittance of building structure, of window (w), of glazing (g), of window frame (f), of floor in contact with ground (gr, f), of wall in contact with ground (gr, w), of slab-on-floor structure (gr, f), of underground floor (gr, u, f), of underground wall (gr, u, w) (W/m^2K)
U'_p	Thermal transmittance of pipe per unit of length (W/mK)
$U_{eq}, U_{\max}$	Equivalent thermal transmittance (eq), maximum permitted thermal transmittance of building structure (max) (W/m^2K)
U_{el}	Voltage (V)
u_c	Uncertainty (1)

(continued)

(continued)

V_e	External (gross) volume of building (m^3)
v	Velocity (m/s)
W	Work (J)
x	Characteristic dimension (1)
Z_{ij}	Element of transformed matrix of the thermal response of building structure (1)
z	Height above terrain (m), depth below terrain (m)

1.2 Energy and Heat

The word "energy" derives from Greek word "energeia" and is attributed to Aristotle. It is best understood as the quantity of work stored in a system, or as the ability of a system to do work. Different systems in nature store energy in different quantities and forms. The study of energies in natural and technical systems is the domain of a scientific discipline known as thermodynamics, developed in the nineteenth century after the invention of the steam engine. Some types of energy are maintained in the same form and quantity for an infinite period; they are called "stored" energies. Others constantly transform to other forms of energy and are known as "transient" energies. An example of the former is the potential energy of a body at a height z above the ground, and an example of the latter is the kinetic energy of a body in motion, free falling from height z. The types of energy considered to be the most important for the supply of energy to buildings are:

- **Electromagnetic radiation**. The most important source of energy for the Earth and for mankind is the Sun. Solar energy in the form of electromagnetic radiation is transformed when it impacts the surface into other forms of energy, for example, environmental heat, the kinetic energy of the wind, and by the process of photosynthesis into organic matter known as biomass (Fig. 1.1).
- **Chemical energy**. All the atoms and molecules in substances are held together by bonds, which represent stored chemical energy. In a chemical reaction the atoms or molecules of a substance transform and the bonds break. If the result is less energy tied up in the bonds, then the reaction is exothermic, and so heat is released. The most important exothermic chemical reaction in terms of the supply of heat is the transformation of fossil fuels and biomass in a combustion reaction.
- **Nuclear energy.** The elementary particles of an atomic nucleus—the protons and neutrons—can store enormous quantities of energy. In the case of the nuclei decay (known as fission) and the nuclei binding (fusion), huge amounts of energy in the form of electromagnetic radiation and kinetic energy are carried by the elementary particles as they are released. The most important nuclear fission reaction takes place naturally at the Earth's core, where the decay of uranium, potassium and

Fig. 1.1 Electromagnetic radiation, emitted by the Sun and received by our planet, is the most important type of energy for mankind (left). Geothermal energy is replenished by the decay of radioactive nuclei in the Earth's core. Geothermal energy is transferred to the surface by conduction and by geothermal water, i.e., in the form of geysers (middle). A reliable supply of electrical energy is crucial for the development of modern societies. Different technologies have been invented for the conversion of renewable energy sources into electrical energy, i.e., photovoltaic modules (right) [1]

thorium atoms replenishes the planet's geothermal energy. The most important nuclear fusion reaction is found at the core of the Sun, where hydrogen atoms join together to produce heavier helium atoms. The mass deficit in the reaction is transformed into electromagnetic radiation, propagating energy at the speed of light.

- **Electrical energy**. Electrons, one of the electrically charged particles of matter, can flow and so transport energy under the influence of an electrical potential. The amount of electrical energy per unit of time depends on the flow of electrically charged particles, known as the electric current I (A), and the difference in the electrical potential, called the voltage U (V);
- **Heat or internal energy**. A measure of the heat in a system is its temperature. At absolute zero (0 K), the internal energy of a body is equal to zero.

1.2.1 Energy Measurement Units

Energy is a scalar quantity that can only be determined indirectly by measuring other quantities, such as the temperature or the velocity of body. To determine the amount of energy, we use units that are prescribed by the International System of Units, or the SI measurement system. These units are length (m), mass (kg), time (s), although other fields of science also use the ampere (A), the kelvin (K), the candela (cd) and the molar quantity of a substance (mol). The units for other quantities can be derived from these units, including that for energy. Some abbreviations of the units are allowed (for example, N newton for force and J joule for energy), and exceptionally, long-established units, for example, yr for year, l for litre, Wh for watt hour, barrel of oil (0.159 m^3) or toe—tonnes of oil equivalent (42×10^9 J).

Case Study Derive the unit of energy J using the SI units (for length, mass and time). Consider the first law of thermodynamics, which states that work is equivalent to energy.

Velocity v is defined as the distance travelled Δs in the time interval Δt:

$$\mathrm{v} = \frac{\Delta \mathrm{s}}{\Delta \mathrm{t}}\left[\frac{\mathrm{m}}{\mathrm{s}}\right]$$

Acceleration a is defined as the change in the velocity Δv in the time interval Δt:

$$\mathrm{a} = \frac{\Delta \mathrm{v}}{\Delta \mathrm{t}}\left[\frac{\mathrm{m}}{\mathrm{s} \cdot \mathrm{s}} = \frac{\mathrm{m}}{\mathrm{s}^2}\right]$$

A force F, acting on a body of mass m, causes this body to accelerate at the rate a:

$$\mathrm{F} = \mathrm{m} \cdot \mathrm{a}\left[\frac{\mathrm{kg} \cdot \mathrm{m}}{\mathrm{s}^2} \equiv \mathrm{N}\right]$$

A force F, acting on a body that travels a distance s, performs the work W, which is equivalent to the inputted energy.

$$\mathrm{W} = \mathrm{F} \cdot \mathrm{s}\left[\frac{\mathrm{kg} \cdot \mathrm{m}^2}{\mathrm{s}^2} \equiv \mathrm{J}\right]$$

1.2.2 Energy Quality

For any form of energy, the rule of the conservation of energy applies, meaning any energy transformation will only change the form and not the amount of energy. However, not all energy forms in the same quantity are of the same quality and usability. The chemical energy of fuel inside a motor vehicle's fuel tank is first converted into heat, then into mechanical work, next into the kinetic energy of vehicle, and ultimately into environmental heat due to the friction between air particles and the cooling of the engine and the brakes. The total amount of energy always remains the same, but it is understandable that we give more value to a full tank of fuel than a slightly higher temperature of the atmosphere, caused by the car travelling certain distance. In thermodynamics, the science of heat and thermal processes, a quantity called exergy was established to provide a measure of the quality of the energy (Fig. 1.2). Exergy is defined as the maximum useful work that can be obtained from a given energy source. Accordingly, the total amount of energy consists of

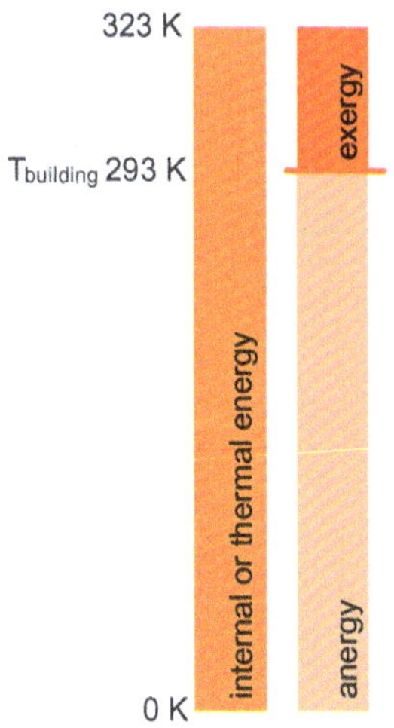

Fig. 1.2 The total quantity of energy for a given energy source consists of its usable part, the exergy, and a non-usable part, the anergy. The boundary between the two is determined by the absolute temperature of the system being supplied with energy. In the case of buildings, this is the temperature of the air inside the building, determined by the requirements of thermal comfort. Nevertheless, some energy forms are pure exergy, for example, electricity

an exploitable part, the exergy, and an ineffective part, the anergy. When assessing the exergy of a heat source, the amount of work that can be done is determined by the temperature. An illustrative example is the internal energy (heat) of outdoor air in wintertime with a temperature of 278 K (5 °C). Although the air temperature is considerably above 0 K, meaning it contains a large amount of energy, the internal energy, or its exergy, available to heat a building is equal to 0 J, since the required temperature of the building is 293 K (20 °C). Both high-exergy and low-exergy sources can be employed to provide comfortable conditions in buildings (Fig. 1.3). High-exergy sources include fossil fuels and electricity, while low-exergy sources include solar energy, geothermal energy as well as the heat and cold of the environment: the atmosphere, soil and groundwater. Low-exergy sources come at a lower cost to the users of a building and cause a considerably smaller environmental burden, but these sources can only be used in energy-efficient buildings. The technologies used to exploit these sources are known as "low exergy" or "low-ex" technologies. The typical applications of these technologies are low-temperature building-heating systems and high-temperature building-cooling systems.

1.2.3 Primary Energy, Final Energy and Useful Energy

The supply of energy in modern societies relies on the utilisation of different energy sources that can be found in nature as primary energy sources. The primary sources

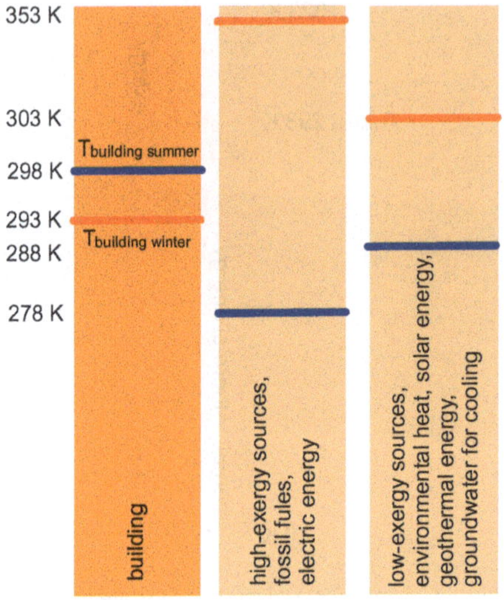

Fig. 1.3 Low-exergy sources can also be employed for indoor thermal comfort in energy-efficient buildings, for example, solar energy, geothermal energy and environmental heat. The exploitation of these sources is less costly to the user, and above all reduces the environmental burden of emissions, which is a characteristic of high-exergy sources such as fossil fuels and electrical energy

of energy are utilised through the chain of energy transformations that have various levels (Fig. 1.4):

- **Primary energy**. This is the energy content of sources in their natural form, e.g., coal in mines, natural gas in subterranean fields, uranium ore, solar radiation, geothermal, biomass, wind and water.

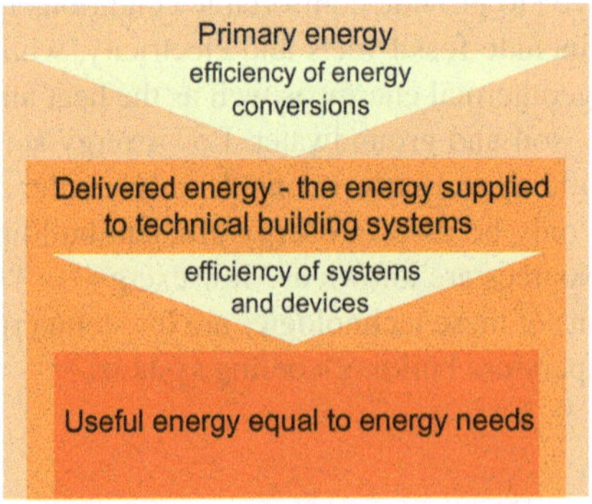

Fig. 1.4 The quantity of energy is measured on three levels: as primary, delivered (or final) and useful energy. They differ in terms of the different efficiencies of energy conversions and the efficiencies of systems and devices used in buildings

- **Final energy (Delivered energy)**. This is the energy that can be transferred by so-called "energy carriers" to the end users, e.g., the heat in a district heating system, coal, wood pellets, oil, natural gas, and electricity.
- **Useful energy**. This is the energy that is emitted by devices in the desired form, e.g., the light from an electric lamp, the heat transferred to the room by a radiator, and the sound from a radio.

The efficiency of the conversion of the primary energy to the final energy depends on the efficiency of the energy systems, and the efficiency of the conversion of the final energy to the useful energy depends on the efficiency of the devices that we use. As energy consumers, we are paying for the final energy, although the total energy consumption of a given system (building, local community or country) is measured at the level of the primary energy. The amount of primary energy also determines the environmental impact caused by the supply and use of energy and is used as a sustainable-development indicator.

1.3 Heat Transfer

Heat is a form of transient energy, i.e., sensible and latent heat. Sensible heat is transferred in any observed system from the regions of higher temperature to the regions of lower temperature until a temperature equilibrium is reached. Latent heat is related to a phase-change process, for example, in the freezing or melting of ice or the evaporation or condensation of water vapour. In this chapter, three basic mechanisms of sensible-heat transfer—by conduction, by convection and by thermal radiation—will be discussed.

1.3.1 Heat Transfer by Conduction

Heat conduction is the transfer of energy between microscopic units of matter, the atoms and molecules, where no chemical or nuclear reactions occur. The process depends on the state of matter (solid, liquid, gaseous) and the microscopic structure of matter. The measure of the quantity of energy transferred among the microscopic particles of matter is their internal energy. The transfer of energy by conduction in liquids and gases is a consequence of random collisions between free molecules. The frequency of collisions depends on the size of the molecules and the number of molecules in a unit of volume. Both the size and the number of molecules are greater in liquids than in gaseous matter, leading to more frequent molecular collisions and vigorous energy transfer. The atoms and molecules in solids are bound together, so the energy is transmitted by the vibration of the atoms and molecules. A special case is metals, which have free electrons in their crystal lattice that greatly increase the transfer of heat. As the free electrons also carry an electric charge, metals are also

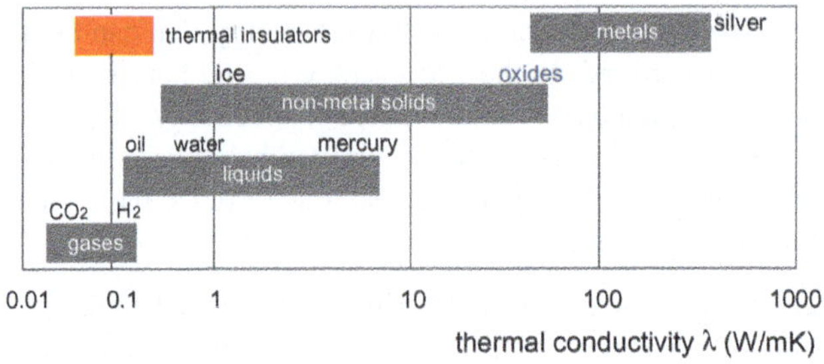

Fig. 1.5 The range of heat conductivities λ for different materials according to their states of matter [2, 3]

good conductors of electrical current. In any case, the internal energy of matter, i.e., the heat, flows from the regions of higher energy (= temperature) to the regions of lower energy in all directions. The temperature difference in matter at a given distance and along a given direction is also known as the temperature gradient. If the heat transfer in the matter is uniform in all directions, the matter is said to be isotropic. In most cases, in the physics of buildings, construction materials are assumed to be isotropic.

Heat transfer by conduction can be described by the thermal conductivity of the matter λ. It is defined as the specific heat flux $\dot{q}$ (W/m^2) passing through a 1-m-thick plane-parallel layer at a temperature difference of 1 K at the boundaries of the layer. Thus, the unit for thermal conductivity is W/mK. Typical values of thermal conductivity are shown in Fig. 1.5.

The thermal conductivities of gases are only low when the gases are steady and the heat is transferred only by the random collisions of molecules. This is only the case when the volume of gas is sufficiently small. Because of that, thermal insulating materials are built of cells and fibres that trap stationary air or some other gases, and this is the reason why the density ρ of these materials is correspondingly small—between 30 and 120 kg/m^3. The most widely used thermal insulating materials with a cellular structure are extruded or expanded polystyrene, as well as natural materials like cork, foam glass, expanded clay and perlite. Fibrous insulating materials include glass and stone wool, and natural materials like wool, coconut fibre and straw. The thermal conductivity of all these materials lies in the range between 0.035 and 0.045 W/mK, but they differ considerably in terms of their fire resistance and in the amount of energy required for their production, also known as embodied energy. In addition to density, the thermal conductivity of construction materials depends on their temperature and moisture content. Some examples are shown in Fig. 1.6.

Heat flux is a vector quantity. It propagates normal to the isotherm, the curve (line) of the same temperature.

The specific heat flux $\dot{q}$ is defined by the ratio between the heat flux $\dot{Q}$ to the surface A (m^2) through which the heat flux is passing:

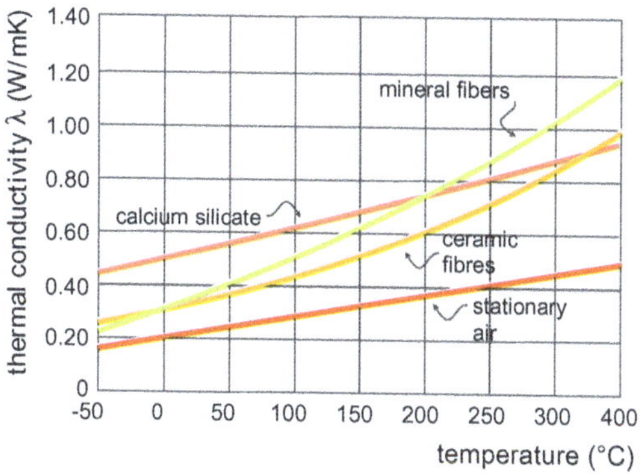

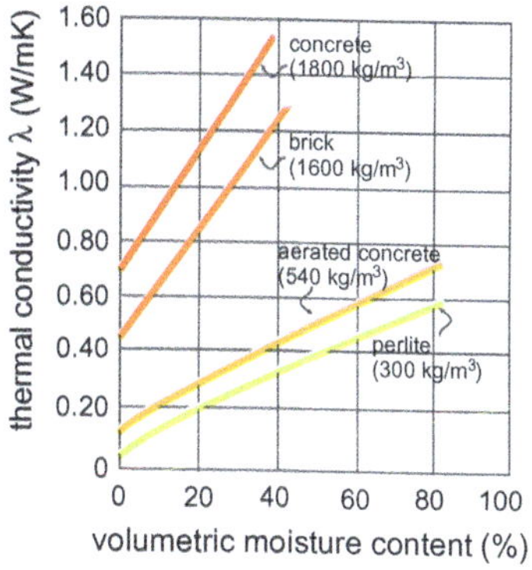

Fig. 1.6 Thermal conductivity λ at different temperatures (left) and versus volumetric moisture content (right) for some building materials [2]

$$\dot{q} = \frac{\dot{Q}}{A}\left[\frac{W}{m^2}\right]$$

Explanation The heat flux $\dot{Q}$ can also be described as the quantity of heat Q (J) transferred through matter by conduction in a unit of time (s).

The specific heat flux conducted through matter is determined by Fourier's law. In the Cartesian coordinate system, the differential form of Fourier's law can be expressed as:

$$\dot{q} = -\lambda_{(\theta,v,n)}\frac{\partial\theta}{\partial n} = -\lambda_{(\theta,v,n)}\left(\frac{\partial\theta}{\partial x} + \frac{\partial\theta}{\partial y} + \frac{\partial\theta}{\partial z}\right)\left[\frac{W}{m^2}\right]$$

In the case that the construction materials are isotropic, and while the ranges of temperatures θ and moisture v of the materials in building constructions are narrow, the parametrically dependent thermal conductivity $\lambda_{(\theta,v,n)}$ can be replaced by a constant value λ. In difference form, Fourier's law can be expressed as follows:

$$\dot{q} = -\lambda\left(\frac{d\theta}{dx} + \frac{d\theta}{dy} + \frac{d\theta}{dz}\right)\left[\frac{W}{m^2}\right]$$

For one-dimensional heat conduction, a characteristic of the majority of single-layer building structures, Fourier's law can be expressed as (Fig. 1.7):

$$\dot{q} = -\lambda\frac{d\theta}{dx} = -\lambda\frac{\theta_1 - \theta_2}{d}\left[\frac{W}{m^2}\right]$$

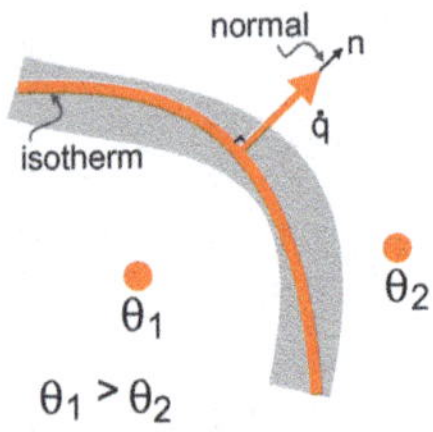

Fig. 1.7 The density of heat flow rate is a vector quantity normal to the isotherms, the curves of equal temperature

1.3.1.1 Determining the Thermal Conductivity of Construction Materials

The thermal conductivity of construction materials is determined by placing a specimen of known thickness d between two parallel plates. One of which is heated and the other one cooled. In this way, a temperature gradient is established in the layer, which is required for a heat flux to conduct through the layer of the construction material. A measuring instrument is used to determine the specific heat flux $\dot{q}$. The instrument consists of a thin layer of a material with known thermal conductivity, enclosed by series of temperature sensors on both sides (Figs. 1.8 and 1.9). The output signal of the meter is the electrical voltage generated by the temperature sensors. For each heat-flux meter the manufacturer indicates the relationship between the heat flux density (W/m^2) and the electrical voltage generated by the meter (mV). This ratio is known as heat flux meter calibration constant. On the heated surface of the sample of building material the temperature is set at $\theta_{s,1}$ and on the cooled side the temperature is set at $\theta_{s,2}$. The thermal conductivity of the construction material λ is calculated by means of the average values of the measured quantities and Fourier's law:

Fig. 1.8 Simple thermal conductivity measuring apparatus consists of temperature sensors on the heated and cooled plates, and a circular heat flux meter (left); the layers of building material enclosing the heat flux meter (middle); and a thermally insulated assembly ensuring the heat flow is transferred through the sample only in one dimension (right)

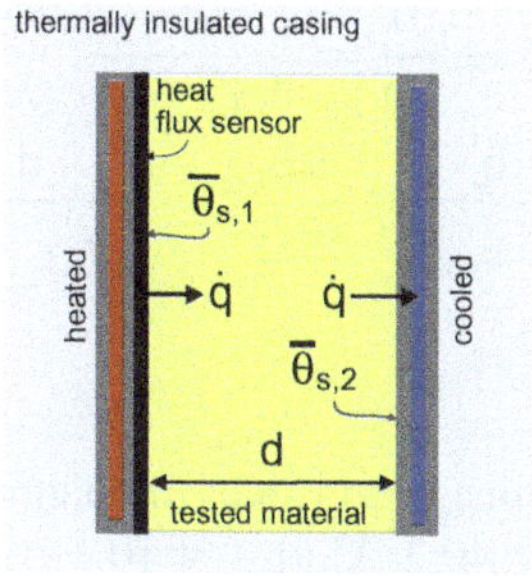

Fig. 1.9 The thermal conductivity λ is determined according to Fourier's law by measuring the specific heat flux $\dot{q}$ and the temperatures ($\theta_{s,1}$ and $\theta_{s,2}$) at the boundary surfaces of the building material layers

$$\dot{q} = \lambda \frac{\theta_{s,1} - \theta_{s,2}}{d} \rightarrow \lambda = \frac{d \cdot \int \dot{q}}{\int (\theta_{s,1} - \theta_{s,2})} = d \cdot \sum_{j=1}^{n} \frac{\frac{\dot{q}_j}{(\theta_{s,1,j} - \theta_{s,2,j})}}{n} \left[\frac{W}{m \cdot K}\right]$$

Each measurement is associated with an error, which is the difference between the measured value of a quantity and its true value. The measurement error consists of a systematic and a random part. The reason for the systematic measurement error is usually due to the measurement method or the design of the device used to conduct the tests. Repeating the experiment may yield the same results, but the values still deviate from the true value. The systematic measurement error can be reduced only by using different measurement methods. The random measurement error causes a spread of the measured values with respect to the average value of the quantity. This difference is known as the measurement uncertainty. When the experiment is repeated, the value of the quantity is always a little bit different, and therefore random. Since the final result of the experiment is most likely obtained based on several measured quantities, the random measurement error is determined using statistical methods. It is also known as total standard uncertainty, expressed as:

$$u_c(y) = \left[\sum_{i=1}^{m} \left(\frac{\partial y}{\partial x_i} u(x_i)\right)^2\right]^{\frac{1}{2}}$$

In the equation, the first element is the partial derivative of the mathematical function used to calculate the end result of the experiment (e.g., Fourier's law), with respect to all m measured quantities x_i. $u(x_i)$ are the measurement uncertainties of individual quantities that are included in the mathematical function. In the case of an experimental determination of the thermal conductivity λ of the layer of a specimen with thickness d, the mathematical function and the partial derivates are equal to:

$$\lambda = \frac{\dot{q} \cdot d}{\theta_{s,1} - \theta_{s,2}} \rightarrow \frac{\partial \lambda}{\partial \dot{q}} = \frac{d}{\theta_{s,1} - \theta_{s,2}}; \frac{\partial \lambda}{\partial d} = \frac{\dot{q}}{\theta_{s,1} - \theta_{s,2}}$$

$$\frac{\partial \lambda}{\partial \theta_{s,1}} = \frac{d \cdot \dot{q}}{\left(\theta_{s,1} - \theta_{s,2}\right)^2}; \frac{\partial \lambda}{\partial \theta_{s,2}} = -\frac{d \cdot \dot{q}}{\left(\theta_{s,1} - \theta_{s,2}\right)^2}$$

Case Study The average measured values of quantities during the experiment are $\dot{q} = 29.5\ \text{W/m}^2$, $\theta_{s,1} = 40.2\ °\text{C}$, $\theta_{s,2} = 10.4\ °\text{C}$. The specimen thickness d is 40 mm. The uncertainty in measuring the heat flux density is 2.5% or 0.73 W/m^2; the uncertainty in measuring the temperature is 0.1 °C; and the uncertainty of the specimen thickness is 0.1 mm. The total standard uncertainty of the thermal conductivity λ is:

$$u_c(\lambda) = \pm\left[\sum_{i=1}^{4}\left(\frac{\partial \lambda}{\partial x_i} u(x_i)\right)^2\right]^{\frac{1}{2}}$$

$$= \left[\left(\frac{\partial \lambda}{\partial \dot{q}} 0.73\right)^2 + \left(\frac{\partial \lambda}{\partial d} 0.0001\right)^2 + \left(\frac{\partial \lambda}{\partial \theta_{s,1}} 0.1\right)^2 + \left(\frac{\partial \lambda}{\partial \theta_{s,2}} 0.1\right)^2\right]^{\frac{1}{2}} \left[\frac{W}{m \cdot K}\right]$$

$$u_c(\lambda) = \pm\left[\left(\frac{d}{\left(\theta_{s,1} - \theta_{s,2}\right)} 0.73\right)^2 + \left(\frac{\dot{q}}{\left(\theta_{s,1} - \theta_{s,2}\right)} 0.0001\right)^2 + \left(\frac{d \cdot \dot{q}}{\left(\theta_{s,1} - \theta_{s,2}\right)^2} 0.1\right)^2 + \left(-\frac{d \cdot \dot{q}}{\left(\theta_{s,1} - \theta_{s,2}\right)^2} 0.1\right)^2\right]^{\frac{1}{2}}$$

$$u_c(\lambda) = \pm\left[\left(\frac{0.04}{(40.2 - 10.4)} 0.73\right)^2 + \left(\frac{29.5}{(40.2 - 10.4)} 0.0001\right)^2 + \left(\frac{0.04 \cdot 29.5}{(40.2 - 10.4)^2} 0.1\right)^2 + \left(-\frac{0.04 \cdot 29.5}{(40.2 - 10.4)^2} 0.1\right)^2\right]^{\frac{1}{2}}$$

$$u_c(\lambda) = \pm 0.019 \frac{W}{m \cdot K}$$

The thermal conductivity λ of the building-material specimen determined by the measurement is:

$$\lambda = \frac{\dot{q} \cdot d}{\theta_{s,1} - \theta_{s,2}} \pm u_c(\lambda) = \frac{29.5 \cdot 0.04}{40.2 - 10.4} \pm 0.019 = 0.0396 \pm 0.019 \frac{W}{m \cdot K}$$

1.3.1.2 Heat Transfer by Convection

Convection is an energy-transfer mechanism combining two phenomena: the transfer of energy by random collisions of molecules and the energy transfer by a common motion of larger groups of molecules, consisting of particles of homogeneous matter. The effects of both mechanisms—the random collisions of molecules at the microscopic level and the motion of matter at the macroscopic level—are summed up and are known by the common name convection. Since the motion of particulate matter is only possible in fluids, convection heat transfer occurs between a solid body and a fluid or between different fluids.

Explanation The term *fluid* is used to describe liquids and gases.

If the movement of the fluid is the result of a temperature gradient, heat is transferred by natural convection and the fluid motion is caused by thermal buoyancy. The density of most fluids depends on the temperature: the higher the temperature, the lower is their density. A particle of matter at a temperature higher than its surroundings is lighter and therefore rises up the fluid due to buoyancy. Fluid motion can also be due to an external energy source, such as wind, a fan or a pump. In this case the convection is forced. In both cases, a velocity boundary layer and a temperature boundary layer develop along to the surface at which the heat transfer is taking place (Fig. 1.10). The velocity boundary layer forms due to the friction among the particles of fluid having different velocities, and the temperature boundary layer develops due to the temperature gradient and the conduction of the heat in the fluid. It follows that the boundary layers depend on the material properties of fluid—the dynamic viscosity μ (Ns/m^2), kinematic viscosity ν (m^2/s), density ρ (kg/m^3), thermal conductivity λ (W/mK) and specific heat capacity c_p (J/kgK).

Explanation The dynamic viscosity of matter μ is a measure of the magnitude of the shear forces acting between fluid particles having different velocities. A higher dynamic viscosity means the fluid is "more viscous". Particularly in liquids, in addition to the microscopic properties of the matter, the dynamic

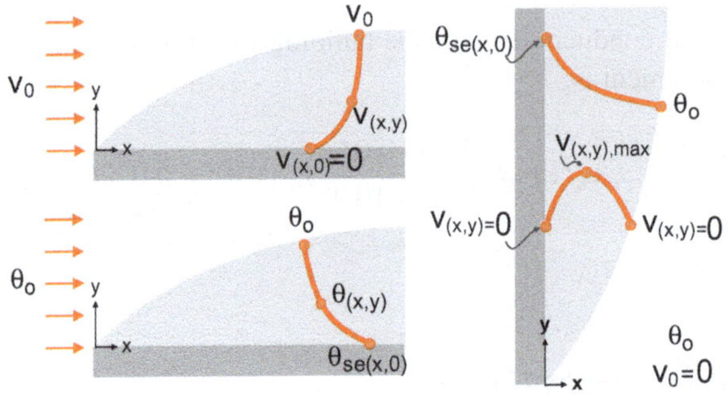

Fig. 1.10 Velocity and temperature boundary layer in the case of forced convection when $\theta_{se(x,0)} > \theta_o$ (left). The velocity and temperature boundary layer in the case of natural convection when $\theta_{se(x,0)} > \theta_o$ (right)

viscosity is also influenced by the temperature—the dynamic viscosity of oils drops from 4 Ns/m^2 at 0 °C to 0.014 Ns/m^2 at 120 °C, meaning the lubrication properties of heated oils is significantly better. The ratio of the dynamic viscosity μ to the fluid density ρ is known as the kinematic viscosity ν. The specific heat capacity c is defined as the amount of heat in J that causes a change in the temperature of 1 kg of substance by 1 K. The total quantity of accumulated heat at a temperature difference of 1 K is equal to the product of the specific heat capacity and the density ($c_p \times \rho$). The ratio between the thermal conductivity of the matter λ to the product of the specific heat capacity and the density ($c_p \times \rho$) is known as the thermal diffusivity or temperature conductivity a measured in m^2/s.

The material properties (Table 1.1) can be combined into dimensionless numbers, for example, the Prandtl number Pr:

$$\mathrm{Pr} = \frac{c_p \cdot \mu}{\lambda} = \frac{c_p \cdot \mu}{\lambda} \cdot \frac{\rho}{\rho} = \underbrace{\frac{c_p \cdot \rho}{\lambda}}_{1/a} \cdot \nu = \frac{\nu}{a}\left[\frac{\mathrm{J} \cdot \overbrace{\mathrm{N}}^{\mathrm{kgm/s^2}} \cdot \mathrm{s} \cdot \mathrm{m} \cdot \mathrm{K}}{\mathrm{kg} \cdot \mathrm{K} \cdot \mathrm{m}^2 \cdot \underbrace{\mathrm{W}}_{\mathrm{J/s}}} = 1\right]$$

In the case of natural convection, the velocity of a fluid at the heat-transfer surface and on the envelope of the velocity boundary layer is zero, while the maximum fluid velocity v_{max} is established inside of the boundary layer. In the case of forced convection, the maximum fluid velocity is found on the envelope of the velocity boundary layer and equal to the velocity in free flow v_o. The temperature boundary layer is confined by an envelope, defined by the isotherm at the surrounding temperature θ_o. As a rule, the velocity and the temperature boundary layers differ. Since the random

Table 1.1 Material properties of air, water and some other gases versus temperature [3]

	θ (°C)	ρ (kg/m³)	λ (W/mK)	cp (kJ/kgK)	μ (Ns/m²) .10⁻⁶	ν (m²/s) .10⁻⁶	a (m²/s) .10⁻⁶	Pr (1)
air	-10	1.334	0.023	1.006	16.61	12.60	17.62	0.717
	0	1.287	0.024	1.006	17.11	13.49	18.94	0.714
	20	1.194	0.026	1.007	18.11	15.27	21.58	0.709
	40	1.118	0.027	1.008	19.07	17.20	24.42	0.705
	60	1.052	0.029	1.008	20.02	19.21	27.38	0.702
	80	0.988	0.030	1.009	20.95	24.54	30.43	0.699
water	0	1000	0.555	4.219	1790	1.79	0.13	13.61
	20	998	0.598	4.182	1008	1.01	0.14	7.063
	40	992	0.627	4.178	653	0.66	0.15	4.385
	60	983	0.651	4.190	470	0.48	0.16	3.077
	80	972	0.669	4.199	354	0.36	0.16	2.285
argon	0	1.784	0.018	0.520	20.99	11.77	19.09	0.616
krypton	0	3.75	0.0094	0.248	23.29	6.21	10.14	0.613
xenon	0	5.9	0.0056	0.158	21.10	3.58	6.05	0.591

collisions of molecules are more frequent in liquids than in gases, the heat transfer is more intense in liquids. The fluid velocity is greater if the fluid motion is caused by an external source, so the heat transfer by forced convection is more intense than by natural convection. The velocity and the temperature boundary layers also depend on the motion of the fluid particles inside the boundary layer (Fig. 1.11). The motion of the particles in laminar flow is more-or-less parallel. Once the ratio of the fluid viscosity to the velocity reaches a certain value, the fluid particles begin to move randomly, and in all directions, while increasing the heat transfer. Such a flow of fluid is known as turbulent flow.

The conditions in which the flow is either laminar or turbulent are determined by the limit values of two dimensionless numbers—the Grashof number *Gr*, in case of natural convection, and the Reynolds number, *Re* in the case of forced convection. As the flow of the fluid is also influenced by the shape of the heat-transfer surface, both dimensionless numbers are determined based on a characteristic dimension *x*. This is, for example, the height of a flat heat-transfer surface in the direction of fluid flow, the diameter of the pipe in the underfloor heating system, or the ratio of the cavity width

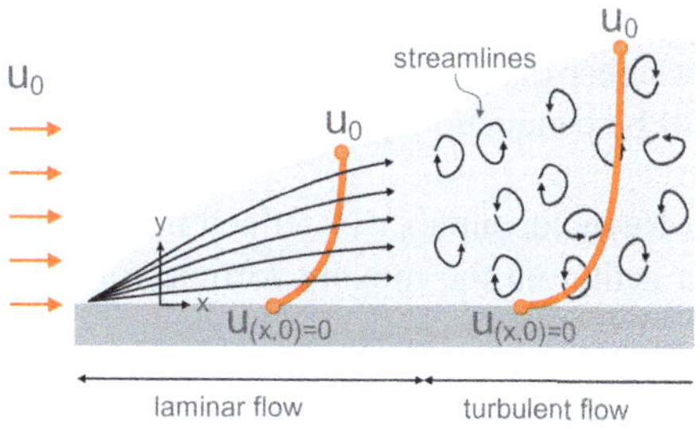

Fig. 1.11 The shape of velocity boundary layer for laminar and turbulent flow

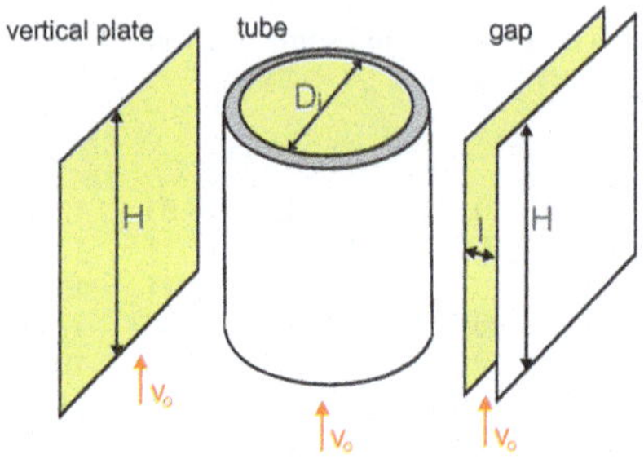

Fig. 1.12 Characteristic dimension x for the determination of the dimensionless Rayleigh and Reynolds numbers; x is height H, diameter D_i and gap width l or height to width ratio H/l

to the length in a mechanically ventilated glass façade. The dimensionless Grashof number Gr determines the relationship between the buoyancy forces and the friction forces, acting upon the buoyancy moving the fluid particles. It is expressed as:

$$\mathrm{Gr} = \frac{g \cdot \beta \cdot (\theta_s - \theta_o) \cdot x^3}{\nu^2}\left[\frac{m \cdot K \cdot m^3 \cdot s^2}{s^2 \cdot K \cdot m^4} = 1\right]$$

where g is the acceleration due to gravity (m/s^2), β is the volumetric thermal expansion coefficient (1/K), θ_s is the heat-transfer surface temperature (°C), θ_o is the fluid temperature outside the temperature boundary layer (°C), ν is the kinematic viscosity of fluid (m^2/s), and x is the characteristic dimension of the heat-transfer surface (Fig. 1.12). The product of the Grashof and Prandtl numbers is known as the Rayleigh number Ra [1]:

$$\mathrm{Ra} = \mathrm{Gr} \cdot \mathrm{Pr} = \frac{g \cdot \beta \cdot (\theta_s - \theta_o) \cdot x^3}{\nu^2} \cdot \frac{\nu}{a} = \frac{g \cdot \beta \cdot (\theta_s - \theta_o) \cdot x^3}{\nu \cdot a}$$

The critical value of Ra at which the type of flow changes from laminar to turbulent is usually assumed to be 10^9. In most cases, the heat transfer in building structures occurs by the natural convection in the laminar velocity boundary layer. In the case of natural convection, the fluid's movement inside the velocity layer only depends on the heat flux passing at the heat-transfer surface. Therefore, the Rayleigh number is also a measure of whether the prevailing heat-transfer mechanism in the case of natural convection will be conduction ($Ra < 1700$) or the motion of fluid particles ($Ra > 1700$).

The dimensionless Reynolds number is defined by the ratio between the forces of inertia and friction acting on the moving particles of fluid, with the following equation:

$$\mathrm{Re} = \frac{\rho \cdot v_0 \cdot x}{\mu} = \frac{v_0 \cdot x}{\nu}\left[\frac{m \cdot m \cdot s}{s \cdot m^2} = 1\right]$$

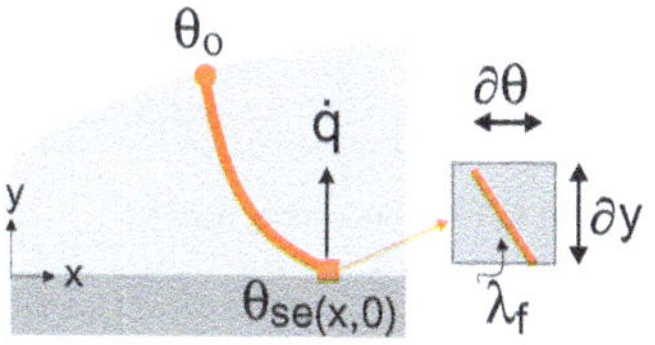

Fig. 1.13 The density of heat flux $\dot{q}$ transferred to the fluid is determined by the thermal conductivity λ_f of the fluid and the temperature gradient $d\theta/dy$ in the infinitesimal particle, which is in contact with the surface

where v_o represents the velocity of the fluid outside the velocity boundary layer, and x is the characteristic dimension of the heat-transfer surface. The critical value of the Re number at which the flow regime changes from laminar to turbulent is most often assumed to be 2300 for heat transfer by convection in pipes and cavities, and 5×10^5 for heat transfer on flat, vertical surfaces.

The heat flux transferred by convection can be described by Newton's law of cooling. Assume that an infinitesimally small fluid particle is right next to the transfer surface, and is therefore steady. In this case the heat is only transferred to it by conduction, but at different rates at every point on the heat-transfer surface (Fig. 1.13).The unknown temperature gradient in the boundary layer is replaced by introducing a local convective surface heat-transfer coefficient $h_{c(x,0)}$, the temperature difference between the heat-transfer surface $\theta_{se(x,0)}$ and the temperature of the fluid outside the temperature boundary layer θ_o. The average convective surface heat-transfer coefficient $\overline{h_c}$ is defined by the heat flux rate that the transfer from or towards 1 m^2 of heat-transfer surface at the temperature difference $(\theta_{se(x,0)} - \theta_o)$ of 1 K. It is measured in W/m^2K. The direction of the fluid flow depends on the temperature difference: if $\theta_{se(x,0)} > \theta_o$, the heat flux flows from the heat-transfer surface into the fluid, and vice versa.

$$\dot{q}_{(x)} = \underbrace{-\lambda_f \frac{\partial\theta}{\partial y}}_{\text{Fourier's law of heat conduction}} \rightarrow -\lambda_f \frac{d\theta}{dy} = \underbrace{h_{c(x,0)}\left(\theta_{se(x,0)} - \theta_o\right)}_{\text{Newton's law of cooling}} \left[\frac{W}{m^2}\right]$$

Often, we must determine the average heat flux transferred by convection across the whole heat-transfer surface. This can be done by integrating the local heat-flux rates $\dot{q}_{(x,0)}$ and replacing the local convective heat-transfer coefficient $h_{c(x,0)}$ and the transfer surface temperature $\theta_{s(x,0)}$ with their average values:

$$\dot{Q} = \int_A \dot{q}_{(x,0)} dA \rightarrow \dot{Q} = \overline{h}_c \cdot A \cdot \left(\overline{\theta}_{se} - \theta_o\right)[W]$$

Examples of convection are heat transfer during the boiling and condensation of liquids (Fig. 1.14). In this case the convective heat-transfer coefficients are consider-

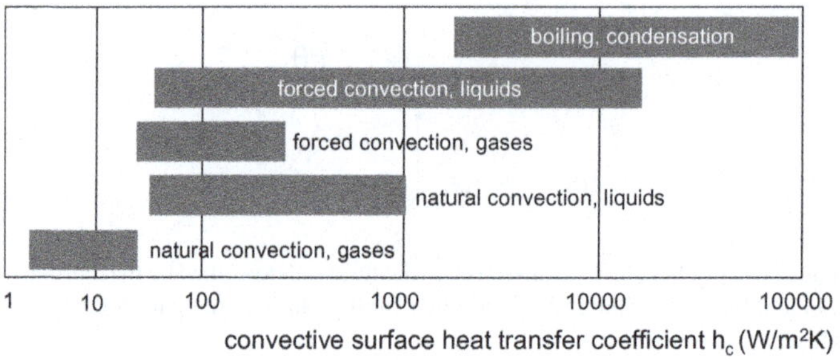

Fig. 1.14 The range of convective heat transfer coefficients according to the mode of convection [3]

ably higher and reach up to 100,000 W/m^{2}K. An example of a device utilising heat transfer by the boiling and condensation of liquids is a heat pipe. In a heat pipe, the heat fluxes are transferred by a latent-heat flux, since the liquid changes the aggregate state at constant temperature. The temperature at which the latent heat flux is transferred inside a heat pipe can be influenced by the choice of the substance that transfers the heat (water, alcohol, organic liquids), or by decreasing the pressure inside the heat pipe. In this case the temperature of the boiling point of the heat-transfer fluid decreases.

1.3.1.3 Determining the Convective Heat-Transfer Coefficient

The convective heat-transfer coefficients are determined with experiments or numerical methods. The experimental approach involves creating a scaled model. In the model the heat flux is supplied to the heat-transfer surface, for example, using an electric heater. By measuring the voltage U_{el} and the electric current I_{el} the heat flux $\dot{Q}$ that impinges on the heat-transfer surface can be determine (Fig. 1.15). At

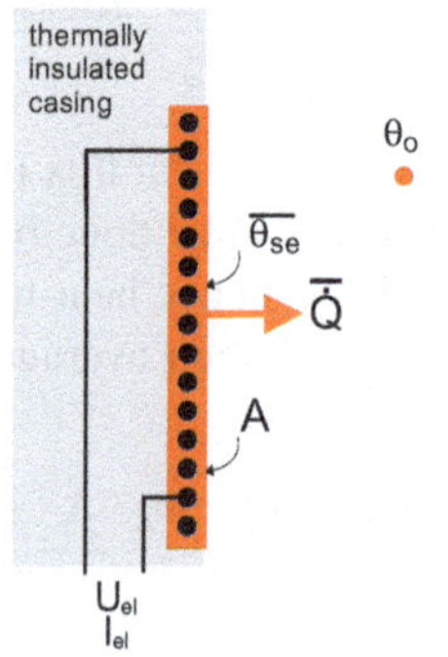

Fig. 1.15 The convective heat-transfer coefficient is determined by experiments on models

the same time, temperature of the heat-transfer surface θ_s and the fluid temperature θ_o outside the boundary layer are measured. The average convective surface heat-transfer coefficient $\overline{h_c}$ can be determined using Newton's law of cooling:

$$\dot{Q} = U_{el}.I_{el} = \overline{h}_c \cdot A \cdot (\theta_{se} - \theta_o) \rightarrow \overline{h}_c = \frac{\dot{Q}}{A \cdot (\theta_{se} - \theta_o)}\left[\frac{W}{m^2 \cdot K}\right]$$

To ensure the experimental results are universally useful, the experiments are repeated for different values of the measured quantities and different fluids, and the dimensionless Nusselt *Nu* number is determined for each experiment. The Nusselt number is defined as the ratio of the convective surface heat-transfer coefficient and the fluid's thermal conductivity:

$$\overline{Nu} = \frac{\overline{h}_c \cdot x}{\lambda}\left[\frac{W \cdot m \cdot K \cdot m}{m^2 \cdot K \cdot W} = 1\right]$$

where x is the characteristic dimension, already known from the definitions of the numbers *Gr* and *Re*. With empirical correlations, the Nusselt number can be associated with other dimensionless numbers: *Ra* in case of the natural convection, and *Re* and *Pr* in case of the forced convection:

$$\overline{Nu} = C \cdot Ra^n[1] \quad \text{natural convection}$$

$$\overline{Nu} = C \cdot Re^m \cdot Pr^n[1] \quad \text{forced convection}$$

After the Nusselt number is known, the convective-surface heat-transfer coefficient $\overline{h_c}$ and the heat flux $\dot{Q}$ at the heat-transfer surface are calculated as:

$$\overline{Nu} = \frac{\overline{h}_c \cdot x}{\lambda} \rightarrow \overline{h}_c = \frac{\overline{Nu} \cdot \lambda}{x} \rightarrow \dot{Q} = \overline{h}_c \cdot A \cdot (\theta_{se} - \theta_o)[W]$$

where A is the area of the heat-transfer surface in m^2.

Case Study Determine the heat flux transferred from the surface of a room heater of height H is 0.8 m and length L is 1.4 m by natural convection. The average temperature of heater surface is θ_{se} is 60 °C, while the air temperature in the room θ_o is 20 °C.

The characteristic dimension is the height of the heater H. The thermodynamic properties of the air at the mean temperature θ_{avg} that is equal to $(\theta_{se} + \theta_o)/2$ can be found in Table 1.1. The volumetric thermal expansion coefficient β is assumed to be the reciprocal value of the absolute air temperature T_o $(273.15 + \theta_o)$, and therefore equal to 0.0032 K^{-1}:

Table 1.2 Informative values of the constants *C*, *m* and *n* for some cases of heat transfer by convection in buildings [3]

	C (1)	m (1)	n (1)	region of validity of empirical equation	characteristic dimension (m)
natural convection on a vertical plate, laminar flow	0.59	0	1/4	$Ra_H < 10^9$	plate height H
natural convection on a vertical plate, turbulent flow	0.10	0	1/3	$Ra_H > 10^9$	plate height H
natural convection on a heated floor	0.54 0.15	0	1/4 1/3	$10^4 < Ra_L < 10^7$ $10^7 < Ra_L < 10^{11}$	ratio L of building structure surface area A to circumference P
natural convection on a cooled ceiling	0.27	0	1/4	$10^5 < Ra_L < 10^{10}$	ratio L of building structure surface area A to circumference P
natural convection on the external surface of a horizontal pipe	0.48	0	1/4	$10^4 < Ra_D < 10^7$	external tube diameter D
natural convection in a closed vertical gap	$Nu = 0.22\left(\frac{Pr}{0.2+Pr}Ra_d\right)^{0.28}\left(\frac{H}{d}\right)^{-1/4}$				gap width d and gap height H; valid for $2 < H/d < 10$, $Pr < 10^5$, $10^3 < Ra_d < 10^{10}$
natural convection in a closed inclined gap	$Nu = 1 + (Nu_{90^\circ} - 1)\sin\beta$ [1]				gap width d and gap height H, gap inclination β valid for 0° (horizontal) < β < 90° (vertical)
natural convection in an open double-skin façade	$Nu = \left(\frac{144}{\left(Ra_d\frac{d}{H}\right)^2} + \frac{2.87}{\left(Ra_d\frac{d}{H}\right)^{1/2}}\right)^{-1/2}$				façade gap width d and façade height H
forced convection on a flat plate, laminar flow	0.66		1/3	$0.6 < Pr < 50$	plate length H

$$Ra = \frac{g \cdot \beta \cdot (\theta_s - \theta_o) \cdot H^3}{\nu \cdot a} = \frac{9.81 \cdot 0.0032 \cdot (60 - 20) \cdot 0.8^3}{17.2 \times 10^{-6} \cdot 24.4 \times 10^{-6}} = 1.53 \times 10^9$$

Since the Rayleigh number exceeds the critical value of 10^9, meaning that fluid flow is turbulent. For such a case, the average Nusselt number is determined by the following empirical extension (Table 1.2):

$$\overline{Nu} = C \cdot Ra^n = 0.10 \cdot Ra^{1/3} = 0.10 \cdot \left(1.53 \times 10^9\right)^{1/3} = 115$$

The convective heat-transfer coefficient $\overline{h_c}$ and the heat-flux rate at the heat-transfer surface are equal to:

$$\overline{h}_c = \frac{Nu \cdot \lambda}{H} = \frac{115 \cdot 0.027}{0.8} = 3.9 \frac{W}{m^2K}$$
$$\rightarrow \dot{Q} = \overline{h}_c \cdot A \cdot (\theta_{se} - \theta_o) = 3.9 \cdot 0.8 \cdot 1.4 \cdot 40 = 173\,W$$

While experiments are suitable for determining the average convective heat-transfer coefficients over the entire heat-transfer surface, the local values can be determined only by numerical methods and a computational fluid dynamics approach.

1.3.2 Heat Transfer by Radiation

The atoms and molecules in matter are not steady. Their level of "activity" depends on the temperature, and only at absolute zero (0 K) would they be at rest. The higher the temperature of matter above absolute zero, the faster is the motion of atoms and molecules, and the more frequently they collide. The electrical charge carried by the particles constantly accelerate and decelerate, generating electromagnetic radiation. One of the characteristics of this radiation is its wavelength λ (m). Often, the wavelength is given in micrometres ($\mu m = 10^{-6}$ m) or in nanometres ($nm = 10^{-9}$ m). Since the electromagnetic radiation propagates at a known velocity ($c_0 = 2.998 \times 10^8$ m/s), the frequency ν (s^{-1}) of electromagnetic radiation is:

$$\nu = \frac{c_0}{\lambda}\left[\frac{m}{s \cdot m} = \frac{1}{s}\right]$$

The electromagnetic radiation with wavelengths between 0.1 and 100 μm is referred to as thermal radiation, and a body at absolute temperature T (K) that emits the largest amount of thermal radiation is called a black-body emitter. The physicist Max Planck was the first to determine the energy transferred by electromagnetic radiation emitted by the black body: it is equal to the product of Planck's constant h with a value of 6.6256×10^{-34} Js, and the frequency of the thermal radiation ν:

$$E = h \cdot \nu = h \cdot \frac{c_0}{\lambda}\left[\frac{Js}{s} = J\right]$$

Since the motion and the collisions of atoms and molecules in a heat emitter are random, the thermal radiation consists of a wide range of wavelengths. For each wavelength of thermal radiation emitted by a black body with an area of 1 m^2 and an absolute temperature T, the emitted specific heat flux $\dot{q}_{(\lambda,T)}$ is determined by Planck's spectral distribution law (Fig. 1.16):

$$\dot{q}_{b,(\lambda,T)} = \frac{C_1}{\lambda^5\left[\exp\left(\frac{C_2}{\lambda \cdot T}\right) - 1\right]}\left[\frac{W}{m^2 \cdot \mu m}\right]$$

where C_1 and C_2 are the first and second radiation constants, respectively, with values of 3.742×10^8 ($W \times \mu m^4/m^2$) and 1.439×10^4 (μm/K). Since $\dot{q}_{b,(\lambda,T)}$ represents the thermal radiation of an emitter with a surface 1 m^2 at a certain wavelength λ, it is also referred to as the specific spectral radiation flux.

The wavelengths of thermal radiation shown in Fig. 1.16 are characteristic of the heat transfer by radiation associated with buildings: the temperature of the photosphere (envelope of the Sun emitting solar radiation) is 5800 K, the temperature of the filament in an incandescent light bulb is 2000 K and the surface temperature of building structures is approximately 300 K. In most engineering problems, thermal radiation is not considered in terms of individual wavelengths, but inside integral

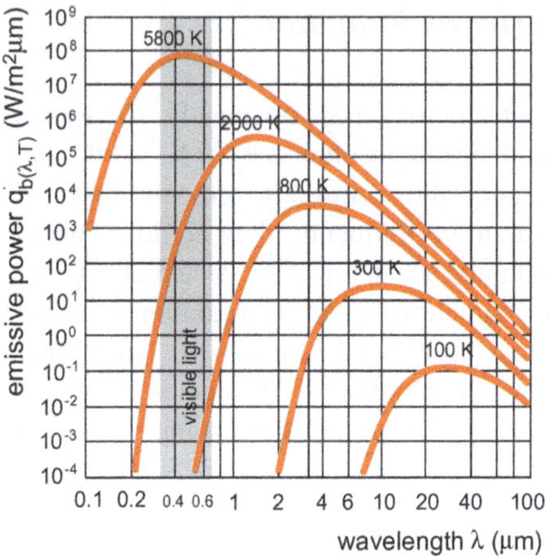

Fig. 1.16 The spectral emissive power of black-body emitters at different absolute temperatures T. An emitter at a higher temperature emits a higher emissive power and radiates at shorter wavelengths

ranges of wavelengths, such as ultraviolet radiation ($0.1 < \lambda < 0.3$ µm), solar radiation ($0.3 < \lambda < 3$ µm), visible light ($0.36 < \lambda < 0.78$ µm) and infrared radiation ($3 < \lambda < 100 +$ µm) (Fig. 1.17).

The total specific radiant heat flux emitted by black body is defined as the integral of the specific spectral radiant heat flux $\dot{q}_{b,(\lambda,T)}$ over the entire range of wavelengths:

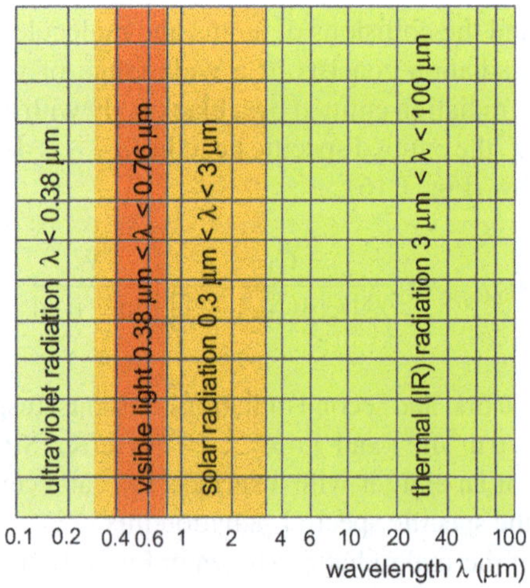

Fig. 1.17 Integral spectral wavelength intervals considered in radiation heat transfer in buildings [4]

$$\dot{q}_b = \int_{\lambda \to 0}^{\infty} \dot{q}_{b,(\lambda,T)} \cdot d\lambda = \sigma \cdot T^4 = 5.67 \times 10^{-8} \cdot T^4 = 5.67 \cdot \left(\frac{T}{100}\right)^4 \left[\frac{W}{m^2}\right]$$

This expression is known as Stefan-Boltzmann's law, where σ is the Stefan-Boltzmann constant with value of 5.67×10^{-8} (W/m^2K^4) and T (K) is the absolute temperature of the surface of a black-body emitter.

Heat transfer by radiation differs from heat transfer by conduction and convection in several ways:

- since electromagnetic radiation propagates in free space, thermal radiation expands in a space without the presence of matter, for example, as solar irradiation from the Sun to the Earth;
- the density of the thermal radiation is proportional to the emitter's absolute temperature to the fourth power, while in case of conduction and convection, the density of the heat flux is proportional to the first power of the temperature difference;
- any body with a temperature above absolute zero emits thermal radiation, which is why a warmer body also receives radiation from a colder one.

1.3.2.1 Black, Grey and Real Emitters

The radiant heat flux defined by the Stefan-Boltzmann law represents the radiant heat flux emitted by a black-body heat emitter at the absolute temperature T. Grey bodies at the same temperature emit a smaller radiation flux (Fig. 1.18). The ratio of the radiation flux emitted by a grey body $\dot{q}$ ($W/m^2\mu m$) to the specific thermal radiation flux emitted by a black body $\dot{q}_b$ ($W/m^2\mu m$) is called the emissivity ε:

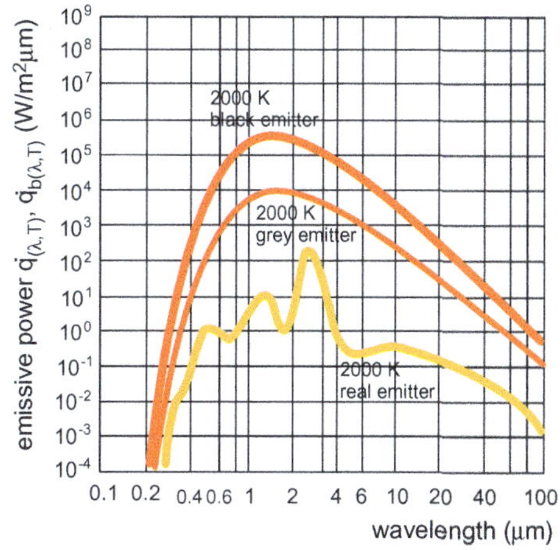

Fig. 1.18 Spectral emissive power radiated by a black, grey and real radiator

$$\varepsilon = \frac{\dot{q}}{\dot{q}_b}[1]$$

For real emitters it is characteristic that their emissivity is different at different emitted wavelengths λ, therefore [1]:

$$\varepsilon_{(\lambda,T)} = \frac{\dot{q}_{(\lambda,T)}}{\dot{q}_{b(\lambda,T)}}$$

Emissivity is a property of the emitter's surface. The emissivity of black bodies equals 1, and the emissivity of grey bodies is somewhere between 0^+ and 1^-. The emissivity of building materials is usually stated as an average value ε over the integral range of wavelengths shown in Fig. 1.17. Thus, the radiant heat flux emitted by building structures is given by the expression:

$$\dot{q} = \int_{\lambda \to 0}^{\infty} \varepsilon_{(\lambda,T)} \cdot \dot{q}_{(\lambda,T)} \cdot d\lambda = \sigma \cdot \varepsilon \cdot T^4 = 5.67 \cdot \varepsilon \cdot \left(\frac{T}{100}\right)^4 \left[\frac{W}{m^2}\right]$$

When the radiant thermal flux reaches another body, part of the radiation that impinges on the body's surface is reflected, while part is absorbed by the body. The ratio between the reflected and the incident spectral radiant thermal flux is called the spectral reflectivity $\rho_{(\lambda)}$, and the ratio between the absorbed and the incident spectral radiant heat flux is the spectral absorptivity $\alpha_{(\lambda)}$ (Fig. 1.19). They are determined by the ratio [1]:

$$\rho_{(\lambda)} = \frac{\dot{q}_{(\lambda),\rho}}{\dot{q}_{(\lambda)}} \quad \alpha_{(\lambda)} = \frac{\dot{q}_{(\lambda),\alpha}}{\dot{q}_{(\lambda)}}$$

If the body is transparent, a part of the incident thermal radiation is transmitted through it. The ratio of transmitted to incident spectral thermal radiant heat flux is known as the spectral transmissivity $\tau_{(\lambda)}$ [1]:

$$\tau_{(\lambda)} = \frac{\dot{q}_{(\lambda),\tau}}{\dot{q}_{(\lambda)}}$$

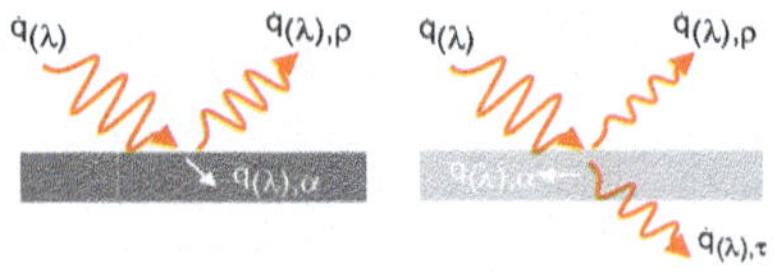

Fig. 1.19 The optical properties of matter—absorptivity, reflectivity and transmissivity—are defined by the law of conservation of energy

The spectral emissivity and spectral absorptivity are related by Kirchhoff's law, which states that the spectral absorptivity of an emitter surface equals its spectral emissivity:

$$\alpha_{(\lambda)} = \varepsilon_{(\lambda)}$$

An emitter with a high spectral emissivity $\varepsilon_{(\lambda)}$ will not only emit a large spectral radiant thermal flux $\dot{q}_{(\lambda)}$, but will strongly absorb the incoming radiation, having the same wavelength λ. The relation between the spectral radiative quantities is derived from the law of the conservation of energy:

$$\alpha_{(\lambda)} + \rho_{(\lambda)} + \tau_{(\lambda)} = 1$$

Kirchhoff's law for opaque bodies yields [1]:

$$\alpha_{(\lambda)} + \rho_{(\lambda)} + \underbrace{\tau_{(\lambda)}}_{0} = \varepsilon_{(\lambda)} + \rho_{(\lambda)} + \underbrace{\tau_{(\lambda)}}_{0} 1 \rightarrow \alpha_{(\lambda)} = 1 - \rho_{(\lambda)} \rightarrow \varepsilon_{(\lambda)} = 1 - \rho_{(\lambda)}$$

Therefore, an emitter with a low spectral emissivity $\varepsilon_{(\lambda)}$ will be highly reflective for the same wavelengths of radiation coming from another emitter. As already stated for emissivity, other radiative quantities are often declared by the average values over the integral wavelength regions and are characterized by the index *s* (for the wavelengths of solar radiation), *IR* (for the wavelengths of thermal radiation) and *vis* (for the wavelengths of light).

In case the radiative properties of an emitter surface differ at different wavelengths of emitted or received radiative thermal flux, the emitter has selective radiative properties. This can be a natural property of the structure, or such properties can be created with a surface treatment of the structure, for example, by coating with thin layer(s). By adapting different forms of selectivity, the overall energy balance of the structure can be significantly improved. Some examples are presented below.

The natural property of glass is that it is selective. Glass transmits the radiant flux with short wavelengths ($\lambda < 3$ μm), for example, the radiation emitted by the Sun), but does not transmit the radiation with long wavelengths ($\lambda > 5$ μm), emitted by objects and people in the building. Instead of transmitting it, the glass absorbs this radiation (Fig. 1.20). This is an excellent property of glass, because buildings can be passively heated and naturally illuminated, while the heat losses of the building decrease. This property of glass is known as the greenhouse effect. According to Kirchoff's law, glass has a high emissivity of long wavelength radiation. Because of that, the radiative properties of glass can be improved even further if a thin layer with a high reflectivity of radiant thermal flux ρ_{IR} is applied to the glass by the process of vacuum deposition or pyrolysis (Fig. 1.21). Oxides like SiO_2 or metal coatings (Ag, Au, Sn) are normally used. Such coatings exhibit a low emissivity of long wavelengths and are known as "low-e" coatings.

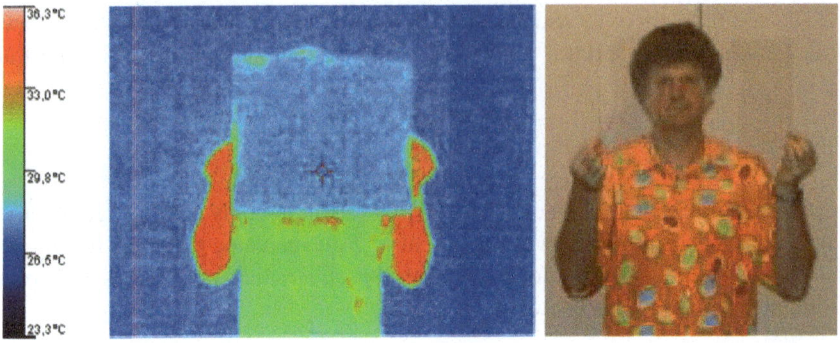

Fig. 1.20 The photograph of the person behind the glass on the right proves the glass transmits well the shortwave radiant thermal flux. As the glass is opaque for the long wavelength radiation emitted by the surface of the skin, a thermal imaging camera only senses the radiation coming from the glass but not from the person behind the glass

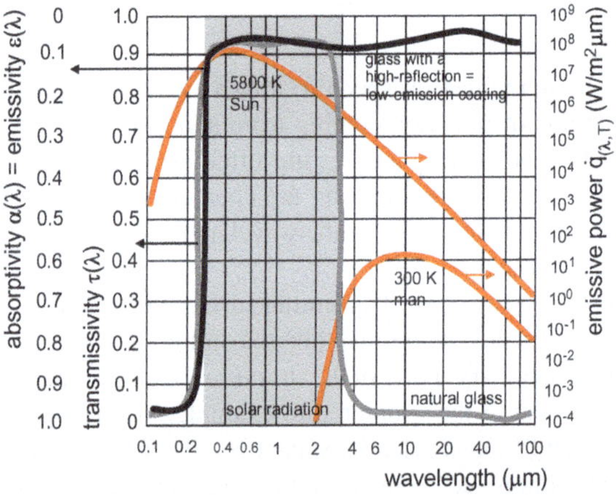

Fig. 1.21 Transmissivity, absorptivity and emissivity of natural glass and glass with a low-emission coating

Ventilated building-envelope structures can be designed to pre-heat the ventilation air by the utilization of the solar energy. The efficiency of solar heating is greater if the absorptance of the solar radiation α_s is close to 1 and the emissivity of the thermal radiation ε_{IR} is as low as possible. The façade shown in Fig. 1.22 is painted with a paint that has an emissivity ε_{IR} equal to 0.3. The air in the ventilated cavity of the façade is heated up over midday on a sunny day to an 8 °C higher temperature compared to the façade painted with high-emissivity paint (ε_{IR} 0.9).

High temperatures on the façade are not, however, always desirable. The reason is the greater temperature expansions and higher mechanical stresses between the final layer of the façade and the thermal insulation. However, by applying a selective

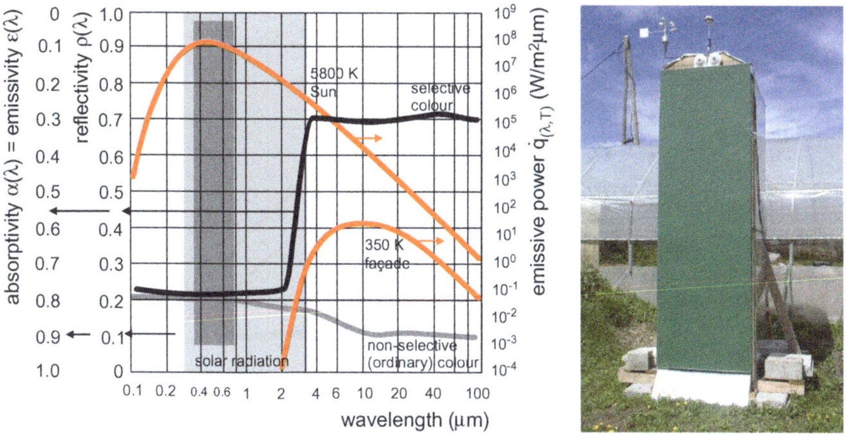

Fig. 1.22 Radiative properties of the surface of the façade painted with a non-selective and a selective paint. Because the reflectivity of the light is the same for both surfaces, the shade of the surface colour will be the same, while the heat flux emitted by the selective surface will be significantly lower due to the lower emissivity of the long-wavelength thermal radiation

coating that reflects the invisible part of the solar radiation (the wavelength 0.38 μm > λ_s > 0.78 μm, Fig. 1.23) and has a high emissivity of long wavelength thermal radiation ε_{IR}, the temperature of the façade decreases considerably. In the case shown in Fig. 1.24, the maximum surface temperatures of the façade on a clear sunny day was reduced by 13 °C.

Fig. 1.23 On a sunny day at noon, the temperature of the facade painted with selective paint was up to 13 °C lower comparing to none selective facade; selective white colour layer with high reflectivity of none visible part of solar radiation was laid first and after green shaded layer having a high thermal radiation emissivity ε_{IR} was added [5]

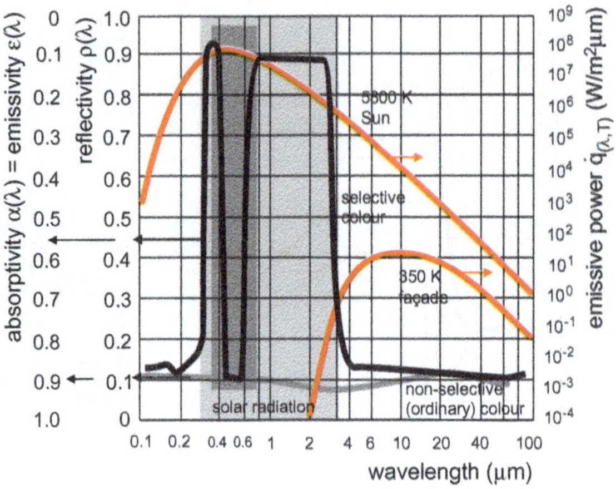

Fig. 1.24 Radiative properties of the surfaces of a façade, coated with a non-selective and with a selective paint. In this case, selective paint is used to reduce the surface temperature and the thermal expansions of the facade

1.3.2.2 Heat Transfer by Radiative Heat Flux Between Surfaces

The surfaces of building structures are grey-body thermal emitters ($\varepsilon_{IR} < 1$ and $\rho_{IR} > 0$) and because of that they reflect part of the received thermal radiation from the surrounding surfaces (Fig. 1.25). The total radiant heat flux $\dot{q}_o$ outgoing from each part of the structure is equal to the sum of the emitted radiant heat flux $\dot{q}$ and the reflected radiant heat flux, which the surface of the structure receives from all the other structures in the surroundings:

$$\dot{q}_o = \dot{q} + \rho_{IR} \cdot \dot{q}_{in} = \sigma \cdot \varepsilon_{IR} \cdot T^4 + (1 - \varepsilon_{IR}) \cdot \dot{q}_{in} \left[\frac{W}{m^2}\right]$$

In enclosed spaces, the outgoing total radiant heat flux $\dot{q}_o$ is distributed to the remaining surfaces in the enclosure. The fraction of the radiant heat flux $\dot{q}_{o,i}$ outgoing

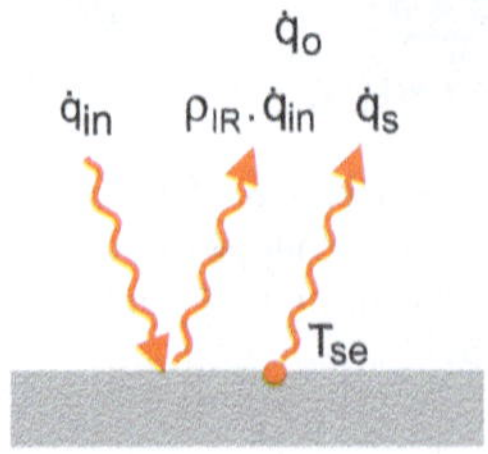

Fig. 1.25 The total radiative thermal flux $\dot{q}_o$ outgoing from the grey body emitter equals the sum of the emitted thermal flux $\dot{q}$ by the emitter and the reflected thermal flux received by the emitter from the surrounding thermal emitters

from the surface i and accepted by the surface j as $\dot{q}_{in,j}$ is in the ratio determined by the radiation-exchange factor (view factor) F_{i-j}:

$$\dot{q}_{in,j} = F_{i-j} \cdot \dot{q}_{o,i} \left[\frac{W}{m^2}\right]$$

The view factor is a geometrical quantity because it depends only on the size and the relative positions of the surfaces that exchange radiant thermal flux. The values of the view factor F_{i-j} for two parallel surfaces are shown in Fig. 1.26. The rule of reciprocity applies:

$$A_i \cdot F_{i-j} = A_j \cdot F_{j-i}$$

The sum of all the view factors of the surfaces forming the enclosure is 1. The view factor for a radiating surface $F_{i\text{-}i}$ is equal to zero.

$$\sum_{j=1}^{n} F_{i-j} = 1 \quad F_{i-i} = 0$$

If two large surfaces i and j are close to one another, they exchange most of the thermal radiation between themselves, so the value of F_{i-j} is close to 1. Typical examples are cavities in glazing and ventilated air gaps in the building structures. In this case, the radiative heat flux between the two boundary surfaces is given by the equation:

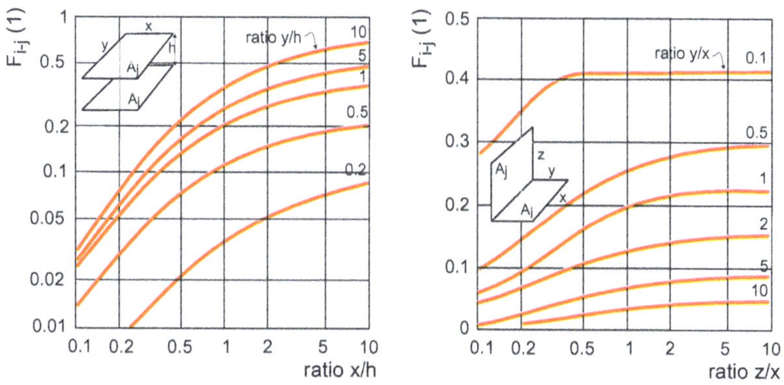

Fig. 1.26 View factor F_{i-j} for two equally sized and parallel surfaces A_i and A_j, for example, a heated or cooled floor or ceiling in a room (left), and for two perpendicular surfaces. The special case is a cube-shaped room ($z/x = y/x = 1$). In this case all the view factors F_{i-j} (i.e. F_{1-2}, F_{1-3}, F_{1-4}, F_{1-5} and F_{1-6}) equal 0.2. Except for the F_{i-i}, which equals 0. In an enclosure the sum $\sum F_{i-j}$ always equals 1 [3 adapted from]

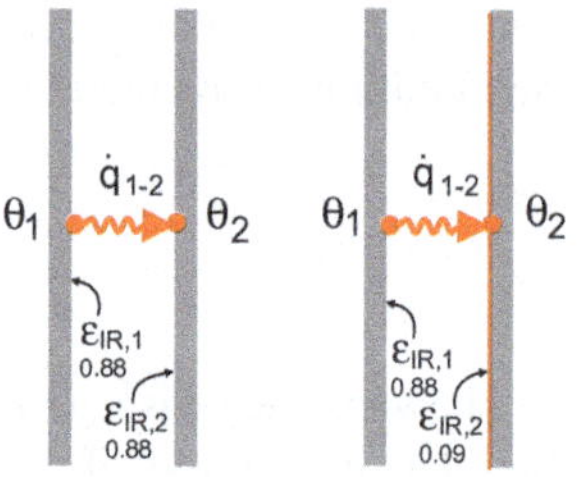

Fig. 1.27 Case study of double pane window glazing

$$\dot{Q}_{1-2} = \frac{\sigma \cdot \left(T_1^4 - T_2^4\right)}{\frac{1-\varepsilon_{IR,1}}{\varepsilon_{IR,1}\cdot A_1} + \frac{1}{A_1\cdot F_{1-2}} + \frac{1-\varepsilon_{IR,2}}{\varepsilon_{IR,2}\cdot A_2}} \left[\frac{W\cdot m^2\cdot K^4}{m^2\cdot K^4} = W\right]$$

where A_1 and A_2 are the surface areas, $\varepsilon_{IR,1}$ and $\varepsilon_{IR,2}$ are the surface emissivities and $F_{1\text{-}2}$ is the radiation-exchange factor for the two surfaces. In the case that both surfaces are black bodies ($\varepsilon_{IR,1} = \varepsilon_{IR,2} = 1$) and close to one another ($F_{1\text{-}2} = 1$), the equation is simplified to the form:

$$\dot{Q}_{1-2} = \frac{\sigma \cdot \left(T_1^4 - T_2^4\right)}{\underbrace{\frac{1-\varepsilon_{IR,1}}{\varepsilon_{IR,1}\cdot A_1}}_{0} + \frac{1}{A_1\cdot F_{1-2}} + \underbrace{\frac{1-\varepsilon_{IR,2}}{\varepsilon_{IR,2}\cdot A_2}}_{0}}$$

$$\rightarrow A_1 \cdot \sigma \cdot \left(T_1^4 - T_2^4\right) = A_1 \cdot 5.67 \cdot \left(\left(\frac{T_1}{100}\right)^4 - \left(\frac{T_2}{100}\right)^4\right) [W]$$

The radiant heat flux passing between the surfaces $\dot{Q}_{1-2}$ is equal to:

$$\dot{Q}_{1-2} = \dot{Q}_1 = -\dot{Q}_2 \text{ at the condition } T_1 > T_2$$

Case Study Determine the specific radiant thermal flux between the glass panes of the double glazing of a window with glass panes made of natural glass ($\varepsilon_{IR,1} = \varepsilon_{IR,2} = 0.88$) and when one of the glass panes has a low-emission coating with the an emissivity $\varepsilon_{IR,2} = 0.09$ (Fig. 1.27). The temperature of the glass towards the interior $\theta_{s,1}$ is 19 °C, and the external glass temperature $\theta_{s,2}$ is −8 °C.

$$\dot{q}_{1-2} = \frac{\dot{Q}_{1-2}}{A} = \frac{\sigma \cdot \left(T_1^4 - T_2^4\right)}{\frac{1}{\varepsilon_{IR,1}} + \frac{1}{\varepsilon_{IR,2}} - 1} = \frac{\sigma \cdot \left(\left(\theta_{s,1}+273\right)^4 - \left(\theta_{s,2}+273\right)^4\right)}{\frac{1}{\varepsilon_{IR,1}} + \frac{1}{\varepsilon_{IR,2}} - 1} \left[\frac{W}{m^2}\right]$$

$$\dot{q}_{1-2} = \frac{\dot{Q}_{1-2}}{A} = \frac{5.67 \cdot 10^{-8} \cdot \left((292)^4 - (265)^4\right)}{\frac{1}{0.88} + \frac{1}{0.88} - 1} = 104.2 \frac{W}{m^2}$$

$$\dot{q}_{1-2} = \frac{\dot{Q}_{1-2}}{A} = \frac{5.67 \cdot 10^{-8} \cdot \left((292)^4 - (265)^4\right)}{\frac{1}{0.88} + \frac{1}{0.09} - 1} = 11.8 \frac{W}{m^2}$$

The low-emissivity coating on one of the glass panes reduces the radiant thermal flux between the glasses by a factor of 8.8. The effect of the additional low-emissivity coating on another glass pane is considerably smaller: the density of radiative thermal flux only decreases from 11.8 to 6.3 W/m^2.

1.3.3 Radiant Heat-Transfer Coefficient

On the surface of building structures the heat is transmitted by convection and by radiation. Since the mechanisms involve a temperature difference with two different powers, the power of 1 for convection and the power of 4 for radiation, the contributions of both mechanisms can be summed only if the temperature difference for the radiation heat transfer is linearized. Since the difference between the temperatures of the surfaces ΔT is small in comparison with T_1 and T_2, a binomial polynomial equation is used for the linearization:

$$T_1 - T_2 = \Delta T \rightarrow T_1 = T_2 + \Delta T \rightarrow T_1^4 = (T_2 + \Delta T)^4$$
$$\rightarrow T_1^4 = T_2^4 + 4 \cdot T_2^3 \cdot \Delta T + \underbrace{6 \cdot T_2^2 \cdot \Delta T^3 + 4 \cdot T_2 \cdot \Delta T^3 + \Delta T^4}_{\text{neglected}}$$

$$\begin{aligned} T_1^4 - T_2^4 &= 4 \cdot T_2^3 \cdot \Delta T \\ &= 4 \cdot T_2^3 \cdot (T_1 - T_2) \approx 4 \cdot \left(\frac{T_1 + T_2}{2}\right)^3 \cdot (T_1 - T_2) \\ &= 4 \cdot T_{avg}^3 \cdot (T_1 - T_2) \end{aligned}$$

The density of the radiant thermal flux $\dot{q}_{1-2}$ between two surfaces is therefore equal to:

$$\frac{\dot{Q}_{1-2}}{A_1} = \dot{q}_{1-2} = \frac{\sigma \cdot \left(T_1^4 - T_2^4\right)}{\frac{1}{\varepsilon_{IR,1}} + \frac{1}{\varepsilon_{IR,2}} - 1} = \underbrace{\frac{4 \cdot \sigma \cdot T_{avg}^3}{\frac{1}{\varepsilon_{IR,1}} + \frac{1}{\varepsilon_{IR,2}} - 1}}_{h_r} (T_1 - T_2) = h_r \cdot (T_1 - T_2) \left[\frac{W}{m^2}\right]$$

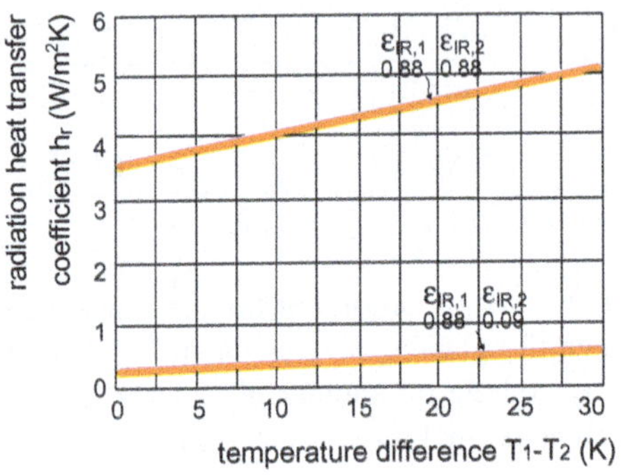

Fig. 1.28 Radiation heat-transfer coefficient h_r versus the temperature difference of the surfaces $(T_1 - T_2)$ exchanging the radiant thermal flux, and their long wavelength emissivites $\varepsilon_{IR,1}$ and $\varepsilon_{IR,2}$

The term h_r before the parenthesis is known as the radiation surface heat-transfer coefficient. Typical values for h_r in the case of heat transfer by radiation in glazing are shown in Fig. 1.28.

1.3.4 Determining the Emissivity of Construction Materials

The devices for determining the emissivity of construction materials differ, depending on whether they measure the integral, i.e., the average emissivity (usually in the wavelength range λ_{IR} between 2.5 and 40 μm), or the spectral emissivity (for each wavelength of thermal radiation). In both cases the measurement of the emissivity is indirect, since a specimen is exposed to the source of thermal radiation and then the reflected thermal radiation is measured. The emissivity is determined with one of the following equations:

$$\bar{\varepsilon}_{\mathrm{IR}} = 1 - \bar{\rho}_{\mathrm{IR}} \quad \varepsilon_{\mathrm{IR},\lambda} = 1 - \rho_{\mathrm{IR},\lambda}\Big|_{2.5\,\mu\mathrm{m}}^{40\,\mu\mathrm{m}}$$

Figure 1.29 shows the principle of the average-emissivity measurement and the device. It consists of a black-body emitter with a semi-hemispherical shape, which is heated to ~100 °C. When the device is placed on the measured surface, part of the radiative heat flux is reflected from the surface. With a lens made of glass that transmits long-wavelength radiation (e.g., zinc selenide glass), the reflected radiant heat flux is directed to a heat-flux sensor, which the device turns into emissivity data. The spectral emissivity is measured in a similar way, except that the sensor is made from a series of filters, each of which is transparent only to a part of the emitted thermal radiation spectrum.

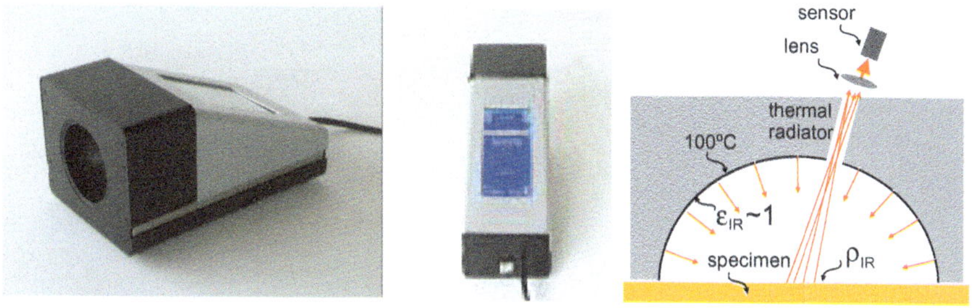

Fig. 1.29 Simple apparatus with a hemispherical heater, used for measuring the integral emissivity of thermal radiation ε_{IR} (2.5 μm < λ < 40 μm) (left, middle) and the measuring principle (right)

1.4 Heat Transfer in Building Structures

Heat transfer in building structures is a consequence of heat-transfer mechanisms, discussed in Sect. 1.3. The heat flux is transferred to the surface of a structure by convection and by radiation, it is conducted through the solid layers of the building structure, and it is transferred from the opposite surface of a building structure by convection and radiation, as shown in Fig. 1.30 (left). In opaque building structures having closed cavities filled with air or gas and in building structures with open air gaps, such as ventilated roofs, the heat flux is transferred between the surfaces of such a layer by convection and by radiation (Fig. 1.30, right). In transparent building structures, such as the window glazing of glass curtain façades, the part of the solar radiation is absorbed by the transparent layers of the structure and converted into heat. Nevertheless, the bulk of the solar radiation passes through the glazing, which enables the natural heating and lighting of buildings. Even more distinctive is the transfer of long-wavelength thermal radiation. Some substances, such as glass, are opaque to such thermal radiation and long-wavelength radiation is absorbed in the glass pane (Fig. 1.31). If a building structure is in contact with the ground, the

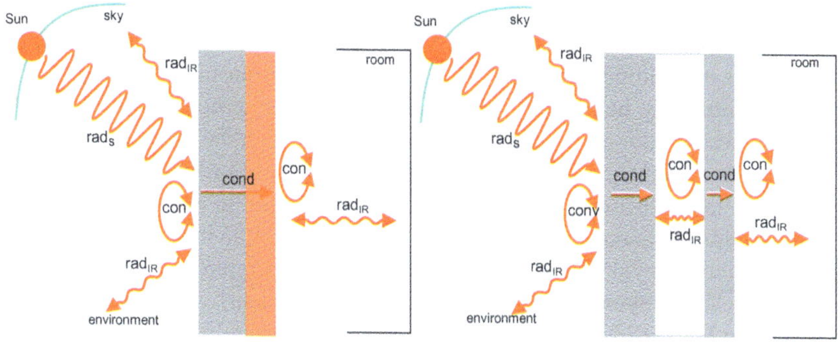

Fig. 1.30 Heat fluxes in opaque building structures (left); heat fluxes in opaque building structure with close cavity (right); cond—conduction, con—convection, rad_s—short wavelength solar radiation, rad_{IR}—long wavelength (IR) radiation

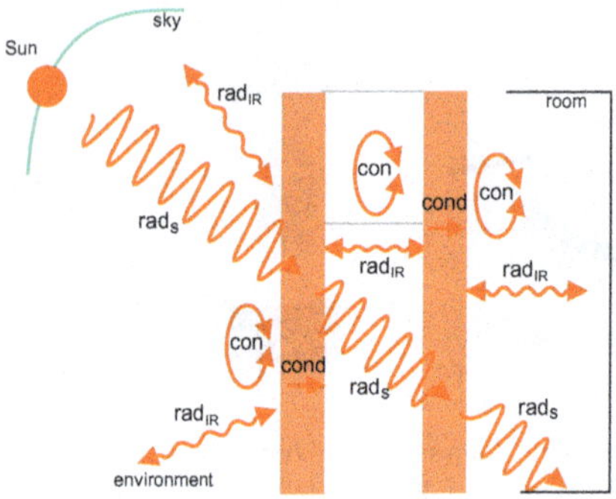

Fig. 1.31 Heat fluxes in transparent building structures with close cavity; cond—conduction, con—convection, rad$_S$—short wavelength solar radiation, rad$_{IR}$—long wavelength (IR) radiation

heat flux is transferred outdoors through the soil layer by conduction. In this case the heat transfer in the building structure depends on thermal conductivity of the soil, the presence of groundwater and on the temperature of the soil, which changes periodically over the year according to the outdoor air temperature.

The heat transfer in building structures also depends on the conditions of the outdoor and indoor environments. The temperature and humidity of the air, the surface temperatures of the surroundings and the sky, the direction and the velocity of the wind, and the solar radiation on the surface of the building structure are most influenced properties of the outdoor environment. Even in the indoor environment, the temperature and humidity of the air are not constant over the time.

The evaluation of heat transfer in building structures differs according to the boundary conditions considered. Thus, we distinguish the following examples:

- **Steady-state heat transfer**. This form of heat transfer occurs when the states of the outdoor and indoor environments do not change with time and the thermal properties of the substances of the building structure are independent of the physical conditions. The heat flux that transmits through the building structure and the temperatures in any part of the structure do not change over time. This is only possible in theory, although it is often used to determine the basic thermal properties of building structures, such as the thermal transmittance.
- **Quasi-stationary heat transfer**. In many cases it is reasonable to replace the steady-state conditions with values that are averaged over the selected time period, taking into account an acceptable deviation from the average values. For example, when the thermal transmittance of a building structure is measured in-situ, the temperature of the outdoor air should not change by more than 2 K. IR scanning of the building envelope can be employed if the temperature of the surroundings does not change by more than 3 K during the experiment.
- **Transient heat transfer**. This is the result of the constantly changing conditions of the outdoor and indoor environments and therefore the most realistic approach. Transient heat transfer can be determined by approximating the boundary conditions using analytical functions, by a single trigonometric function or by a series

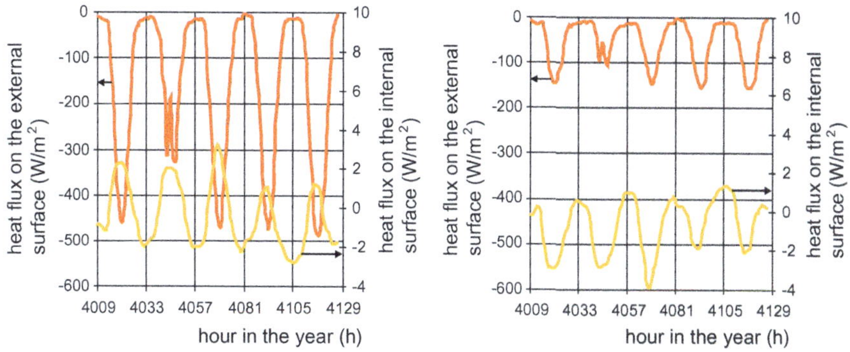

Fig. 1.32 Example of transient heat transfer in a heavy-weighted (left) and a light-weighted (right) vertical, south-orientated building structure. Both structures have the same thermal transmittance $U = 0.280$ W/m^{2}K, absorptivity of solar irradiation $\alpha_s = 0.6$ and emisivity of the long wavelength radiation $\varepsilon_{IR} = 0.9$. The outdoor boundary conditions for the period between 4009th and 4129th hour in the year were taken from the TRY data for the city of Ljubljana

of trigonometric functions in the form of a Fourier series in the case of more complex quantities (e.g. the daily solar radiation). Instead of the analytical function, discrete values of the outdoor environment's quantities in the form of an hourly historical database called the "typical reference year (TRY)" are often used in engineering practice (Fig. 1.32).

The heat transfer in building structures in a real environment is always transient. Such a mode of heat transfer will be discussed in Sect. 1.7. An analysis of the heat transfer in the steady state is usually sufficient for the verification of the thermal insulation properties of a building structure. In this case, the most informative property of a building structure is its thermal transmittance U. It is defined by the heat flux $\dot{Q}$ (W) that passes through the building structure, having a surface area of 1 m^2 for a difference in the air temperature on both sides of the structure equal to 1 K. Therefore, the unit of thermal transmittance is equal to W/m^{2}K.

1.5 Thermal Transmittance of Building Structures

The methods used to determine the thermal transmittance U of building structures are divided into simplified ones, where the heat transfer is assumed to be one-dimensional, and advanced methods, where the thermal transmittance is determined on the basis of the three-dimensional temperature field in the building structures using numerical methods and computer-based tools. An example is shown in Sect. 1.5.6.3. With a known temperature distribution in the structure, the heat fluxes at the boundary surfaces are determined, together with the thermal transmittance of the building structure. In engineering practice, simplified methods are common; however, the thermal transmittance of building structures can also be determined on the basis of experiments performed "in-situ" (see Sect. 1.5.6).

The simplified methods are summarized in [6], a mandatory standard in many EU countries. With respect to the specifics of heat transfer in building structures and according to the provisions of the standards, the structures are divided into:

- homogeneous,
- simple non-homogeneous,
- structures with a closed gap,
- structures with an open, buoyancy or forced ventilated gap,
- greened building structures,
- structures in contact with the ground.

1.5.1 Thermal Transmittance of Homogeneous Building Structures

Homogeneous building structures are made of homogeneous building materials, materials with properties that are assumed to be constant throughout the entire volume and independent of the temperature and the moisture content. All the layers in the building structures as well as the inner and outer surfaces of the structure are parallel. The heat flux passes through the structure in the direction normal to the surface of the structure (Fig. 1.33). We can conclude that almost no building structure fulfils the requirement, because, for example, bricks are bound with mortar and the thermal conductivity of fibrous materials depends on the orientation of the fibres. Since these anomalies have a relatively minor impact on heat transfer, most of the building structures can be considered as homogeneous. In this case, the thermal transmittance is determined by defining the thermal resistances R_λ between temperature nodes placed on each of the surfaces of the layers. The resistance R_λ of each layer is determined by the thickness d (m) of the layer and the thermal conductivity of the material λ

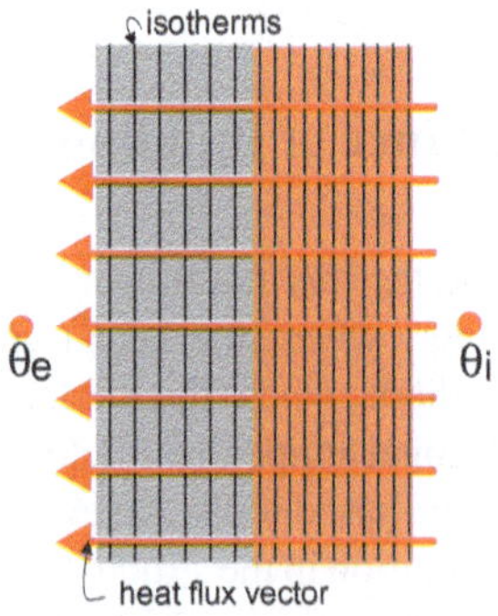

Fig. 1.33 Isotherms in homogeneous building structures are parallel and the heat-flux vector coincides with the surface normal ($\theta_i > \theta_e$)

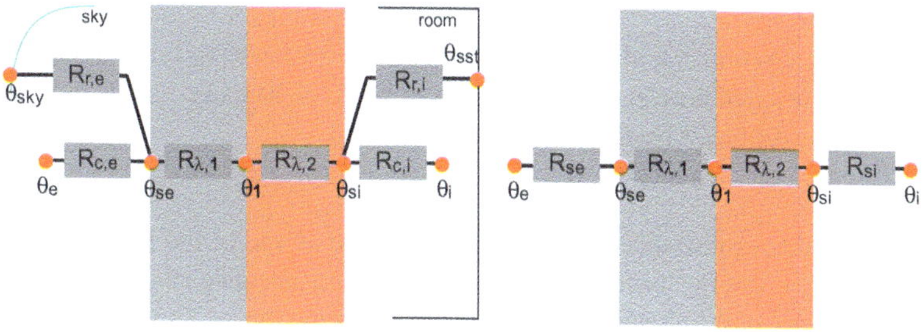

Fig. 1.34 Building structure thermal transmittance U is determined as the reciprocal value of the sum of all resistances R to the heat transfer (left). The surface thermal resistances to convective and radiative heat transfer are replaced by the combined values R_{si} and R_{se} (right)

(W/mK):

$$R_{\lambda} = \frac{d}{\lambda}\left[\frac{m^2 \cdot K}{W}\right]$$

The resistance to heat transfer by convection and radiation at the surface of the external building structure is replaced by the combined surface heat-transfer resistances R_{si} and R_{se} (m^2K/W) (Fig. 1.34), which are determined as the reciprocals of the heat-transfer coefficients:

$$R_{si} = \frac{1}{h_{c,i} + h_{r,i}}, R_{se} = \frac{1}{h_{c,e} + h_{r,e}}\left[\frac{m^2 \cdot K}{W}\right]$$

where the *si* and *se* indexes indicate the internal and external surfaces of the building structure. In this way, the surface heat-transfer resistances are simplified, since the different temperature nodes in the convection and radiation heat transfers are not taken into account. The values shown in Table 1.4 were adopted from [6] with respect to the direction of the heat flux in the building structure. Since all the resistances to heat transfer in the homogeneous building structure are ordered in series, the total thermal resistance R_{tot} (m^2K/W) of the building structure is determined using the equation:

$$R_{tot} = R_{si} + \sum_{j=1}^{n} R_{\lambda,j} + R_{se} = R_{si} + \sum_{j=1}^{n} \frac{d_j}{\lambda_j} + R_{se}\left[\frac{m^2 \cdot K}{W}\right]$$

where j is the numerator of the building structure's layers. Finally, the building structure's thermal transmittance U is equal to the reciprocal value of the total thermal resistance R_{tot}:

$$U = \frac{1}{R_{tot}} = \frac{1}{R_{si} + \sum_{j=1}^{n} \frac{d_j}{\lambda_j} + R_{se}}\left[\frac{W}{m^2 \cdot K}\right]$$

Explanation When calculating the thermal transmittances of internal building structures, situated between heated spaces or between a heated and an unheated space, R_{se} is replaced by R_{si}.

The thermal transmittance of a building structure is one of the most pronounced indicators of the energy efficiency of buildings, and because of that, the thermal transmittances of building structures should not exceed the maximum value $U_{\max}$:

$$U \leq U_{\max} \left[\frac{W}{m^2 \cdot K} \right]$$

The values of $U_{\max}$ differ regarding the type of building structure. The values of $U_{\max}$ are defined by different values in various countries. As a case study, the requirements according $U_{\max}$ adopted in Slovenia in 2002, 2008 and 2010 [7–9] are given in Table 1.3. It is clear that that the requirements have become increasingly stringent in the past decade.

Case Study Determine the thermal transmittance U of the external wall and check whether it is lower than the permissible value $U_{\max}$ 0.28 W/m^{2}K. The composition of the wall structure is given in Table 1.5. The surface heat resistances are $R_{se} = 0.04$ m^{2}K/W and $R_{si} = 0.13$ m^{2}K/W (Table 1.4).

Table 1.3 An example of the evolution of the maximum permissible thermal transmittances of building structures $U_{\max}$, as required by the Slovenian legislation over the past two decades. Similar values were adopted in other EU countries [7–9]

	maximum permissible building structure thermal transmittance (W/m^{2}K)		
	2002	2008	2010
outer walls and walls enclosing unheated rooms	0.60	0.28	0.28
ceiling under an unheated attic	0.35	0.20	0.20
walls between heated rooms	1.60	0.90	0.90
ceiling structure between heated apartments	1.35	1.35	0.90
pitched and flat roofs	0.25	0.20	0.20

Table 1.4 The combined surface resistances to heat transfer by convection and radiation at the surface of the external building structure. In the case of an internal building structure, the value R_{si} is applied to both surfaces [6]

heat flux direction	R_{se} (m^2K/W)	R_{si} (m^2K/W)
bottom-up	0.04	0.10
horizontal	0.04	0.13
top-down	0.04	0.17

Table 1.5 External wall layers from the calculation example

	d (cm)	λ (W/mK)
cement mortar	2	1.40
hollow brick	19	0.52
rock wool	15	0.041
façade mortar	2	0.70

$$R_{tot} = R_{si} + \frac{d_1}{\lambda_1} + \frac{d_2}{\lambda_2} + \frac{d_3}{\lambda_3} + \frac{d_4}{\lambda_4} + R_{se}\left[\frac{m^2 \cdot K}{W}\right]$$

$$R_{tot} = 0.13 + \frac{0.02}{1.4} + \frac{0.19}{0.52} + \frac{0.15}{0.041} + \frac{0.02}{0.7} + 0.04 = 4.237\frac{m^2 \cdot K}{W}$$

$$U = \frac{1}{R_{tot}} = \frac{1}{4.237} = 0.236\frac{W}{m^2 \cdot K} < U_{max} = 0.28\frac{W}{m^2 \cdot K}$$

Solution: The building structure's thermal transmittance U is less than the permissible value U_{max}.

Case Study Determine the required thickness of the thermal insulation layer so that the thermal transmittance of the building structure from the previous example will not exceed 0.15 W/m^2K.

$$R_{tot} = R_{si} + \frac{d_1}{\lambda_1} + \frac{d_2}{\lambda_2} + \frac{d_{ti}}{\lambda_{ti}} + \frac{d_4}{\lambda_4} + R_{se} \rightarrow d_{ti}$$

$$\rightarrow d_{ti} = \left(R_{tot} - \left(R_{si} + \frac{d_1}{\lambda_1} + \frac{d_2}{\lambda_2} + \frac{d_4}{\lambda_4} + R_{se}\right)\right) \cdot \lambda_{ti}[m]$$

$$R_{tot} = \frac{1}{U} = \frac{1}{0.15} = 6.667\frac{m^2 \cdot K}{W}$$

Fig. 1.35 Thick thermal insulation layer is needed to fulfil the criteria of permissible thermal transmittance U_{max}

$$d_{ti} = \left(6.667 - \left(0.13 + \frac{0.02}{1.4} + \frac{0.19}{0.52} + \frac{0.02}{0.7} + 0.04\right)\right) \cdot 0.041$$
$$= 0.249 \rightarrow 25\,\text{cm}$$

Solution: The thermal insulation layer must be 25 cm thick (Fig. 1.35).

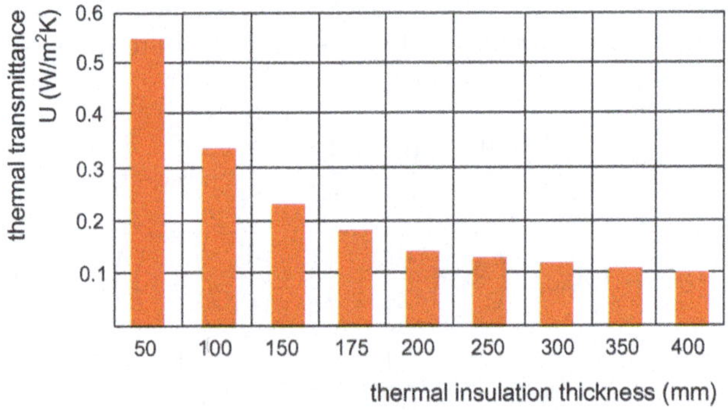

Fig. 1.36 Thermal transmittance of a building structure U does not decrease linearly with the thermal insulation thickness!

Table 1.6 Thermal conductivity λ and the embodied energy of different thermal insulating materials. The values are informative and depend on the density of the material [10]

thermal insulating material	thermal conductivity λ (W/mK)	embodied energy (kWh/m^3)
cellulose	0.040	20
sheep wool	0.040	30
coconut fibre	0.042	60
cork	0.041	90
expanded clay	0.055	170
perlite	0.045	200
glass wool	0.040	250
mineral wool	0.040	270
expanded polystyrene	0.039	600
extruded polystyrene	0.038	810

Case Study Draw a diagram showing the required thickness of the thermal insulation layer on the brick wall of the previous example, so that the thermal transmittance of the wall will is between 0.6 and 0.1 W/m^2K.

Solution is shown in Fig. 1.36.

Important conclusion: The thermal transmittance of a building structure does not decrease linearly with the thickness of the thermal insulation (Fig. 1.36). Therefore, the optimum thickness of the thermal insulation cannot be determined solely on the basis of the heat-transfer physics; economic and environmental criteria must be taken into account, together with the expected lifespan of the building. For example, the embodied energy (Table 1.6), which represents the amount of primary energy needed in the production of a certain insulating material, has an important role as well. Insulating materials made from natural and locally available raw materials will generally have less embodied energy. To avoid the long-term unreliability of assessment methods, the maximum allowable thermal transmittances U_{max} of building structures were imposed. These values are determined based on the best-available technologies (BATs) and indicate a country's commitment to efficient energy use.

1.5.1.1 Temperatures in Homogeneous Building Structures

In steady-state heat-transfer conditions the heat flux entering a building structure equals the heat flux exiting the building structure, and the heat flux is therefore one

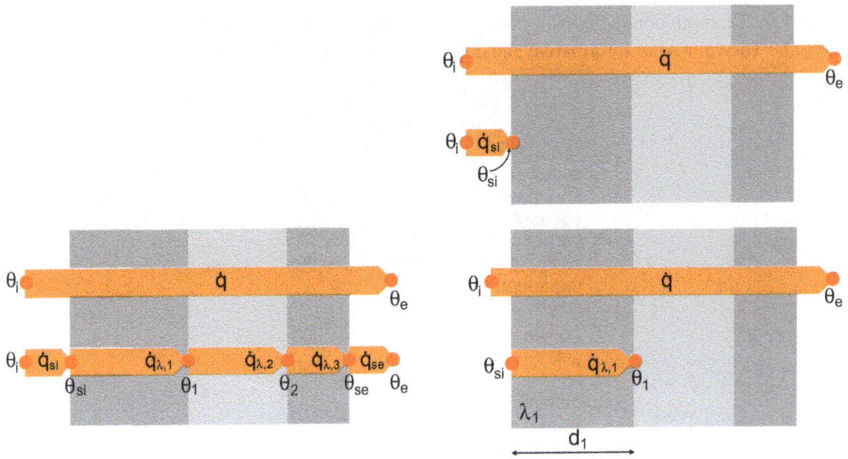

Fig. 1.37 In the case of steady-state heat transfer in homogeneous building structures the heat flux that enters is equal to the heat flux that exits the building structure, and is equal to the heat flux that conducts through any section of the building structure ($\theta_i > \theta_e$) (left). The heat fluxes involved in the determination of the interior surface temperature θ_{si} and the temperature θ_1 at the contact of the first and second layers (right)

dimensional. As consequence, the heat flux conducted through each of n layers is equal to the heat flux entering the building structure. The density of the heat flux per 1 m^2 that enters, is conducted through and exits the building structure, is equal to (Fig. 1.37, left):

$$\dot{q} = U \cdot (\theta_i - \theta_e) = \dot{q}_{si} = \dot{q}_{\lambda,1} = \ldots = \dot{q}_{\lambda,n} = \dot{q}_{se} \left[\frac{W}{m^2}\right]$$

where U is the thermal transmittance of the building structure, and θ_i and θ_e are the indoor and outdoor air temperatures. Because the heat flux is equal in any section of the building structure, it is easy to determine the temperature of any point in the building structure. For example, the internal surface temperature θ_{si} (Fig. 1.37, top right):

$$\dot{q} = U \cdot (\theta_i - \theta_e) = \dot{q}_{si} = h_{si} \cdot (\theta_i - \theta_{si}) \rightarrow \theta_{si} = \theta_i - \frac{\overbrace{U \cdot (\theta_i - \theta_e)}^{\dot{q}}}{h_{si}} [^\circ C]$$

or the temperature θ_1 at the junction between the first and second layers of the building structure (Fig. 1.37, bottom right):

$$\dot{q} = U \cdot (\theta_i - \theta_e) = \dot{q}_{\lambda,1} = \frac{\lambda_1}{d_1} \cdot (\theta_{si} - \theta_1) \rightarrow \theta_1 = \theta_{si} - \frac{\overbrace{U \cdot (\theta_i - \theta_e)}^{\dot{q}} \cdot d_1}{\lambda_1} [^\circ C]$$

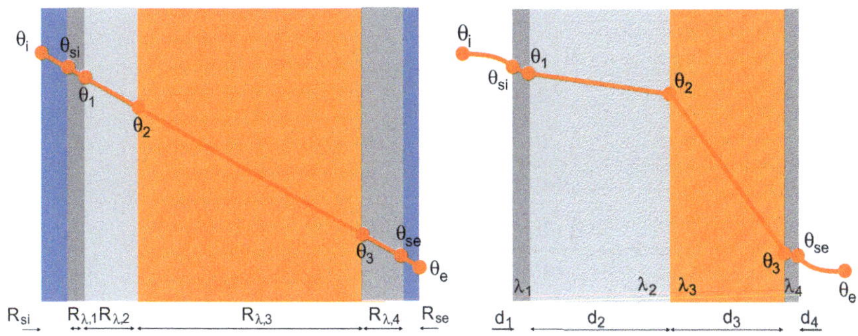

Fig. 1.38 Temperature distribution in a homogeneous building structure when the thicknesses of the layers are replaced by their resistances to heat transfer (left). The actual temperature distribution in the same building structures (right)

The temperatures in the building structure can also be determined graphically if the thickness of the layer of the building structure is replaced by the heat conductance resistance $R_{\lambda,j}$ and by adding the combined surface resistances R_{si} and R_{se} to both sides of building structure. The distribution of the temperatures inside the building structure's layer shown on the graph will be linear (Fig. 1.38).

Case Study Determine the temperatures in the building structures with the layers shown in Table 1.5. The indoor-air temperature θ_i is 20 °C and the outdoor-air temperature is θ_e is − 10 °C. The thermal transmittance of the structure U is 0.236 W/m^{2}K. The combined resistances to surface heat transfer (Table 1.3) R_{si} and R_{se} are equal to 0.13 and 0.04 m^{2}K/W, respectively.

The density of the heat flux $\dot{q}$ that transfers through the wall equals:

$$\dot{q} = U \cdot (\theta_i - \theta_e) = 0.236 \cdot (20 - (-10)) = 7.08 \frac{W}{m^2}$$

The surface heat-transfer coefficients h_{si} and h_{se} are equal to:

$$h_{si} = \frac{1}{R_{si}} = \frac{1}{0.13} = 7.7 \frac{W}{m^2 \cdot K} \quad h_{se} = \frac{1}{R_{se}} = \frac{1}{0.04} = 25.0 \frac{W}{m^2 \cdot K}$$

$$\theta_{si} = \theta_i - \frac{U \cdot (\theta_i - \theta_e)}{h_{si}} = \theta_i - \frac{\dot{q}}{h_{si}} = 20 - \frac{7.08}{7.7} = 19.1°C$$

$$\theta_1 = \theta_{si} - \frac{\overbrace{U \cdot (\theta_i - \theta_e)}^{\dot{q}} \cdot d_1}{\lambda_1} = \theta_{si} - \frac{\dot{q} \cdot d_1}{\lambda_1} = 19.1 - \frac{7.08 \cdot 0.02}{1.4} = 19.0°C$$

$$\theta_2 = \theta_1 - \frac{\dot{q} \cdot d_2}{\lambda_2} = 19.0 - \frac{7.06 \cdot 0.19}{0.52} = 16.4^\circ C$$

$$\theta_3 = \theta_2 - \frac{\dot{q} \cdot d_3}{\lambda_3} = 16.4 - \frac{7.06 \cdot 0.15}{0.041} = -9.4^\circ C$$

$$\theta_{se} = \theta_3 - \frac{\dot{q} \cdot d_4}{\lambda_4} = -9.4 - \frac{7.06 \cdot 0.02}{0.7} = -9.6^\circ C$$

1.5.2 Thermal Transmittances of Simple Nonhomogeneous Building Structures

Non-homogeneous building structures consist of various non-homogeneous layers of different thickness. These are, for example, roof structures with supporting beams and thermal insulation or exterior walls with concrete pillars. As a rule, the heat-flux vectors are still oriented in the direction of the normal to the surface of the building structure, but the heat flux is not the same over the building structure's surface (Fig. 1.39). The building structure is divided into the number of apparent homogeneous segments in the perpendicular (T, transverse) and parallel (II, longitudinal) directions. For each apparent layer, the substitutional thermal conductivities λ' (T) and λ'' (II) and the associated resistances of the building structure R_T and R_{II} are determined, taking into the account the fraction f_j of the surface of each j-th segment, relative to the total surface of the building structure. The sum of all the fractions $\sum f_j$ is equal to 1. The thermal transmittance of the structure U is determined

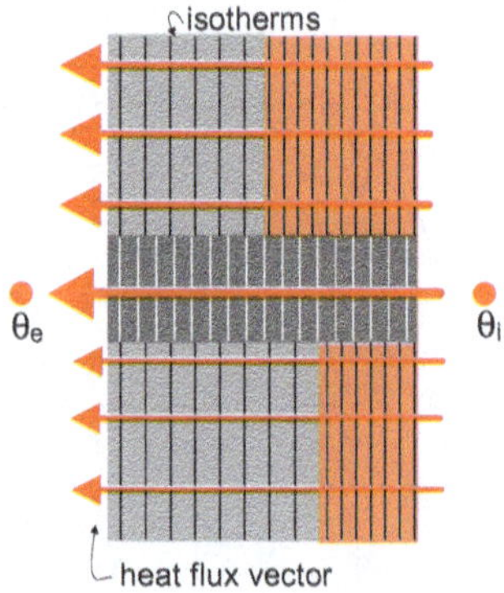

Fig. 1.39 Directions of the heat fluxes in the simple nonhomogeneous building structure coincide with the direction of the surface normal, but the magnitude of the heat-flux transfer through the sections defer ($\theta_i > \theta_e$)

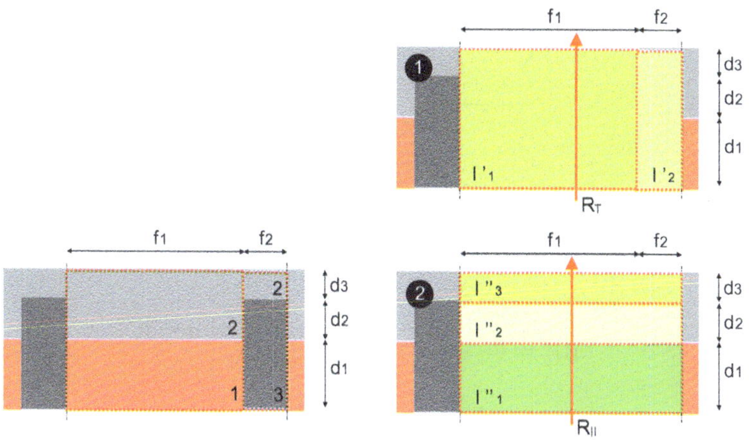

Fig. 1.40 Non-homogeneous building structure consist of two parallel layers, partially pierced by the beam. The total resistance R_{tot} is the average resistance between the resistance in the perpendicular and parallel cross-section of the building structure. The virtual thermal conductivities λ'_j and λ''_j transform the nonhomogeneous structure into a homogeneous one

as the reciprocal value of the average thermal resistance R_T and R_{II}. The procedure for the structure in Fig. 1.40 is as follows:

❶ the building structure is divided into two apparent homogeneous segments f_1 and f_2. The substitutional thermal conductivities λ'_1 and λ'_2 of both segments and the heat-transfer resistance in the normal direction R_T equals:

$$\lambda' = \frac{\sum_{j=1}^{n} \lambda_j \cdot d_j}{\sum_{j=1}^{n} d_j} \rightarrow \lambda'_1 = \frac{\lambda_1 \cdot d_1 + \lambda_2 \cdot d_2 + \lambda_2 \cdot d_3}{d_1 + d_2 + d_3} \left[\frac{W}{m \cdot K}\right]$$

$$\rightarrow \lambda'_2 = \frac{\lambda_3 \cdot d_1 + \lambda_3 \cdot d_2 + \lambda_2 \cdot d_3}{d_1 + d_2 + d_3} \left[\frac{W}{m \cdot K}\right]$$

$$R_{T,1} = R_{si} + \frac{d_1 + d_2 + d_3}{\lambda'_1} + R_{se}$$

$$R_{T,2} = R_{si} + \frac{d_1 + d_2 + d_3}{\lambda'_2} + R_{se}$$

$$R_T = \frac{1}{\frac{f_1}{R_{T,1}} + \frac{f_2}{R_{T,2}}} \left[\frac{m^2 \cdot E}{W}\right]$$

❷ the substitutional thermal conductivities λ''_1 and λ''_2 and the heat-transfer resistance in the parallel direction R_{II}:

$$\lambda'' = \frac{\sum_{j=1}^{n} \lambda_j \cdot f_j}{\underbrace{\sum_{j=1}^{n} f_j}_{1}} \rightarrow \lambda_1'' = \lambda_1 \cdot f_1 + \lambda_3 \cdot f_2 \left[\frac{W}{m \cdot K}\right]$$

$$\lambda_2'' = \lambda_2 \cdot f_1 + \lambda_3 \cdot f_2 \left[\frac{W}{m \cdot K}\right]$$

$$\lambda_3'' = \lambda_2 \cdot (f_1 + f_2) \left[\frac{W}{m \cdot K}\right]$$

$$R_{II} = R_{si} + \frac{d_1}{\lambda_1''} + \frac{d_2}{\lambda_2''} + \frac{d_3}{\lambda_3''} + R_{se} \left[\frac{m^2 \cdot K}{W}\right]$$

❸ the thermal transmittance of the non-homogeneous structure U is equal to the reciprocal value of the average heat-transfer resistances R_T and R_{II}:

$$U = \frac{2}{R_T + R_{II}} \left[\frac{W}{m^2 \cdot K}\right]$$

Case Study Determine the thermal transmittance of the non-homogeneous structure shown in Fig. 1.40. The thicknesses of the segments are d_1 15 cm, d_2 5 cm and d_3 15 cm. The fractions of the surfaces are f_1 0.75 and f_2 0.25. The thermal conductivity λ of segment one is 1 W/mK, of segment two is 0.04 W/mK and of segment three is 0.14 W/mK. The combined resistances to surface heat transfer R_{si} and R_{se} are equal to 0.13 m²K/W and 0.04 m²K/W (Table 1.4).

$$\lambda_1' = \frac{\lambda_1 \cdot d_1 + \lambda_2 \cdot d_2 + \lambda_2 \cdot d_3}{d_1 + d_2 + d_3} = \frac{1 \cdot 0.15 + 0.04 \cdot 0.05 + 0.04 \cdot 0.15}{0.15 + 0.05 + 0.15} = 0.45 \frac{W}{m \cdot K}$$

$$\lambda_2' = \frac{\lambda_3 \cdot d_1 + \lambda_3 \cdot d_2 + \lambda_2 \cdot d_3}{d_1 + d_2 + d_3} = \frac{0.14 \cdot 0.15 + 0.14 \cdot 0.05 + 0.04 \cdot 0.15}{0.15 + 0.05 + 0.15} = 0.097 \frac{W}{m \cdot K}$$

$$\lambda_1'' = \lambda_1 \cdot f_1 + \lambda_3 \cdot f_2 = 1 \cdot 0.75 + 0.14 \cdot 0.25 = 0.785 \frac{W}{m \cdot K}$$

$$\lambda_2'' = \lambda_2 \cdot f_1 + \lambda_3 \cdot f_2 = 0.04 \cdot 0.75 + 0.14 \cdot 0.25 = 0.065 \frac{W}{m \cdot K}$$

$$\lambda_3'' = \lambda_2 = 0.04 \frac{W}{m \cdot K}$$

$$R_{T,1} = R_{si} + \frac{d_1 + d_2 + d_3}{\lambda_1'} + R_{se}$$

$$= 0.13 + \frac{0.15 + 0.05 + 0.15}{0.45} + 0.04 = 0.948 \frac{m^2 \cdot K}{W}$$

$$R_{T,2} = R_{si} + \frac{d_1 + d_2 + d_3}{\lambda_2'} + R_{se}$$

$$= 0.13 + \frac{0.15 + 0.05 + 0.15}{0.097} + 0.04 = 3.778 \frac{m^2 \cdot K}{W}$$

$$R_T = \frac{1}{\frac{f_1}{R_{T,1}} + \frac{f_2}{R_{T,2}}} = \frac{1}{\frac{0.75}{0.948} + \frac{0.25}{3.778}} = 1.166 \frac{m^2 \cdot K}{W}$$

$$R_{II} = R_{si} + \frac{d_1}{\lambda_1''} + \frac{d_2}{\lambda_2''} + \frac{d_3}{\lambda_3''} + R_{se}$$

$$= 0.13 + \frac{0.15}{0.785} + \frac{0.05}{0.065} + \frac{0.15}{0.04} + 0.04 = 4.880 \frac{m^2 \cdot K}{W}$$

$$U = \frac{2}{R_T + R_{II}} = \frac{2}{1.166 + 4.880} = 0.331 \frac{W}{m^2 \cdot K}$$

The thermal transmittances of the non-homogeneous structures U must also not be greater than the permissible value U_{max}. These are the same as for the homogeneous building structures and are listed in Table 1.3.

1.5.3 Heat Transfer and Thermal Transmittances of Building Structures with a Cavity or Gap

Building structures with an air layer are common. Depending on the purpose, the building structure could consist of one or more closed cavities with one or more ventilated air gaps. Closed cavities improve the sound insulation and fire resistance, as well as reducing the thermal transmittance of the building structure. Ventilated air gaps are most commonly on the external side of the building structure, just below the final façade layer or roof cover. This allows the transfer of water vapour, which enters the building structure from the indoor environment, to the air in the surroundings without risk of the interstitial condensation of water vapour and consequently increased moisture in the building structure. The impact of ventilated air gaps on the heat transfer is relatively small, especially if the air flow in the gap is due to the thermal buoyancy and the air velocity does not exceed 0.3 m/s (Fig. 1.41).

A special case is building structures with forced ventilated gaps. In this case, using a mechanical device (e.g., a fan), the velocity of air flow is increased and influences the rate of heat transfer to the air. This helps to keep the roofing cold and to lessen the melting of snow on the roof, as well as making it possible to use solar energy for pre-heating the ventilation air and for the natural heating of buildings (Fig. 1.42).

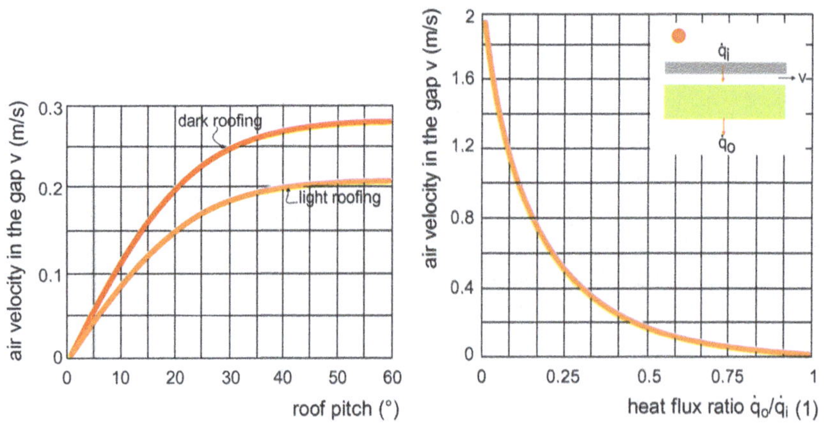

Fig. 1.41 Air velocity in a thermal buoyancy ventilated gap in the roof depends mainly by the roofing temperature and the roof pitch (left). In typical conditions found in a naturally ventilated gap ($v < 0.3$ m/s), the heat-flux transfer into the building $\dot{q}_i$ is only slightly decreased (right). A significant decrease of $\dot{q}_i$ can only be realised by mechanical ventilation of the air gap with air velocities above 1.5 m/s [2]

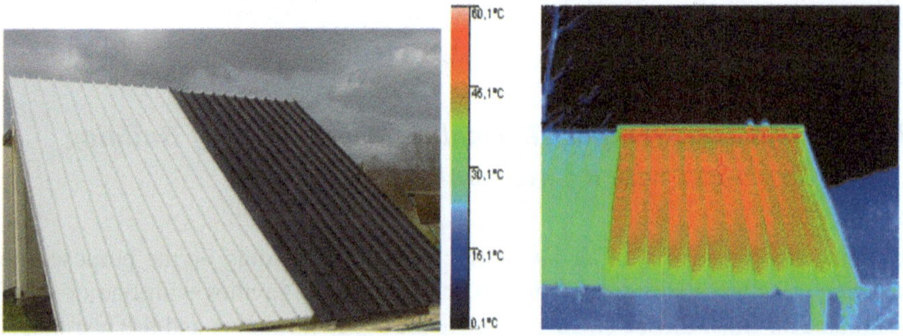

Fig. 1.42 Building structures with ventilated air gaps can be used for pre-heating of the ventilation air or space heating. The difference in the temperature of the white and black ventilated roof structures is shown

Fig. 1.43 Even in the case of a low thermal transmittance of the roof structure, the heat flux exiting the building can cause the formation of dangerous icicles (left). Such conditions can often be eliminated by mechanical ventilation of the gap between the thermal insulation and the roofing. Ventilation with flow rate above 50 m^3/h per each metre of roof width must be provided [11]. However, the formation of icicles is prevented more frequently by installing electric heaters in the rainwater drainage

In this way, for example, we can reduce the melting of snow on roofs, because the roof cover remains cold, or enable the use of solar energy for pre-heating the air for ventilation and the natural heating of buildings (Fig. 1.43).

The heat transfer in building structures with a closed cavity or ventilated air gap can be modelled in two ways:

- **Empirical method.** It is possible to use empirical expressions to determine the heat flux that is transferred by convection on the surfaces of the gap. The expressions include the dimensionless Prandtl (Pr), Grashof (Gr) and Rayleigh (Ra) numbers to model the heat transfer in a closed gap, and the Reynolds number (Re) for the heat transfer in a ventilated gap (see Sect. 1.3.2). In both cases the heat flux inside the cavity or the gap surfaces is also transmitted by the radiant heat flux (see Sect. 1.3.3).
- **Energy conservation method.** By using the law of energy conservation, the heat fluxes and the temperatures in the building structure with an air cavity or gap can be determined. The heat fluxes and the temperatures in the building structure with an air cavity or gap can be determined from the law of the conservation of energy. Here, the heat transfer is usually treated as a one-dimensional problem. This method also calls for a rough simplification of the geometrical shape of the building structure and the shape of the air gap, particularly the inlet and outlet openings of the ventilated air gap.
- **Computational fluid dynamics.** The development of computer technologies and software has enabled us to solve the heat-transfer problems in building structures using computational fluid dynamics (CFD). In this approach, the building structure is divided into a finite number of 2D or 3D elements. For each element, interrelated quantities of the state—the temperature, the pressure and the velocity of the matter in the element—are determined by the laws of the conservation of mass, energy and momentum. With known temperature and velocity fields,

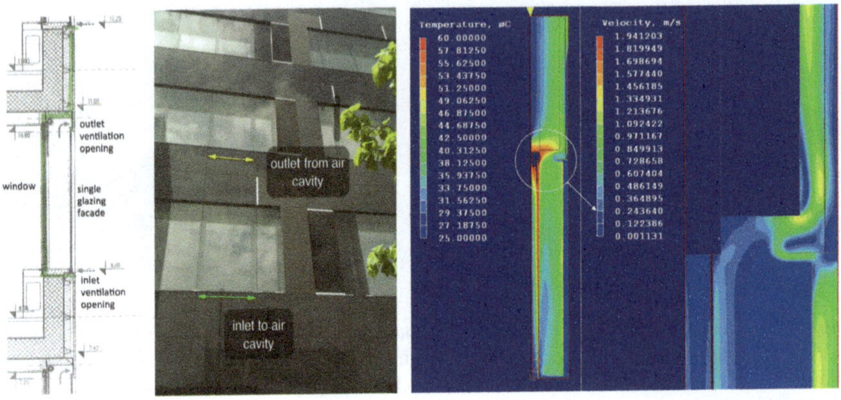

Fig. 1.44 The temperatures and velocities in the ventilated gap of a double-skin façade with shades drawn in the interstitial space simulated with PHONICS CHAM code. The natural flow of air in the gap depends on the solar irradiation absorbed by the façade elements. At the external temperature $\theta_e = 25$ °C and the room temperature $\theta_i = 26$ °C, the average temperature of the internal surface of glazing is 32.2 °C, the density of heat flow rate entering the room is 49.6 W/m^2, and the air flow rate in the ventilated gap is 140 m^3/h [12]

the heat fluxes and the air-flow rate in the cavity or ventilated gap can be determine. CFD numerical methods of the heat-transfer modelling make it possible to analyse building structures with very complex details. It is increasingly popular in engineering practice (Figs. 1.44, 1.45).

However, despite the rapid development of CFD techniques, simpler methods are still employed in engineering practice to determine the thermal transmittance of building structures with a closed cavity or ventilated air gap.

1.5.3.1 Thermal Transmittances of Building Structures with a Closed Cavity

In a closed cavity, the heat flux is transferred by convection and radiation. The heat-transfer resistance is combined into the alternate resistance of the heat transfer of a closed air cavity R_a. In [6] R_a is specified depending on the direction of the heat flux (i.e., the position of the structure), the thickness of the cavity d_a, and the thermal emissivity of the surfaces $\varepsilon_{IR,a}$ enclosing the gap (Fig. 1.46). The values are summarized in Table 1.7.

When calculating the total thermal resistance of a building structure, the air cavity is treated like all the other homogeneous layers. The thermal transmittance U is equal to the reciprocal of the total thermal resistance R of the building structure:

$$\mathrm{U} = \frac{1}{\mathrm{R}} = \frac{1}{\mathrm{R_{si}} + \sum_{\mathrm{j}=1}^{n} \frac{\mathrm{d_j}}{\lambda_\mathrm{j}} + \mathrm{R_a} + \mathrm{R_{se}}} \left[\frac{\mathrm{W}}{\mathrm{m}^2 \cdot \mathrm{K}}\right]$$

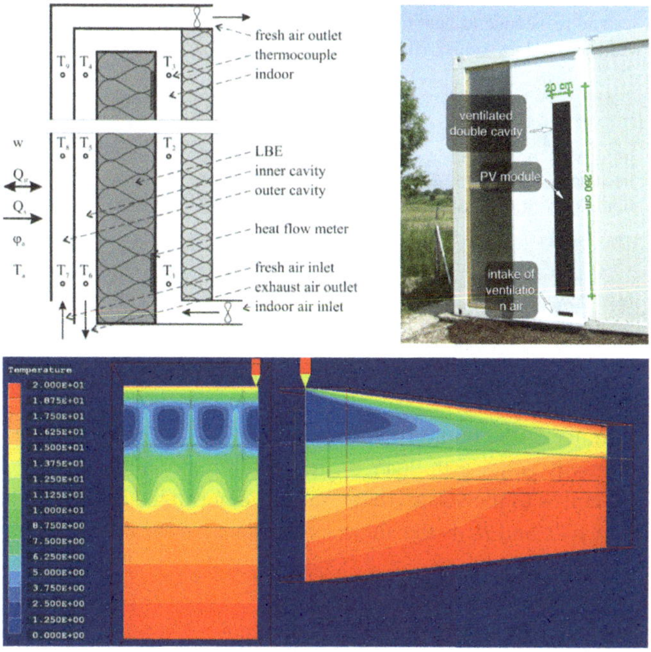

Fig. 1.45 A 2.4-m-long double-site counter-flow mechanically ventilated gap in the thermal insulation panel enables heat transfer between the exhaust and the supply air (acting as a heat exchanger) and the utilisation of solar energy for pre-heating the supply air. In the example shown, the solar energy absorbed by the external surface of the insulation panel is 400 W/m^2, the external temperature is $\theta_e = 0$ °C and the room temperature is $\theta_i = 20$ °C. The airflow in both ventilated gaps is 80 m^3/h. The density of the heat-flow rate entering the room is 1.19 W/m^2, which means the thermal transmittance U of the insulation panel at such conditions is 0.06 W/m^{2}K. The heat flux transferred into the room by the supply air is 480 W [13]

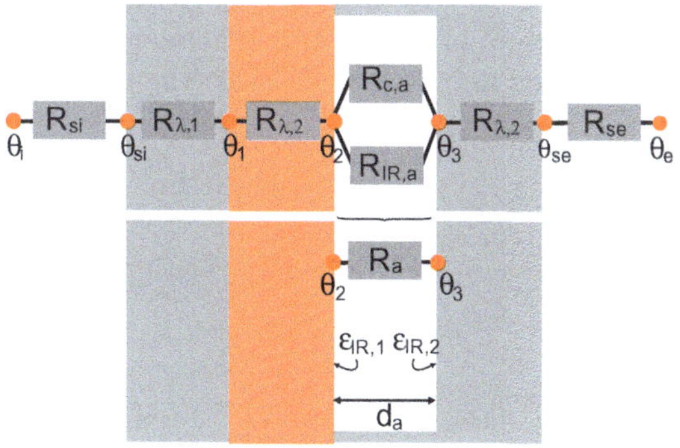

Fig. 1.46 Resistance to heat transfer by radiation and convection in the closed gap is substituted by the alternate resistance R_a

Table 1.7 Alternate resistance to heat transfer R_a of a closed air gap vs. the direction of the heat flux, the gap width d_a and the emissivity of parallel surfaces on the boundary of the air gap $\varepsilon_{IR,1}$ and $\varepsilon_{IR,2}$ [6]

R_a	gap width d_a (mm)						
$\varepsilon_{IR,1} = \varepsilon_{IR,2} = 0.9$	5	7	10	15	25	50	100
heat flow - bottom-up	0.11	0.13	0.15	0.16	0.16	0.16	0.16
heat flow - top-down	0.11	0.13	0.15	0.17	0.18	0.18	0.18
heat flow - horizontal	0.11	0.13	0.15	0.17	0.19	0.22	0.23
R_a	gap width d_a (mm)						
$\varepsilon_{IR,1} = \varepsilon_{IR,2} = 0.05$	5	7	10	15	25	50	100
heat flow - bottom-up	0.19	0.26	0.37	0.46	0.46	0.46	0.46
heat flow - top-down	0.19	0.26	0.37	0.53	0.68	0.68	0.68
heat flow - horizontal	0.19	0.26	0.37	0.53	0.81	1.37	2.31

where j is a counter of the homogeneous layers, d_j and λ_j are the thickness and the thermal conductivity of the j-th layer of the building structure, R_{si} and R_{se} are the combined resistances to the surface heat transfer determined by the direction of the heat flux (Table 1.4), and R_a is the alternate resistance to heat transfer of the closed air cavity (Table 1.7).

1.5.3.2 Thermal Transmittances of Building Structures with Ventilated Gap

The methodology of calculating the thermal transmittances of building structures with a ventilated air gap is given in [6]. The standard distinguishes structures with slightly, moderate and well-ventilated gaps (Fig. 1.47). In all cases it is assumed that the building structure is naturally ventilated by thermal buoyancy. The type of ventilation depends on the total free cross-section of the inlet openings (A_i) and the air outlets (A_o). If it exceeds an area of 1500 mm^2 per metre of length in vertical structures, or 1500 mm^2 per m^2 of surface area in horizontal or slightly tilted structures, the air gap is assumed to be well ventilated. The inlet and outlet openings'

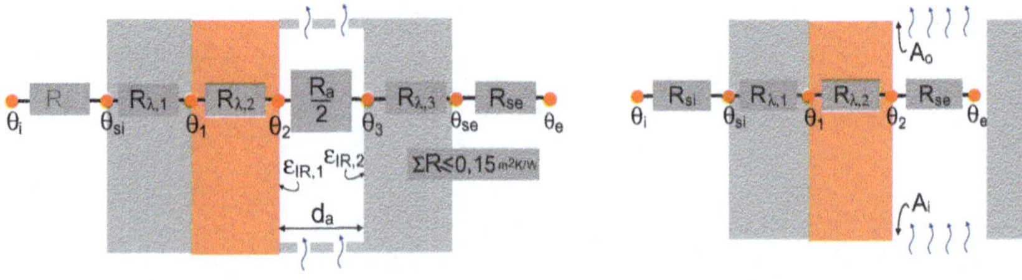

Fig. 1.47 Thermal transmittances U of building structures with a slightly ventilated gap and with a well-ventilated gap are determined differently. In the first case (left) the thermal transmittance is determined by taking into account half of the alternate resistance R_a of the closed gap (Table 1.7). In the second case (right) only the layers from the interior to air gap are considered

Fig. 1.48 Ventilation openings above the apartment ceiling. The same number of equally sized ventilation openings is also situated on the opposing north façade

cross-sections must be the same. The thermal transmittance of a building structure with a ventilated gap is determined as:

- **For a slightly ventilated gap**. The thermal transmittance U of a building structure with a slightly ventilated gap is determined in the same way as the thermal transmittance of a structure with a closed cavity, whereby the alternate resistance to heat transfer in the gap R_a from Table 1.7 is halved, and the total resistance to heat transfer between the air gap and the exterior is no more than 0.15 m^2K/W (Fig. 1.47);
- **For a well-ventilated gap.** The thermal transmittance U of a well-ventilated building structure is determined by calculating the total thermal resistance R_{tot} of the structures, not taking into account any layer between the gap and the exterior, and replacing the exterior surface thermal resistance R_{se} with R_{si}.

Case Study Find whether the cross-ventilated flat-roof structure above an apartment (see Fig. 1.48) is well ventilated with respect to the size of the ventilation openings. The surface area of the horizontal roof structure A_{roof} is 74.4 m^2. The free cross-section is 60% of the grille area (0.2 × 0.3 m^2), and therefore the total free cross-section of the ventilation openings is equal to:

$$A_i + A_o = 2 \cdot 5 \cdot 0.6 \cdot (0.2 \cdot 0.3) = 0.36\,m^2 = 360\,000\,mm^2$$

The ratio of the openings and roof structure areas is equal to:

$$\frac{A_i + A_o}{A_{roof}} = \frac{360\,000}{74.4} = 4839\frac{mm^2}{m^2} > 1500\frac{mm^2}{m^2}$$

It can be concluded that roof structure can be treated as a well-ventilated structure and must be considered as such in a calculation of the thermal transmittance U.

1.5.3.3 Heat Transfer in Transparent Building Structures with a Cavity

The study of the heat transfer in transparent building structures with one or multiple cavities, for example, the window glazing, glazed façades or double skin façades, distinguishes two main cases: the heat transfer due to the difference in temperature between the building and the outdoor environment and the transfer of heat flux resulting from the penetration of solar radiation.

The transfer of the heat flux between the building and the exterior due to temperature differences takes place with all the mechanisms of heat transfer (Fig. 1.49). If the transparent panes are made from glass, the thermal radiation with wavelength λ greater than 3 μm cannot pass the glass and is absorbed in panes due to the greenhouse effect of glass.

In this case the heat transfer is similar to that of opaque building structures with a closed cavity, and the thermal transmittance is determined by taking into account the alternate resistance to heat transfer in a closed cavity R_a. This means that the thermal transmittance of the glazing can be decreased by increasing the number of cavities, and by using low-emission coatings on the surfaces of the glass panes.

The most common transparent building structures with one or more cavities are the windows. Due to the complex geometrical shape and the related evaluation of heat-transfer mechanisms, the thermal transmittance of the windows U_w is determined on the basis of the thermal transmittance of the glazing U_g, the thermal transmittance

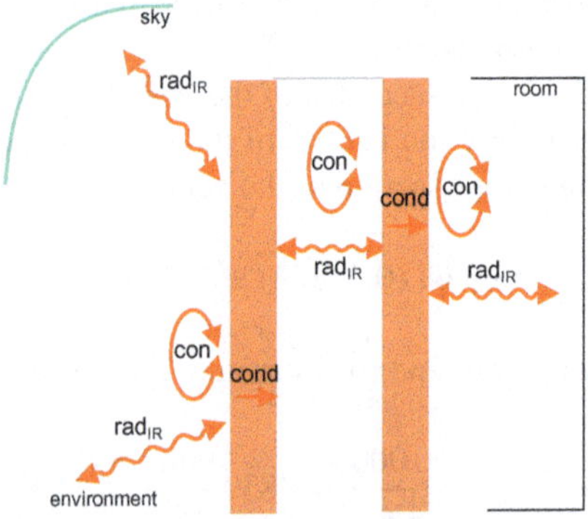

Fig. 1.49 Heat-transfer mechanisms in double glazing. The thermal transmittance of the glazing is influenced above all by the convective (con) and longwave radiant heat flux (rad$_{IR}$) between the window panes

of the window frame U_f and the linear thermal transmittance of the thermal bridge Ψ_{spacer} attributed to the glazing spacers:

$$U_w = \frac{U_g \cdot A_g + U_f \cdot A_f + \Psi_{spacer} \cdot l_{spacer}}{A_g + A_f} \left[\frac{W}{m^2K}\right]$$

where A_g and A_f (m^2) are the glazing and window frame areas, and l_{spacer} (m) is the total length of the circumference of the spacer in the glazing (Fig. 1.50).

The thermal transmittance of the window glazing U_g depends above all on the width d_a and the height of the cavity H_a, the gas filling the cavity and the emissivity of glass surfaces enclosing the cavity. The window glazing cavities are usually filled with the noble gases, i.e., argon, krypton and xenon, which are known to have a considerably larger density ρ and smaller thermal conductivity λ than air (Table 1.8). These material properties and the aspect ratio of the cavity (the ratio of cavity height to width $A_a = d_a/H_a$), influence the heat transfer between the window panes by convection. The convective surface heat-transfer coefficient $h_{c,a}$ (W/m^{2}K) in the cavity in relation to the width of the cavity d_a and the gas filling is shown in Fig. 1.51. As the convective heat transfer involves the conduction of heat in the gas

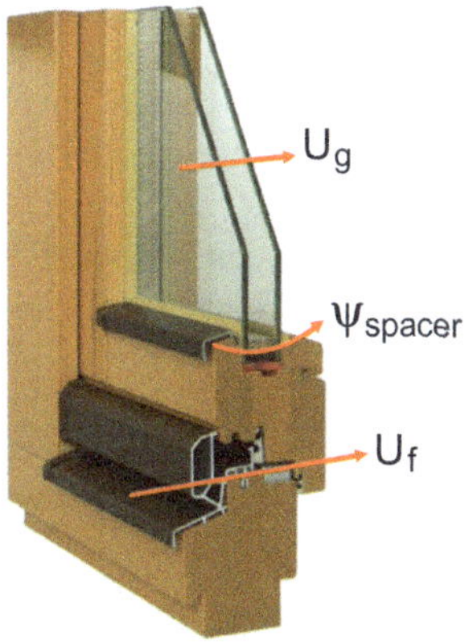

Fig. 1.50 Thermal transmittance of a window is determined by the thermal transmittance of the glazing U_g, the frame U_f, and the linear thermal bridge transmittance of the spacer ψ_{spacer}

Table 1.8 Relative density and thermal conductivity of noble gases compared to the air

gas	density ratio ρ (1)	thermal conductivity ratio λ (1)
air	1	1
argon	1.39	0.74
krypton	2.91	0.39
xenon	4.58	0.24

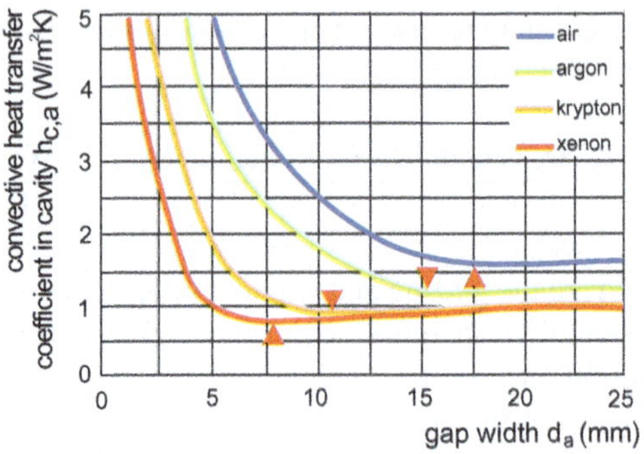

Fig. 1.51 Convective surface heat-transfer coefficient $h_{c,a}$ inside the window glazing depends on the cavity width d_a and the gas filling. The convective heat transfer in glassing filled by argon is the lowest at d_a ~ 16 mm (for aspect ratio A_a ~ 75) and in glazing filled by krypton at d_a ~ 12 mm (for aspect ratio A_a ~ 100). This means that the gaps in the krypton-filled glazing can be narrowed, which is especially important in case of triple-glazed windows

and simultaneously the movement of gas particles, the local minimum values of the convective heat-transfer coefficient are found for different aspect ratios of the gap. In Sect. 1.3 we noted that glass is not transparent for long-wavelength thermal radiation and absorbs it instead. Because of the high thermal emissivity of natural glass, the radiative heat transfer is the prevailing heat-transfer mechanism between the window panes. This deficiency of glass can be improved by implementing a low-emission coating (Fig. 1.52).

Low-emission coatings on the glass can be "hard" or "soft". Hard coatings are thin metal oxide films (for example, tin oxide SnO_2), having a thickness of the order

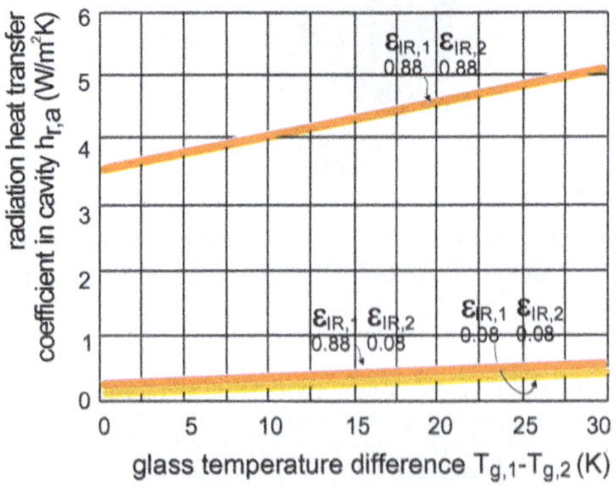

Fig. 1.52 Radiation surface heat-transfer coefficient $h_{r,a}$ between glass panes in the window glazing. Natural glass has a high long-wavelength thermal emissivity (ε_{IR} ~ 0.88). With a thin-film low-emissivity coating, the emissivity can be reduced to ~ 0.05 and the radiative heat transfer between the glass panes significantly decreases, although a low-emissivity coating on the second glass has a negligible impact on the radiative surface heat-transfer coefficient

Fig. 1.53 Changed colour of the flame's reflection reveals the glazing surface with a low-emission coating (left). The holes on the spacer bare enable the water vapour molecules to enter the adsorbent. The basic parameters of the glazing are displayed on the spacer bar (right)

of tens of nanometres, deposited in a hot glass surface by a spraying or pyrolysis process. The metal in a liquid compound (for example, tin chloride $SnCl_2$) is sprayed with propellant gas onto the hot glass surface where the high temperature triggers a chemical reaction. The "soft" low-emission coatings are applied to the glass in a vacuum chamber by the sputtering of atoms from metal electrodes. The presence of oxygen causes metal oxides to form on the surface of the glass. This technology can be used to manufacture multilayer thin films (for example, $TiO_2/Ag/TiO_2$), with a thickness of the order of a few μm to a few tens of μm. By adapting the thicknesses of each layer, the emissivity at certain wavelengths can be modified. Hard low-emission coatings exhibit better mechanical durability, but they also have a higher emissivity of thermal radiation (e_{IR} between 0.1 and 0.15). Soft coatings with an emissivity e_{IR} under 0.05 are used in modern glazing, so it suffices if only one of the window panes surrounding the gap is coated (Fig. 1.53).

The thermal transmittance of glazing is also be reduced by multiple cavities. The indicative values of the thermal transmittance for different glazing are listed in Table 1.10.

Window frames are mainly made of wood, plastic or metal (aluminium). The choice of materials depends on the mechanical strength, the required maintenance, the cost, and the required thermal transmittance U_f of the frame. Wooden window frames are built with a solid wooden profile that is 70–100 mm thick and have the lowest thermal transmittance U_f. The thermal transmittance of plastic and metal can be decreased by dividing the frame into partitions or chambers, and by the interruption of the thermal bridges formed on the frame profile's ribs. Nevertheless, the thermal transmittance of plastic and metal frames exceeds the thermal transmittance of wooden frames (Fig. 1.54). Windows suitable for installation into energy-efficient buildings have frames packed with layers of thermal insulation, so the thermal transmittance of the frame is not greater than the thermal transmittance of the glazing

Fig. 1.54 Window frame made of plastic. The frame profile is divided into sections and chambers to increase the resistance to heat transfer. The depicted frame has 5 chambers [14] (left). A window suitable for installation into an energy-efficient building. Triple glazing with a low-emission coating filled with krypton. A window frame with additional thermal insulation layer covered by an aluminium profile on the external side that ensures low maintenance cost and long durability [14] (right)

U_g. For example, the glazing's and the frame's thermal transmittance for windows installed in a passive buildings must be under 0.8 W/m^{2}K.

The spacer between the window panes has several functions. The primary and secondary seals make sure the glazing gaps stay tight. The spacers are also filled with water vapour sorption material (i.e., silica gel) which binds the molecules of water vapour that pass the glazing by diffusion. This prevents the formation of water droplets on the glass surfaces inside the cavities at low outdoor-air temperatures. Once the solar radiation heats up the window and the sorbent, the molecules of water vapour are released without any risk of window fogging. Because of the spacer, a thermal bridge is formed, which is evaluated by the linear bridge thermal transmittance Ψ_{spacer}. It depends on the material from which the spacer is made. For an ordinary aluminium spacer, the Ψ_{spacer} is approximately 0.10 W/mK, and for a plastic spacer it is only 0.03–0.04 W/mK. Note that spacer thermal bridge is defined per length of 1 m.

The thermal transmittance of windows U_w can be determined using numerical methods and computer tools. Such tools require some simplification of the geometric model of the glazing, frame and spacer. Because of that, in engineering practice, the thermal transmittance of glazing and windows is commonly determined experimentally, as shown in Fig. 1.55. The glazing or a complete window is placed between two temperature-controlled rooms, in which different temperatures θ_i and θ_e are established. The density of the heat flux transferred in steady-state conditions is measured as well as the temperatures of the air and the surfaces. The ratio of the heat flux density $\dot{q}_i$ that passes the glazing or the whole window and the differences in the

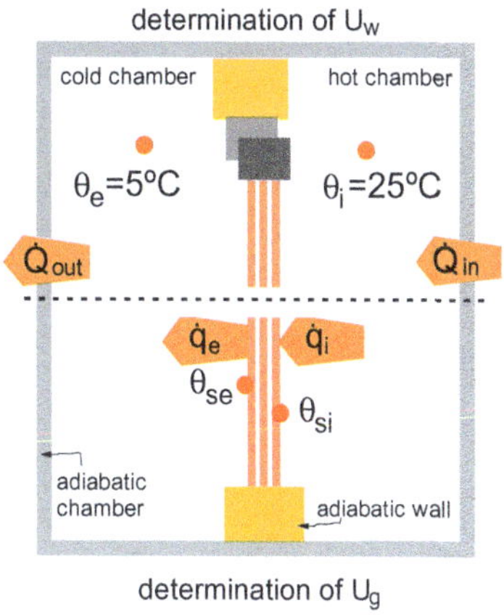

Fig. 1.55 Thermal transmittance of the windows U_w, the glazing U_g and the frames U_f can be determined by experiments, installing the windows in the adiabatic wall between two temperate spaces. By measuring the density of the heat flux that transfer the window or the glazing and temperatures of the air in the spaces and the temperature on the surface of the glazing. Numerical methods and computer tools can be used instead of experiments to determine the thermal properties of windows

air temperatures in the temperature-controlled spaces $(\theta_i - \theta_e)$, the thermal transmittances U_w and U_g are determine. The same method could be used for an in-situ evaluation of the thermal properties of windows in quasi-steady-state conditions.

The typical values of windows, glazing and frames are given in Table 1.10. The table also shows how the size of the window influences the thermal transmittance U_w. In some countries, instead of thermal transmittance, requirements for windows are stated according to the all-year energy balance—the ratio between the heat losses and the solar heat gains. An example of such a balance is shown in Table 1.10, taking into account the climate conditions in the city of Ljubljana. A positive energy balance indicates that the solar gains during the heating period prevail over the heat losses.

As is the case with other building structures, the thermal transmittance of the windows must not exceed the permissible thermal transmittance. As an example, the values for $U_{g,\max}$, $U_{f,\max}$ and $U_{w,\max}$ as defined by the law in Slovenia, are listed in Table 1.9 [9].

The glazing is transparent to short-wavelength solar irradiation with wavelengths between 0.3 and 3 μm. The solar irradiation is transferred to the building as a direct beam irradiation if the glazing is transparent or as diffuse irradiation if the glazing is translucent. In both cases part of the solar irradiation is reflected at the boundary separating the two materials of different density (glass-air or air-glass), while a part of the solar irradiation is absorbed when passing through the glass. The internal glass pane is therefore heated to a temperature θ_{si}, which can be higher than the building's indoor-air temperature θ_i, causing the heat flux to pass from the surface of the glass into the room by convection and long-wavelength radiation (Fig. 1.56). The total specific heat flux $\dot{q}_i$ entering the room through 1 m^2 of glazing due to the

Table 1.9 Development of the maximum permitted thermal transmittance of windows, glazing and frames as set in the Slovenian requirements regarding to the energy efficiency of buildings; the similar values were adopted in other EU countries [7–9]

	Permitted thermal transmittance $U_{g,max}$, $U_{f,max}$, $U_{w,max}$ (W/m²K)		
	2002	2008	2010
glazing	1.4	1.1	1.1
wooden or plastic frame	1.6	1.3	1.3
metal frame	1.8	1.6	1.6
windows, glass walls	1.4	1.3	1.3
roof windows	-	1.3	1.4
suntunnels, skylights	-	2.7	2.4

Table 1.10 Thermal properties of different windows and glazing [15–17]

glazing	U_g (W/m²K)	g (1)	τ_{vis} (1)
single	5.80	0.85	0.90
double	2.80	0.80	0.80
triple	2.10	0.72	0.65
double/ Ar	1.40	0.78	0.80
double / Ar + 1 low-e	1.00 - 1.20	0.64	0.78
double / Kr + 1 low-e	0.80 - 1.10	0.64	0.78
triple / Ar + 2 low-e	0.80	0.50 – 0.60	0.71
triple/ Kr + 2 low-e	0.50	0.50 – 0.60	0.72
quadruple/ Kr + 4 low-e	0.30	0.40	0.60
solar control	1.10	0.12 – 0.40	0.40 – 0.65
vacuum insulated double/triple *laboratory value	0.50 – 0.30	0.65 – 0.60	0.70 – 0.80
frame	**share of frame surface area is 25% of window area**		
	U_g (W/m²K)	U_f (W/m²K)	U_w (W/m²K)
wooden	0.80	1.50 – 0.80	1.10 – 0.80
plastic	0.80	1.30 – 1.10	1.10 – 0.80
aluminium	0.80	1.50 – 1.20	1.10 – 1,00
window size	**window thermal transmittance U_w (W/m²K)**		
	U_g 1.3 (W/m²K)	U_g 0.9 (W/m²K)	U_g 0.4 (W/m²K)
0.6 . 0.6 m	1.50	1.15	0.60
1.0 . 1.0 m	1.45	1.10	0.55
1.0 . 2.0 m	1.40	1.00	0.45
window	**energy balance of a vertical south window in Ljubljana climate (kWh/m² an)**		
	south	east	north
U_w 1.40 W/m²K, g 0.68	+ 1.1	- 41.0	- 68.3
U_w 0.90 W/m²K, g 0.50	+ 1.7	- 25.2	- 35.6

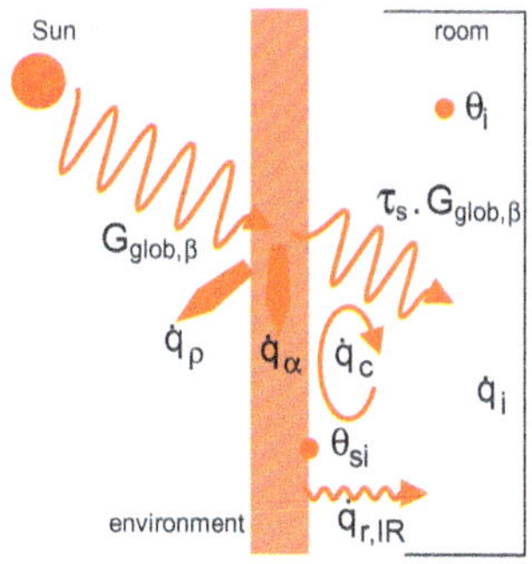

Fig. 1.56 Total heat flux q_i entering the room due to the insolation of the window equals the sum of the transmitted solar radiation $\tau_s.G_{glob,\beta}$, the convective heat flux $\dot{q}_c$ and the long-wave heat flux $\dot{q}_{r,IR}$ transferred into the room from the internal surface of the glazing. The ratio of these heat fluxes determines the total solar energy transmittance of the glazing g

solar irradiation $G_{glob,\beta}$ equals the sum of the transmitted solar irradiation and the convective $\dot{q}_{c,}$ and radiant $\dot{q}_{r,IR}$ specific heat flux entering the room from the internal surface of the glazing. The transmitted solar irradiation equals the solar irradiation $G_{glob,\beta}$ multiplied by the transmittance of the solar irradiation of the glazing ($\tau_s \bullet G_{glob,\beta}$). The ratio of the total specific heat flux $\dot{q}_i$ to the solar irradiation received by the external surface of glazing $G_{glob,\beta}$ is referred to as the total solar energy transmittance g:

$$g = \frac{\tau_s \cdot G_{glob,\beta} + \dot{q}_c + \dot{q}_{r,IR}}{G_{glob,\beta}} = \frac{\dot{q}_i}{G_{glob,\beta}} \left[\frac{W \cdot m^2 \cdot K}{m^2 \cdot K \cdot W} = 1 \right]$$

The total solar energy transmittance g is a property of the glazing, or the system comprising of the glazing and the window blinds or shading devices. It is an indicator that can be used to evaluate the risk of building overheating in the summer. The total solar energy transmittance of the solar control glazing should be well below 0.25. The thin coatings on the glass can be adjusted to modify the transmission of the thermal (IR) part of the solar radiation in the wavelengths above 0.8 μm and to a lesser extent the transmission of visible light (0.38 μm < λ < 0.76 μm). The solar radiation transmittance τ_s can, as a result, differ from the visual light transmittance τ_{vis}. This must be considered when planning the natural lighting of rooms. However, since such glazing affects the natural heating of buildings, a minimum permissible total solar energy transmittance is required for energy-efficient buildings. In passive buildings, the value of g for the glazing must be greater than 0.5, and the risk of building overheating is mitigated with shading devices. To be efficient, they must be installed on the external side of the glazing, must have a high reflective surface for short-wavelength solar irradiation, and must be offset from the plane of the window so they can be cooled by convection through the ventilated gap between the window and the shading device. The efficiency of the shading is expressed by the combined

Fig. 1.57 Shading devices should have a high solar irradiation reflectance and must be offset from the façade, so they cool by circulating air on both sides and enable daylighting (left). Shades should be controlled by a building management system (BMS) (middle), nevertheless they must be designed so that the occupants can adjust the position of the shades to personal needs (right)

total solar energy transmittance of the shades and the glazing. For the cases shown in Fig. 1.57, the combined total solar energy transmittance is below 0.05.

1.5.4 Heat Transfer in Greened Building Structures

Greened building structures have an important role in mitigating the site's microclimate and in decreasing the heat transfer into the building during the summer as well as during the winter (Fig. 1.58). Furthermore, greened building structures retain the precipitation in the vegetative blanket. Plants absorb CO_2 and other pollutants from the air, retain particulate matter on the surfaces of the leaves and absorb the urban noise (Fig. 1.59).

Leaves have a high absorptivity for solar radiation and the emissivity of thermal radiation, and are naturally cooled by evapo-transpiration (Fig. 1.60). Therefore, leaves act as very efficient micro sunshades. Evapo-transpiration is a process that involves the evaporation of water from the soil below the pant and the transpiration of water moves through a plant and evaporates from the leaves, stems and flowers. The result of evapo-transpiration is the evaporative cooling of air in the vegetation blanket and in the canopy of the trees.

The density of the heat flux due to evapotranspiration $\dot{q}_{ET}$ (W/m^2) is determined by the amount of water that evaporates and the latent heat of water evaporation r (J/kg) required to transform one kilogram of water into water vapour. At an average temperature of the plants, r is 2500 kJ/kg or 0.68 kWh/kg. The evaporative heat-flux density is:

$$\dot{q}_{ET} = r \cdot ET = r \cdot ET_0 \cdot LAI \left[\frac{W}{m^2} \right]$$

Fig. 1.58 Trees and plants influence the heat transfer in building structures and the thermal comfort of outdoor environment by shading the surfaces that would otherwise be exposed to solar irradiation

Fig. 1.59 Temperature of the vegetation blanket is significantly lower than the surface temperature of ordinary building structures. For this reason greened roofs and facades regulate the microclimate in the urban environment and mitigate the overheating of the buildings

where ET (mm/m^{2}h) is the amount of water that evaporates from 1 m^2 of greened area per hour, ET_0 (mm/m^{2}h) is the reference amount of evaporated water per hour from 1 m^2 of meadow with a grass height of 12 cm and LAI (m^2/m^2) is the leaf-area index (Fig. 1.61). ET_0 depends on the soil moisture in the area of the root system, the solar

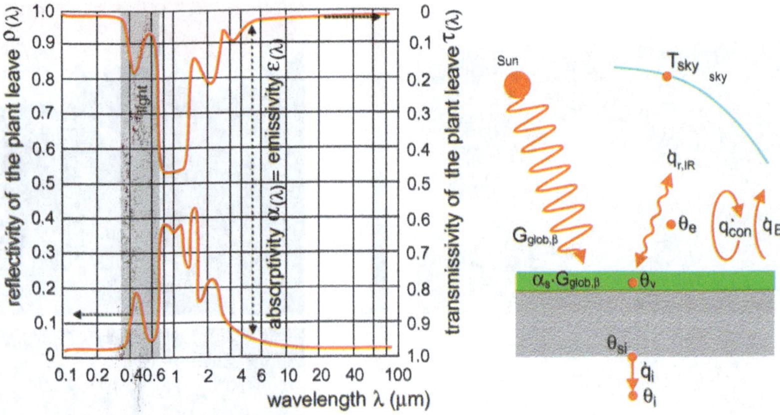

Fig. 1.60 Radiative properties of plant leaves are adapted to the absorption of wavelengths involved in photosynthesis (~ 0.43 and ~ 0.66 μm), while the leaves have a high reflectivity with respect to the thermal part of the solar radiation ($0.78 < \lambda < 3$ μm) and a high emissivity of thermal radiation ($\lambda > 3$ μm), which reduces the overheating of the leaves (left). The vegetation blanket temperature θ_v is determined by the energy balance of the heat fluxes. The greened structures differ with respect to the others by the evapo-transpiration heat flux $\dot{q}_{ET}$, with which a major part of the absorber solar irradiation by the leaves ($\alpha_s \times G_{glob,\beta}$) is transferred from the plant to the surroundings. Because of this, the temperature of the vegetation blanket is very close to the temperature of the outdoor air θ_e even in case of high solar irradiation (right)

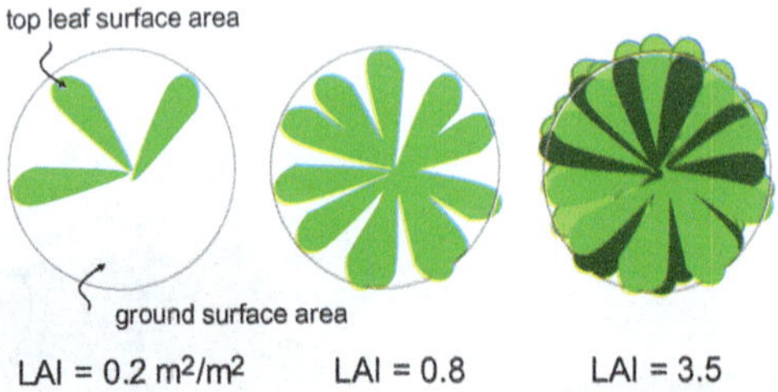

Fig. 1.61 Leaf area index *LAI* is the ratio of the total plant leaf top surface area to the surface area of the ground beneath the plant. Typical values are between 0 for barren ground and 8 m^2/m^2 for thick woods. The *LAI* of grassy land is 1 m^2/m^2

radiation absorbed by the vegetative blanket, the wind speed, and the temperature and relative humidity of the surrounding environment. It is determined with an empirical expression that depends on the outdoor-air temperature θ_e (°C), the relative humidity of the outdoor air φ_e (1) and the absorbed solar irradiation $\alpha_s \times G_{glob,\beta}$ (W/m^2K) [16]:

$$\mathrm{ET_o} = 0.0299 + 0.005 \cdot \theta_e - 0.0022 \cdot \varphi_e + 0.00058 \cdot \left(\alpha_s \cdot G_{glob,\beta}\right)\left[\frac{\mathrm{mm}}{\mathrm{m^2 \cdot h}}\right]$$

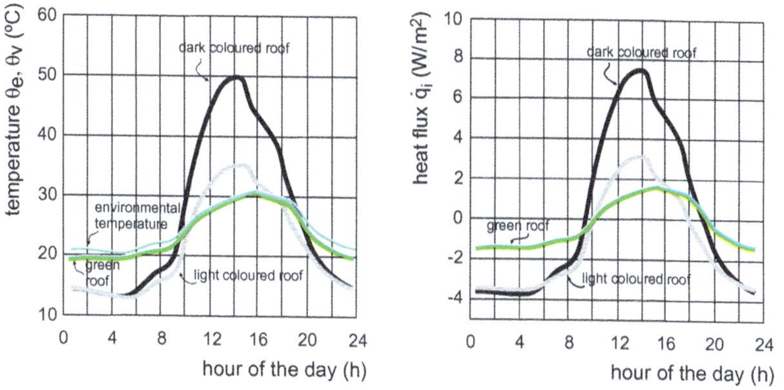

Fig. 1.62 Extremal surface temperatures θ_e of ordinary flat roofs with different absorptivity of solar irradiation (dark and light surfaces) and the vegetation blanket of the greened flat roof θ_e with a thermal transmittance U 0.300 W/m^{2}K over a clear summer's day (left). The density of the heat flux that enters (+) or exits (−) the building over the same summer day (right)

In general, the average daily value of ET_o in places with dry and moderate climate is between 3 and 6 mm/m^{2}h at an outdoor-air temperature θ_e ~ 20 °C, and 6–10 mm/m^{2}h at an outdoor-air temperature θ_e ~ 30 °C. Nevertheless, this will be the case if enough water in the soil is available to the plant's root system. Various plants can be used for the greening of building structures. Since the size of the leaf affects the evapo-transpiration process, the ratio between the total top surfaces of the plant leaves to the surface of the soil beneath the plant or the leaf area index *LAI* is defined as a property of the plants. For barren land, the *LAI* is 0, for a dense forest it is up to 8 m^2/m^2. Extensive plants like moss, grass and herbs, which do not require special care, are mainly used for the greening of the building structures. Their *LAI* is less than 1.

With an iterative solution of the energy balance of heat fluxes shown in Fig. 1.60, we determine the temperature of the vegetation blanket θ_v and the heat flux density that enters and exits the building $\dot{q}_i$. From an example shown in Fig. 1.62, the respiration of the plants during the night increases the temperature of the vegetation blanket and the radiant cooling of the greened roof is much less pronounced.

1.5.4.1 Thermal Transmittance of Greened Building Structures

The thermal transmittance of a greened building structure is determined by considering the resistances to heat conduction of all the layers that are not in contact with precipitation water (Fig. 1.63). The penetration of the precipitation water is prevented by a waterproofing layer and it is assumed that all the layers above the waterproofing are wet and have no resistance to heat transfer. The exception is waterproof thermal insulation materials (i.e., extruded polystyrene foam XPS), which are taken into account in the calculation of the thermal transmittance, despite being placed on the

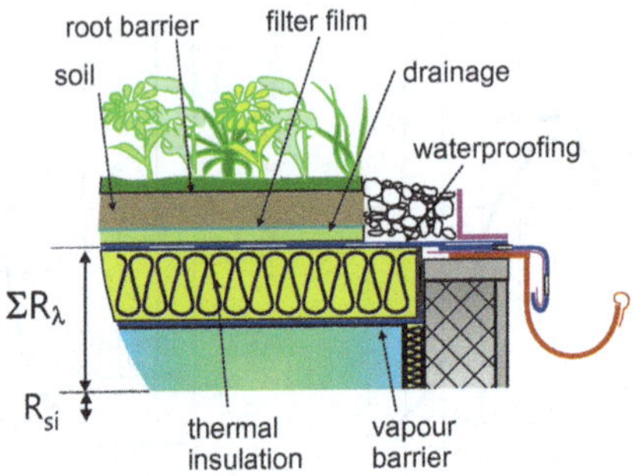

Fig. 1.63 Only the layers from the interior up to the waterproofing should be considered when determining the thermal transmittance of greened building structures

outside of the waterproofing. The thermal transmittance of the greened structure is determined with the equation:

$$U = \frac{1}{R_{tot}} = \frac{1}{R_{si} + \sum_{j=1}^{n} \frac{d_j}{\lambda_j}} < U_{max}\left[\frac{W}{m^2 \cdot K}\right]$$

where j is the number of layers of the building structure from the inside to the first wet layer. The combined resistance to convective and radiative heat transfer on the inner surface R_{si} is presented in Table 1.3. As an example, Table 1.11 presents the permissible thermal transmittance of a greened building structure, as required by Slovenian regulations on the energy efficiency of the buildings.

Table 1.11 Thermal transmittances U_{max} for greened building structures, as defined by Slovenian legislation on energy-efficient buildings [7–9]

	maximum thermal transmittance of vegetated building structures U_{max} (W/m^2K)		
	2002	2008	2010
residential building roofs	0.25	0.20	0.20
other building roofs	0.25	0.35	0.20
residential building façades	0.60	0.28	0.28
other building façades	0.60	0.35	0.28

1.5.5 *Heat Transfer in Building Structures in Contact with the Ground*

Over the course of a year, the soil temperature changes much less than the outdoor air temperature. The impact of the absorbed solar irradiation on the surface of the ground and the periodic variation of the outdoor-air temperature over the year in dry soil can only be noticed up to ~ 20 m beneath the surface. Above this depth, the soil temperature varies periodically over the period of one year, and only at depths exceeding 50 m does the soil temperature start to rise because the geothermal heat flux flows from the hot Earth's core.

Since buildings are usually built either on the surface of the ground or dug up to a depth of some 10 m into the soil, the heat losses of the building structures in contact with the ground depend on how the soil temperature changes over the year. The transient soil temperature θ_{soil} depends on a set of three meteorological variables:

- average annual outdoor-air temperature $\overline{\theta}_{e,an}$ (°C);
- amplitude of the outdoor-air temperature variation over a year $\widehat{\theta}_{e,an}$ (°C); amplitude is half the difference between the highest and lowest outdoor-air temperatures during the year;
- number of days Δn from 1st of January to the day when the lowest annual outdoor-air temperature is measured.

Indicative values for different climate zones are listed in Table 1.12.

The thermal properties of soil are expressed by the thermal diffusivity a_{soil} (Table 1.13), which combines the thermal conductivity of the soil λ_{soil} and the density ρ_{soil} and the specific heat capacity $c_{p,soil}$:

$$a_{soil} = \frac{\lambda_{soil}}{\rho_{soil} \cdot c_{p,soil}} \left[\frac{W \cdot m^3 \cdot kg \cdot K}{m \cdot K \cdot kg \cdot J} = \frac{W \cdot m^3 \cdot kg \cdot K}{m \cdot K \cdot kg \cdot W \cdot s} = \frac{m^2}{s} \right]$$

Table 1.12 Indicative values of the meteorological variables that influence the heat flux that passes the through building structures in contact with the ground. Heating degree days DDs, designed winter outdoor-air temperature $\theta_{e,p}$, average annual outdoor-air temperature $\overline{\theta}_{e,an}$, the amplitude $\widehat{\theta}_{e,an}$ of the outdoor-air temperature over the year and the number of days Δn after the beginning of the year when the lowest daily average outdoor-air temperature occurs

climate	DD (Kday)	$\theta_{e,p}$ (°C)	$\theta_{e,av}$ (°C)	A_e (°C)	Δn (days)
mediterranean	1880	- 6	13.8	9.0	37
	2700	- 12	10.8	9.5	37
continental	2980	- 18	9.7	10.5	35
	3200	- 21	9.2	11.0	35
alpine	3500	- 24	8.2	11.0	32

Table 1.13 Thermal conductivity λ_{soil} and thermal diffusivity a_{soil} of different ground types

ground type	λ_{soil} (W/mK)	a_{soil} (m^2/s)
fine dry sand, desert sand	0.27	$2.23 \cdot 10^{-7}$
clay	1.30	$1.01 \cdot 10^{-6}$
granite	2.79	$1.37 \cdot 10^{-6}$
quartz sand	5.38	$1.84 \cdot 10^{-6}$

Explanation The low thermal diffusivity α_{soil} of desert sand is the reason why less heat is stored in the desert ground and as consequence days in the deserts are hot, and nights are cold. Meanwhile, the high thermal diffusivity of quartz sand is the reason why ground heat exchangers are embedded in a layer of quartz sand to increase the heat transfer between the soil and such heat exchangers.

The average daily soil temperature at a depth Z (m) beneath the surface of the ground for a given day n is determined as a periodic function (Fig. 1.64):

$$\theta_{soil,(Z,n)} = \overline{\theta}_{e,an} - \hat{\theta}_{e,an} \cdot e^{K} \cdot \cos\left[\frac{2 \cdot \pi}{365}(n - \Delta n) - \frac{Z}{2}\left(\frac{365}{\pi \cdot a^{*}_{soil}}\right)^{1/2}\right] [°C]$$

where the exponent K defines the periodic variation of the amplitude $\hat{\theta}_{e,an}$ over a period of 1 year [1]:

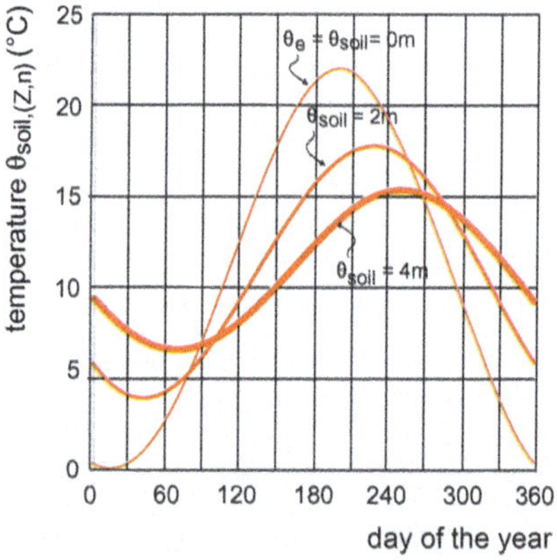

Fig. 1.64 Temperature $\theta_{soil(Z,n)}$ in dry clay ground at different distance Z beneath the surface determines for a location with $\overline{\theta}_{e,an}$ 9.7 °C and $\hat{\theta}_{e,an}$ 10.5 °C

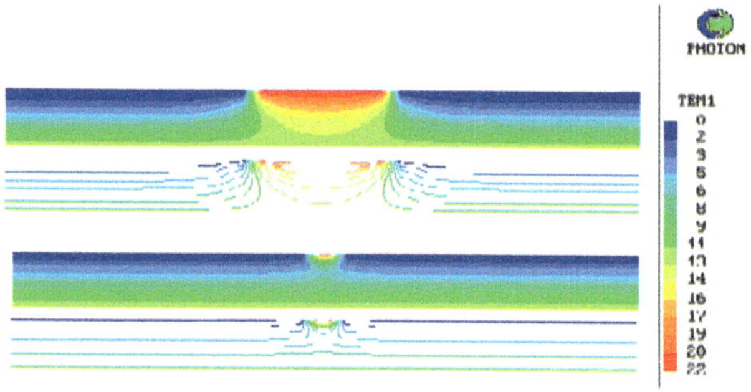

Fig. 1.65 Temperature distribution in the ground under a 50 × 50 m^2 and a 10 × 10 m^2 building, down to a depth of 50 m. The thermal losses through the building structures in contact with the ground are considerably smaller than the losses into the air through building structures at the building envelope. Due to a large affected zone, the numerical simulation of heat transfer through the structures in contact with the ground must consider the ground volume, a multiple of the building volume

$$K = -Z \cdot \left(\frac{\pi}{365 \cdot a^*_{soil}} \right)^{1/2}$$

To match the units, the thermal diffusivity of soil a_{soil} (m^2/s) must be converted into $a_{soil}{}^*$ (m^2/day):

$$a^*_{soil} = a_{soil} \cdot 60 \cdot 60 \cdot 24 \left[\frac{m^2}{s} \frac{s}{min} \frac{min}{h} \frac{h}{day} = \frac{m^2}{day} \right]$$

The heat losses of the building structure in contact with the ground such as a floor on the ground, a submerged floor, a wall dug into the ground or an underground ceiling depends not only on its structure, but also on the properties of the soil and the size of the structure. As shown in Fig. 1.65, during the winter period, the temperature of the soil is higher if the area of the structure on the envelope of the building is larger. Because of that, the impact of the size of the building structure in contact with the ground is modelled by characteristic dimension B′:

$$B' = \frac{A_f}{0.5 \cdot P} [m]$$

where A_f (m^2) is the area of the slab-on-the-ground building structure, P (m) is the perimeter of the structure.

1.5.5.1 Thermal Transmittance U_{gr} and the Steady-State Heat-Transfer Coefficient H_{gr} of Building Structures in Contact with the Ground

The methodology for calculating the thermal transmittance of building structures in contact with the ground is provided in [18]. The methodology differs depending on the type of building structure, which can be a slab-on-the-ground structure or a structure below ground level.

Slab-on-the-ground building structures can be either with or without continuous thermal insulation (Fig. 1.66) or with edge horizontal or vertical thermal insulation (Fig. 1.69). The thermal transmittance of these structures depends on the characteristic dimension of floor B' (m) and the total equivalent thickness d_f (m) of the building structure defined by the expression:

$$d_f = d_w + \lambda_{soil}(R_{si} + R_c + R_{se})(m)$$

where d_w (m) is the full thickness of the external wall joined with a slab-on-the-ground structure, λ_{soil} (W/mK) is the thermal conductivity of soil, R_{si} and R_{se} (m^{2}K/W) are the computational surface resistance to heat transfer on the internal and external surfaces of the structure (Table 1.14), and R_c (m^{2}K/W) is the thermal resistance of the floor slab. The calculation of R_c accounts for the layer of thermal insulation, while

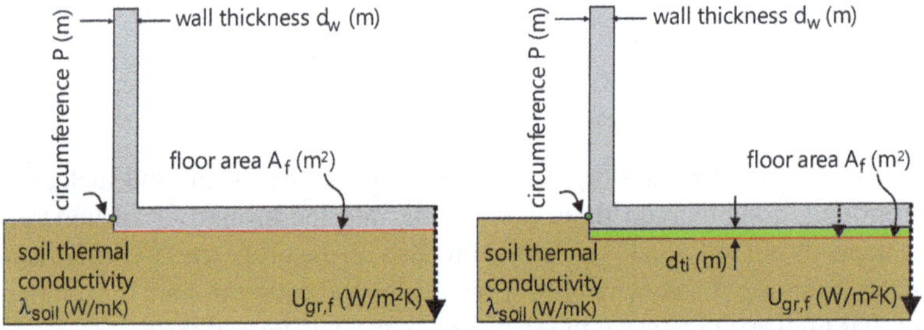

Fig. 1.66 Slab-on-the-ground building structures without thermal insulation (left) and with continuous thermal insulation layer (right). The parameters involved in the determination of thermal transmittance $U_{gr,f}$ are indicated

Table 1.14 Computational surface heat-transfer resistances of the building structures in contact with the ground

surface and direction of heat flow	resistance to heat transfer R_{si}, R_{se} (m^{2}K/W)
internal, top-down	0.17
internal, horizontal	0.13
internal, bottom-up	0.10
external, all directions	0.04

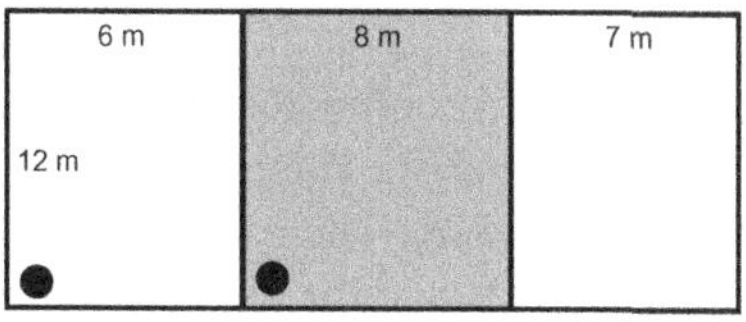

Fig. 1.67 Ground-floor plan of a row of houses. The external dimensions of the floor structure are shown

the dense concrete slab and the floor covering can (but not necessarily) be neglected. It follows that R_c of uninsulated floor structures can be 0 m^{2}K/W if surface resistances are neglected. Two examples of such structures are shown in Fig. 1.66. The thermal transmittance $U_{gr,f}$ is determined as:

$$U_{gr,f} = \frac{2 \cdot \lambda_{soil}}{\pi \cdot B' + d_f} \ln\left(\frac{\pi \cdot B'}{d_f} + 1\right) \quad \text{for} \quad d_f < B' \left[\frac{W}{m^2 \cdot K}\right]$$

$$U_{gr,f} = \frac{\lambda_{soil}}{0.457 \cdot B' + d_f} \quad \text{for} \quad d_f \geq B' \left[\frac{W}{m^2 \cdot K}\right]$$

Case Study Determine the thermal transmittance of a slab-on-the-ground structure for the final ($U_{gr,f1}$) and the middle ($U_{gr,f2}$) row house with the size of the floor slabs shown in Fig. 1.67. Both slabs have a continuous thermal insulation layer of thickness d_{ti} 10 cm with a thermal conductivity λ_{ti} of 0.04 W/m^{2}K (R_c = 2.5 m^{2}K/W). The thickness of the external wall d_w at the edge of the floor structure is 45 cm. The thermal conductivity of the soil λ_{soil} is 2 W/m^{2}K. The values of surface heat-transfer resistances R_{si} and R_{se} are given in Table 1.14.

$$d_f = d_w + \lambda_{soil} \cdot (R_{si} + R_c + R_{se}) = 0.45 + 2 \cdot (0.17 + 2.5 + 0.04) = 5.87\,m$$

$$B'_1 = \frac{A_{f,1}}{0.5 \cdot P_1} = \frac{6 \cdot 12}{0.5 \cdot (6 + 12 + 6)} = 6\,m$$

$$B'_2 = \frac{A_{f,2}}{0.5 \cdot P_2} = \frac{8 \cdot 12}{0.5 \cdot (8 + 8)} = 12\,m$$

$$d_f < B'_1 \quad U_{gr,f,1} = \frac{2 \cdot 2.0}{\pi \cdot 6 + 5.8} \ln\left(\frac{\pi \cdot 6}{5.8} + 1\right) = 0.235 \frac{W}{m^2 \cdot K}$$

$$d_f < B'_2 \quad U_{gr,f,2} = \frac{2 \cdot 2.0}{\pi \cdot 12 + 5.8} \ln\left(\frac{\pi \cdot 12}{5.8} + 1\right) = 0.185 \frac{W}{m^2 \cdot K}$$

The thermal transmittances of slab-on-the-ground building structures $U_{gr,f}$ without thermal insulation and with a continuous thermal insulation layer are shown in Fig. 1.68, according to the characteristic dimensions of the floor B' and the thicknesses of the thermal insulation layer d_{ti}. The values are determined for the thermal conductivity of the soil λ_{soil} 2 W/m^{2}K and the thickness of the external wall d_w, i.e., 45 cm.

A slab-on-the-ground building structure with an edge horizontal thermal insulation layer and structure with a vertical thermal insulation layer along the foundation are shown in Fig. 1.69. Such an element of thermal insulation can be installed in the case of uninsulated as well as insulated structures as an additional thermal insulation layer that decreases the impact of edge thermal bridge. The thermal transmittance $U_{gr,f}$ of such structures is determined by the construction thermal transmittance $U_{gr,fc}$ and the linear thermal transmittance of the thermal bridge $\Delta\Psi_f$(W/mK) at the edge junction between the floor and the external wall or the floor and the foundations:

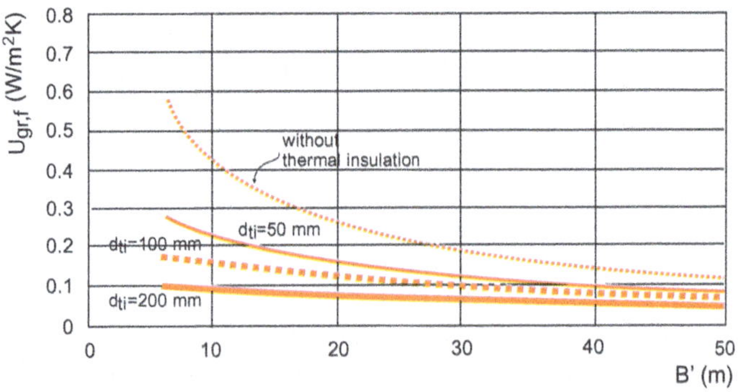

Fig. 1.68 Thermal transmittance $U_{gr,f}$ of a slab-on-the-ground building structure with and without a continuous thermal insulation layer with a thickness d_{ti} versus the characteristic dimensions of the floor B' [18]

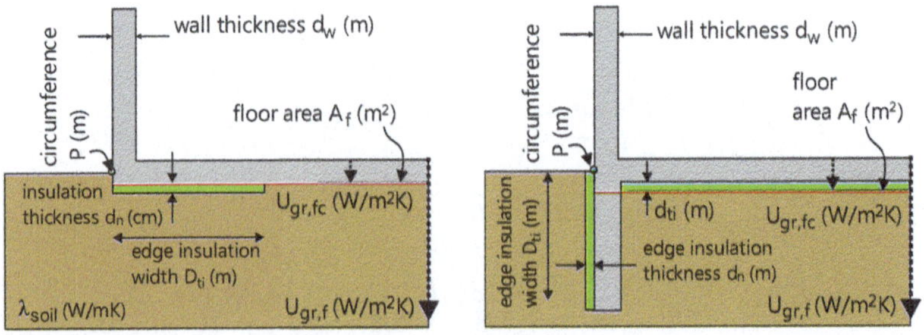

Fig. 1.69 Slab-on-the-ground building structures with horizontal edge thermal insulation layer (left) and vertical edge thermal insulation layer on the foundation (right). The parameters involved in determining the thermal transmittance $U_{gr,f}$ are designated

$$U_{gf,f} = U_{gr,fc} + 2 \cdot \frac{\Delta\psi_f}{B'} \left[\frac{W}{m^2 \cdot K} \right]$$

When the thermal transmittance of slab-on-the-ground building structures is compared to the required maximum value, the corrected value $U_{gf,f}$ must be compared. The edge thermal bridge linear thermal transmittance $\Delta\Psi_f$ for a typical structure's design is given in Fig. 1.70, taking into account that the edge thermal insulation is added to the continuous thermal insulation layer with the thickness d_{ti}. The edge thermal bridge linear thermal transmittance $\Delta\Psi_f$ depends on the thickness d_n and width D_{ti} of the edge thermal insulation layer and is always negative

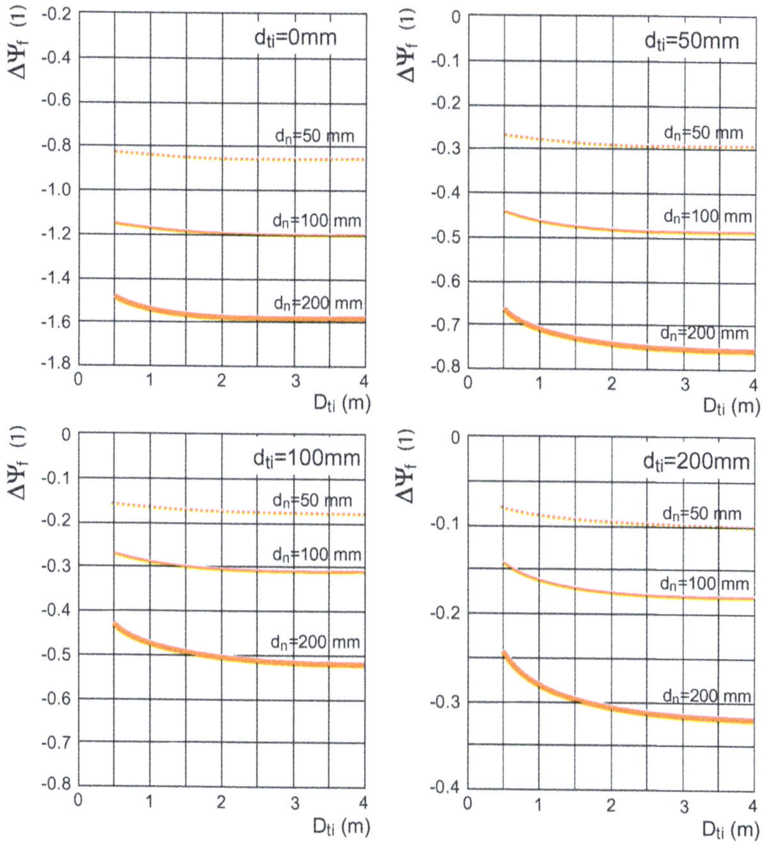

Fig. 1.70 Edge thermal bridge linear thermal transmittance $\Delta\Psi_f$ versus the thickness of the continuous thermal insulation layer d_{ti} (d_{ti} equals 0 mm means that structure has no continuous thermal insulation layer) and thickness $d_n$and width D_{ti} of the edge thermal insulation layer are as shown in Fig. 1.69 [19]

as the edge thermal insulation decreases the structure's thermal transmittance. From the diagrams in Fig. 1.70, it is clear that the optimum width D_{ti} is approximately 2 m for thicknesses of the edge thermal insulation layer d_n up to 100 mm and up to 3 m if the thickness of the edge thermal insulation layer d_n is 200 mm. The edge thermal insulation has a larger impact in the case of a slab-on-the-ground structure without a continuous thermal insulation layer (d_{ti} equals 0 mm). In the case of large buildings, like shopping centres or industrial facilities with $B' > 40$, the edge thermal insulation layer is usually sufficient to fulfil the requirements of permissible thermal transmittance criteria for such building structures.

Case Study Using the diagrams in Figs. 1.68 and 1.70 determine the thermal transmittance of the slab-on-the-ground structure $U_{gr,f}$ for the row-building shown in Fig. 1.67. Consider the building as a single object with a floor 21 m long and 12 m wide. The thickness of the uniform thermal insulation layer d_{ti} is 100 mm, while the foundation has an additional edge thermal insulation layer with thickness d_n 100 mm and width D_{ti} 150 cm. The characteristic dimension of the floor B' is:

$$B' = \frac{A_f}{0.5 \cdot P} = \frac{21 \cdot 12}{0.5 \cdot (21 + 12 + 21 + 12)} = 7.6\,\mathrm{m}$$

From the diagram in Fig. 1.68 it is clear that the centre thermal transmittance of the floor structure ($d_{ti} = 100$ mm, $B' = 7.6$ m) $U_{gr,fc}$ equals 0.185 W/m^{2}K and from the diagram in Fig. 1.70, bottom left, that the edge thermal bridge linear thermal transmittance $\Delta\Psi_f$ is—0.3 W/mK. The thermal transmittance of the floor slab $U_{gr,f}$ is therefore equal to:

$$U_{gr,f} = U_{gr,fc} + 2 \cdot \frac{\Delta\psi_f}{B'} = 0.175 + 2 \cdot \frac{-0.3}{7.6} = 0.096 \frac{W}{m^2 \cdot K}$$

The heat transfer in underground structures in the thermal envelope of the building also depends on the depth z (m) of the structure beneath ground level. Examples of such structures are shown in Fig. 1.71. The thermal transmittance of a basement's floor slab (Fig. 1.71, left bottom) or underground ceiling (Fig. 1.71 left top) $U_{gr,u,f}$ (W/m^{2}K) is determined like the thermal transmittance of the slab-on-the-ground, the difference being that the equivalent thickness d_f of the building structure is increased by half of the basement depth z, both in the equations and in the conditions that define which equation must be used:

$$U_{gr,u,f} = \frac{2 \cdot \lambda_{soil}}{\pi \cdot B' + d_f + \frac{1}{2} \cdot z} \ln\left(\frac{\pi \cdot B'}{d_f + \frac{1}{2} \cdot z} + 1\right) \quad \text{for} \quad d_f + \frac{1}{2} \cdot z < B' \left[\frac{W}{m^2 \cdot K}\right]$$

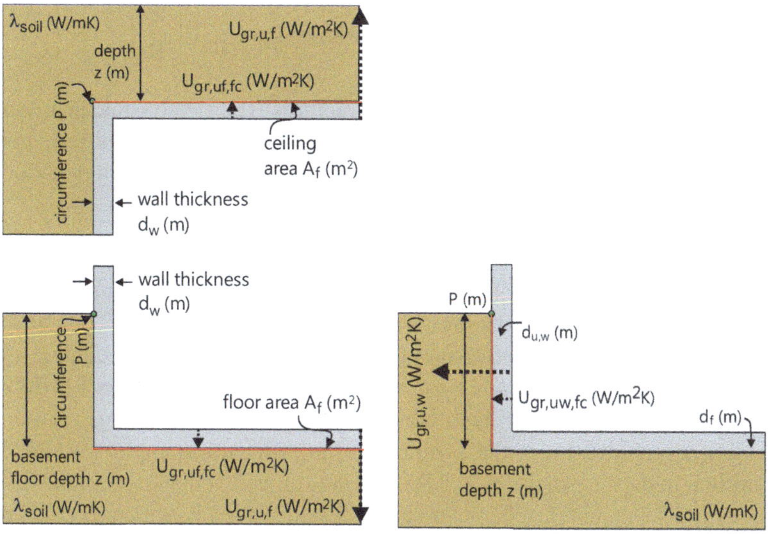

Fig. 1.71 Floor and ceiling building structures of an underground basement (left) and a basement's underground wall (right). The parameters involved in determining the thermal transmittance $U_{gr,u,f}$ and $U_{gr,u,w}$ are designated

$$U_{gr,u,f} = \frac{\lambda_{soil}}{0.457 \cdot B' + d_f + \frac{1}{2} \cdot z} \quad \text{for} \quad d_f + \frac{1}{2} \cdot z \geq B' \left[\frac{W}{m^2 \cdot K}\right]$$

The thermal transmittance of the underground wall $U_{gr,u,w}$ (Fig. 1.71, left) is determined by:

$$U_{gr,u,w} = \frac{2 \cdot \lambda_{soil}}{\pi \cdot z}\left(1 + \frac{0.5 \cdot d_f}{d_f + z}\right) \ln\left(\frac{z}{d_{u,w}} + 1\right)\left[\frac{W}{m^2 \cdot K}\right]$$

The equation is valid for $d_{u,w} \geq d_f$, otherwise $d_f = d_{u,w}$. In the equation R_c is the thermal resistance of all the wall layers, and R_{si} and R_{se} are the resistances to the surface heat transfer, listed in Table 1.4, d_f is the total equivalent thickness of the basement underground floor and $d_{u,w}$ is the total equivalent thickness of the basement underground wall defined by the expression:

$$d_{u,w} = \lambda_{soil}(R_{si} + R_c + R_{se})[m]$$

The steady-state heat-transfer coefficient $H_{gr,f}$ (W/K) of the slab-on-the-ground structure (Figs. 1.66 and 1.69), the steady-state heat-transfer coefficient $H_{gr,u,f}$ (W/K) of the underground floor or ceiling slab and the steady-state heat-transfer coefficient $H_{gr,u,w}$ of the underground wall structure (Fig. 1.71) are given by the expressions:

$$H_{gr,f} = A_f \cdot U_{gr,f} = A_f \cdot U_{gr,fc} + P \cdot \Delta\Psi_f$$

$$H_{gr,u,f} = U_{gr,u,f} \cdot A_f \quad H_{gr,u,w} = U_{gr,u,w} \cdot z \cdot P \left[\frac{W}{K}\right]$$

where A_f is the external surface area of the structures. The steady-state heat-transfer coefficient determines the heat flux that transmits the building structure in contact with the ground per 1 K temperature difference between the indoor- and outdoor-air temperatures.

Case Study The building shown in Fig. 1.67 has a heated basement with a floor area of A_f 252 m^2 and a perimeter P of 66 m, which is z 2.5 m beneath the ground level. The surface area of the underground walls $A_{,u,w}$ equals $z \times P$, which is 122 m^2. The thermal resistance of the basement walls R_c is 1 m^{2}K/W. Determine the steady-state heat-transfer coefficient H_{gr} of the underground building structures. From the solution of the previous example we already know that in the case that λ_{soil} is 2 W/mK, the equivalent thickness of the floor slab d_f is 5.87 m and the characteristic dimension of the basement B *is* 7.6 m.

The equivalent thickness of the basement wall structure is:

$$d_{u,w} = \lambda_{soil}(R_{si} + R_c + R_{se}) = 2 \cdot (0.13 + 1 + 0.04) = 2.34\,m$$

The thermal transmittance of the basement floor structure is:

$$d_f + \frac{1}{2} \cdot z \geq B' \rightarrow U_{gr,u,f} = \frac{2}{0.457 \cdot 7.62 + 5.87 + \frac{1}{2} \cdot 2.5} = 0.188 \frac{W}{m^2 \cdot K}$$

and the thermal transmittance of the basement wall structure is:

$$U_{gr,u,w} = \frac{2 \cdot 2}{\pi \cdot 2.5}\left(1 + \frac{0.5 \cdot 5.87}{5.87 + 2.5}\right) \ln\left(\frac{2.5}{2.34} + 1\right) = 0.499 \quad \frac{W}{m^2 \cdot K}$$

The steady-state heat-transfer coefficient of the underground basement building structure is:

$$H_{gr} = H_{gf,u,f} + H_{gr,u,w} = U_{gr,u,f} \cdot A_{gr,u,f} + U_{gr,u,w} \cdot z \cdot P =$$
$$= 0.188 \cdot 252 + 0.499 \cdot 2.5 \cdot 66 = 129.7 \left[\frac{W}{K}\right]$$

1.5.5.2 Maximum Heat Flux Rate Through the Building Structure in Contact with the Ground

Designing the heating load of the buildings requires data about the maximum heat flux passing through the building envelope structure under the design conditions—the indoor air θ_i, and the outdoor air $\theta_{e,p}$ design temperatures. The heat accumulates in the ground, so the maximum heat flux $\Phi_{gr,\max}$ (W) through building structures in contact with the ground does not coincide with the maximum heat flux through other building structures adjacent to the outdoor air. Therefore, $\Phi_{gr,\max}$ should be determined as the maximum monthly value (max $\Phi_{gr,\max,m}$), taking into account the transient component of the heat flux:

$$\Phi_{\mathrm{gr,max,m}} = \mathrm{H_{gr}} \cdot \left(\overline{\theta}_{\mathrm{i,m}} - \overline{\theta}_{\mathrm{e,m}}\right) + \hat{\mathrm{H}}_{\mathrm{gr}} \cdot \mathrm{A}_{\theta\,\mathrm{e}}[\mathrm{W}]$$

where H_{gr} (W/K) is the steady-state heat-transfer coefficient, $\widehat{H}_{gr}$ (W/K) is the transient heat-transfer coefficient, $\overline{\theta}_{i,m}$ (°C) and $\overline{\theta}_{e,m}$ (°C) are the average monthly indoor and outdoor air temperatures and $A_{\theta e}$ (°C) is the amplitude of the monthly outdoor-air temperature. For the slab-on-the-ground building structure shown in Fig. 1.66, the $\Phi_{gr,f,\max}$ can be approximated with the expression [19]:

$$\Phi_{\mathrm{gr,f,max}} \approx \left(\mathrm{a_f} \cdot \mathrm{B}'^{\left(-\frac{2}{3}\right)}\right) \cdot \mathrm{U_{gr,fc}} \cdot \mathrm{A_f} \cdot \left(\theta_{\mathrm{i}} - \theta_{\mathrm{e,p}}\right)[\mathrm{W}]$$

where a_f (−) is the approximation coefficient shown in Table 1.15, $U_{gf,fc}$ (W/m^{2}K) and A_f(m^2) are the thermal transmittance and the external area of the structure and θ_i (°C) and $\theta_{e,p}$ (°C) are the indoor air and the design heating period outdoor-air temperature. Similarly, the maximum heat flux $\Phi_{gr,u,\max}$ in the case of an underground floor and wall building structures shown in Fig. 1.71, can be approximated [19]:

$$\Phi_{\mathrm{gr,u,max}} \approx \left(\mathrm{b_f} \cdot \mathrm{B}'^{\left(-\frac{3}{5}\right)} \cdot \mathrm{U_{gr,uf,fc}} \cdot \mathrm{A_{u,f}} + \mathrm{c_f} \cdot \mathrm{U_{gr,uw,fc}} \cdot \mathrm{A_{w,f}}\right) \cdot \left(\theta_{\mathrm{i}} - \theta_{\mathrm{e,p}}\right)[\mathrm{W}]$$

where b_f (−) and c_f (−) are approximation coefficients shown in Table 1.16, $U_{gf,uf,fc}$ (W/m^{2}K) and $A_{u,f}$ (m^2) are the thermal transmittance and external area of the underground floor or ceiling structure and $U_{gf,uw,fc}$ (W/m^{2}K) and $A_{w,f}$ (m^2) are the thermal transmittance and the external area of the underground wall.

Table 1.15 Approximation coefficients a_f for different thermal transmittances of a slab-on-the-ground structure $U_{gr,cf}$ and thermal conductivities of the soil λ_{soil} [19]

af	Ugr,fc (W/m²K)			
λsoil (W/mK)	0.3	0.4	0.5	0.6
1.5	0.97	0.83	0.72	0.65
2.0	1.15	1.00	0.89	0.80
3.5	1.43	1.27	1.12	1.04

Table 1.16 Approximation coefficients b_f for an underground floor and ceiling structures (left) and c_f for underground wall structures (right) determined for different thermal transmissivities $U_{gr,uf,fc}$ and $U_{gr,uw,fc}$ of structures, conductivities of the soil λ_{soil} and the level beneath the ground z [19]

bf		$U_{gr,uf,fc}$ (W/m²K)			
λ_{soil} (W/mK)	z (m)	0.3	0.4	0.5	0.6
1.5	1.5	0.62	0.55	0.49	0.44
	3.0	0.60	0.53	0.47	0.42
	4.5	0.58	0.49	0.44	0.40
	6.0	0.55	0.48	0.43	0.39
2.0	1.5	0.78	0.70	0.63	0.57
	3.0	0.75	0.68	0.61	0.55
	4.5	0.72	0.63	0.59	0.53
	6.0	0.70	0.62	0.58	0.52
3.5	1.5	1.02	0.94	0.87	0.81
	3.0	0.99	0,92	0.85	0.79
	4.5	0.97	0.89	0.82	0.77
	6.0	0.95	0.88	0.81	0.76

cf		$U_{gr,uw,fc}$ (W/m²K)			
λ_{soil} (W/mK)	z (m)	0.3	0.4	0.5	0.6
1.5	1.5	0.33	0.31	0.29	0.27
	3.0	0.23	0.21	0.20	0.19
	4.5	0.19	0.17	0.16	0.15
	6.0	0.17	0.15	0.13	0.12
2.0	1.5	0.39	0.37	0.35	0.33
	3.0	0.28	0.26	0.24	0.22
	4.5	0.23	0.21	0.19	0.18
	6.0	0.20	0.18	0.16	0.15
3.5	1.5	0.48	0.45	0.44	0.42
	3.0	0.33	0.31	0.30	0.29
	4.5	0.28	0.26	0.24	0.23
	6.0	0.24	0.22	0.21	0.20

Case Study Determine the maximum heat flux rate $\Phi_{gr,u,\max}$ passing through an underground floor and the walls of a heated basement with the air temperature θ_i 18 °C and the floor area A_f 450 m² (30 × 15 m). The thermal transmittances of the floor structure and the walls are $U_{gr,uf,fc}$ 0.3 W/m²K and $U_{gr,uw,fc}$ 0.4 W/m²K, respectively. The design temperature of the outdoor air $\theta_{e,p}$ at the building's geographical location is −13 °C. The basement depth under ground level z is 3 m, and the thermal conductivity of the ground λ_{soil} is 1.5 W/mK.

The characteristic dimension of the basement B' is equal to:

$$B' = \frac{A_f}{0.5 \cdot P} = \frac{30 \cdot 15}{0.5 \cdot (2 \cdot 30 + 2 \cdot 15)} = 10.0\,\mathrm{m}$$

The approximation coefficients taken from Table 1.16 are: b_f 0.60 and c_f 0.21, and the maximum heat flux rate $\Phi_{gr,u,\max}$ is equal to:

$$\begin{aligned}\Phi_{gr,u,\max} &\approx \left(b_f \cdot B'^{\left(-\frac{3}{5}\right)} \cdot U_{gr,uf,fc} \cdot A_{u,f} + c_f \cdot U_{gr,uw,fc} \cdot A_{w,f}\right) \cdot (\theta_i - \theta_{e,p}) \\ &= \left(0.60 \cdot 10^{\left(-\frac{3}{5}\right)} \cdot 0.3 \cdot 450 + 0.21 \cdot 0.4 \cdot 270\right) \cdot (18 - (-13)) \\ &= 1334[\mathrm{W}]\end{aligned}$$

1.5.6 *Heat Transfer in Building Structures with Thermal Bridges*

A thermal bridge is a part of the building structure where the thermal transmittance of the structure differs compared to the value determined assuming that the structure is homogeneous. The main reasons for the formation of thermal bridges are:

- complete or partial penetration of the structure of the building envelope with an element that has a different thermal conductivity (for example, supporting column, wooden beam),
- difference in the thickness and shape of the structure (for example, edge of window, opening in the wall),
- different size of the interior and exterior surfaces (for example, wall corners, wall and floor junctions, structures in contact with the ground).

At the sites of thermal bridges the heat transfer is more intense. In the winter this means that the temperatures on the external surface of the outer wall will be higher compared to the structure without thermal bridges (Fig. 1.72). Because of that, thermal bridges can be easily detected on thermographic images (see Sect. 1.5.1). On the inner surface of the structures with a thermal bridge, the temperatures are lower, which increases the risk of surface condensation and the conditions for the formation of mould and the growth of harmful organisms appears (Fig. 1.73). Apart from greater heat losses, this is the main reason why the impact of the thermal bridges must be reduced as much as possible.

In the part of the building structures with a thermal bridge, the isotherms (lines connecting points of equal temperatures) are no longer parallel, as the heat flux does not pass through the structure only in the direction normal to the surface, but is two (2D) or three dimensional (3D). The thermal bridges are divided into 2D- and 3D-bridges, accordingly.

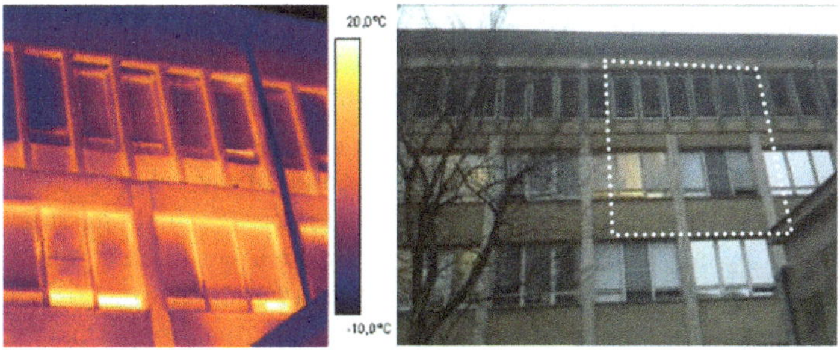

Fig. 1.72 During the winter the temperatures of the external surface of the building structures with thermal bridges are higher and such details on the building envelope can be detected on thermographic images

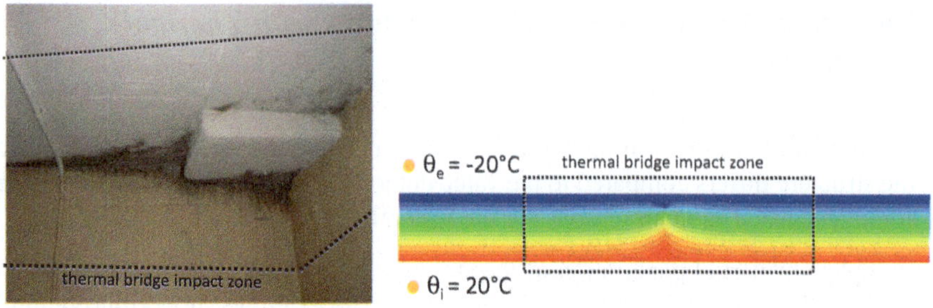

Fig. 1.73 Mould formation on the wall as the consequence of a thermal bridge at the junction of an uninsulated outer wall and a ceiling slab. The piece of thermal insulation installed by the occupants is not big enough to prevent mould formation since the influence area of the thermal bridge is much larger (left). Isotherms inside a lightweight building structure with an I-shaped metal support. The area where the isotherms are not parallel is the thermal-bridge-affected zone (right)

1.5.6.1 Linear and Point Thermal Transmittances of Thermal Bridges

Analysing all the thermal bridges in a building with a single numerical model is usually impossible or too complex. Therefore, the building structures must be divided into several smaller elements, corresponding to the zone of influence of the individual thermal bridge. The rules of partitioning the building structures into smaller elements are defined in [20]. If there are no adjacent thermal bridges, it is most often assumed that the zone influenced by the thermal bridge are not to be larger than 1 m.

The influence of thermal bridges on the thermal transmission of building structures is evaluated by correction terms. These factors account for the difference between the heat flux passing through the building structure with the thermal bridge and the heat flux that passes through the same structure on the section without the thermal bridge.

The effect of thermal bridges on the size and direction of the heat flux in a building structure is taken into account by the linear thermal transmittance Ψ ("psi") of a two-dimensional thermal bridge or by the point thermal transmittance χ ("chi") of a three-dimensional thermal bridge (Fig. 1.74). Linear thermal transmittances are normalised to 1 m of the length of the thermal bridge (for example, 1 m of the junction between a partition wall and an external wall, the circumference of a door or window openings in the external wall, the length of the balcony slab joined to the external load-bearing structure). Since the surface of a building structure can be determined by internal or external dimensions, the linear thermal transmittances can be expressed based on internal (Ψ_i) or external (Ψ_e) dimensions. The impact of 3D thermal bridges is given by the number of point thermal bridges in the building structure. In this case too, the surface area of a building structure can be determined using the internal or external dimensions. The impact of 3D thermal bridges is given by the point thermal transmittance of thermal bridges χ and their number in the building structure.

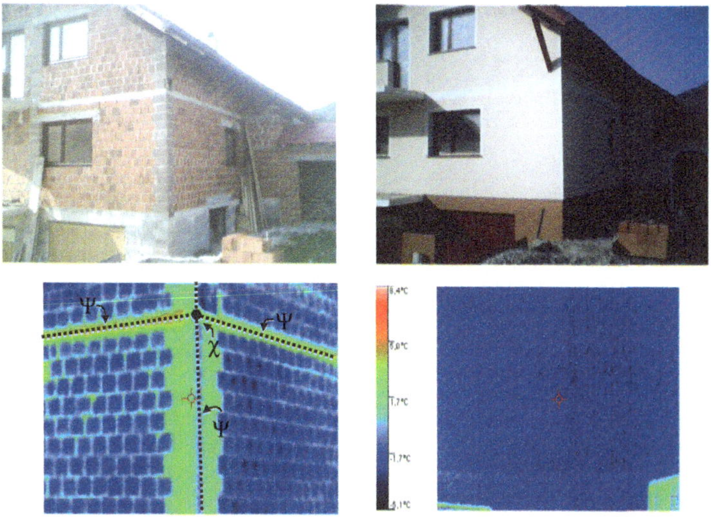

Fig. 1.74 Thermal bridge can be 2 or 3 dimensional. A two-dimensional (2D) thermal bridge on the corner of the external walls, and a three-dimensional (3D) thermal bridge on the junction of a corner and a floor slab can be seen on the thermographic image (left). Properly designed thermal insulation significantly decreases the intensity of thermal bridges (right)

Two methods for determining the thermal bridge correction terms have appeared in engineering practice as a result of the rapid development of computer hardware and software, as well as an increasing number of predefined construction details the catalogue methods and numerical modelling.

1.5.6.2 Methods for Evaluating Thermal Bridges

The catalogue method is employed for the evaluation of the thermal bridges, associated to predefined construction details (Fig. 1.75). Some examples are building structures made of light structural elements and C-, S- and T-shaped joining elements of predefined dimensions (Fig. 1.76), specific window installation details, predefined details of thermal insulation installation, light structural element assembly, and so forth.

Computer methods are based on the numerical solution of temperature fields in the building structure. Special-purpose computer tools (for example, Physibel Trisco) or computer tools for computational fluid dynamics (CFD) are used, such as Phoenics or Fluent. In this case, the course of determining the linear and point thermal transfers of thermal bridges is as follows [example is shown in Fig. 1.78]:

- **Segmentation**. The building structures are partitioned into segments with thermal bridges.

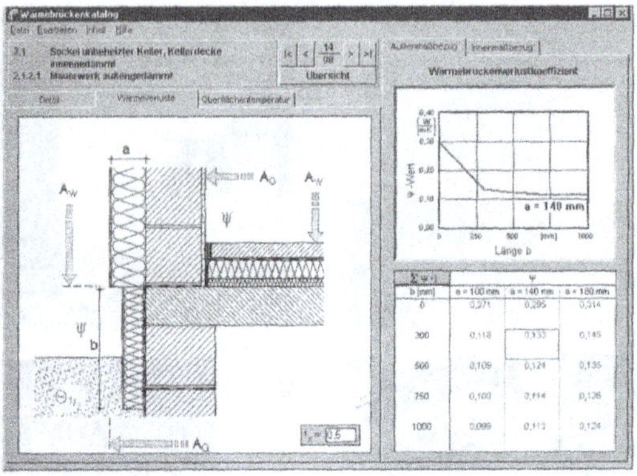

Fig. 1.75 "Wärmebrücken-kata-log" is an extensive database of building-structure details and the corresponding thermal transmittances of thermal bridges. The figure shows the construction detail of external wall and values of Ψ for different thermal insulation-layer thicknesses (**a**) and thermal insulation lengths along the foundation (**b**) [2]

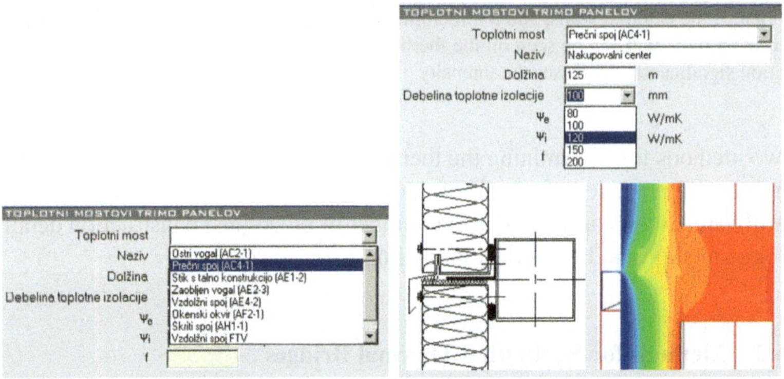

Fig. 1.76 Light weighted structural elements TRIMO are joined in different ways. The catalogue of thermal bridges is made and uploaded to the computer code for evaluation of the energy efficiency of buildings, TRIMOExpert [21]

- **Boundary conditions**. The boundary conditions are defined as a state that remains unchanged during the process of numerical solving. For example, the indoor-air temperature θ_i, the outdoor-air temperature θ_e, the surface combined convection and radiation heat-transfer coefficients h_{si} and h_{se}, the boundary conditions assumed in the calculations are prescribed (e.g. θ_i, 20 °C, θ_e −10 °C, h_{si} 8 W/m^{2}K or h_{se} 23 W/m^{2}K);
- **Meshing**. The building structure can be meshed into a large number of small elements, for example, finite volumes. The accuracy of the calculation depends on the size of the structural elements and can be improved by halving the finite

volumes until the temperatures obtained in two consecutive iterations differ by no more than a predefined value (e.g. 0.05 °C).

The final results of the numerical simulations are the temperatures of the finite elements on the surface and inside the building structure, with the heat fluxes between finite elements and the thermal transmittances of the linear and point thermal bridges are determined.

1.5.6.3 Determination of Heat Bridges' Linear and Point Thermal Transmittance

The corner of a homogeneous building structure considered as a linear thermal bridge is shown in Fig. 1.77. The thermal transmittances of the homogeneous building structures are U_1 and U_2 (W/m^{2}K). A numerical analysis is used to determine the heat fluxes Q_1 and Q_2 (W) passing through the structure with length l (m). The specific heat flux L^{2D} that passes through the structure is determined with the expression:

$$L^{2D} = \frac{\dot{Q}_1}{\theta_i - \theta_e} + \frac{\dot{Q}_2}{\theta_i - \theta_e} = \frac{\dot{Q}^{2D}}{\theta_i - \theta_e}\left[\frac{W}{K}\right]$$

In Fig. 1.77 the width of the influencing area of the thermal bridge is denoted by the internal dimensions ($d_{i,1}$ and $d_{i,2}$) and the external dimensions ($d_{e,1}$ and $d_{e,2}$). Depending on the chosen system of measurements, the linear thermal transmittances Ψ_i and Ψ_e differ, especially in the case of a thick thermal insulation layer. As an example, the linear thermal transmittance of the thermal bridge Ψ_e of the detail on the figure expressed by the external dimensions is:

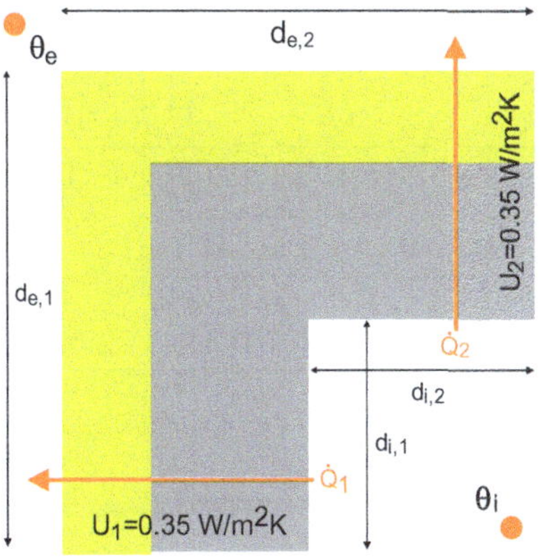

Fig. 1.77 Example of a two-dimensional thermal bridge; $d_{i,1}$ and $d_{i,2}$ are internal and $d_{e,1}$ and d_{e2} are external dimensions of the impact zone of thermal bridge. At the boundary of the detail the thermal transmittances U_1 and U_2 as for homogeneous structure

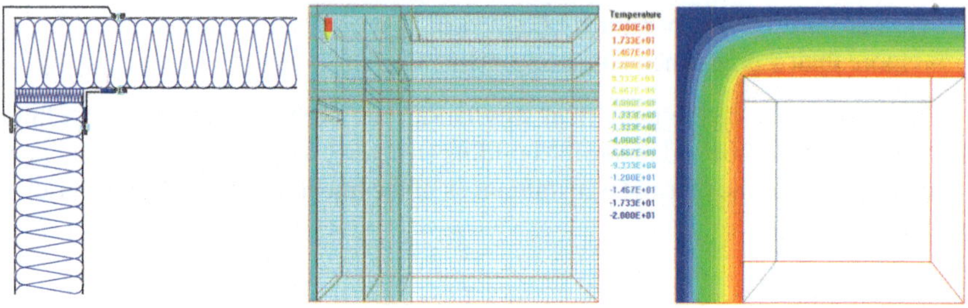

Fig. 1.78 Corner building structure made of light-weighted structural elements (left), 2D meshing of the building structure (middle) and isotherm in the building structure (right) [21]

$$\Psi_e = \frac{L^{2D} - \sum_{j=1}^{2} U_j \cdot A_{e,j}}{l} = \frac{L^{2D} - \left(U_1 \cdot d_{e,1} \cdot l + U_2 \cdot d_{e,2} \cdot l\right)}{l} \left[\frac{W}{mK}\right]$$

where l is the length of the linear thermal bridge (in this case the length of the corner of the building structure) and $d_{e,1}$ and $d_{e,2}$ (m) are the external dimensions of the thermal bridge impact zone.

Case Study Numerical modelling was used to determine the heat fluxes $\dot{Q}_1$ and $\dot{Q}_2$ for the corner building structure shown in Fig. 1.77. Since building structures that form the corner are the same. The thermal transmittance of the building structures that form the corner U_1 and U_2 are 0.35 W/m^{2}K. The internal widths of the thermal bridge impact area $d_{i,1}$ and $d_{i,2}$ are 1 m and the external widths of the thermal bridge impact area $d_{e,1}$ and $d_{e,2}$ are 1.3 m. The length l of the corner is 2.6 m. The heat fluxes $\dot{Q}_1$ and $\dot{Q}_2$ were determined at an indoor-air temperature θ_i of 20 °C and at an outdoor-air temperature θ_e of −15 °C [22] and equal 37.7 W ($\dot{Q}^{2D} = 75.4$ W) (Fig. 1.78).

$$L^{2D} = \frac{\dot{Q}_1}{(\theta_i - \theta_e)} + \frac{\dot{Q}_2}{(\theta_i - \theta_e)} = \frac{37.7 + 37.7}{(20 - (-15))} = 2.15\frac{W}{K}$$

$$\Psi_i = \frac{L^{2D} - \left(U_1 \cdot d_{i,1} \cdot l + U_2 \cdot d_{i,2} \cdot l\right)}{l} = $$

$$= \frac{2.15 - (0.35 \cdot 1 \cdot 2.6 + 0.35 \cdot 1 \cdot 2.6)}{2.6} = 0.127\frac{W}{mK}$$

$$\Psi_e = \frac{L^{2D} - \left(U_1 \cdot d_{e,1} \cdot l + U_2 \cdot d_{e,2} \cdot l\right)}{l}$$

$$= \frac{2.15 - (0.35 \cdot 1.3 \cdot 2.6 + 0.35 \cdot 1.3 \cdot 2.6)}{2.6} = -0.083\frac{W}{mK}$$

The negative value of ψ_e is a consequence of the fact that the heat flux is entering the corner building structure at its internal surface and exiting at the larger external surface. The density of the heat flux is therefore lower. For structures with the same size of internal and external surfaces, ψ_i and ψ_e are equal.

For some details on the building envelope, the temperature field is distinctly three-dimensional. Such examples are the final corner, a thermal insulation anchor, the joint of a partition wall, a floor panel and an external wall. In this case, a point thermal bridge appears in the 3D form of the building structure, which is evaluated by the point thermal transmittance χ (W/K). A system of internal measures is used to determine the length of the j-th linear thermal bridge $l_{i,j}$, and the system of external dimensions for determining the surfaces of the k-th building structure $A_{e,k}$. Accordantly, χ is determined by:

$$L^{3D} = \frac{\dot{Q}^{3D}}{(\theta_i - \theta_e)} \rightarrow \chi = L^{3D} - \sum_{j=1}^{m} \Psi_{e,j} \cdot l_{i,j} - \sum_{k=1}^{n} U_k \cdot A_{e,k} \left[\frac{W}{K}\right]$$

where $\dot{Q}^{3D}$ (W) and L^{3D} (W/K) are the heat flux and the specific heat flux that passes the impact area of the thermal bridge section of the building structure. $\Psi_{e,j}$ (W/mK) is the linear thermal transmittance of the linear thermal bridges with length $l_{i,j}$. $U_k$and $A_{e,k}$ are the thermal transmittances and the surface areas of the homogeneous parts of the building structures, m is total number of linear thermal bridges and n is the total number of building structures that form the structure with a point thermal bridge.

Case Study Determine the point thermal transmittance χ of the corner formed by the three light-weight structures shown in Fig. 1.79. The indoor-air temperature θ_i is 20 °C and the outdoor-air temperature θ_e is −15 °C.

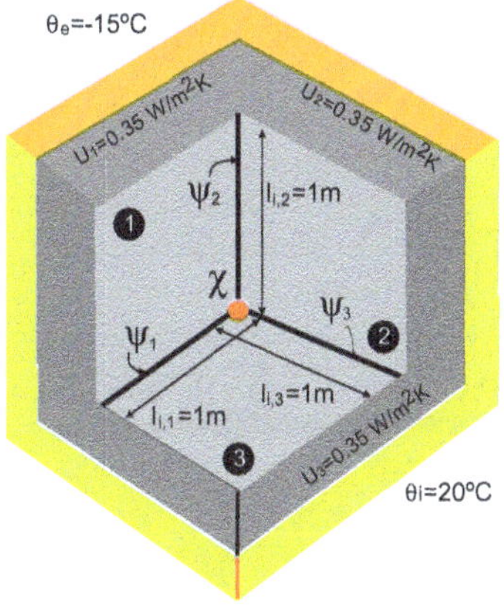

Fig. 1.79 Building structure analysed in the case study

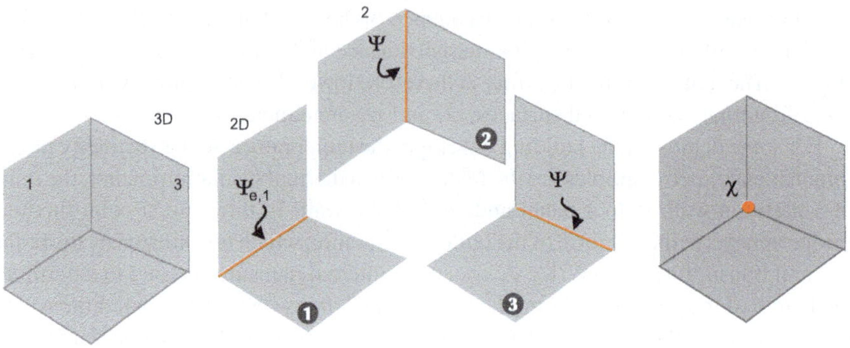

Fig. 1.80 Calculation procedure for determining the point thermal transmittance χ. In the first step the specific heat flux L^{3D} that passes through the analysed detail is determined (left). Next, the linear thermal transmittances for each joint of the building structure are determined (middle). Finally, the point thermal bridge transmittance χ is calculated (right)

The procedure for determining the point thermal transmittance χ is presented in Fig. 1.80. In the first step the heat flux $\dot{Q}^{3D}$ (W) that passes through the corner building structure on the total surface area $\sum A_{e,k}$ is calculated. The numerical simulation shows that the total heat flux Q^{3D} through the building structure's segments 1, 2 and 3 is 105.7 W. The specific heat flux L^{3D} passing the building structure is equal to:

$$L^{3D} = \frac{\dot{Q}^{3D}}{(\theta_i - \theta_e)} = \frac{105.7}{(20 - (-15))} = 3.02\frac{W}{K}$$

In the second step the building structure is divided into elements with linear thermal bridges and heat flux $\dot{Q}^{2D}{}_j$ that passes each of the j-th elements and the linear thermal transmittance $\Psi_{e,j}$ is calculated. For the presented case study, it was found that $\Psi_{e,1} = \Psi_{e,2} = \Psi_{e,3}$ and equals −0.08 W/mK.

The point thermal transmittance χ of the corner of the building structure is equal to:

$$\chi = L^{3D} - \sum_{j=1}^{m} \Psi_{e,j} \cdot l_{i,j} - \sum_{k=1}^{n} U_k \cdot A_{e,k} =$$

$$= 3.02 - \sum_{j=1}^{3} (-0.083 \cdot 1) \cdot 3 - \sum_{k=1}^{3} 0.350 \cdot (1.3 \cdot 1.3) \cdot 3 = 0.488\frac{W}{K}$$

1.5.6.4 Surface-Temperature Factor of the Thermal Bridge

In addition to higher heat losses, the thermal bridges increase the risk of condensing water vapour from the air in the room and the formation of harmful micro-organisms on a cold inner surface of the building. The risk can be evaluated with the method of the surface-temperature factor f_{Rsi}. It is determined from the calculated minimum temperature on the building structure's internal surface $\theta_{si,\min}$, the indoor-air temperature θ_i and outdoor-air temperature θ_e. It can be assumed, that there is no risk of water vapour condensation on the building structure's surface and associated growth of harmful organisms if $f_{Rsi.\min}$ is kept over 0.8. The temperature factor f_{Rsi} is given by:

$$f_{Rsi} = \frac{(\theta_{si,\min} - \theta_e)}{(\theta_i - \theta_e)} > f_{Rsi,\min} \equiv 0.8[1]$$

The method of evaluation based on the thermal bridge temperature factor is elaborated in Sect. 2.6.

1.5.6.5 The Influence of Thermal Bridges on the Thermal Transmittance of the Building Structure

The catalogues and numerical modelling of thermal bridges are used to determine their influence on the thermal transmittance of the building structure. A common rule is that linear thermal transmittance of thermal bridges Ψ_i must be less than 0.2 W/mK to avoid the risk of surface condensation. If all the linear thermal transmittances Ψ_e are less than 0.01 W/mK, the building is considered to be thermal bridge free. Otherwise, the effect of thermal bridges on the thermal transmittance of the building structure is evaluated using the expression:

$$U_{eq} = \frac{U \cdot A + \sum_{j=1}^{m} \psi_{e,j} \cdot l_{e,j} + \sum_{g=1}^{l} \chi_g \cdot n_g}{A} = \frac{U \cdot A + \Delta\psi}{A} \left[\frac{W}{m^2K}\right]$$

where U_{eq} is equivalent thermal transmittance of the building structure with surface A, j and g are numerators of the linear and point thermal bridges in the construction, $l_{e,j}$ is the length of the j-th thermal bridge and n_g is the number of g-th point thermal bridges in observed building structure and $\Delta\Psi$ *(W/K)* is thermal bridge factor of building structure.

Case Study The case study shown in Fig. 1.79 was solved with the Physibel Trisco computer tool. The structure has an area of the external surface equal to 5.07 m^2. Figure 1.81 shows the inner-surface temperatures of the corner building structure at an indoor-air temperature θ_i of 20 °C and and outdoor-air

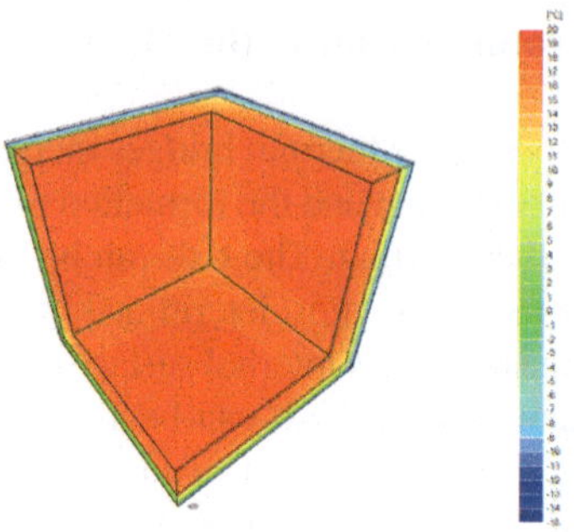

Fig. 1.81 Inner surface temperature of case study corner building structure at indoor-air temperature θ_i 20 °C and outdoor-air temperature θ_e −15 °C

temperature θ_e of −15. The thermal bridges factor of the building structure influenced by the thermal bridge $\Delta\Psi$ is −0.305 W/K. Taking into account the external surface area of the structure and the thermal transmittance of the homogeneous structures is 0.350 W/m^{2}K, the equivalent thermal transmittance of the building structure U_{eq} is 0.290 W/m^{2}K. The total heat flux that transmits through the construction, influenced by the thermal bridges, can be determined by replacing the thermal transmittances of homogeneous building structures U with U_{eg}, taking into account the external dimensions of the building structure without taking into account any of the thermal bridges.

1.5.6.6 Simplified Calculation of the Impact of the Thermal Bridges

An accurate evaluation of thermal bridges' effects on the heat transfer through the building envelope is quite time consuming and can be, in engineering practice, simplified by involving the empirical values of the thermal bridges factor of the building envelope $\Delta\Psi_{env}$ (W/m^{2}K). It is also common that the thermal bridges factor of the building envelope is added to the specific thermal heat-transfer coefficient H'_T of the building envelope (see Sect. 1.5.7), rather than to the thermal transmittance of the individual building structure. For massive and poorly insulated buildings $\Delta\Psi_{env}$ is typically in the range 0.1–0.14 W/m^{2}K, while for low-energy buildings and buildings built of wooden structures, it is in the range 0.03 and 0.04 W/m^{2}K.

1.5.6.7 The Influence of Thermal Bridges on the Use of Heat Required for Space Heating

Thermal bridges usually cause an increase in the heat transfer through the building envelope and as a consequence the transmission heat loss increases. Nevertheless, the energy requirements for the space heating also depend on the type of ventilation, the internal heat sources and the passive solar heating. This must be taken into account when determining the impact of thermal bridges on the energy efficiency of the buildings. Table 1.17 shows the energy needs as a specific value, normalized to 1 m^2 of useful area of the building A_u for a single-family building with the minimum required thermal protection of the building structures and low-energy buildings with additional thermal insulation equipped with a mechanical ventilation system that has heat recovery. It is obvious that the relative influence of the thermal bridges in buildings with a low energy requirement for heating is much greater.

1.5.7 Overall Thermal Transmittance of the Building Envelope

The envelope of the building is the boundary between the temperature-controlled indoor environment and the surroundings (outdoor air, soil or water) or unheated spaces (attic, staircase, garage). The shape of the envelope is characterized by the shape factor of the building f_o (m^{-1}):

$$f_o = \frac{A_{env}}{V_e} \left[\frac{m^2}{m^3} = m^{-1} \right]$$

where A_{env} (m^2) is the total external surface area of the building envelope and V_e (m^3) is the gross heated volume of building, which is bounded by the surface of the building envelope A_{env}. The building shape factor has a major impact on the transmission

Table 1.17 Specific energy needs for the heating of a single-family building according to the level of thermal insulation and the value of the normalized thermal bridges factor of the building envelope $\Delta\Psi_{env}$

	Energy needs for heating $Q_{H,nd}$		
	no thermal bridges	$\Delta\psi_{env}$ = 0.1 (W/m²K)	$\Delta\psi_{env}$ = 0.06 (W/m²K)
building with minimum thermal insulation	105 kWh/m²a	116 kWh/m²a +10%	122 kWh/m²a + 16%
energy-efficient building with mechanical ventilation with heat recovery	43 kWh/m²a	53 kWh/m²a +23%	59 kWh/m²a + 37%

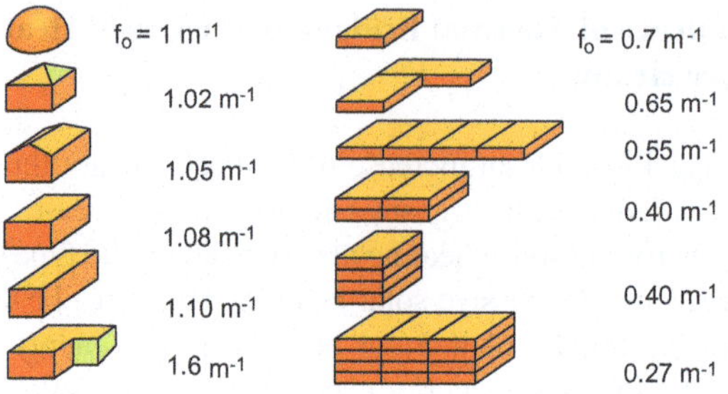

Fig. 1.82 Shape factor f_o of single-family buildings with equal useful area of the building A_u (left) and the apartment blocks consist of virtual living units with size of 10 × 20 × 5 m (right)

thermal losses of a building and buildings should be designed to minimize their shape factor. Figure 1.82 shows how the architectural design of a building influences its shape factor.

The building envelope consists of different building structures, which differ by thermal transmittances and the surface areas. Because of that, the overall thermal transmittance of the building envelope H'_T (W/m^{2}K) is one of the key indicators of the energy efficiency of the building, and often its value is limited by the maximum permitted value stated in the national legislation. It is determined using the expression:

$$H'_T = \frac{\sum_{i=1}^{n} b_i \cdot U_i \cdot A_i + \sum_{j=1}^{m} \psi_j \cdot l_j + \sum_{k=1}^{l} \chi_k \cdot n_k}{\sum_{i=1}^{n} A_i} \left[\frac{W}{m^2K}\right]$$

where U_i (W/m^{2}K) and A_i (m^2) are the thermal transmittance and the surface of the i-th opaque or transparent structure on the building envelope, b_i (−) is the weighting factor with a value equal to 1 for structures towards the outdoor environment, including the ground, lower than 1 for structures towards unheated spaces and above 1 for structures with floor heating, Ψ_j (W/mK) and l_j (m) are the linear thermal transmittance of the j-th 2D thermal bridge and its length and X_k (W/K) and n_k (−) are the point thermal transmittance of the k-th 3D thermal bridge and their number.

For the purposes of simplification, the thermal transmittance of the heat bridges Ψ_j and X_k can be replaced by the empirical thermal bridge factor of the building $\Delta\Psi_{env}$ (W/m^{2}K). The typical values of $\Delta\Psi_{env}$ are presented in Sect. 1.5.6.6. In this case the overall thermal transmittance of the building envelope H'_T is:

$$H'_T = \frac{\sum_{i=1}^{n} b_i \cdot U_i \cdot A_i}{\sum_{i=1}^{n} A_i} + \Delta\psi_{env} \left[\frac{W}{m^2K}\right]$$

1.5.8 Determination of the Thermal Transmittances of the Building Structures on Built Objects

The thermal transmittances of building structures on built objects can deviate from the calculated values due to the differences in the properties of the materials or the method of construction. The compliance with the theoretical values is checked with the test defined in [23]. The same sensors as described in Sect. 1.3.1.1 are used during the test. The measurement takes place over a longer period of time in which the heat flux $\dot{q}_j$ at time j $\dot{q}_j$ that passes through the building structure and the air temperature in the room $\theta_{i,j}$ and the surroundings $\theta_{e,j}$ are measured at the same time intervals. The measured thermal transmittance of the building structure $U_{in\text{-}situ}$ is determined from the average values of the measured quantities:

$$U_{in\text{ - }situ} = \overline{U_j} = \frac{\sum_{j=1}^{n} \dot{q}_j}{\sum_{j=1}^{n} (\theta_{i,j} - \theta_{e,j})} \left[\frac{W}{m^2 \times K}\right]$$

The test can be ended when the thermal transmittance $U_{in\text{-}situ}$ after the end of the 24-h period does not differ by more than $\pm$ 5% from the value calculated 24 h ago.

Explanation For lightweight building structures, according to the [23], these are the structures with a specific thermal capacity C of less than 20 kJ/m^{2}K, the values of q, θ_i and θ_e that are measured over a period of 1 h after the solar sunset in the evening and until sunrise the next morning are taken into consideration. The only lightweight building structures that fit the criteria in practice are the window glazing and the thin insulation panels. In the case of building structures with a higher thermal capacity, the thermal transmittance is averaged over 24-h periods.

Under real conditions, the heat flux and the temperatures in the building structure change over time due to the weather conditions and the unsteady sources of heat in the building. Therefore, the heat transfer in the building structures on site is always transient and depends on the heat accumulated in the building structure. This requires a long running time of the experiment. The experiment can be shortened if the measured heat flux $\dot{q}_j$ is corrected with the heat-accumulation factors F_i and F_e:

$$F_i = \sum_{k=1}^{n-1} F_{i,k} \left[\frac{J}{m^2 \cdot K}\right] \quad F_e = \sum_{k=1}^{n-1} F_{e,k} \left[\frac{J}{m^2 \cdot K}\right]$$

where n is the number of layers of the building structure, and $F_{i,k}$ and $F_{e,k}$ are the heat-accumulation factors of the individual layer k determined from the inside and outside sides of the building structure:

$$F_{i,k} = C_k\left(\frac{R_{e,k}}{R} + \frac{R_k^2}{3R^2} - \frac{R_{i,k}R_{e,k}}{R^2}\right)\left[\frac{J}{m^2 \cdot K}\right]$$

$$F_{e,k} = C_k\left[\frac{R_k}{R}\left(\frac{1}{6} + \frac{R_{i,k} + R_{e,k}}{3R}\right) + \frac{R_{i,k}R_{e,k}}{R^2}\right]\left[\frac{J}{m^2 \cdot K}\right]$$

where R_{tot} is the total thermal resistance of the building structure, $R_{i,c}$ is the resistance to the heat transfer on the internal and $R_{e,c}$ on the external side of the building structure. An example for the case of a construction consisting of three layers ($n = 3$) is shown in Fig. 1.83. The specific heat capacity of the k-th layer of the building structure C_k is:

$$C_k = d_k \cdot \rho_k \cdot c_{p,k}\left[\frac{J}{m^2 \cdot K}\right]$$

The thermal transmittance $U_{in\text{-}situ.cor}$, determined by taking into account the heat accumulated in the building structure, is defined by the expression:

$$U_{in-situ,cor} = \frac{\sum_{j=1}^{n} \dot{q}_j - \frac{F_i \cdot \Delta\bar{\theta}_{i,24} + F_e \cdot \Delta\bar{\theta}_{e,24}}{\Delta t}}{\sum_{j=1}^{n} (\theta_{i,j} - \theta_{e,j})}\left[\frac{W}{m^2 \cdot K}\right]$$

The term Δt (s) is the time period between the j and j + 1-th reading of the measured quantities, $\Delta\theta_{i,24}$ and $\Delta\theta_{e,24}$ are the differences between the average temperatures in the building and the outdoor during the period of the first 24 h and the last 24 h of the experiment, which must be performed for at least 96 h. The test can be completed when the corrected heat transmittance $U_{in\text{-}situ,cor}$ after the end of the 24-h period does not differ by more than ± 5% from the value before 24 and 48 h.

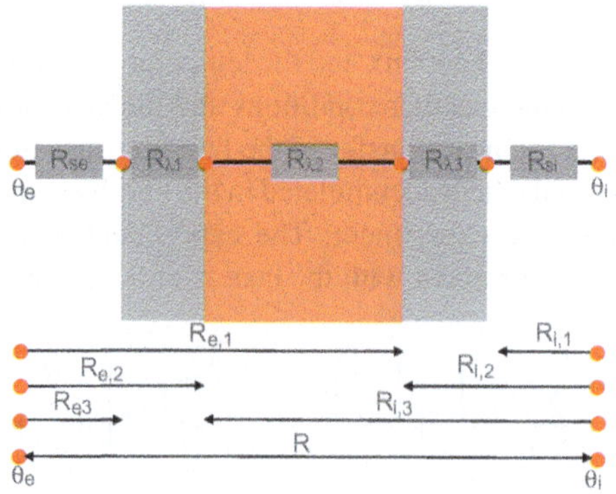

Fig. 1.83 The principle for the determination of the resistance to the heat transfer $R_{i,c}$ on the inner part and $R_{e,c}$ on the external part of the building structure in the case of building structure with 3 layers

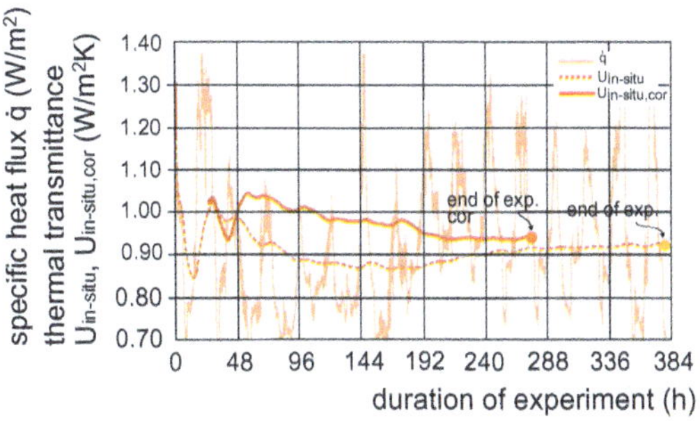

Fig. 1.84 Measured heat flux and the values of the calculated thermal transmittance of the wall without ($U_{in\text{-}situ}$) and with heat-accumulation factors ($U_{in\text{-}situ,cor}$)

Table 1.18 Layers of the building structure for which we determine the thermal transmittance by in-situ experiment

layer in the wall	d (cm)	λ (W/mK)	ρ (kg/m³)	c (kJ/kgK)	R_k (m²K/W)	C_k (J/m²K)
inner plaster	2	0.81	1600	1.050	0.025	33600
solid brick	50	0.58	1400	0.920	0.862	644000
outer plaster	3	0.81	1600	1.050	0.037	50400

Case Study The thermal transmittance of the brick wall with three layers should be determined on the built object. The properties of the materials, the thermal resistances and the specific heat capacities of the layers are listed in Table 1.18. The resistance of the surface heat transfers R_{si} and R_{se} were taken from Table 1.4. The total thermal resistance of the wall R is 1.094 m²K/W, and the total specific heat capacity of the building structure C is 728,000 J/m²K.

The heat-accumulation factors determined for the internal F_i and the external F_e sides of the wall are equal to:

$$R_{i,1} = \frac{1}{\alpha_i} = 0.13\frac{m^2 \cdot K}{W}$$

$$R_{i,2} = \frac{1}{\alpha_i} + \frac{d_1}{\lambda_1} = 0.13 + \frac{0.02}{0.81} = 0.155\frac{m^2 \cdot K}{W}$$

$$R_{i,3} = \frac{1}{\alpha_i} + \frac{d_1}{\lambda_1} + \frac{d_2}{\lambda_2} = 0.13 + \frac{0.02}{0.81} + \frac{0.50}{0.58} = 1.017\frac{m^2 \cdot K}{W}$$

$$R_{e,1} = \frac{1}{\alpha_e} + \frac{d_3}{\lambda_3} + \frac{d_2}{\lambda_2} = 0.04 + \frac{0.03}{0.81} + \frac{0.5}{0.58} = 0.939\frac{m^2 \cdot K}{W}$$

$$R_{e,2} = \frac{1}{\alpha_e} + \frac{d_3}{\lambda_3} = 0.04 + \frac{0.03}{0.81} = 0.077 \frac{m^2 \cdot K}{W}$$

$$\begin{aligned} F_{e,1} &= C_1 \left[\frac{R_1}{R} \left(\frac{1}{6} + \frac{R_{i,1} + R_{e,1}}{3R} \right) + \frac{R_{i,1} \cdot R_{e,1}}{R^2} \right] \\ &= 33600 \left[\frac{0.025}{1.094} \left(\frac{1}{6} + \frac{0.13 + 0.939}{3 \cdot 1.094} \right) + \frac{0.13 \cdot 0.939}{1.094^2} \right] \\ &= 3805.1 \frac{J}{m^2 \cdot K} \end{aligned}$$

$$\begin{aligned} F_{e,2} &= C_2 \left[\frac{R_2}{R} \left(\frac{1}{6} + \frac{R_{i,2} + R_{e,2}}{3R} \right) + \frac{R_{i,2} \cdot R_{e,2}}{R^2} \right] \\ &= 644000 \left[\frac{0.862}{1.094} \left(\frac{1}{6} + \frac{0.155 + 0.077}{3 \cdot 1.094} \right) + \frac{0.155 \cdot 0.077}{1.094^2} \right] \\ &= 126863.1 \frac{J}{m^2 \cdot K} \end{aligned}$$

$$\begin{aligned} F_{e,3} &= C_3 \left[\frac{R_3}{R} \left(\frac{1}{6} + \frac{R_{i,3} + R_{e,3}}{3R} \right) + \frac{R_{i,3} \cdot R_{e,3}}{R^2} \right] \\ &= 50400 \left[\frac{0.037}{1.094} \left(\frac{1}{6} + \frac{1.017 + 0.04}{3 \cdot 1.094} \right) + \frac{1.017 \cdot 0.04}{1.094^2} \right] \\ &= 2262.1 \frac{J}{m^2 \cdot K} \end{aligned}$$

$$\begin{aligned} F_{i,1} &= C_1 \left(\frac{R_{e,1}}{R} + \frac{R_1^2}{3R^2} - \frac{R_{i,1} \cdot R_{e,1}}{R^2} \right) \\ &= 33600 \left(\frac{0.939}{1.094} + \frac{0.025^2}{3 \cdot 1.094^2} - \frac{0.13 \cdot 0.939}{1.094^2} \right) \\ &= 25418.3 \frac{J}{m^2 \cdot K} \end{aligned}$$

$$\begin{aligned} F_{i,2} &= C_1 \left(\frac{R_{e,2}}{R} + \frac{R_2^2}{3R^2} - \frac{R_{i,2} \cdot R_{e,2}}{R^2} \right) \\ &= 644000 \left(\frac{0.077}{1.094} + \frac{0.862^2}{3 \cdot 1.094^2} - \frac{0.155 \cdot 0.077}{1.094^2} \right) \\ &= 172178.9 \frac{J}{m^2 \cdot K} \end{aligned}$$

$$F_{i,3} = C_3 \left(\frac{R_{e,3}}{R} + \frac{R_3^2}{3R^2} - \frac{R_{i,3} \cdot R_{e,3}}{R^2} \right)$$

$$= 50400\left(\frac{0.04}{1.094} + \frac{0.037^2}{3 \cdot 1.094^2} - \frac{1.017 \cdot 0.04}{1.094^2}\right)$$

$$= 148.9\frac{\mathrm{J}}{\mathrm{m^2 \cdot K}}$$

$$\mathrm{F_e} = \sum_{i=1}^{3} \mathrm{F_{e,i}} = 132,930\frac{\mathrm{J}}{\mathrm{m^2 \cdot K}}$$

$$\mathrm{F_i} = \sum_{i=1}^{3} \mathrm{F_{i,i}} = 197,746\frac{\mathrm{J}}{\mathrm{m^2 \cdot K}}$$

The measured heat flux $\dot{q}_j$ during the experiment and the evaluated thermal transmittances $U_{in\text{-}situ}$ and $U_{in\text{-}situ,cor}$ of the wall are given in Fig. 1.84. It is clear that the in-situ experiment could be shorten by approximately 100 h if heat-accumulation factors are involved in the evaluation of in-situ determined thermal transmittance of the building structure.

1.5.9 Determination of Thermal Irregularities in the Envelope for Built Buildings

An increased thermal transmittance and thermal bridges in building structures cause a rise in the temperatures on their external surfaces during the heating period. Any irregularities in the thermal insulation of the building envelope can therefore be determined by measuring the surface temperature of the building structures. A direct measurement using temperature sensors would be impractical, so contact-free methods are employed to determine the temperature distribution on large areas of the building envelope. The method is called thermography or thermal imaging and is based on measuring the radiant heat flux emitted from the building structure's surface. The thermography devices are called thermal imaging cameras (Fig. 1.85). A thermal imaging camera converts the thermal radiation, which is invisible to our eyes, into colour images. The following natural properties and laws are exploited:

- An atmospheric window is an optical property of the atmosphere, making it possible for thermal radiation in the wavelength range between 8 and 14 μm to pass without being absorbed by the air.
- The thermal radiation emitted from the surfaces at ambient temperature (in the range between 263 and 323 K) is the largest in the wavelength range of the atmospheric window;
- Most construction materials have a large emissivity of thermal radiation ($\varepsilon_{IR} >$ 0.90), meaning that they emit radiant heat flux almost like black bodies.

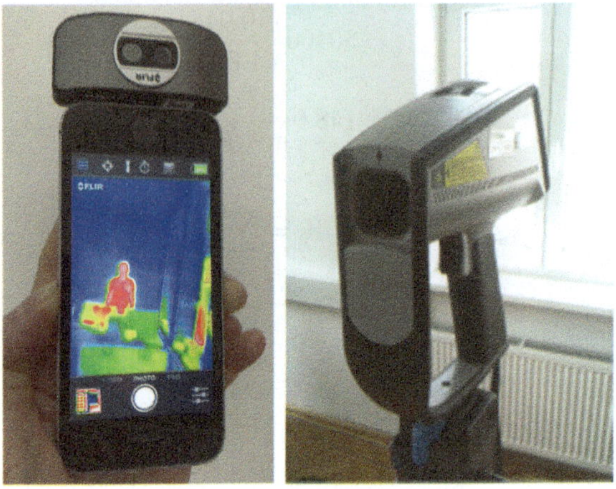

Fig. 1.85 Thermal imaging cameras are available for smartphones

- Most construction materials act as Lambert radiators, i.e., the radiant heat flux emitted from the building envelope surfaces is proportional to the cosine of the angle between the observer's line of sight and the surface normal.

The conditions for the thermal imaging of buildings are defined in [24].

One of the most important conditions is that the thermal imaging of the building is carried out when there is a cloudy sky, because otherwise radiative cooling affects the surface temperature of the building structure. At least 12 h prior to the thermal imaging, the building structure must not be exposed to direct solar irradiation, and the indoor and outdoor air temperatures must not change by more than ± 2 °C while the thermal imaging is conducted.

The thermal imaging of buildings is a common technique for assessing the energy efficiency of buildings, including the building service systems and a range of other applications related to energy conversion (Fig. 1.86).

1.6 Thermal Insulation of Technical Installations

The efficiency of energy use in buildings also depends on the effectiveness of the operation of the building service systems. Heat losses in heating and cooling distribution systems must be considered, including pipes, ducts and storage tanks. The thermal transmittance of thermally insulated pipelines is determined as the reciprocal of the sum of the total resistances to heat transfer between the fluid and the pipe surrounding the ambient air (or soil, for example, in the case of district heating and cooling).

Because the pipelines are round in shape, the density of the heat flux transferred on to an external surface of the thermal insulation of the pipeline depends on the square of the radius. Therefore, the thermal transmittance of the pipelines is defined as the

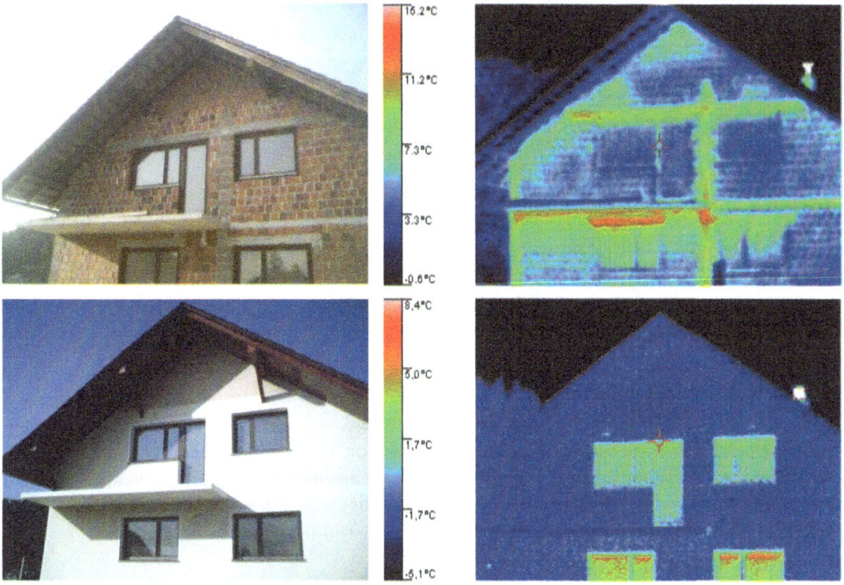

Fig. 1.86 Photograph of the facade before and IR images after the thermal renovation of the façade. The surface temperatures drop significantly because the thermal insulation of the facade wall is much lower. A much lower impact of the thermal bridge scan be seen as well

normalized value of the thermal transmittance U'_p (W/mK) per metre of pipeline:

$$R'_{p,tot} = R_{p,si} + R_{p,\lambda} + R_{p,se} \left[\frac{m \cdot K}{W}\right]\left[\frac{W}{m \cdot K}\right]$$

$$R'_{p,tot} = \underbrace{\frac{1}{h_{si} \cdot \pi \cdot D_i}}_{0} + \frac{1}{2 \cdot \pi \cdot \lambda_{ti}} \ln \frac{D_e}{D_i} + \frac{1}{h_{se} \cdot \pi \cdot D_e} \rightarrow U'_p = \frac{1}{R'_{p,tot}}\left[\frac{W}{m \cdot K}\right]$$

The expression is shown for single layer of thermal insulation with thermal conductivity λ_{izol} and layer thickness $(D_e - D_i)/2$. The resistance to heat conduction in the pipe wall is not taken into account because of the high thermal conductivity of the pipe material. In the expression $R_{p,si}$ is the resistance to the surface heat transfer between the fluid and the pipe wall, $R_{p,\lambda}$ is resistance to heat conduction in the thermal insulation, $R_{p,se}$ is resistance to surface heat transfer by convection and radiation from the surface of the pipe thermal insulation towards the exterior (and vice versa), and D_i and D_e (m) are the internal and external diameters of the thermal insulation layer. $R_{p,si}$ may be ignored due to the high surface heat-transfer coefficient h_{si} between the fluid and the pipe wall (Fig. 1.87).

The combined surface heat-transfer coefficient on the external surface of the heat insulation layer h_{se} depends on the position of the pipe (horizontal, vertical), the wind

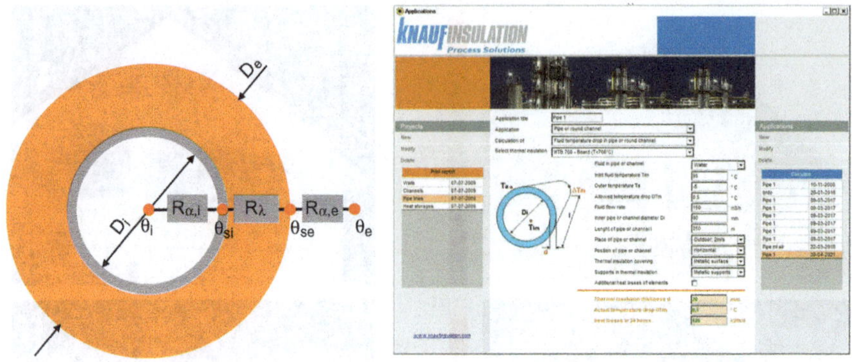

Fig. 1.87 Thermal resistances involved in determining the thermal transmittance U'_p of pipes with a layer of thermal insulation in the case that the resistance to conduction heat transfer in the pipe is neglected (left). Software tools such as KnaufInsulation Technical Applications Calculator [25], accessible at www.knaufinsulation.com, can be used for an analysis of the different applications of technical installations in buildings: flat walls, pipelines, air ducts, cylindrical and spherical hot-water storage tanks. The methodology was adopted from [22] (right)

speed and the emissivity of the surface. The values h_{se} for the pipelines installed in the building are between 4.5 and 6 W/m^{2}K, in the the case of external pipelines and wind speeds of 1 m/s, they are between 12 and 15 W/m^{2}K, and at wind speeds of 5 m/s, the values are between 40 and 55 W/m^{2}K. The values depend on the external diameter of the heat-insulation layer D_e as well. The specific heat flux $\dot{q}_p{}'$ passing through the thermal insulated pipe at the temperature of the liquid θ_i and the external air temperature θ_e per 1 m of pipeline is equal to:

$$\dot{\mathrm{q}}'_{\mathrm{p}} = \mathrm{U}'_p \cdot (\theta_{\mathrm{i}} - \theta_{\mathrm{e}}) \left[\frac{\mathrm{W}}{\mathrm{m}}\right]$$

It is common that the minimum depth of the pipeline thermal insulation layer instead of the maximum normalized value of thermal transmittance $U'_p$is required. As an example, Table 1.19 shows the requirements as in [9].

1.7 Transient Heat-Transfer in Building Structures

In Sect. 1.4 we defined the transient heat transfer as a phenomenon resulting from a constant change in the state of the outdoor and indoor environments, separated by the building structure. The causes of transient heat transfer are the temperature-dependent material properties or changes in the humidity of the materials in the building structure. However, these causes have a relatively low impact on the thermal response of the building structures and will not be considered in the following chapters. Otherwise, the building materials that consist of a phase-change material have

Table 1.19 Requirements regarding the thermal insulation of technical installations as set out in [9]

requirement	criterion
thermal insulation of hot water pipes	$d_{insul} > D_i/2$ for $D_i < 100$ mm $d_{insul} > 100$ mm for $D_i > 100$ mm
thermal insulation of pipes covered with cement plaster	$d_{insul} > 6$mm
thermal insulation of pipes for the distribution of chilled water (5-7 °C) in water cooling systems	$d_{insul} > 13$ mm for $D_i < 40$ mm $d_{insul} > 38$ mm for $D_i < 200$ mm the water vapour must not condense from the air on the surface of thermal insulation layer at air temperature θ_i and air humidity φ_i
hot water storage tanks* *this requirement is not listed in the Rules on efficient energy use in buildings	water at temperature 60 °C may not cool by more than 3 K in 24 hours

a great impact on the transient heat transfer in building structures due to the storage of latent heat.

1.7.1 Modelling Transient Heat Transfer in Building Structures

Modelling transient heat transfer in homogeneous building structures combines the modelling of boundary conditions and the modelling of the transient thermal response of the building structure. The boundary conditions are defined by the temporal variation of the temperature or heat flux on the surface of the building structure. Transient heat transfer in the building structure is modelled using a mathematical formulation of Fourier's law of heat conduction in a differential form. Although the heat transfer is 3D, in the following examples only the modelling of the one-dimensional transient heat conduction through an isotropic material with a constant thermal conductivity λ will be shown. The building structure is divided into differential control volumes with a rectangular shape and a size dx, dy and dz, as shown in Fig. 1.88. The incoming heat flux $\dot{q}_x$ in the direction x, is equal to:

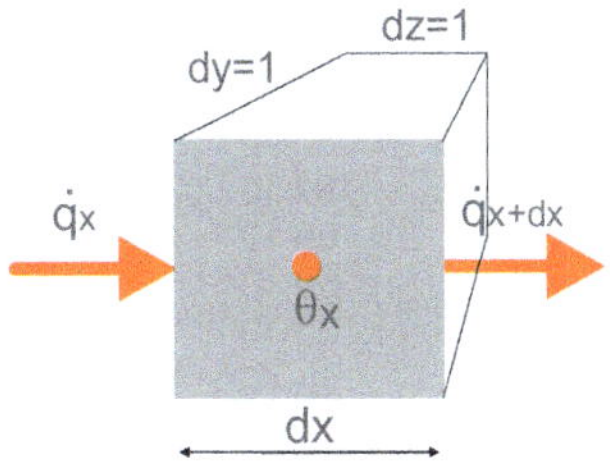

Fig. 1.88 Temperature change in the differential control volume $dx \times dy \times dz$ is proportional to the difference between the incoming and outgoing heat flux transferred in a unit of time

$$\dot{q}_x = -\lambda \cdot \frac{\partial \theta}{\partial x}$$

The outgoing heat flux $\dot{q}_{x+dx}$ at the position $x + dx$ is determined using a Taylor series, omitting the derivatives of higher order:

$$\dot{q}_{x+dx} = \dot{q}_x + \frac{\partial \dot{q}_x}{\partial x} \cdot dx$$

Explanation The Taylor series defines the value of a continuous function in the vicinity of a function-independent variable (in this case x), where the value of the function f_x (in this case, the heat flux $\dot{q}_x$) is known:

$$f_{x+dx} = f_x \pm dx \cdot \left(\frac{\partial f}{\partial x}\right)_x + \frac{dx^2}{2!} \cdot \left(\frac{\partial^2 f}{\partial x^2}\right)_x \pm \frac{dx^3}{3!} \cdot \left(\frac{\partial^3 f}{\partial x^3}\right)_x + \ldots + \frac{dx^m}{m!} \cdot \left(\frac{\partial^m f}{\partial x^m}\right)_x$$

The difference between the incoming $\dot{q}_x$ and the outgoing heat flux $\dot{q}_{x+dx}$ during the time period dt changes the amount of heat stored in the differential control volume $dx \times dy \times dz$, in the case of 1D $dy = dz = 1$. The change in the amount of heat stored in the control volume in an infinitesimal time interval dt is given by (Table 1.20):

$$\underbrace{q_x - \left(q_x + \frac{\partial q_x}{\partial x} \cdot dx\right)}_{\text{heat flux difference}} = \underbrace{\rho \cdot c_p \cdot \frac{\partial \theta}{\partial t} \cdot dx}_{\text{stored heat difference}}$$

If the difference between $\dot{q}_x$ and $\dot{q}_{x+dx}$ is negative, the temperature inside the volume will decrease, and vice-versa. Using Fourier's law, the expression can be rearranged to obtain:

Table 1.20 Thermal diffusivity a and thermal effusivity b of some construction materials [2, 26]

material	a (m^2/s)	b ($kJ/m^2Ks^{0.5}$)
aluminium	$8230 \cdot 10^{-8}$	22.5
stone wool	$167 \cdot 10^{-8}$	0.03
concrete	$68 \cdot 10^{-8}$	2.18
brick	$59 \cdot 10^{-8}$	1.17
aerated concrete	$26 \cdot 10^{-8}$	0.25
water	$14 \cdot 10^{-8}$	1.56
wood	$14 \cdot 10^{-8}$	0.38

$$-\frac{\partial q_x}{\partial x} = -\frac{\partial}{\partial x}\left(-\lambda\frac{\partial \theta}{\partial T}\right) = \lambda\frac{\partial^2 \theta}{\partial x^2} = \rho \cdot c_p \cdot \frac{\partial \theta}{\partial t} \rightarrow \frac{\partial \theta}{\partial t} = \underbrace{\frac{\lambda}{\rho \cdot c_p}}_{a}\frac{\partial^2 \theta}{\partial x^2}$$

Thus, we derive the differential equation of a one-dimensional transient heat conduction. The properties of a material that conducts the heat flux are combined in the thermal diffusivity a (m^2/s) and thermal effusivity b ($kJ/m^2Ks^{0.5}$) (Table 1.20), the ratio between the thermal conductivity λ (W/mK) and the heat-accumulation capacity, which depends on the density ρ (kg/m^3) and the specific heat capacity at constant pressure c_p (J/kgK).

$$a = \frac{\lambda}{\rho \cdot c_p}\left[\frac{W \cdot m^3 \cdot kg \cdot K}{m \cdot K \cdot kg \cdot J} = \frac{W \cdot m^3 \cdot kg \cdot K}{m \cdot K \cdot kg \cdot W \cdot s} = \frac{m^2}{s}\right]$$

While it can be assumed that the thermal conductivity λ and the density ρ of construction materials are independent of the temperature of the building structures, the specific heat capacity depends on whether the heat is accumulated in the material as sensible or as latent heat. In the first case, the heat is stored in the material by increasing its temperature, while in the second case by changing the state of matter of the substance, e.g., from solid to liquid. Such materials are called phase-change materials (PCMs). The phase change in pure materials takes place at a constant temperature. The amount of heat needed to change the state of matter of 1 kg of substances from solid to liquid is known as the enthalpy of fusion r (J/kg) and the same amount of heat is released upon solidification. Since both, the sensible and the latent heat can be stored in the substance, the total amount of accumulated heat is indicated by the enthalpy h (J/kg). It is defined by the term:

$$h = c_p \cdot \theta\left[\frac{J \cdot K}{kg \cdot K} = \frac{J}{kg}\right]$$

Explanation The paraffin mixtures commonly used for latent-heat storage in building structures do not change state at a constant temperature, rather in a range of melting temperature, which should be close to the diurnal indoor-air temperature swing. This is a common feature of organic mixture PCMs and because of that, the temperatures of a building structure that contain a PCM rise only slightly until the melting is completed.

Case Study Determine the quantity of heat stored in 1 kg of water, 1 kg of concrete and 1 kg of PCM paraffin mixture Rubitherm RT20, if the temperature

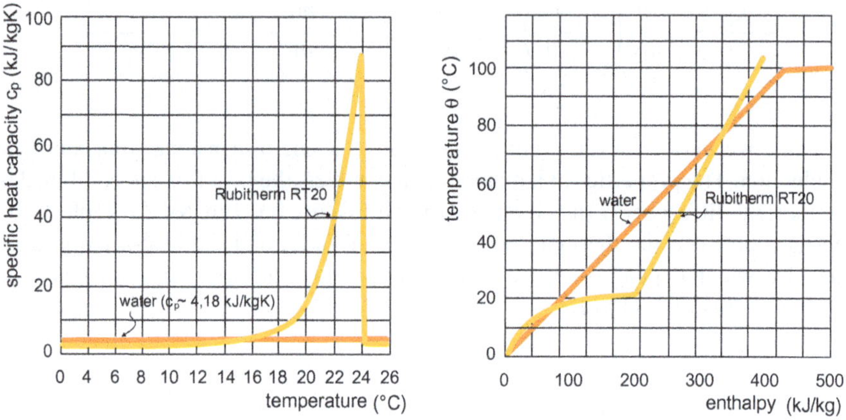

Fig. 1.89 The specific heat capacity of water and a phase change material (PCM), the paraffin mixture Rubitherm RT 20 (left). The melting point of this PCM lies in the range 16–24 °C. The average specific heat capacity of the PCM in the melting temperature range is 20.99 kJ/kg, which is almost five times the specific heat capacity of water [27, 28]. The enthalpy of the sensible and latent heat stored in 1 kg of construction materials (right). Latent-heat storage has significant advantages over sensible-heat storage in the case of a low-temperature swing of the heat source. Such an example is a periodic change in the indoor-air temperature during the day

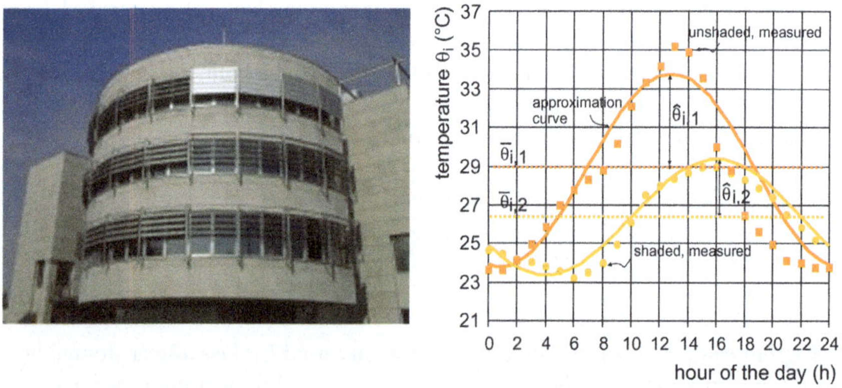

Fig. 1.90 The course of temperature θ_i in the room of an office building (left) with and without shades (right) (Table 1.22)

of those materials changes the same as the indoor-air temperature θ_i from 18 to 24 °C. The total amount of heat stored is equal to the difference in the enthalpy Δh shown in Fig. 1.89. The density ρ of water is 1000 kg/m^3, of concrete 2500 kg/m^3 and of PCM 450 kg/m^3. The results are listed in Table 1.21.

Table 1.21 Amount of the total heat stored in 1 kg and 1 m^3 of the mater. The total stored heat consists of the sensible and the latent heat and is equal to the enthalpy

	water	concrete	Rt20
enthalpy at 18°C (kJ/kg)	76.7	15.1	86.3
enthalpy at 24°C (kJ/kg)	93.7	20.1	197.2
stored heat=Δh (kJ/kg)	17.0	5.1	110.9
stored heat=Δh (kJ/m^3)	17000	11000	77630

Table 1.22 Average diurnal indoor-air temperature $\overline{\theta}_i$ and amplitude $\hat{\theta}_i$ in the office shown in Fig. 1.90 without and with the shaded windows

	$\overline{\theta}_i$ (°C)	$\hat{\theta}_i$ (°C)	Δt (s)
unshaded windows	28.9	5.3	3600
shaded windows	26.5	3.0	14700

1.7.2 *Causes of Transient Heat Transfer in Building Structures*

Regarding indoor thermal comfort and the thermal response of buildings, transient heat transfer appears in following conditions: when we walk barefoot over a heated floor;

- when heat is accumulated in internal building structures due to a periodic change in the indoor-air temperature $\tilde{\theta}$. The value of $\tilde{\theta}_i$ changes during the day as a consequence of the time-varying internal heat sources, solar
- radiation gains and transmission and ventilation heat losses. Under such conditions, heat is stored at higher indoor-air temperatures (during the day-time) and is realised at lower indoor-air temperatures (during the night-time) and as consequence the overheating risk and the energy demand for providing the thermal comfort in the building is reduced. A periodic variation of the indoor-air temperature $\tilde{\theta}_{i,(t)}$ in time t (s) for the duration of the period t_p 86,400 s (24 h) can be modelled with the expression:

where $\overline{\theta}_i$ (°C) is the average diurnal indoor-air temperature, $\hat{\theta}_i$ (°C) is the indoor-air temperature amplitude and Δt (s) is the time shift of the amplitude with regards to 12 o' clock (solar time) in a day.

- when the external surface of a building structure is exposed to the on-site climate boundary conditions, the temperature of the external surface θ_{se} of the structure is changing over the course of the day as a response to the absorbed short-wavelength solar irradiation, the emitted long-wavelength thermal radiation from the surface of the structure, the received long-wavelength thermal radiation from the sky and external environment surfaces, and the

- heat flux that is conducted through the structure towards the building's interior. The outdoor boundary conditions can be modelled by introducing the equivalent outdoor-air temperature $\theta_{e,eq}$ (°C). This is the apparent outdoor-air temperature at which the same heat flux transfers at the external surface of a building structure by convection only, as it is actually at the current meteorological conditions. It is defined using the expression:

$$\theta_{e,eq} = \theta_e + \frac{\overbrace{\alpha_{s,se} \cdot G_{glob,\beta}}^{\text{absorbed solar radiation}} + \overbrace{\sum_{i=1}^{n} \left(\varepsilon_{IR,i} \cdot \sigma \cdot F_i \cdot T_{se,i}^4\right)}^{\text{long - wave emissive power from the neighbouring surfaces}} + \overbrace{\varepsilon_{IR,sky} \cdot F_{sky} \cdot \sigma \cdot T_{sky}^4}^{\text{long - wave emissive power from the sky}} - \overbrace{\varepsilon_{IR,se} \cdot \sigma \cdot T_{se}^4}^{\text{long - wave emissive power from the structure}}}{h_{c,e}}$$

The quantities included in the expression are described in Table 1.23.

The calculation procedure can be simplified by assuming that the thermal emissivity $e_{IR,i}$ of all the external environment surfaces are the same, and that the temperatures of all the external environment surfaces $T_{se,i}$ are equal to the outdoor-air temperature T_e. The geometric view-factors F_i of the building structure towards the external surfaces are equal to $1 - F_{sky,i}$ for the vertical surfaces and equal to 0 for the horizontal surfaces:

Table 1.23 Quantities defining the equivalent outdoor-air temperature $\theta_{e,eq}$ at which the actual heat flux at the external surface of the building structure can be expressed using the convective surface heat-transfer coefficient $h_{c,e}$

	quantity	unit	description
	T_e, θ_e	(K), (°C)	outdoor air temperature
	$h_{c,se}$	(1)	convective heat transfer coefficient between the building structure's external surface and the environment; determined using empirical expressions, for example as a function of wind speed v (m/s): $h_{c,se} = 4 + 4.4 \cdot v$ (W/m²K)
absorbed solar radiation	α_{sse}	(1)	absorptivity of short-wave solar radiation
	$G_{glob,\beta}$	(W/m²)	total / global solar irradiation of external building structure surface with inclination β and orientation ψ
long-wave emissive power from the neighbouring surfaces	i, i = 1 to n		index of neighbouring surfaces, n = total number of surfaces
	$\varepsilon_{IR,i}$	(1)	long-wave thermal emissivity, tipical versions of ε_i are in the range between 0.8 and 0.9
	σ	(W/m²K⁴)	Stefan-Boltzmann constant; $5.67 \cdot 10^{-8}$ W/m²K⁴
	F_i	(1)	geometric view factor for i-th surface in the environment and surface of the building surface, $F_i = 0$ for horizontal surfaces
	T_i	(K)	temperature of i-th surface in the environment
long-wave emissive power from the sky	$\varepsilon_{IR,sky}$	(1)	long-wave thermal emissivity for the sky
	F_{sky}	(1)	geometric view factor for the sky and the surface of the building surface, EN ISO 13791 standard defines the following values of F_{sky} for vertical surfaces: $F_{sky} = 0.33$ for urban environments $F_{sky} = 0.41$ for suburban environments $F_{sky} = 0.45$ for the countryside $F_{sky} = 1$ for horizontal surfaces
	T_{sky}	(K)	sky temperature; EN ISO 13791 defines the following expression for the determination of T_{sky}: $T_{sky} = 55.3 \cdot 10^{-3} \cdot T_e^{3/2}$ (K)
long-wave emissive power from the building structure	$\varepsilon_{IR,se}$	(1)	long-wave thermal emissivity of building structure's surface
	T θ	(K), (°C)	building structure surface temperature

$$\theta_{e,eq} = \theta_e + \frac{\alpha_{s,se} \cdot G_{glob,\beta} + \overbrace{\varepsilon_{IR} \cdot \sigma \cdot (1 - F_{sky}) \cdot T_e^4}^{=0\text{ for horizontal surfaces}} + \overbrace{\varepsilon_{IR,sky}}^{\sim 1} \cdot F_{sky} \cdot \sigma \cdot T_{sky}^4 - \varepsilon_{IR,se} \cdot \sigma \cdot T_{se}^4}{h_{c,e}}$$

Explanation Sol–air temperature $\theta_{sol\text{-}air}$ is often used in engineering practice in place of the equivalent outdoor-air temperature $\theta_{e,eq}$. The temperatures differ in the model of long-wave heat transfer by radiation.

The global irradiation on an arbitrary surface $G_{glob,\beta}$ (W/m^2) is the sum of the direct, diffuse and reflected irradiation (Fig. 1.91). Direct irradiation reaches the building structure's surface directly from the sun's photosphere as beam radiation and depends on the incident angle of the sun's rays i (°). This is the angle between the sun's ray and the direction of the normal of the sunlit surface. Diffuse irradiation is received by the surface of the building structure after a part of the beam irradiation is scattered from particles and molecules in the atmosphere. It depends on the atmospheric pollution, the type of clouds and the proportion of the sky that is covered by clouds. The reflected solar irradiation is reflected from the external environment surfaces and depends on the reflectivity (albedo) of those surfaces. The basis for the modelling of $G_{glob,\beta}$ is the meteorological data formed in the "Typical Reference Year", files with whole-year hourly values of the global solar irradiation on the horizontal plane $G_{glob,o}$. as well as other meteorological quantities. Figure 1.91

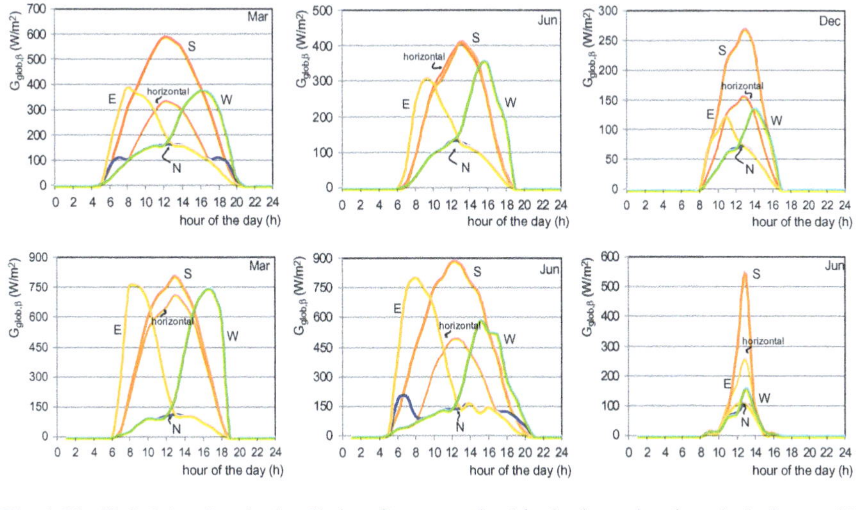

Fig. 1.91 Global (total) solar irradiation $G_{glob,\beta}$ received by horizontal and vertical planes with different celestial orientations on an average day in December, March and June (top) and a clear day on the 3rd of December, 23th of March and 21st of June in Ljubljana (L 46°03′, λ 14°30′) (bottom)

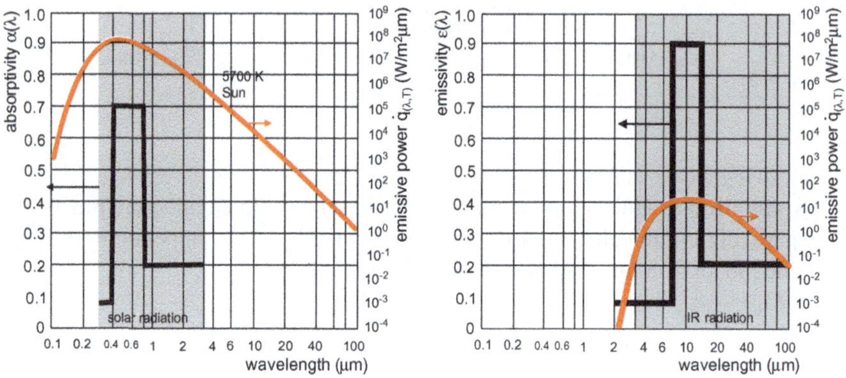

Fig. 1.92 Spectral absorptivity of the surface analysed as case study 1 (left). The spectral emissivity of the surface analysed as case study 2 (right)

shows an example of the global solar irradiation received by the horizontal $G_{glob,o}$ and vertical planes $G_{glob,\beta}$ with different celestial orientations on average and a clear day in December, March and June in Ljubljana.

Figure 1.92 presents selective radiative properties of matter related to the radiative heat transfer. It is shown that, for example, the spectral absorptivity $\alpha_{r,\lambda}$ of the short-wavelength solar irradiation (0.3 μm < λ < 3 μm) is not the same for all wavelengths of the solar irradiation. Selective surfaces can also have a different spectral emissivity $\varepsilon_{IR,\lambda}$ of the long-wavelength (3 μm < λ < 100 μm) thermal radiation at different wavelengths. Since the emitted thermal radiative heat flux depends on the absolute temperature of the emitter's surface, the average absorptivity α_s of the solar irradiation and the average emissivity ε_{IR} of the selective surfaces must be determined by taking into account the spectral radiative heat flux, as shown in the following case study.

Case Study 1 Determine the average absorptivity of the solar irradiation of the surface with the following spectral radiative properties: $\alpha_{s,0,3\ \mu m<\lambda<0,4\ \mu m}$ equal to 0.1, $\alpha_{s,0,4\ \mu m<\lambda<0,8\ \mu m}$ equal to 0.7 and $\alpha_{s,0,8\ \mu m<\lambda<3\ \mu m}$ equal to 0.2 (Fig. 1.92, left). Assume that the sun is a thermal emitter with a temperature of T 5700 K. The proportion factors $F_{0\text{-}\lambda}$ of the total radiative heat flux for a thermal emitter at the temperature T (K) between the wavelength 0 μm and a selected wavelength λ μm are presented in Table 1.24.

The average absorptivity of the solar irradiation should be used as $\alpha_{s,se}$ in the determination of equivalent outdoor-air temperature $\theta_{e,eq}$.

$$\overline{\alpha}_s = \int\limits_{\lambda=0,3}^{\lambda=3} \alpha_\lambda \cdot \left(F_{0-3} - F_{0-0,3}\right) \cdot d\lambda$$

Table 1.24 Proportion factors $F_{0-\lambda}$ of the total emissive power for a thermal radiator at the temperature T (K) between the wavelength 0 μm and a selected wavelength λ μm

λ·T (μm·K)	$F_{(0-\lambda)}$	λ·T (μm·K)	$F_{(0-\lambda)}$	λ·T (μm·K)	$F_{(0-\lambda)}$	λ·T (μm·K)	$F_{(0-\lambda)}$
200	0.0000	3200	0.3181	6400	0.7692	12000	0.9451
400	0.0000	3400	0.3617	6600	0.7832	13000	0.9551
600	0.0000	3600	0.4036	6800	0.7961	14000	0.9629
800	0.0001	3800	0.4434	7000	0.8081	15000	0.9700
1000	0.0003	4000	0.4809	7200	0.8192	16000	0.9738
1200	0.0021	4200	0.5160	7400	0.8295	18000	0.9809
1400	0.0078	4400	0.5488	7600	0.8391	20000	0.9856
1600	0.1971	4600	0.5793	7800	0.8480	25000	0.9922
1800	0.0393	4800	0.6076	8000	0.8563	30000	0.9953
2000	0.0667	5000	0.6337	8500	0.8746	40000	0.9980
2200	0.1009	5200	0.6590	9000	0.8900	50000	0.9990
2400	0.1403	5400	0.6804	9500	0.9031	75000	0.9997
2600	0.1831	5600	0.7010	10000	0.9142	100000	0.9999
2800	0.2279	5800	0.7202	10500	0.9237		
2898	0.2501	6000	0.7378	11000	0.9319		
3000	0.2732	6200	0.7541	11500	0.9400		

$$\approx \overline{\alpha}_{\lambda_1-\lambda_2}\left(F_{0-\lambda_2} - F_{0-\lambda_1}\right)$$
$$+ \overline{\alpha}_{\lambda_2-\lambda_3}\left(F_{0-\lambda_3} - F_{0-\lambda_2}\right) + .. + \overline{\alpha}_{\lambda_{n-1}-\lambda_n}\left(F_{0-\lambda_n} - F_{0-\lambda_{n-1}}\right)$$

$$\lambda_1 \cdot T_s = 0.3 \cdot 5700 = 1710[\mu\,m \cdot K] \rightarrow F_{0-0.3} = 0.0305$$

$$\lambda_2 \cdot T_s = 0.4 \cdot 5700 = 2280[\mu\,m \cdot K] \rightarrow F_{0-0.4} = 0.1166$$

$$\lambda_3 \cdot T_s = 0.8 \cdot 5700 = 4560[\mu\,m \cdot K] \rightarrow F_{0-0.8} = 0.5732$$

$$\lambda_4 \cdot T_s = 0.3 \cdot 5700 = 17100[\mu\,m \cdot K] \rightarrow F_{0-0.8} = 0.9777$$

$$\overline{\alpha}_s \approx 0.1 \cdot (0.1166 - 0.0305) + 0.7 \cdot (0.5732 - 0.1166)$$
$$+ 0.2 \cdot (0.9777 - 0.5732) = 0.41$$

Case Study 2 Determine the average emissivity of a building structure's surface having the following spectral emissivity: $\varepsilon_{IR,03\,\mu m<\lambda<8\,\mu m} = 0.1$, $\varepsilon_{IR,8\,\mu m<\lambda<12\,\mu m} = 0.9$ and $\varepsilon_{IR,12\,\mu m<\lambda<100\,\mu m} = 0.2$ (Fig. 1.92, right). The proportion factors $F_{0\text{-}\lambda}$ of the total radiative heat flux for a thermal emitter at the temperature T (K) between the wavelength 0 μm and a selected wavelength λ μm are presented in Table 1.24.

The average emissivity of the long-wavelength irradiation should be used as $\varepsilon_{IR,se}$ in the determination of the equivalent outdoor-air temperature $\theta_{e,eq}$.

$$\bar{\varepsilon}_{IR} = \int_{\lambda=3}^{\lambda=100} \varepsilon_\lambda \cdot (F_{0-100} - F_{0-3}) \cdot d\lambda \approx$$

$$\approx \bar{\varepsilon}_{\lambda_1-\lambda_2}\left(F_{0-\lambda_2} - F_{0-\lambda_1}\right) + \bar{\varepsilon}_{\lambda_2-\lambda_3}\left(F_{0-\lambda_3} - F_{0-\lambda_2}\right) + .. + \bar{\varepsilon}_{\lambda_{n-1}-\lambda_n}\left(F_{0-\lambda_n} - F_{0-\lambda_{n-1}}\right)$$

$$\lambda_1 \cdot T = 2 \cdot 300 = 600[\mu\,m \cdot K] \rightarrow F_{0-2} = 0.0$$

$$\lambda_2 \cdot T = 8 \cdot 300 = 2400[\mu\,m \cdot K] \rightarrow F_{0-8} = 0.1403$$

$$\lambda_3 \cdot T = 12 \cdot 300 = 3600[\mu\,m \cdot K] \rightarrow F_{0-12} = 0.4036$$

$$\lambda_4 \cdot T = 100 \cdot 300 = 30000[\mu\,m \cdot K] \rightarrow F_{0-100} = 0.9953$$

$$\bar{\varepsilon}_{IR} \approx 0.1 \cdot (0.1403 - 0.0) + 0.9 \cdot (0.4036 - 0.1403) + 0.2 \cdot (0.9953 - 0.4036) = 0.37$$

Case Study 3 Determine the equivalent air temperature $\theta_{e,eq}$ during a sunny day for a vertical south-oriented building structure with the average absorbance of short-wavelength solar irradiation $\alpha_{s,se}$ 0.4 (bright surface) and 0.8 (dark surface) and the average emissivity of long-wavelength thermal radiation $\varepsilon_{IR,se}$ 0.9. The diurnal values of the global solar irradiation $G_{glob,\beta}$ on the structure surface, the sky temperature θ_{sky} and the outdoor-air temperature θ_e are shown in Fig. 1.93. Other values of the influencing quantities are included in Table 1.24.

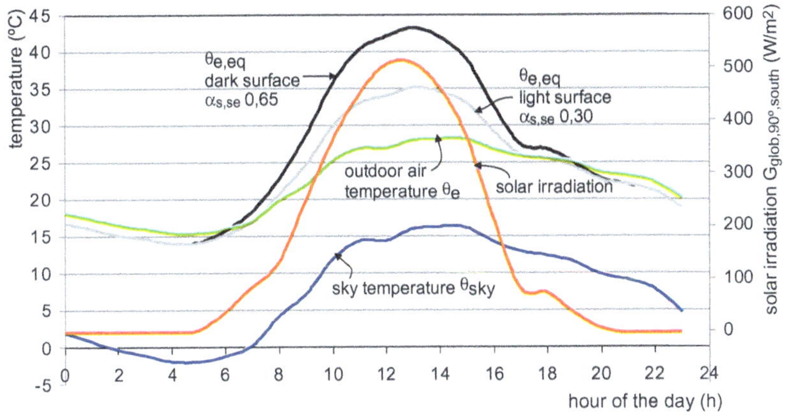

Fig. 1.93 Equivalent outdoor air temperatures $\theta_{e,eq}$ over a clear summer day for a vertical south-facing wall with different absorptivities of the solar irradiation $\alpha_{s,se}$

1.7.3 *Modelling of the Transient Heat Transfer in Building Structures*

1.7.3.1 Response of Building Structures to a Step Change in the Surface Temperature

An example of transient heat transfer in building structures is the case when the cold surface of a building structure is touched by a bare part of the body, for example, when walking on the bathroom floor. As the temperature of a foot is normally higher than the building structure, the heat flux will transfer from the body to the building structure. The transient heat transfer occurs as a consequence of a step change in the temperature on the surface of the building structure. The time-dependent temperature at a depth x of the building structure $\theta(x, t)$ is defined by an expression that includes the "error function" erfc (Fig. 1.94):

$$\theta(x, t) = \theta_0 + (\theta_1 - \theta_0) \cdot erfc\underbrace{\left(\frac{x}{\sqrt{4 \cdot a \cdot t}}\right)}_{z}$$

where θ_0 is the initial steady-state temperature on the surface and within the building structures before the surface temperature of the structure changed, θ_1 is the increased surface temperature after the step change, x (m) is the depth of the building structure ($x = 0$ m on the surface), a is the thermal diffusivity (m^2/s, Table 1.20), and t is the time since the step change (s). The specific heat flux at an arbitrary depth x is determined by Fourier's law (for the one-dimensional case):

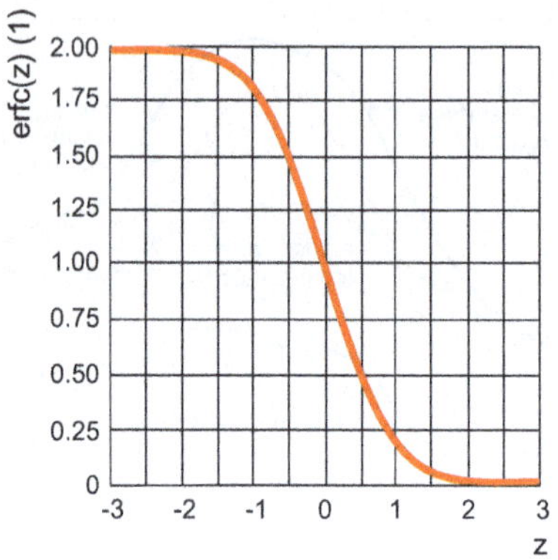

Fig. 1.94 Erfc—"error function", determining the temporal variation of the temperature at the depth of the building structure $\theta(x,t)$ as a result of a step temperature change from θ_0 to an elevated temperature θ_1 at the surface of the building structure

$$\dot{q} = -\lambda \frac{\partial \theta(x, t)}{\partial x}$$

The temperature derivative $\partial\theta(x,t)/\partial x$ is obtained by introducing a new variable z:

$$\frac{\partial \theta(x, t)}{\partial x} = \underbrace{\frac{d}{dz} erfc(z)}_{-\frac{2 \cdot e^{-z^2}}{\sqrt{\pi}}} \cdot \underbrace{\frac{dz}{dx}}_{\frac{1}{\sqrt{4 \cdot a \cdot t}}} \rightarrow \dot{q}(x, t) = -\lambda \cdot \left(-\frac{2 \cdot e^{-z^2}}{\sqrt{\pi}} \right) \cdot \frac{1}{\sqrt{4 \cdot a \cdot t}} \rightarrow$$

$$\dot{q}(x, t) = \lambda \cdot \frac{2}{\sqrt{\pi}} \cdot e^{-\left(\frac{x}{\sqrt{4 \cdot a \cdot t}}\right)^2} \cdot \frac{1}{\sqrt{4 \cdot a \cdot t}} \rightarrow \dot{q}(0, t) = \frac{\lambda}{\sqrt{\pi \cdot a \cdot t}} \cdot (\theta_1 - \theta_0)$$

The final equation states the time-dependent heat flux transferred into the building structure from its surface, after the time t from the moment of the temperature step change. The heat flux $\dot{q}(0,t)$ can also be expressed with the thermal effusivity b (Table 1.20):

$$\dot{q}(0, t) = \frac{\lambda}{\underbrace{\sqrt{a}}_{\sqrt{\frac{\lambda}{\rho \cdot c_p}}}} \frac{1}{\sqrt{\pi \cdot t}} \cdot (\theta_1 - \theta_0) = \frac{1}{\underbrace{\sqrt{\lambda \cdot \rho \cdot c_p}}_{b}} \frac{1}{\sqrt{\pi \cdot t}} \cdot (\theta_1 - \theta_0) \left[\frac{W}{m^2} \right]$$

If two bodies with the temperature θ_1 and θ_0 are in contact and $\theta_1 > \theta_0$, the heat flux transferred from body 1 is equal to the heat flux entering body 0 with the lower temperature (Fig. 1.95). The temperature of the contact surface θ_{10} at the new

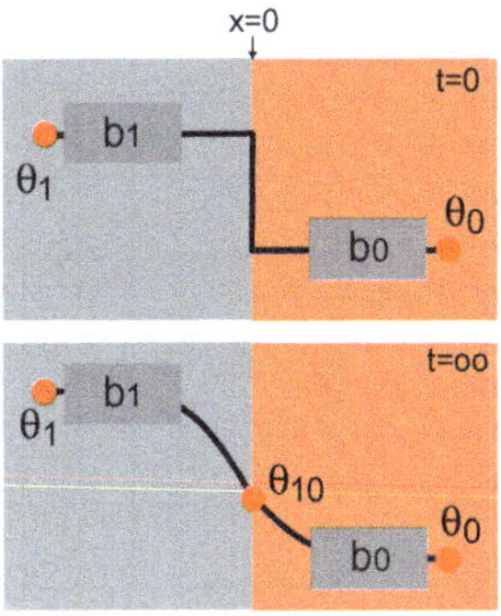

Fig. 1.95 Building-structure response to a step change of the surface temperature. The temperature distribution in bodies in contact at initial state at $t = 0$ s is shown above, the temperature distribution at infinite time $t = \infty$ is shown below. Such a state will be reached in typical building materials after 100+ s

stationary state (after infinity long time) can be determined by considering that the heat fluxes are equal:

$$\dot{q}(0^-, t_{\to\infty}) = \dot{q}(0^+, t_{\to\infty}) \to \frac{1}{b_1} \frac{1}{\sqrt{\pi \cdot t}} \cdot (\theta_1 - \theta_{10}) = \frac{1}{b_0} \frac{1}{\sqrt{\pi \cdot t}} \cdot (\theta_{10} - \theta_0)$$

$$\to \theta_{10} = \frac{b_1 \cdot \theta_1 + b_0 \cdot \theta_0}{b_1 + b_0} [^\circ C]$$

Until the stationary state is reached, the density of the heat flux passing from body 1 to body 0 changes with time and is determined using the expression:

$$\dot{q}(0, t) = \frac{b_1 \cdot b_0}{b_1 + b_2} \frac{1}{\sqrt{\pi \cdot t}} \cdot (\theta_1 - \theta_0) \left[\frac{W}{m^2}\right]$$

Case Study Determine the heat flux transferred from the sole of a foot into a concrete floor. The thermal effusivity of the human body b_1 is 1470 $Ws^{1/2}/m^2K$, and the thermal effusivity of concrete b_o is 1800 $Ws^{1/2}/m^2K$. The temperature of the foot θ_1 is 33 °C, the temperature of the floor θ_0 is 18 °C. Calculate the contact temperature after the heat flux stabilizes. Compare the results for a concrete floor with the results for a wooden one with a thermal effusivity b_0 = 310 $Ws^{1/2}/m^2K$. What should the temperature of the surface of the concrete floor be so that after 2 s the person would not feel any difference between the concrete and wooden floors.

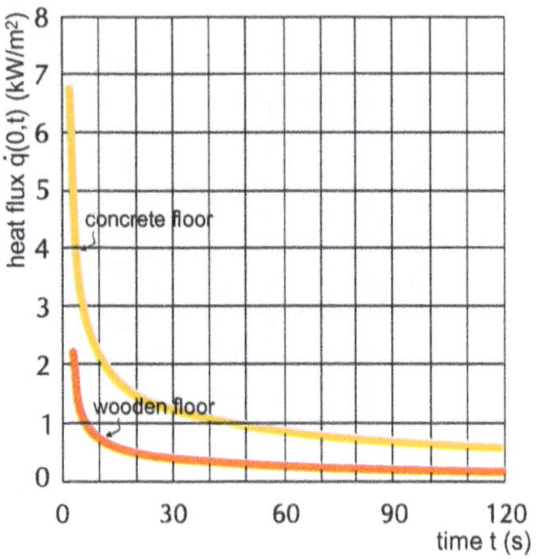

Fig. 1.96 Density of the heat flux that transfers from the bare foot into the cooler concrete or wood floor

The temperatures on the contact surfaces are independent of the time t and are equal to:

$$\theta_{foot/wood} = \frac{b_l \cdot \theta_l + b_0 \cdot \theta_0}{b_l + b_0} = \frac{1470 \cdot 33 + 310 \cdot 18}{1470 + 310} = 30.4°C$$

$$\theta_{foot/concrete} = \frac{b_l \cdot \theta_l + b_0 \cdot \theta_0}{b_l + b_0} = \frac{1470 \cdot 33 + 1800 \cdot 18}{1470 + 1800} = 24.7°C$$

A lower temperature of the contact surface causes us to perceive a warm surface when we step on a wooden floor and a cold surface when step on a concrete floor. The thermal diffusivity of wood *a* is equal to 14×10^{-8} (Table 1.20), and the heat flux 2 s after touching the wooden floor is:

$$\dot{q}(0,t) = \frac{b_l \cdot b_0}{b_l + b_0}\frac{1}{\sqrt{\pi \cdot t}} \cdot (\theta_l - \theta_0) = \frac{310 \cdot 1470}{310 + 1470}\frac{1}{\sqrt{\pi \cdot 2}} \cdot (33 - 18)$$
$$= 1532\frac{W}{m^2}$$

and to perceive the concrete floor when treading barefoot as pleasant as wooden one, the concrete floor must be heated to the temperature (Fig. 1.96):

$$\dot{q}(0,t) = \frac{b_l \cdot b_0}{b_l + b_0}\frac{1}{\sqrt{\pi \cdot t}} \cdot (\theta_l - \theta_0) \rightarrow \theta_0 = \theta_l - \dot{q}(0,t)\frac{b_l + b_0}{b_l \cdot b_0}\sqrt{\pi \cdot t} =$$
$$= 33 - 1532\frac{1470 + 1800}{1470 \cdot 1800}\sqrt{\pi \cdot 2} = 28.3°C$$

1.7.3.2 Modelling the Thermal Response of an Infinitely Thick Single-Layer Building Structure with a Periodic Change in the Temperature on the Surface

An example of the thermal response of a building structure to a periodic change in the temperature on its surface appears as a periodic process of storing and releasing the heat in building structures. In this case the indoor-air temperature is modelled as a periodic function $\tilde{\theta}_i\ (t)$, as shown in Sect. 1.7.2. The assumption that there is no thermal resistance to the surface-heat transfer between the indoor air and the surface of the structure, results in a surface temperature of the structure $\theta_{si}\ (0,t)$ equal to the indoor-air temperature $\theta_i(t)$. For a one-dimensional case and a homogeneous single-layer construction with infinite thickness, the solution to the differential equation for transient heat transfer (Sect. 1.7.1) is:

$$\frac{\partial \theta}{\partial t} = a \frac{\partial^2 \theta}{\partial x^2}$$

and the solution of the differential equation assuming a periodic boundary temperature condition is:

$$\theta(x,t) = \bar{\theta}_i + \underbrace{\hat{\theta}_i \cdot e^{-x\sqrt{\frac{\pi}{t_p \cdot a}}}}_{\text{periodic response}} \cdot \underbrace{\sin\left(\frac{2\cdot\pi}{t_p} t - x\sqrt{\frac{\pi}{t_p \cdot a}}\right)}_{\text{amplitude time shift}} [^\circ\text{C}]$$

where $\theta(x,\ t)$ is the temperature at a depth x of the building structure at a time t, $\bar{\theta}_i$ (°C) and $\hat{\theta}_i$ (°C) are the average surface temperature of the building structure and the amplitude of the temperature in the period t_p, a is the thermal diffusivity (m^2/s) and t is the time (s). The temperature of the building structure at a depth x will be the highest when the part defining the time shift of the amplitude is 0:

$$\theta(x)_{max} = \bar{\theta}_i + \hat{\theta}_i \cdot e^{-x\sqrt{\frac{\pi}{t_p \cdot a}}}$$

$$\theta(x)_{min} = \bar{\theta}_i - \hat{\theta}_i \cdot e^{-x\sqrt{\frac{\pi}{t_p \cdot a}}} [^\circ\text{C}]$$

The thermal response zone of the building structure is defined by the maximum thickness of the building structure in which the temperature change is observed, is an important design parameter for the passive heating and cooling of the buildings. The zone defines the amount of the heat stored during half of the period t_p (24 h or 86,400 s) and which is realised during the second half-period duration. The periodic heat-penetration depth d_p (m) is the depth of the structure at which the amplitude of the temperature $\hat{\theta}_{(x=dp)}$ decreases by a factor of 1/e—to 0.367 of the amplitude of the indoor-air temperature $\hat{\theta}_{i,(x=0)}$. The periodic heat-penetration depth is defined using the expression:

$$\hat{\theta}_{i,(x=0)} \cdot \frac{1}{e} = \theta(d_p)_{max} - \bar{\theta}_i = \hat{\theta}_i \cdot e^{-dp\sqrt{\frac{\pi}{tp \cdot a}}} \rightarrow d_p = \overbrace{\sqrt{\frac{tp \cdot a}{\pi}}}^{tp=84600\,s} = 165.8 \cdot \sqrt{a}[m]$$

The double ($2 \cdot d_p$) and triple ($3 \cdot d_p$) periodic heat-penetration depths are the thicknesses of the building structure, whereby the amplitude of the temperature of the building structure decreases by a factor $1/(2 \cdot e)$—to 0.184 of the amplitude $\hat{\theta}_{i,(x=0)}$, or by a factor of $1/(3 \cdot e)$—to 0.122 of the amplitude $\hat{\theta}_{i,(x=0)}$ (Figs. 1.97, 1.98). When designing the heat storage capacity of the building structure for passive heating and cooling, the double periodic penetration depth $2 \cdot d_p$ is usually chosen,

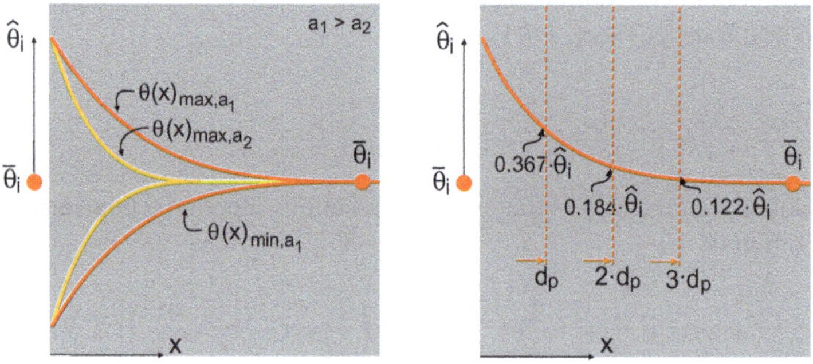

Fig. 1.97 Thermal response zone of a building structure with infinite thickness to a periodic change of the surface temperature depends on the thermal diffusivity of matter *a*. The envelope curves $\theta(x)_{max,a}$ and $\theta(x)_{min,a}$—showing the cases where a_1 is greater than a_2—determine the volume of the building structure inside which the temperature over the period t_p periodically changes (left). The periodic heat-penetration depth d_p (m) is the depth of the structure at which the amplitude of the temperature decreases by a factor 1/e of the amplitude of the indoor-air temperature $\hat{\theta}_{i,(x=0)}$, while at a double periodic heat-penetration depth the amplitude of the temperature decreases by a factor $1/(2 \times e)$ and at a triple heat-periodic penetration depth by the factor $1/(3 \times e)$ of the amplitude of the indoor-air temperature $\hat{\theta}_{i,(x=0)}$; (right)

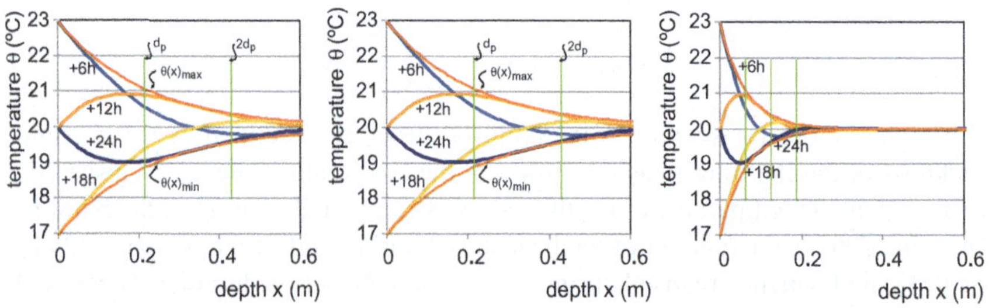

Fig. 1.98 Thermal response of a thermal insulation layer (thermal diffusivity $a = 167.10^{-8}$ m^2/s, left), a concrete ($a = 68.10^{-8}$ m^2/s, middle) and a wooden ($a = 14.10^{-8}$ m^2/s, right) building structure of infinite thickness to a periodic change in the surface temperature; $\theta_{se} = 20$ °C, $\theta_a = 3$ K, $t_0 = 0$ h, $t_0 = 24$ h

because a greater thickness of the building structure will have a relatively small impact on the amount of heat or cold stored in the half-period $t_p/2$. The amount of heat stored in the building structure in half of the day ($t_p/2$) and then released into the room in the second half of the day is determined by integrating the difference in the temperature between $\theta(x)_{max}$ and the average structure temperature equal to $\overline{\theta}_i$ with the expression:

$$q = 2 \cdot \sqrt{\frac{t_p}{2 \cdot \pi}} \cdot \lambda \cdot c \cdot \rho \cdot \hat{\theta}_i = 2 \cdot \overbrace{\sqrt{\frac{t_p}{2\pi}}}^{t_p = 86400\,s} \cdot b \cdot \theta_a = 234.53 \cdot b \cdot \hat{\theta}_i \left[\frac{kJ}{m^2}\right]$$

where b is the thermal effusivity ($Ws^{0.5}/m^2K$) of the homogeneous building structure. The values of the thermal effusivities for some building materials are given in Table 1.20.

Explanation While the thermal response zone of the building structure is determined by the thermal diffusivity a, the amount of accumulated heat is determined by the thermal effusivity b of the structure.

Case Study Determine the required thickness and the size of a concrete and wooden building structure for diurnal heat storage. The material properties of concrete and wood are given in Table 1.20. The daily heat gain for solar radiation is 2 kWh and this causes a periodic change in the indoor-air temperature for the amplitude $\hat{\theta}_{i,(x=0)}$ 3 K. It can be assumed that θ_i is equal to θ_{se}.

$$d_{p,concrete} = 165.8 \cdot \sqrt{a_{concrete}} = 165.8 \cdot \sqrt{68 \cdot 10^{-8}} = 0.137\,m$$

$$d_{p,concrete} = 165.8 \cdot \sqrt{a_{wood}} = 165.8 \cdot \sqrt{14 \cdot 10^{-8}} = 0.064\,m$$

$$q_{concrete} = 234.53 \cdot b_{concrete} \cdot \theta_a = 234.53 \cdot 2.18 \cdot 3$$
$$= 1534 \frac{kJ}{m^2} = \frac{1534}{3600} = 0.429 \frac{kWh}{m^2}$$

$$q_{wood} = 234.53 \cdot b_{wood} \cdot \theta_a$$
$$= 234.53 \cdot 0.38 \cdot 3 = 267 \frac{kJ}{m^2} = \frac{267}{3600} = 0.074 \frac{kWh}{m^2}$$

Solution: The required building structure thickness for the storage of the heat is assumed to be equal to 2 × d_p or ~ 25 cm for concrete, and ~ 12 cm for the wooden building structure. The surface areas of the building structures that accumulate 2 kWh of daily solar radiation are ~5 m^2 in the case of concrete and ~27 m^2 in the case of the wooden structure.

1.7.3.3 Modelling the Thermal Response of Multi-layered Building Structures Under Arbitrary Boundary Conditions

An analytical solution of the thermal response of a building structure is only possible in the case of simple boundary conditions and homogeneous structures like one discussed in Sects. 1.7.3.1 and 1.7.3.2. In practice, the constructions consist of several layers with different material properties and have a finite thickness. Furthermore, the boundary conditions on both sides of the building structure, for example, the equivalent temperature $\theta_{e,eq}$ on the external side or the indoor-air temperature in passive heated or cooled spaces, changes with time. For such cases, the thermal response of the building structure must be determined using numerical methods, for example, the finite-difference method. The method of finite differences is based on the transformation of differential into difference equations. The partial differential equation of the transient heat transfer (one in the case of a 1D problem) is transformed into a system of linear algebraic equations, the solution of which is the temperature at the surface and at a predefined final number of points inside the building structure. A numerical solution of the differential equation of the transient one-dimensional heat conduction is as follows:

$$\frac{\partial \theta}{\partial t} = a\frac{\partial^2 \theta}{\partial x^2}$$

and is found by the approximate derivative of the differential of the temperature $\partial\theta$ over the differential of thickness ∂x and derivative of the differential of temperature $\partial\theta$ over the differential of time ∂t in form (Fig. 1.99):

$$\frac{\partial \theta}{\partial x} = \frac{\theta^t_{x+1} - \theta^t_x}{\Delta x} \quad \frac{\partial \theta}{\partial t} = \frac{\theta^{t+1}_x - \theta^t_x}{\Delta t}$$

and by the approximate second derivative of the differential of the temperature $\partial\theta^2$ over the differential of the thickness ∂x^2 (Fig. 1.99):

$$\frac{\partial \theta^2}{\partial^2 x} = \frac{\left.\frac{\partial \theta}{\partial x}\right|_{x+\frac{1}{2}} - \left.\frac{\partial \theta}{\partial x}\right|_{x-\frac{1}{2}}}{\Delta x}$$

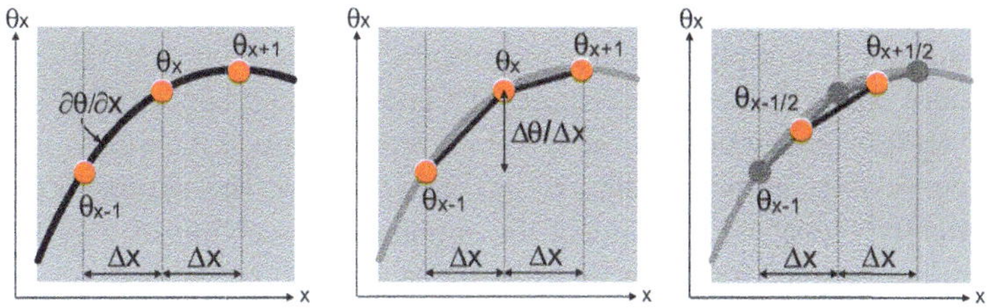

Fig. 1.99 Procedure of the discretization of the differential equation of 1D transient heat conduction. The space descretization (left), the aproximation of the first (midle) and the second derivate of temperature θ (right) in the vicinity of depth x

$$= \frac{\left(\theta^t_{x+1} - \theta^t_x\right) - \left(\theta^t_x - \theta^t_{x-1}\right)}{\Delta x \cdot \Delta x} = \frac{\theta^t_{x+1} - 2 \cdot \theta^t_x + \theta^t_{x-1}}{\Delta x^2}$$

The differential equation of 1D heat conduction is then replaced by the equation involved finite-differences as an algebraic equation:

$$\begin{aligned}
\frac{\partial \theta}{\partial t} &= a\frac{\partial^2 \theta}{\partial x^2} \rightarrow \frac{\theta^{t+1}_x - \theta^t_x}{\Delta t} \\
&= a\frac{\theta^t_{x+1} - 2 \cdot \theta^t_x + \theta^t_{x-1}}{\Delta x^2} \rightarrow \theta^{t+1}_x - \theta^t_x \\
&= \overbrace{\frac{a \cdot \Delta t}{\Delta x^2}}^{Fo}\left(\theta^t_{x+1} - 2 \cdot \theta^t_x + \theta^t_{x-1}\right)
\end{aligned}$$

The term combining the time step Δt, the spatial step Δx and the thermal diffusivity of matter a is known as the dimensionless Fourier number. The numerical solution of above algebraic equation is numerically stable when $Fo < 0.5$. This numerical solving method is also known as the explicit scheme (Fig. 1.100). The method always calculates only one new temperature ($\theta_x{}^{t+1}$) at a time.

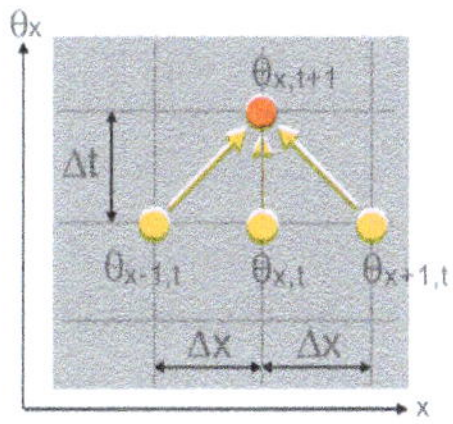

Fig. 1.100 An explicit numerical method is used to determine the temperature at depth x inside the building structure in the next time step $\theta_x{}^{t+1}$ using the temperatures $\theta_{x-1}{}^t$, $\theta_x{}^t$ and $\theta_{x+1}{}^t$ at x and in the vicinity of x at the present time (t)

Explanation The implicit method can also be used as the numerical solution procedure, where the new temperature θ_x^{t+1} is determined from the temperature at the depth x at the current time t, and the temperature at the time $t + \Delta t$ in the vicinity of the point at depth x ($\theta'_{x-1}{}^{t+t}$ and $\theta_{x+1}{}^{t+1}$). This method is always numerically stable, but it requires the simultaneous solving of a system of algebraic equations, for example, using the inverse matrix method.

The composition and the shape of building structures is, as a rule, complex, and the heat conduction through building structures is three-dimensional. Therefore, numerical computer tools are used to analyse the transient heat conduction in building structures. An example of a computer tool is described in Sect. 1.5.6.3.

1.7.3.4 Evaluation of the Dynamic Thermal Properties of Building Structures

The thermal transmittances of building structures U presented in Sect. 1.5 is a property of the structure presented as constant and therefore does not take into account the transient external and/or internal environment conditions. In nature, the thermal response of building structures changes over time and because of that, indicators of the dynamic thermal properties of building structures were introduced. They are based on an analytical solution of the differential equation of the transient heat conduction in the building structure, which is only possible for limited cases of the boundary conditions, for example, in the case of periodic temperature of the external $\tilde{\theta}_{se}(t)$ or internal $\tilde{\theta}_{si}(t)$ surfaces or the periodic heat fluxes $\tilde{\dot{q}}_{se}(t)$ at the external or at the internal $\tilde{\dot{q}}_{si}(t)$ surface of the building structure. The unsteady temperature $\theta(x,t)$ in the building structure and the density of the heat flux $\dot{q}(x,t)$ at thickness x and at time t can be solved by involving hyperbolic functions and complex numbers [29]. The solution can be found assuming sinusoidal boundary conditions in the form of a heat-transfer matrix:

$$\begin{vmatrix} \tilde{\theta}_{si}(t) \\ \tilde{\dot{q}}_{si}(t) \end{vmatrix} = \begin{vmatrix} Z_{11} & Z_{12} \\ Z_{21} & Z_{22} \end{vmatrix} \cdot \begin{vmatrix} \tilde{\theta}_{se}(t) \\ \tilde{\dot{q}}_{se}(t) \end{vmatrix}$$

where the elements of the heat-transfer matrix Z are complex numbers, described by the modulus $|Z_{ij}|$ that represents the magnitude, and the argument $\varphi_{ik} = \arg(Z_{ik})$ that represents the time. The elements of the matrix have the following physical meanings:

- $|Z_{11}|$ (1): is the ratio of the temperature amplitude $\hat{\theta}_{se}$ (at $x = 0$) and the temperature amplitude $\hat{\theta}_{si}$ (at $x = d$);

- φ_{11} (s): is the phase difference of the surface temperatures, the phase difference is the time interval between the moment of maximum θ_{se} (at $x = 0$) and maximum θ_{si} (at $x = d$) in the observed period;
- $|Z_{12}|$ (Km²/W): is the ratio of the amplitude of the surface temperature amplitude $\hat{\theta}_{se}$ (at $x = 0$) and the amplitude of the heat flux $\hat{\dot{q}}_{si}$ at the inner surface of the structure (at $x = d$);
- φ_{12} (s): is the phase difference (time interval) between the moment of maximum θ_{se} (at $x = 0$) and the moment when the maximum heat flux at the inner surface of the structure $\dot{q}_{si}$ occurs (at $x = d$);
- $|Z_{21}|$ (W/m²K): is the ratio between the amplitude of the heat flux $\hat{\dot{q}}_{se}$ at the external surface of the building structure (at $x = 0$) and the amplitude of the inner surface temperature of the structure $\hat{\theta}_{si}$ (at $x = d$);
- φ_{21} (s): is the phase difference, the time interval between the moment of maximum heat flux $\dot{q}_{se}$ (at $x = 0$) and the maximum inner surface temperature θ_{si} (at $x = d$);
- $|Z_{22}|$ (1): is the ratio between the amplitude of the heat flux at the external surface of the structure $\hat{\dot{q}}_{se}$ (at $x = 0$) and the amplitude of the heat flux $\hat{\dot{q}}_{si}$ at the inner surface of the structure (at $x = d$);
- φ_{22} (s): is the phase difference, the time interval between the moment of maximum heat flux at the external surface $\dot{q}_{se}$ (at $x = 0$) and the maximum heat flux at the inner surface of structure $\dot{q}_{si}$ (at $x = d$).

In the case of a multi-layer building structure, the heat-transfer matrix of the structure $|Z|$ consists of the matrixes of the individual layers $|Z_N|$, where N is the number of homogeneous layers taking into account that $|Z_1|$ is the matrix on the interior layer of the structure:

$$\begin{vmatrix} Z_{11} & Z_{12} \\ Z_{21} & Z_{22} \end{vmatrix} = \begin{vmatrix} Z_{N,11} & Z_{N,12} \\ Z_{N,21} & Z_{N,22} \end{vmatrix} \cdot \begin{vmatrix} Z_{N-1,11} & Z_{N-1,12} \\ Z_{N-1,21} & Z_{N-1,22} \end{vmatrix} \cdot \ldots \cdot \begin{vmatrix} Z_{1,11} & Z_{1,12} \\ Z_{1,21} & Z_{1,22} \end{vmatrix}$$

The entire heat-transfer matrix of a multi-layer building structure, taking into account the thermal resistances to the heat transfer on the inner R_{si} and outer surfaces R_{se}, is written in the form:

$$\begin{vmatrix} Z_{ii} & Z_{ie} \\ Z_{ei} & Z_{ee} \end{vmatrix} = \begin{vmatrix} 1 & -R_{se} \\ 0 & 1 \end{vmatrix} \cdot \begin{vmatrix} Z_{11} & Z_{12} \\ Z_{21} & Z_{22} \end{vmatrix} \cdot \begin{vmatrix} 1 & -R_{si} \\ 0 & 1 \end{vmatrix}$$

The standard [28] specifies that the dynamic thermal properties of building structures are indicated by two factors:

- **Decrement factor** f (−). This is the ratio between the dynamic and the stationary thermal transmittance U and exemplifies ability of a structure to dampen the amplitude of the heat flux $\hat{\dot{q}}_{si}$ at the inner surface of a building structure caused by sine-changing external surface temperature with amplitude $\hat{\theta}_{se}$. f is between 0^+ for high damping and 1^- for low damping of the heat flux. Structures with a high damping of the heat flux are usually thermally insulated massive structures, while lightweight building structures usually have a low damping ability. The coefficient of damping is defined by the expression (Table 1.25):

$$f = \frac{\left|\hat{q}_{si}\right|}{\left|\hat{\theta}_{se}\right| U} = \frac{1}{|Z_{12}| U}[1]$$

- **Time shift Δt *(s)*.** This is the period of time between the moment when the outdoor-air temperature is at amplitude $\hat{\theta}_{se}$ and the moment when the heat flux at the inner surface of the structure is equal to the amplitude v $\hat{q}_{si}$. Lightweight building structures usually have a shorter time shift, while thermally insulated massive structures usually have a longer time shift. The time shift Δt is defined by the expression:

$$\Delta t = \frac{\overbrace{t_p}^{86400\ s}}{2 \cdot \pi} \arg(Z_{12})[s]$$

Figure 1.102 shows an example of the thermal response of a lightweight (1) and a massive (2) building structure with the same static thermal transmittance U 0.467 W/m^{2}K. The thermal response is presented by the heat flux at the inner surface of the structures $\dot{q}_{si,1}$ and $\dot{q}_{si,2}$ caused by the amplitude of the outdoor-air temperature $\hat{\theta}_{se}$ 1 K.

Table 1.25 The composition of a light and a massive structure analysed in the case study

light structure (1)	d (cm)	λ (W/mK)
Fe sheet metal	0.06	43
stone wool	8	0.041
Fe sheet metal	0.06	43
massive structure (2)	d (cm)	λ (W/mK)
stone wool	8	0.041
brick	8	0.580

Table 1.26 Dynamic thermal response properties of analysed building structures. All the structures, shown in Fig. 1.103 have the same static thermal transmittance U (0.310 W/m^{2}K)

	damping factor f (-)	time lag Δt (h:min)
structure 1	0.281	7:07
structure 2	0.470	6:07
structure 3	0.480	7:14
structure 4	0.065	7:55

Case Study 1 Determine the damping factor f and the time shift Δt of the building structures with the composition shown in Table 1.25. The average heat fluxes $\dot{q}_{si,1}$ and $\dot{q}_{si,2}$ at the inner surfaces of the building structures are equal to $U \times \hat{\theta}_{se}$ (0.467 W/m^2) and are the same in both structures because they have equal static thermal transmittance U.

The decrement factors f_1 *and* f_2 are:

$$f_1 = \frac{\left|\hat{\dot{q}}_{si,1}\right|}{\left|\hat{\theta}_e\right| \cdot U} = \frac{\left|\dot{q}_{si,min,1} - \bar{\dot{q}}_{si,1}\right|}{\left|\hat{\theta}_e\right| \cdot U} \frac{|-0.930 - 0.467|}{|1| \cdot 0.467} = 0.99[-]$$

$$f_2 = \frac{\left|\hat{\dot{q}}_{si,2}\right|}{\left|\hat{\theta}_e\right| \cdot U} = \frac{\left|\dot{q}_{si,min,2} - \bar{\dot{q}}_{si,2}\right|}{\left|\hat{\theta}_e\right| \cdot U} \frac{|-0.537 - 0.467|}{|1| \cdot 0.467} = 0.15[-]$$

and the time shift Δt_1 and Δt_2 is (Fig. 1.102):

$$\Delta t_1 = 0:58[h] \quad \Delta t_2 = 10:04[h]$$

The case study shows that the dynamic thermal properties of building structures differ considerably, despite the same static thermal transmittance U.

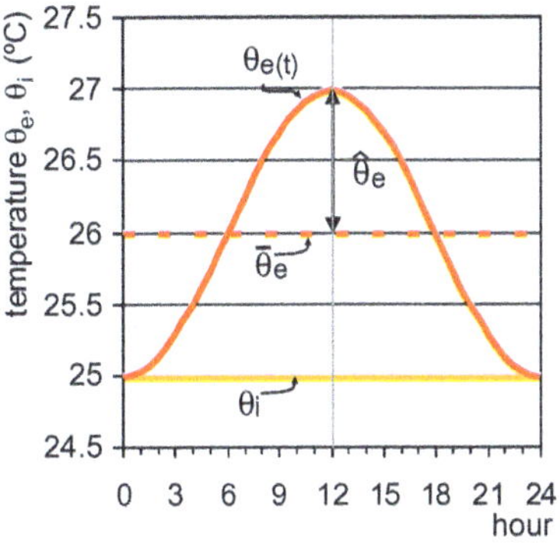

Fig. 1.101 Boundary conditions used to determine the indicators of the dynamic thermal properties for a building structure, as defined in [28]. The outdoor temperature θ_e is a sine function with the amplitude $\hat{\theta}_e$ 1 K. The mean $\overline{\theta}_e$ is 26 °C. The indoor temperature θ_i is constant at 25 °C. The resistance to heat transfer at the external surface R_{se} 0.04 m^{2}K/W and at the inner surface R_{si} 0.13 m^{2}K/W

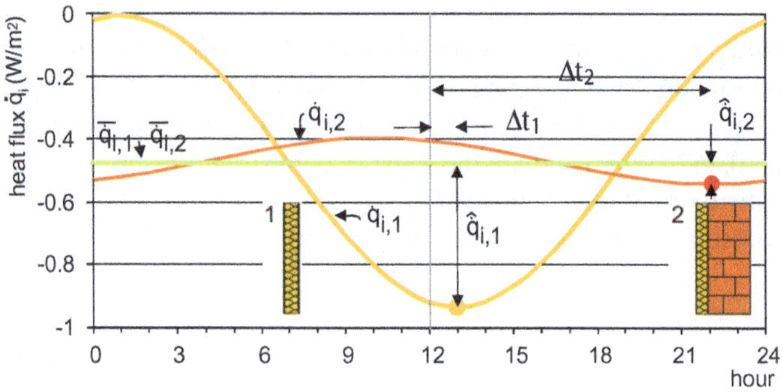

Fig. 1.102 Thermal response of a lightweight (1) and a massive (2) building structure having the same static thermal transmittance U (0.467 W/m^{2}K) taking into account the boundary conditions presented in Fig. 1.101 heat fluxes at inner surface of the structures ($\dot{q}_{si,1}$ and $\dot{q}_{si,2}$), amplitudes of heat flux ($\hat{\dot{q}}_{si,1}$ and $\hat{\dot{q}}s_{si,2}$) and time shifts (Δt_1 and Δt_2) are shown [29]

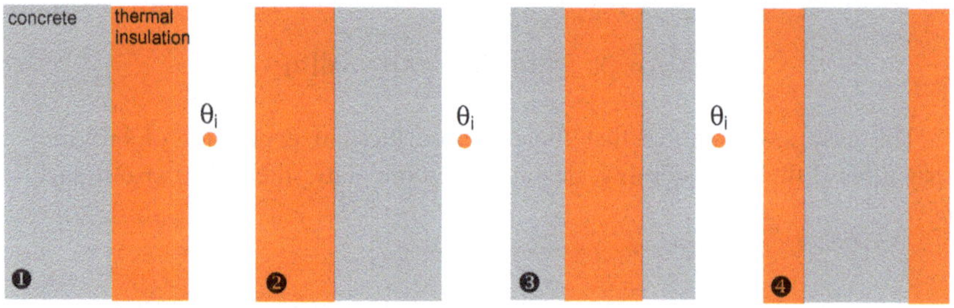

Fig. 1.103 Arrangement of thermal insulation layer(s) and the concrete layer(s) in the external wall analysed in the case study

Case Study 2 Use the "BTR Building Thermal Response" software tool (accessible at www.knaufinsulation.com) [30] to determine how the positions of the thermal insulation and the concrete layers affect the dynamic thermal response of the building structure. The total thickness of the thermal insulation is 10 cm (as one of two separate layers) and the total thickness of the concrete wall is 15 cm. The arrangement of the layers is shown in Fig. 1.103. The thermal insulation has a density of ρ_{ti} 50 kg/m^3, a thermal conductivity λ_{ti} 0.035 W/mK and a specific heat capacity $c_{p,ti}$ 840 J/kg/K. The concrete layer has a density of ρ 1800 kg/m^3, a thermal conductivity λ 0.93 W/mK and a heat capacity $c_p = 960$ J/kg/K. The static thermal transmittance of the building structures is 0.310 W/m^{2}K, no matter how the structures are built (Figs. 1.102 and 1.103).

Solution: The arrangement of the layers has a significant impact on the dynamic thermal properties of the building structure. The maximum damping of the heat flux (maximum damping factor f) and the maximum time lag of the heat-flux at the inner surface of the building structure Δt is achieved if the massive concrete wall is thermally insulated on both sides ($f_4 = 0.065$, $\Delta t_4 = 7{:}55$ h). The damping factor f is the lowest if the thermal insulation is installed on the outside part of the structure, rather than on the internal side ($f_2 = 0.281, f_1 = 0.470$), and the time lag of the heat flux is greater if the thermal insulation is installed on the inside of the massive layer ($\Delta t_1 = 7{:}07$ h, $\Delta t_2 = 6{:}07$ h). The division of the massive part of the wall (3) and the installation of the thermal insulation in the core (4) of the structure are comparable with the installation of thermal insulation on the external side of the building structure (Table 1.26).

1.8 Indoor Thermal Comfort

In terms of thermodynamics the human body is a kind of heat engine that converts the chemical energy of ingested nutrients through metabolism into heat and mechanical work with chemical, biological and physical processes. With its metabolism, the body also maintains the necessary temperature for survival, since it ensures that the vital internal organs and the brain (the body's core) have an almost constant temperature of approximately 37 °C (between 36.8 °C at rest and 37.8 °C with increased physical activity), while a seated person has a skin temperature of between 33 and 34 °C. The condition when the body's core temperature decreases below 28 °C, known as hypothermia, causes heart failure, while at elevated temperatures, or hyperthermia, damage occurs to the cells in the brain. At low ambient temperatures, the body seeks to maintain an unchanged temperature of the body core as long as possible. Therefore, the blood flow to the limbs, especially to the fingers, is reduced and they become strongly hypothermic. At high temperatures, the body must be cooled rapidly to avoid overheating of the internal organs. The body transfers the heat from the body core to the skin with blood, and the heat flux is controlled by the flow of blood. If necessary, blood flow to the skin can be increase by up to 15 times.

As any other heat engine, the human body can only convert a relatively small amount of consumed energy into mechanical work and the surplus of energy must be transferred into the environment in the form of heat. The body emits heat from the skin's surface with sensible and latent-heat fluxes with the following heat-transfer mechanisms (Fig. 1.104):

- by convective heat flux from the skin's surface to the surrounding air;
- by radiative heat flux from the skin's surface to the neighbouring surfaces;
- by the conduction of heat into the building structures that the body is in contact with (for example, from bare feet or through the surface of a seat);

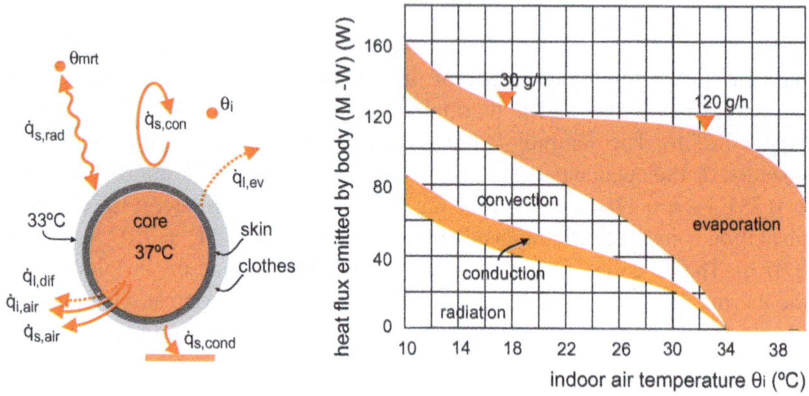

Fig. 1.104 Our body regulates its temperature with the transfer of sensible and latent heat from the body's core and skin towards the environment with the convection, radiation, conduction and evaporation of water vapour (left). The amount of the transferred heat depends on the metabolism (met) of the person, while the share between different heat fluxes depends on the air temperature in the surroundings of the body (right)

- by latent heat flux upon the evaporation of water from the skin's surface when sweating and by the diffusion of water-vapour molecules through the skin;
- by the heat removed from body with a sensible and latent heat flux with exhaled heated and moistened air by breathing.

The energy balance for a naked body can be written using the law of the conservation of energy:

$$M - W = \underbrace{\dot{q}_{s,con} + \dot{q}_{s,rad} + \dot{q}_{s,cond} + \dot{q}_{s,air}}_{\text{sensible heat fluxes}} + \underbrace{\dot{q}_{l,ev} + \dot{q}_{l,dif} + \dot{q}_{l,air}}_{\text{latent heat fluxes}} \left[\frac{W}{m^2}\right]$$

where M is the amount of heat generated by the metabolic and W is the amount of mechanical work produced by body in a unit of time, $\dot{q}_{s,con}$, $\dot{q}_{s,rad}$, and $\dot{q}_{s,cond}$ are the sensible heat fluxes of heat convection, radiation and conduction, $\dot{q}_{s,air}$ is the heat flux transferred from the body by the air heated inside the breathing organs, and $\dot{q}_{l,ev}$, $\dot{q}_{l,dif}$ and $\dot{q}_{l,air}$ are the latent heat fluxes emitted by the skin and the moistened air that is exhaled. All these quantities are expressed per 1 m^2 of naked-body surface area. The amount of work carried out during most of our time spent indoors is close to 0, therefore the sum of the heat fluxes that the body must remove is approximately equal to the metabolism rate M. The total heat flux transferred from the body is determined by taking the actual surface area of a naked body, which is commonly determined using the expression of DuBois [31], accounting for the body mass and the height. The typical body surface area for an adult person is 1.8 m^2. The relationship between the individual heat fluxes emitted by the body depends on the ambient temperature (Fig. 1.104, right). At a low air temperature the heat transfer by sensible heat fluxes prevails, while at higher temperatures the latent heat fluxes are dominant.

Table 1.27 Metabolic heat emitted by a person doing different tasks expressed in the activity rate met (left). The metabolic heat rate 1 met corresponds to the heat flux 58 W per m^2 of the skin area. The thermal resistance of clothes expressed with the unit clo. The total number of clo is determined as the sum of all clothes worn, e.g., underwear + socks + shirt + long trousers + jacket + shoes = 0.02 + 0.02 + 0.25 + 0.28 + 0.02 = 0.84 clo [30, 31] (right)

activity	met
sleeping	0.80
sitting person	1.00
sitting person doing office work	1.20
teacher	1.50
person doing light house work (having a wash, dressing oneself)	1.70
walking, 2 km/h	1.90
person doing harder house work (doing laundry, ironing)	2.90
walking, 5 km/h	3.40
ice skating, 10 km/h	6.20
running	8.00

clothes	clo
underwear	0.02
singlet	0.01
socks	0.02
t-shirt	0.09
shirt	0.25
trousers, long	0.06
trousers, short	0.25
skirt	0.10
jacket	0.28
jacket, winter	0.35
anorak	1.13
coat	0.70
shoes	0.02
gloves	0.05

The metabolism M depends on the level of the person's activity. The average adult at rest creates a metabolic heat output of 100 W, a person doing office work between 125 and 170 W, a person who performs heavy physical work up to 500 W. That is why a relative scale was introduced to define the metabolic heat output and the unit of activity is met (from "metabolism"). One met corresponds to a heat flux of 58 W per m^2 of body surface area. For a seated person the activity is 1.0–1.2 met, the activity of a walking person is 2–4 met, and the activity while dancing is up to 5 met. Such a (standard) person emits a heat flux of 5 met $\times$ 58 W/m^2 $\times$ 1.8 m^2 which is equal to 522 W. People usually wear clothes to thermally insulate their bodies. A relative scale and quantity clo (from clothes) was introduced to measure the insulation performance of clothing. The naked body has 0 clo, 1 clo corresponds to the insulation of a business suit, and winter clothing has 2 clo (Table 1.27) [32].

Explanation The data on design activity (met) and clothing (clo) of inhabitants in different building and room categories are listed in the international standards ISO 7730 [33] and CR 1752 [34].

1.8.1 *Indoor Thermal Comfort Requirements*

Besides visual comfort, indoor air quality, room acoustics and noise protection, indoor thermal comfort is one of the segments for assessing the indoor environment quality (IEQ). All the segments of IEQ are measured by quality classes. The

EN 16798-1 [35] standard defines four quality classes (I, II, III, IV) that apply to all areas of assessment. Analogously, EN ISO 7730 [32] defines four quality classes and designates them as A, B, C and D (Table 1.28).

The state of the indoor environment in which, regardless of the current activity or clothing, the heat transfer between the body and the environment is not perceived by the person as uncomfortable, is known as the state of thermal comfort. An individual's perception of thermal indoor environment is subjective and depends on the nature of the physical properties and other factors, such as age and gender, and the psycho-physical properties of the person. In engineering practice, the thermal indoor environment is defined by the physical quantities of the state of the indoor environment that can be measured. Because of the subjective perception of indoor thermal comfort, the acceptable levels of IEQ indicator are given by the value range that suits statistically the majority of people. The following physical parameters of the indoor environment are the most important for thermal comfort:

- **The indoor-air temperature θ_i(°C)**. A comfortable indoor-air temperature in accommodation spaces is between 20 and 22 °C during the winter, and between 25 and 26 °C during the summer due to the body's natural adaptation (Fig. 1.105). During the heating season, the air temperature in rooms where no clothes are worn should be between 22 and 24 °C, the air in rooms where light physical activity is conducted should be between 20 and 21 °C, and the air temperature in rooms where people are doing hard physical activity should be between 16 and 18 °C in order to facilitate the cooling of the skin's surface. During a period of extremely high external temperatures, the indoor temperature can also be somewhat higher, but should not exceed 28 °C. Note that the temperatures above represent the indoor operative temperatures (see Sect. 1.8.2.1). However, in a well thermally insulated building the difference between the operative and air temperature is less than 0.5 °C.
- **The mean radiant temperature θ_{mrt} (°C)**. This is the average temperature of all the surfaces in a closed space with which our body exchanges heat by radiation.

Table 1.28 IEQ quality classes as defined in EN 16798-1 [35]; in the similar way in EN ISO 7730 [33] the IEQ quality classes are expressed as A, B, C and D

Category	Explanation
I (A)	High level of expectation, recommended for spaces occupied by very sensitive persons (very young children, sick persons, elderly persons, etc.)
II (B)	Normal level of expectation for new building and renovations
III (C)	An acceptable, moderate level of expectation, for existing buildings; common defined as reference quality category of IEQ
IV (D)	Acceptable in case of limited duration of buildings exploitation over the year; not acceptable for permanent living

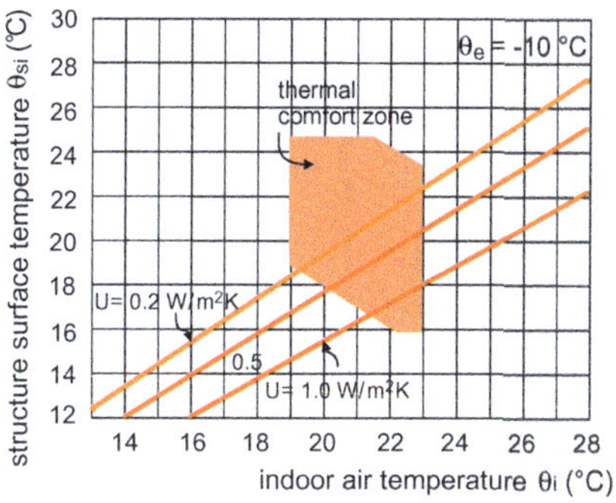

Fig. 1.105 The thermal comfort zone limited by the range of indoor air temperatures θ_i and the mean radiant temperature $\theta_{mrt;;}$the lines drawn for different thermal transmittance U determine the building structure inner surface temperature θ_{si} at different indoor air temperatures θ_i; it can be seen that building structures with U 1 W/m^{2}K has, at indoor-air temperature θ_I; 21 °C, the surface temperature 16 °C; at that conditions of thermal comfort is not achieved

θ_{mrt} depends on the surface temperature, the emissivity of the thermal radiation ε_{IR} and the areas of the surfaces that surround us in a room. θ_{mrt} also depends on the building structure's thermal transmittance, the exposure to solar irradiation and the temperature and the size of the heat exchangers for space heating or cooling. The mean radiant temperature of the room enclosing surfaces is determined using the radiation-exchange factors F_{i-j}, presented in Sect. 1.2.2;

- **Humidity of the indoor air**. This is referred to as the relative humidity φ_i (%) (see Sect. 2.3). A relative humidity of the indoor air between 35% in 70% in living spaces is assumed as appropriate. Nevertheless, a relative air humidity in accommodation spaces between 40 and 60% is optimal (Fig. 1.106, left). At lower humidity, airborne particles and aerosols are caught in the air that force

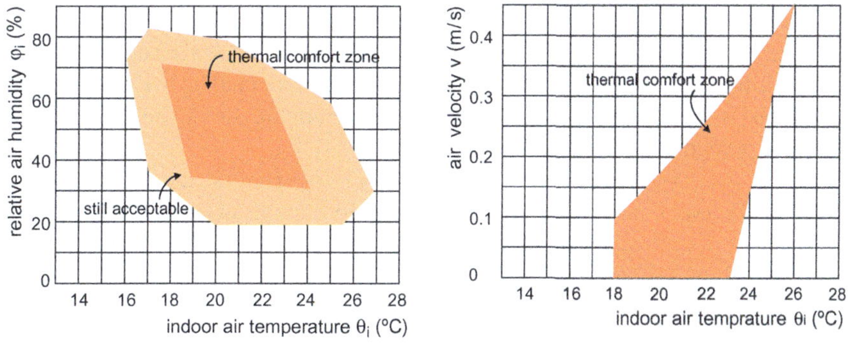

Fig. 1.106 Range of indoor air temperatures θ_i and relative humidity φ_i at which the thermal comfort will be achieved (left). Thermal comfort zone with respect to the indoor-air temperature θ_i and the air velocity v (right)

Fig. 1.107 A ceiling fan does not cool the air, but perceived temperature in the room is lower due to the higher air velocity. The perceived temperature for an air velocity 1 m/s next to the skin is 3 °C lower compared to still-air conditions

us to cough, and when there is higher humidity, there is the risk of developing harmful micro-organisms.

- **Air velocity** v**(m/s)**. The movement of air in buildings is a consequence of the thermal buoyancy, the air pressure difference between the outdoor environment and the interior of the buildings due to wind or mechanical devices (fans). At lower indoor-air temperatures (θ_i ~ 20 °C), air movement at a velocity above 0.15 m/s causes the convection from the body to be perceived as uncomfortable during the heating period, while in the summer a higher velocity of air helps the body to cool. Therefore, in a room with a temperature of 26 °C, the indoor air velocity that does not exceed 0.65 m/s will be perceived as comfortable (Figs. 1.106, right and 1.107).

All these four, so-called "global", parameters influence the thermal comfort simultaneously, so there is a significant link between them. At a higher temperature of the enclosing surfaces, i.e., in the case of lower thermal transmittances of building structures or in the case of floor heating, the indoor-air temperature can be reduced by 1–2 °C, while maintaining the same thermal comfort level, although heat losses through the exterior building structures and therefore the energy demand for space heating decrease. At higher indoor-air temperatures the perceived IEQ will be adequate if the indoor-air humidity is lower, while the velocity of the air could be higher. However, the greatest impact on the thermal comfort comes from the activity and the clothing of people in relation to the purpose of the space.

Even if the global parameters of thermal comfort provide the highest degree of indoor thermal comfort, we may feel uncomfortable if any part of the body is exposed to a "local" condition of thermal discomfort. The most important causes of local thermal discomfort are:

- **Radiant heat flux.** A large radiant heat flux emitted or received by different parts of the body causes radiant asymmetry. The reasons for radiant asymmetry can be, for example, the surface of a heated or cooled ceiling, a cold window surface or a sunlit inner window shading. The thermal comfort requirements stated in [32] or [36], define the assessment method of the radiant temperature asymmetry according to the difference in the temperature of two opposite surfaces or the asymmetry of emitted radiant heat fluxes. Table 1.29 (left) shows the acceptable asymmetry in the surface radiant temperature θ_{mrt} of opposite surfaces in a room. It was also found that people are more sensible to the excessive radiant flux ($\dot{q}_{rad} > 12$ W/m^2) from the heated ceiling towards the head and less by the radiant flux received from heated vertical surfaces or to the high radiant flux emitted from the body towards a cold ceiling. These facts should be taken into account when a floor/ceiling/walls heating or cooling system is designed. Note that the difference between the radiant θ_{mrt} and the surface temperature θ_{si} of the building structure depends on the surface area and especially on the thermal emissivity ε_{IR} of the structure's surface.

- **Air temperature gradient**. A large air-temperature gradient, i.e., the difference in air temperature with the height of a space, can cause local thermal discomfort, even if the body is in thermal equilibrium with the indoor environment, for example, a seated person feels the difference in the air temperature at the height of the head (1.1 m above the ground) and the ankles (0.1 m above the ground) as local thermal discomfort, if it exceeds 3 K. A large air-temperature gradient is characteristic of rooms with a local furnace, while the temperature gradient does not appear in the case of floor space heating. Conversely, people are less disturbed by the inverse temperature gradient, i.e., when the air at the head level is cooler than the air at the feet. This is also confirmed by the lower sensitivity to radiation asymmetry in the case of a cooled ceiling.
- **Draught and turbulence of the air**. Feeling a draught is a consequence of too intense cooling of body parts. Humans are particularly sensitive to sudden changes in the velocity and direction of air. Local draught discomfort is therefore evaluated based on the time-averaged air velocity v and the turbulence Tu (%) for a period of at least 180 s. Tu is defined by the ratio of the standard deviation of instantaneous air velocities to the average air velocity v over a given period of time and measured

Table 1.29 Maximum radiant asymmetry of two opposite building structures that will not cause thermal discomfort (left). The thermal effusivity b of some building materials used as flooring. The materials having b less than 400 Ws$^{0.5}$/m^{2}K are warm to the touch, while materials with b over 4000 Ws$^{0.5}$/m$_2$K are cold to the touch (right)

opposing surfaces in the room	$\Delta\theta_{sst}$ (K)
heated ceiling-floor	<5K
cooled ceiling-floor	<14K
cooled wall-opposing wall	<10K
heated wall-opposing wall	<23K

material	b (W s$^{0.5}$/m^{2}K)
expanded polystyrene	32
mineral wool	44
needled floor covering	200
aerated concrete	250
wood	380
rubber	522
brick	900
glass	1250
concrete	1800
granite	2800
stainless steel	8000

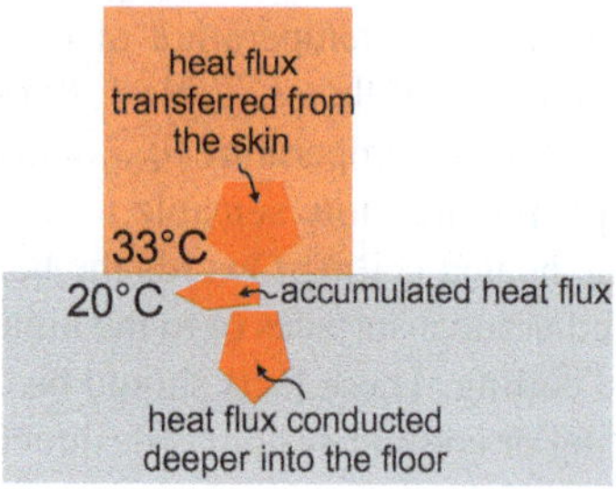

Fig. 1.108 Thermal effusivity b defines whether the material will be perceived as warm or cold in contact with the skin. This phenomenon depends on the ability of the substance to store the heat and its thermal conductivity

in percent. The values of acceptable air velocities in the space are reduced to half if the turbulence increases from 0 to 50%;

- **Temperature of surfaces**. A too high or a too low temperature of building structure surfaces that we touch with unprotected parts of the body is most often the result of walking barefoot on the floor. The perceived temperature of the structure depends on the relationship between the heat flux that passes from the skin and is accumulated in the building structure close to the surface and the heat flux that is being conducted into the depth of the building structure (Fig. 1.108). As already explained in Sect. 1.7.3.1, the amount of the accumulated heat depends on the density of the material ρ and its specific heat capacity c_p, while the heat flux that is conducted into the interior of the building structure depends on the thermal conductivity of the material λ. These material properties are combined in the thermal effusivity b (see Sect. 1.7.3.1, Table 1.29, right). Materials with a b of less than 400 $Ws^{0.5}/m^2K$ are perceived as warm to the touch, and those with b greater than 4000 $Ws^{0.5}/m^2K$ feel cold. In order to avoid local heat discomfort when walking barefoot, the floor must be heated. Thus, the acceptable temperature of a floor with a needled covering is 21 °C, meanwhile the acceptable temperature of a floor with marble tiles is 29 °C.

1.8.2 Combined Thermal Comfort Indicators

1.8.2.1 Effective or Operative Temperature

The effective or operative temperature θ_{op} is a combined indicator of the influence of the air temperature θ_i and the mean radiant temperature θ_{mrt} in a room. It is the temperature perceived when all other physical parameters of the thermal comfort are ideal. It is defined by the expression:

$$\theta_{op} = a \cdot \theta_i + (1 - a) \cdot \theta_{mrt} [°C]$$

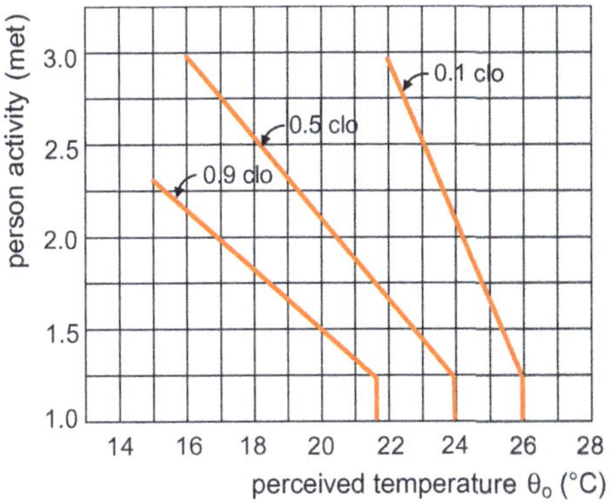

Fig. 1.109 Optimal operative temperature θ_{op}according to the activity and clothing of the person occupying the room

The constant *a* in the expression is a weighting factor, which takes into account the air velocity in the room. At an air velocity up to 0.2 m/s, *a* is equal to 0.5, at an air velocity up to 0.6 m/s, 0.6 and at air velocity up to 1 m/s, 0.7. The optimal operative temperature is also dependent on the activity (met) and the clothing (clo) of occupants, as shown in Fig. 1.109. The operative temperature is used as a reference indoor temperature when determining the energy demand for space heating and cooling.

1.8.2.2 Mean Predicted Vote of the Thermal Comfort—The *PMV* Index

In the 1970s the method of a combined indicator of thermal comfort was developed, based on the subjective perception of the basic physical quantities of the thermal comfort of the respondents. The indicator, called the "Predicted Mean Vote" (PMV), has seven descriptive values for the perceived thermal comfort [very cold (−3), cold (−2), pleasantly cold (−1), neutral (0), pleasantly warm (+1), warm (+2), hot (+3)]. The PMV scale is shown in Fig. 1.110. The scale is also called the Fanger's scale of thermal comfort after the Danish professor Fanger, who developed this method [40]. The method is based on the assumption that there is a linear relationship between

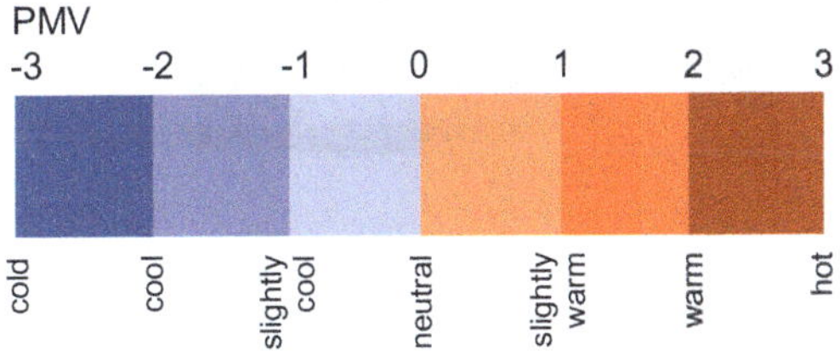

Fig. 1.110 Descriptive values of the PMV index defined in the standard [32]

the average temperature of the skin, sweating and metabolism, based on which the sensation of thermal comfort can be predicted. The predicted degree of thermal comfort is determined analytically by solving the energy balance of the human body for the selected values of the basic physical parameters of thermal comfort (θ_i, θ_{mrt}, φ_i and v), and considering the activity (met) and clothing (clo) of the occupants. The mathematical expression is given in [33]. The objective is the design of the building and the building service systems that ensure such a combination of values θ_i, θ_{sst}, φ_i and v that the *PMV* index in the rooms is as close as possible to 0, thus ensuring a neutral indoor thermal environment and the highest thermal comfort category. In existing buildings, the *PMV* index is measured to verify the actual quality of the thermal comfort (Fig. 1.111).

The *PMV* index can also be related to the expected percentage of people who are dissatisfied with the current indoor thermal environment. This number is known as the *PPD* index (Predicted Percent of Dissatisfied). The link between the *PMV* index and the *PPD* index is shown in Fig. 1.112. The model assumes that people feel equally uncomfortable in a too-warm or too-cold environment, so the dependence of the *PPD* index and *PMV* index is symmetrical. It is also assumed that 5% of people are dissatisfied even in a thermally neutral environment. This percentage increases with the increase or decrease in the *PMV* index. If the *PMV* in a room is ± 1, it is expected that 28% of occupants will be dissatisfied with the thermal comfort, and 77% of occupants if the *PMV* in the room is ± 2. The criteria for the design of indoor thermal comfort are stated in [36]. When designing the indoor thermal comfort, the *PMV* index must be at least in the range between − 0.7 and + 0.7 to provide the minimum requested thermal comfort category III (C) and in such a case the *PPD* index will not be higher than 15%. The ranges of the *PMV* and *PPD* values for other categories of the quality of thermal comfort are summarized in Table 1.30.

The *PPD* index was also developed to determine the acceptance of indoor thermal conditions of "non-reference" people. Figure 1.112 (right) shows how the operative temperature θ_{op} impacts on the percentage of dissatisfied (*PPD*) at different occupant

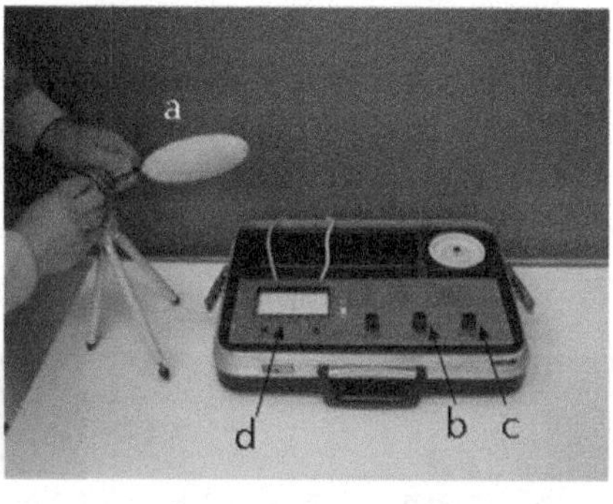

Fig. 1.111 Thermal comfort measurement device with sensor for detection θ_i, θ_{mrt}, φ_I and v (a) and two analogue bottoms for setting met (b) and clo (c). *PMV* is shown on an analogue scale with the range − 3 to + 3 (d), making it possible to direct the assessment of the *PMV*

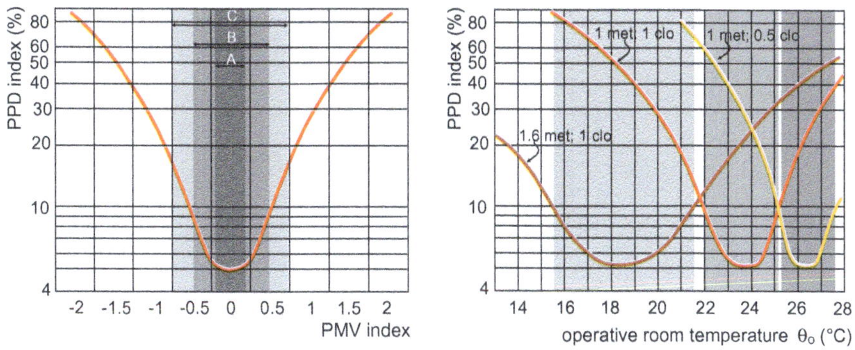

Fig. 1.112 *PPD* index vs. *PMV* index as determined by Prof. Fanger in the 1980s (left). It is assumed that even in a thermally neutral environment (*PMV* 0), 5% of people will be dissatisfied with their indoor thermal comfort. The curve in the figure on the left is given for a seated person (1.2 met) wearing light clothes (1.0 clo). The range of the operative temperature θ_o at which the *PPD* index will be lower than 10% at different activity *met* and clothing *clo* of the person in the room (right). The *PPD* is determined for v 0.15 m/s and a relative air humidity φ_i 50% [33]

Table 1.30 Indoor thermal comfort categories and the corresponding PMV and PPD index ranges [33] and [34]

thermal comfort class		predicted mean vote PMV	percentage of dissatisfied PPD
A	I	-0.2 to + 0.2	< 6 %
B	II	-0.5 to + 0.5	< 10 %
C	III	-0.7 to +0.7	< 15 %
D	IV	<-0.7 or >+0.7	> 15 %

activities and clothing. The relationship between local thermal comfort indicators and the percentage of dissatisfied occupants was also studied. Figure 1.113 shows such examples for radiant asymmetry, floor temperature and air temperature gradient. The figures show the values of the physical quantities that must be provided to establish the A, B or C category of indoor thermal comfort.

1.8.2.3 Effective and Physiological Effective Temperature

Instead of a relative *PMV* scale, the effective air temperature θ_{ET} can be used to estimate the thermal comfort. The θ_{ET} is the substitute room air temperature, determined assuming that all the other physical parameters of thermal comfort are optimal. θ_{mrt} is equal to θ_i at an air velocity of 0.1 m/s and a relative humidity φ_i of 50%.

By taking into account the heat transfer between the core of the body and the skin, and the skin and the environment, and taking into account the actual skin temperature and sweating in different parts of the body, the so-called physiologically equivalent temperature θ_{PET} was established. Since the solar irradiation and the

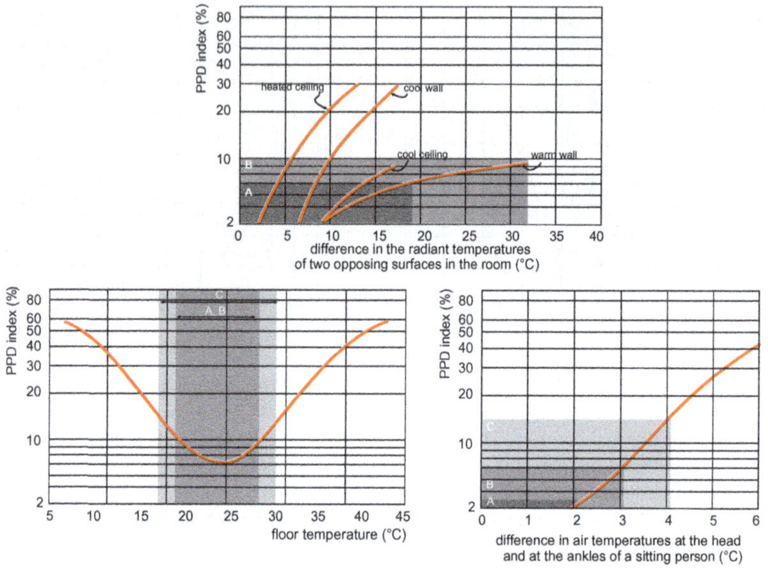

Fig. 1.113 *PPD* index versus selected parameters of local thermal comfort—radiant asymmetry, temperature gradient and floor temperature; the zones of thermal comfort categories are shaded; in all cases the sensation of thermal environment is given for a sitting person (activity 1.2 met) and wearing light clothes (1 clo); θ_i equals θ_{sst}, air velocity $v = 0.15$ m/s, air humidity $\varphi_i = 50\%$ [32]

thermal radiation of the environment on the body is taken into account in the energy-balance equations, θ_{PET} can be used to evaluate the outdoor thermal comfort and the thermal stress in the external environment. Similar to θ_{ET}, θ_{PET} is the substitute air temperature in the outdoor or indoor environment when all the remaining physical parameters of the thermal comfort are ideal (Table 1.31). Since the determination of θ_{PET} requires the use of iterative numerical methods, it is determined by computer tools. An example of such a tool is listed in the literature [37].

Case Study To increase the use of public transport, it is very important how the stops are planned. The bus-stop shelters must protect passengers against sun irradiation and adverse weather events. As evidenced, the physiological effective temperature θ_{PET} perceived by a passenger waiting for a bus at a bus

Table 1.31 Values of the physiological effective temperature θ_{PET} in different climate conditions

room/environment	θ_i (°C)	θ_{sst} (°C)	v (m/s)	φ_i (%)	θ_{PET} (°C)
appartment	20	20	0,1	50	20
hot summer day, in the sun	30	60	0,5	50	45
hot summer day, in the shade	30	30	0,5	50	24
cold winter evening	-10	-15	3,0	50	-18

Fig. 1.114 Modern bus stop shelter that protects passengers against adverse weather events has a great impact on the use of public transportation in an urban environment. From Fig. 1.115 it is clear that the physiological effective temperature θ_{PET} can be lowered by 20–30 °C when protecting people from direct solar irradiation by shading the bus stop

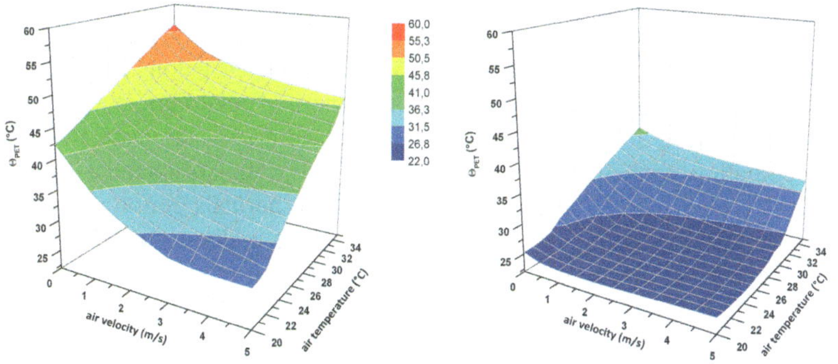

Fig. 1.115 Physiological effective temperatures θ_{PET}, sensed by two passengers waiting for the bus at an unprotected bus stop (left) and at a modern Ljubljana Public Transport bus stop shown in Fig. 1.114 [37]

stop without and with modern shelter shown in Fig. 1.114 at a summer sunny day should be compared. The passenger activity is assumed to be 1.5 met, the clothing 0.8 clo and the wind velocity 1.2 m/s. The results are presented in Fig. 1.115 for an unshaded (left) and a shaded bus stop (right).

1.8.3 Subjective Sensation of Thermal Environment

The above-presented indicators of thermal comfort define the pleasant conditions of indoor thermal environment for an "average person". However, because people differ both physiologically and psychologically, their perceived thermal comfort also

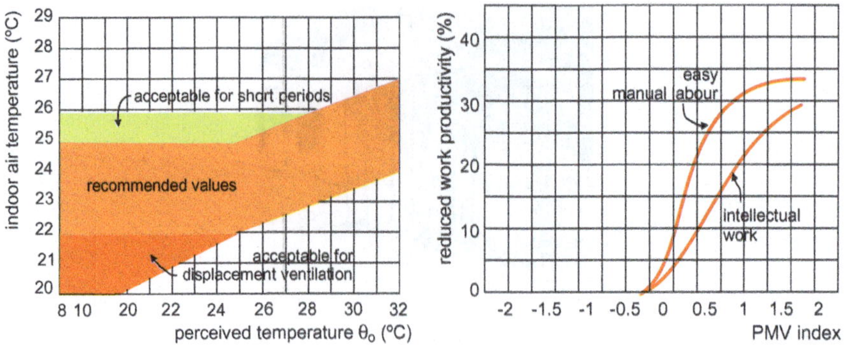

Fig. 1.116 Humans adapt to the conditions of the external environment, and because of that, the acceptable perceived indoor temperature depends on the outdoor conditions (left) [36]. If the *PMV* index deviates from neutral thermal comfort condition (*PMV* 0), productivity decreases significantly and therefore the cost of labour increases (right) [38]

depends on several subjective criteria. A person's sex and age belong among the most influential parameters of subjective sensation of thermal environment. Older people are characterized by lower activity and their bodies have a diminished ability for thermoregulation, so they are more sensitive to the deviation of the physical parameters of the environment. The physiological aspect of thermal comfort sensation can also be noticed in the summertime. The occupants of naturally ventilated and natural cooled buildings accept, as not discomforting, temperatures up to 3 °C higher than the occupants of air-conditioned buildings. People adapt to the state of the external environment and the state of indoor thermal comfort could be evaluated according to the outdoor-air temperature (Fig. 1.116, left). For shorter time periods, temperatures that are 3–6 °C lower than the highest daily outdoor-air temperature will consequently not be perceived as disturbing [31].

1.8.4 Thermal Comfort, Health and Productivity

In the 1980s, the World Health Organization (WHO) drew attention to the increasing number of occupants of buildings with health problems. The phenomenon is called "sick building syndrome" or SBS. Common symptoms are seen as mucosal irritation, allergies and malaise. Although SBS is often related to polluted air in buildings, due to inadequate ventilation, one of the causes might also be an inadequate global and local indoor thermal environment. An inadequate thermal environment also affects productivity at work. Both illnesses and reduced productivity affect the economic efficiency of the staff, and in many cases, it has been proven that the cost to compensate for the consequences of inappropriate indoor thermal comfort considerably exceed the costs of energy for the operation of the buildings (Fig. 1.116, right). Among the others, Roelofsen [38] proposed an approximation model for the determination of

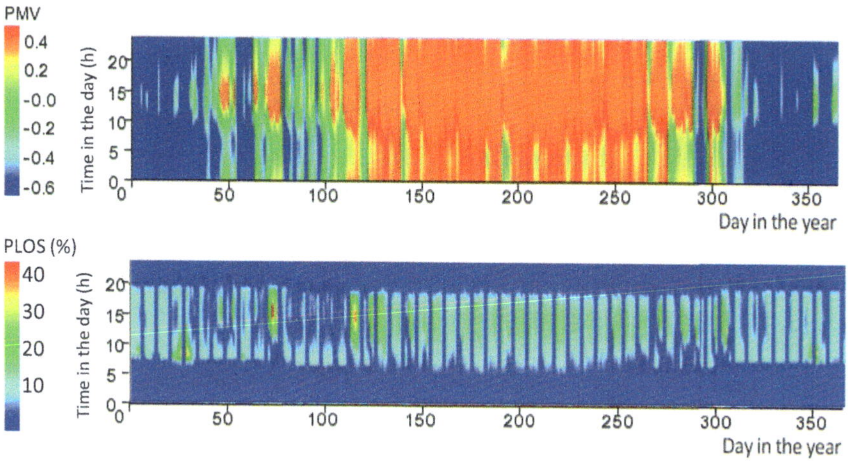

Fig. 1.117 Example of PMV and PLOS determined for an office space during working hours (8:00–18:00) over a year

the Productivity Loss index (*PLOS*) based on the predicted mean vote (*PMV*) in the form:

$$\mathrm{PLOS} = \sum_{i=0}^{6} b_i \cdot \mathrm{PMV}^i [\%]$$

where b_i are the approximation coefficients with values: for $-3 < PMV < 0$: b_0 1.2807, b_1 15.9954, b_2 31.507, b_3 11.7549, b_4 1,4737 and $b_5 = b_6 = 0$; and for $0 < PMV < 3$: b_0 -0,520, b_1 3.8820, b_2 25.1764, b_3 −26.6413, b_4 13.1101, b_5 −3.1298 and b_6 0.2926 [38]. Figure 1.117presents the *PLOS* (%) over the year determined by *PMV* for working hours (8:00–18:00) for office an modelled with the IDA-ICE building-simulation tool [39].

References

1. Behling, S., Behling, S.: Sol Power: Die Evolution der solaren Architecture. Prestel, D (1996)
2. Bauphysik Kalender: Ernst & Sohn Verlag, D (2001)
3. Incropera, F.P., DeWitt, D.P.: Fundamentals of heat and mass transfer. Wiley, USA (1996)
4. Medved, S., Novak, P.: Environment protection and renewable energy sources. University of Ljubljana, Faculty of Mechanical Engineering (2000).(in Slovenian language)
5. Medved, S., Arkar, C., Keršič, P., Božič, N.: Research on the surface temperatures of thermal insulated façades with selective coatings JUB. University of Ljubljana, Faculty of Mechanical Engineering (2008).(in Slovenian language)
6. EN ISO 6946:2017 Building components and building elements—Thermal resistance and thermal transmittance—Calculation methods (2017)

7. Rules on thermal insulation and efficient energy use in buildings; Ur.l. RS 42/2002 (in Slovenian language)
8. Rules on efficient use of energy in buildings; Ur.l. RS 93/2008 (in Slovenian language)
9. Rules on efficient use of energy in buildings with a technical guideline; Ur.l. RS 52/2010 (in Slovenian language)
10. Schmitz-Gunther, T.: Lebensräume: Der Grosse Ratgeber für ökologisches Bauen und Wohnen, Könemann, D (1998)
11. Tobiasson, W., Buska, J., Greatorex, A.: Ventilated attics to minimize icing at eaves. Energy and Building **21**, 229–234 (1994)
12. Medved, S., Leskovšek, U.: Numerical modelling of thermal response of double glazed façade on ROTONDA office building. University of Ljubljana, Faculty of Mechanical Engineering, HypoBank (2009).(in Slovenian language)
13. Medved, S., Arkar, C.: Heat transfer and optimization of compact double side ventilated gap. University of Ljubljana, Faculty of Mechanical Engineering, TRIMO (2006).(in Slovenian language)
14. www.glass.de (1.1.2014)
15. www.refleks.si (1.1.2014).
16. Glücklich, D.: Ökologische Bauen: Von Grundlagen zu Gesamtkonzepten. Deutsche Verlags München, D (2005)
17. Hegger, M., et al.: Energy Manual: Sustainable Architecture. Birkhäuser, D (2008)
18. EN ISO 13370:2017 Thermal performance of buildings—Heat transfer via the ground—Calculation methods; ISO 2017 (2017)
19. Medved, S., Černe, B.: A simplified method for calculating heat losses to the ground according to EN ISO 13370 standard. Energy and Buildings **34**, 5 (2002)
20. EN ISO 10211:2017: Thermal bridges in building constructions—Heat flows and surface temperatures—Detailed calculations; ISO 2017 (2017)
21. Medved, S., Arkar, C., Iamanaev, N., Domjan, S.: TrimoExpert CAD, Building Physic simulation tool; TRIMO, Trebnje (2005)
22. EN ISO 12241:2008: Thermal insulation for buildings equipment and industrial installations—Calculation rules (2008)
23. ISO 9869-1:2014: Thermal insulation—building elements—In-situ measurement of thermal resistance and thermal transmittance—Part 1: Heat flow meter method; ISO 2014 (2014)
24. EN 13187:2000 Thermal performance of buildings—Qualitative detection of thermal irregularities in building envelopes—Infrared method; ISO 2000 (2000)
25. Medved, S., Arkar, C., Iamanaev, N.: KnaufInsulation Technical Applications Calculator; KNAUF INSULATION, Škofja Loka (2008)
26. Keller, B.: Bautechnologie III: Bauphysik, ETH, CH (2002)
27. Arkar, C., Medved, S.: Influence of accuracy of thermal property data of a phase change material on the result of a numerical model of a packed bed latent heat storage with spheres. Thermochim. Acta **438**, 1–2 (2005)
28. Medved, S., Arkar, C.: Correlation between the local climate and the free-cooling potential of latent heat storage. Energy and Buildings **40**, 4 (2008)
29. EN ISO 13786: Thermal performance of buildings components—Dynamic thermal characteristics—Calculation method (2007)
30. Medved, S., Arkar, C., Černe, B.: BTR Building Thermal Response Software Tool (accessible at www.knaufinsulation.com), KNAUF INSULATION, Škofja Loka (2009)
31. McQuiston, F., Parker, J., Spliter, J.: Heating, Ventilating and Air Conditioning, Analysis and Design. Wiley, USA (2005)
32. Davis, M.G.: Building Heat Transfer. Wiley, USA (2004)
33. ISO 7730:2005: Ergonomics of the thermal environment—Analytical determination of thermal comfort using calculation of the PMV and PPD indices and local thermal comfort; ISO 2005 (2005)
34. CR 1752:1998: Ventilation of buildings—Design criteria for the indoor environment; CEN 1998 (1998)

35. EN 16798-1:2019: Energy performance of buildings—Ventilation of buildings—Part 1: Indoor environmental input parameters for design and assessment of energy performance of buildings addressing indoor air quality, thermal environment, lighting and acoustics, CEN (2019)
36. ASHRAE Standard 55—Thermal environmental conditions for human occupancy; ASHRAE 2017 (2017)
37. Vidrih, B., Medved, S.: Outdoor environment modelling. University of Ljubljana, Faculty of Mechanical Engineering (2009).(in Slovenian language)
38. Kosonena, R., Tanb, F.: Assessment of productivity loss in air-conditioned buildings using PMV index. Energy and Buildings **36**, 10 (2004)
39. IDA-ICE, IDA Indoor Climate and Energy 4.8. EQUA Simulation AB, Solna (2018)
40. Fanger, P.O.: Thermal Comfort. Danish Technical Press, Copenhagen (1970)

Chapter 2
Ventilation and Energy Performance of Buildings

Abstract There are many sources of air pollutants in buildings, such as people, appliance and built materials, which cause the air in buildings to be more pollutant than in the outdoor environment. The increased content of pollutants in the indoor air can affect the well-being, work performance and even the health of the occupants. Such conditions can be avoided by dilution of pollutants with adequate ventilation of buildings, the process of the air exchange between building interior and outdoor. Buildings can be ventilated naturally due to the difference in air pressure caused by the difference in the air temperature and exposure of the buildings to the wind or forced ventilated by mechanical systems. Because ventilation of the buildings increases the energy use, the uncontrolled air exchange due to the air infiltration must be avoided as much as possible with consistent airtightness of the building envelope. In the chapter the principles of planning and the evaluation methods for indoor air quality comfort are presented. The impact of the ventilation on energy needs is explained, including the ventilation related measures for increasing of the energy efficiency of buildings.

Learning objectives In this chapter you will learn about:

- Significance of building's ventilation for indoor air quality (IAQ);
- sick building syndrome (SBS) and building-related illnesses (BRI);
- principles of natural ventilation;
- requirements and indicators of indoor air quality;
- methods for design of the ventilation air supply
- modelling of the time-dependant concentrations of indoor air pollutants;
- in-situ evaluation and airtightness requirements for buildings;
- heat losses due to ventilation and impact of the ventilation on a building's energy efficiency.

S. Medved, *Building Physics*, Springer Tracts in Civil Engineering,
https://doi.org/10.1007/978-3-030-74390-1_2

2.1 Symbols for Quantities and Units

A, A_{env}, A_{floor}	Area, area of external surface of the building thermal envelope (env), of floor (m^2)
A_i, A_o	Free surface of inlet (i), of outlet (o) ventilation opening (m^2)
A_u, A_{env}	Useful area of building (u), of building envelope (env) (m^2)
a	Airtightness factor of building block (window/door) ($m^3/hmPa^{2/3}$)
α	Hellmann's exponent (1)
β	Volumetric thermal expansion coefficient (1/K)
$C_i, C_e, C_{i,\max}, C_0, C_{oo}$	Concentration of the pollutant in the indoor air (i), in the outdoor air (e), maximum permissible concentration of the pollutant in the indoor air (max), at time 0, at infinite time (steady state) (g/m^3, mg/m^3, $\mu g/m^3$)
C_p, C_d	Wind pressure coefficient, pressure loss coefficient (1)
c_{olf}	Specific emission of olfs (olf/m^2)
χ	Point thermal bridge transmittance (W/K)
d	Thickness (m)
dp	Decipol (1)
$E, E_P, E_{P,nren}, E_{P,ren}, E'_P$	Energy, primary energy (P), non-renewable primary energy (P, nren), renewable primary energy (P, ren) (J), specific primary energy (kWh/m^2an)
η_{rec}	Heat recovery effciency of mechanical ventilation system (1, %)
ε_v	Efficiency of ventilation (1, %)
$\Phi_{NH}, \Phi_{NC}, \Phi'_{NH}, \Phi'_{NC}$	Heat load (NH), cooling load (NC) (W), specific heat load, specific cooling load (W/m^2, W/m^3)
G	Mass source of pollutant (g/h)
g	Gravitational acceleration (m/s^2)
H, h, H_a	Height of air gap (m)
$H_T, H_V, H_{V,n}, H_{v,m}$	Transmission heat-loss coefficient (T) (W/K), ventilation heat-loss coefficient (V) (W/K), natural ventilation heat-loss coefficient (V,n) (W/K), mechanical ventilation heat-loss coefficient (V,m) (W/K)
k	Air permeability (m^2), decay-rate of pollutant (h^{-1})
l	Length of building block (window/door) joint (m)
Ψ	Linear thermal bridge transmittance (W/mK)
M	Molar mass (kg/kmol)
MPC	Maximum permissible concentration ($\mu g/m^3$, mg/m^3)
m	mass (kg)
$\dot{m}$	Mass flow rate (kg/s)
μ	Dynamic viscosity (Ns/m^2, kg/ms)
N	Number of occupancies (person)

(continued)

(continued)

n, n_{inf}, n_{50}	Air-exchange rate, air-exchange rate caused by infiltration, air-exchange rate at a pressure difference of 50 Pa (h^{-1})
olf	Amount of odours emitted by the source (1)
p_{atm}, Δp, $\Delta p_{i\text{-}e}$, Δp_t	Air pressure, air-pressure difference, air-pressure difference between indoor and outdoor, stagnation pressure difference (Pa)
ρ, ρ_i, ρ_e	Density, of indoor air (i), of outdoor air (e) (kg/m^3)
Q, Q'	Heat demand per year (J/a, kWh/a), spacific heta demand per year (J/m^2a, kWh/m^2a)
Q_{nd}, Q_f	Energy needs, final energy for operation of building service systems (J, kWh)
θ, θ_i, θ_e, $\theta_{e,p}$	Temperature, of indoor air (i), of outdoor air (e), design winter outdoor temperature (e, p) (°C)
T, T_i, T_e	Absolute temperature, of indoor (i) air, of outdoor (e) air (K)
t	time
U	thermal transmittance of building structure (W/m^2 K)
V	Internal (nett) volume of building (m^3)
$\dot{V}$, $\dot{V}'$, $\dot{V}_\varepsilon$ $\dot{V}_{per}$	Volume air flow rate (m^3/h), specific volume air flow rate (m^3/hm^2), specific volume air flow rate per person (m^3/h per), corrected volume air flow rate (m^3/h)
$\dot{V}_{100,A,\max}$, $\dot{V}_{100,l,\max}$	Permissible permeability volumetric air flow rate at air pressure difference 100 Pa, per area of building block (window or door) ($m^3/hm^2\ Pa^{2/3}$), per length of joint ($m^3/hm\ Pa^{2/3}$)
v	Velocity (m/s)
z	Height above groud (m)

2.2 Building Ventilation

2.2.1 *Significance of Building Ventilation*

In the modern world, people spend most of their time in buildings. In addition to ensuring the requirements of indoor thermal comfort in apartments and at working places, adequate indoor air quality in the buildings through effective ventilation must be provided. Ventilation is a process of diluting the pollutants generated inside the building with outdoor air. Since fresh air must be heated in the winter and cooled in the summer, ventilation is associated with energy use. In 1984, the World Health Organization (WHO) drew attention to the fact that indoor air quality (IAQ) in at least one-third of new and renovated buildings is inadequate and a cause of several diseases. The reason is in the regulations and requirements for increased energy savings for space heating and the increased energy efficiency of buildings in general.

The symptoms of malaise and ailments due to excessive concentrations of pollutants in the indoor air are divided into two categories that indicates unhealthy buildings:

- sick building syndrome (SBS),
- building-related illnesses (BRIs).

The sick building syndrome describes situations in which building occupants experience health problems such as headache, poor concentration, dry throat, irritation of mucous membranes and nausea, linked to the time spent in the building. The exact cause of the malaise is difficult to identify, since soon after the occupant leaves the building the symptoms disappear. The long-term exposure to air contaminants in buildings can lead to a number of disorders, for example, respiratory infections, viral and bacterial diseases, and links between air quality and cardiovascular disorders have been identified.

2.2.1.1 Air Pollutants in Buildings

The pollutants are formed in the building or enter the building from the external environment. The most important process and the sources of pollutants inside buildings are breathing (Table 2.1), human and animal metabolism, organic matter decay, building materials and devices. When evaluating the IAQ, it is important to consider:

- which airborne pollutants that pose a health risk are present (Fig. 2.1, left);
- the maximum allowable concentration of separate pollutants that do not represent a risk to health. The values, known as the Maximum Permissible Concentrations (*MPC*), are defined mostly by World Health Organization (WHO) and take into account not only the concentration of the pollutant but the duration of the exposure as well (Fig. 2.1, right; Table 2.2).
- the synergistic effects of several pollutants. Often, the overall effects of two pollutants on human health are greater than "their sum". For example, respiratory diseases due to the inhalation of asbestos dust are 50–70 times more common in smokers than non-smokers.

Among the most typical air pollutants originating in buildings are carbon dioxide CO_2, carbon monoxide CO, nitrogen compounds (for example, ammonia), ozone

Table 2.1 Amount of the pollutant exhaled by an adult person during the day [2, 3]

pollutant	exhaled quantity of pollutant (g/day)
carbon dioxide CO_2	250–400
carbon monoxide CO	0.3–1
ethanol	0.01-0.05
methane	0.04-0.1
acetone	0.03–0.06

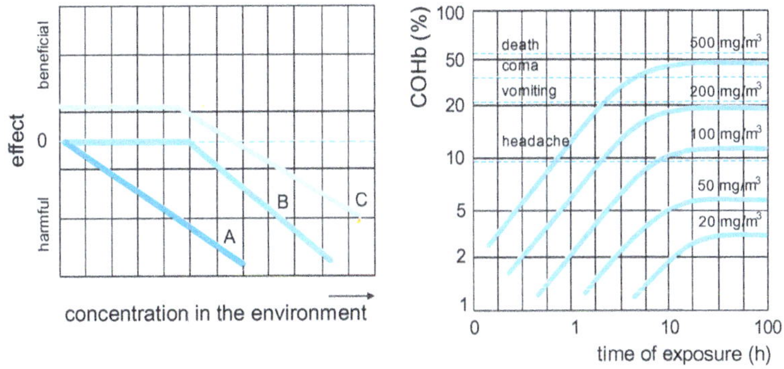

Fig. 2.1 Impact of pollutants on health depend on the concentration of the pollutants in the environment. Some pollutants are harmful at any concentration (A, e.g., carbon monoxide CO), while other are only harmful when a limit value is exceeded (B, e.g., carbon dioxide CO_2). Some of the pollutants are beneficial for the development of living organisms, but become harmful at larger concentrations (C, e.g., heavy metals) [1] (left). The impact of pollutants on well-being and health is related to the level of contamination in the environment and to the duration of the exposure [1]. The physiological effects of carboxyhemoglobin (COHb) forming in human bloodstream when body is exposed to CO in the air according to duration of exposure (right)

Table 2.2 Maximum-permissible concentrations ($MPC = C_{i,\max}$) of some indoor air pollutants and average concentrations of those pollutants in the outdoor air. Note that the concentrations of air pollutants in the outdoor air depend on anthropogenic sources as well as climate conditions and weather [1]

pollutant	maximum permissible concentration of pollutant in indoor air MPC, $C_{i,max}$	average concentration of pollutant in the outdoor air C_e
CO_2	1000 to1500 mg/m³	450 mg/m³
CO	10 μg/m³	10 μg/m³
radon	<200 Bq/m³	10-50 Bq/m³
water vapour	10 - 12 g/m³	2 - 8 g/m³

O_3, airborne particles smaller than 10 μm and hundreds of different volatile organic compounds (VOCs). The latter is a group of hazardous carbon compounds that cause photochemical reactions in the air. The main sources in buildings are paints, solvents, and cleaning agents. They are grouped into halogen-containing compounds containing chlorine, aromatic compounds, for example, benzene, and aliphatic compounds such as formaldehyde (CH_2O). These substances cause irritation of the mucous membrane, damage to the internal organs, and some substances are carcinogenic.

Air pollutants that enter the building from the outdoor environment are primarily substances generated by the combustion of fossil fuels. These are carbon dioxide CO_2 and monoxide CO, nitrogen oxides NO_x, sulphur dioxide SO_2 as well as heavy metals, for example, lead and mercury, volatile organic matter and solid particles, while airborne micro-organisms and ozone O_3, are created in the outdoor air by biological and photochemical reactions. Among the important pollutants entering the building, particularly from soil, is radon.

2.2.2 Requirements for the Ventilation of Buildings

Taking into account that outdoor air is typically less polluted than indoor air in buildings, the exchange of indoor with outdoor air by ventilation is common practice for maintaining the IAQ. Several empirical methods are used to determine the amount of outdoor air needed to ventilate the building, such as:

- for residential buildings, in which occupants are the main source of pollutants, the method of the air exchange rate *n* (h^{-1}), which defines how many times in one hour the air in the building is replaced by outdoor air:

$$\mathrm{n} = \frac{\dot{\mathrm{V}}}{\mathrm{V}}\left[\frac{\mathrm{m}^3}{\mathrm{h} \cdot \mathrm{m}^3} = \mathrm{h}^{-1}\right] \rightarrow \dot{\mathrm{V}} = \mathrm{n} \cdot \mathrm{V}\left[\frac{\mathrm{m}^3}{\mathrm{h}}\right]$$

where $\dot{V}$ (m^3/h) is required volume air-flow rate and V (m^3) is the nett volume of the room or building, determined by the interior dimensions of the premises. Because airborne pollutants are also emitted by building materials and devices, we distinguish the minimum air-exchange rate during the period of the unoccupied and occupied building. An example of the required air-exchange rates *n*, is given in Table 2.3 (left);

- in the case of larger premises in commercial and public buildings and for buildings with a high density of occupancies, such as cinema halls or congress rooms, the total required amount of fresh air $\dot{V}$ (m^3/h) is defined by considering the occupants physiological needs:

$$\dot{\mathrm{V}} = \mathrm{N} \cdot \dot{\mathrm{V}}_{\mathrm{per}}\left[\frac{\mathrm{m}^3}{\mathrm{h}}\right]$$

  where $\dot{V}_{per}$ (m^3/h pers) is the required amount of fresh air per occupant and N (pers) is the number of people in a fully occupied space. The recommended values of $\dot{V}_{per}$ are shown in Table 2.3 (right);
- for the buildings in which non-occupant sources of airborne pollutants prevail and the type and quantity of pollutants are known, the required amount of outdoor air

for ventilation is determined based on the maximum permissible concentration *MPC* of the given pollutant. The required amount of outdoor air for ventilation $\dot{V}_j$ is determined for the *j*th pollutant and the maximum value $\dot{V}_{j,\max}$ is selected as the design-volume fresh-air flow rate:

$$\dot{V}_j = \frac{G_j}{MPC_j - C_{e,j}} \quad \dot{V} = \dot{V}_{j,\max} \left[\frac{g \cdot m^3}{h \cdot g} = \frac{m^3}{h} \right]$$

where G_j (g/h) is the mass of the *j*th pollutant generated in the room in a unit of time unit, MPC_j is the permissible concentration of the *j*th pollutant in the indoor air, and $C_{e,j}$ is the concentration of *j*th pollutant in the outdoor air. MPC_j and $C_{e,j}$ are given as mass fractions, i.e., in g, mg or µg per m^3 of air [6].

Case Study 1 A sitting person in an office emits 100 g of water vapour per hour, increasing the content of water vapour in the indoor air, which affects the indoor-air quality. The acceptable content of water vapour in the indoor air is $C_{i,H_2O,\max}$ of 11 g/m^3. Determine the required amount of outdoor air to ventilate the building if the humidity of the outdoor air is 4.5 g/m^3.

$$\dot{V} = \frac{G}{C_{i,H_2O,\max} - C_e} = \frac{100}{11 - 4.5} = 15.4 \left[\frac{g \cdot m^3}{h \cdot g} = \frac{m^3}{h} \right]$$

Case Study 2 A seated person exhales approx. 25 g of CO_2 per hour. The volume fraction of CO_2 in the outdoor air is 550 ppm, or 550×10^{-6} m^3 of CO_2 in 1 m^3 of air. Determine the required amount of outdoor air for ventilation

Table 2.3 Required air-exchange rate *n* during the building's unoccupied and occupied periods (left). Necessary quantity of outside air for ventilation per person $\dot{V}_{per}$ in different rooms [4, 5] (right)

room	air-exchange rate n (h^{-1}) unoccupied	air-exchange rate n (h^{-1}) occupied
living rooms and bedrooms	0.2	0.7
toilets, bathrooms	0.2	4 to 6 occasionally
kitchens	0.2	0.5 to 25 occasionally

room	volume air flow rate $\dot{V}_{per}$ (m^3/h per person)
residential buildings	25 – 30
bathrooms	60
offices	40 – 60
conference halls	40
sports halls	30
shops	30
classrooms	30
restaurants	30

so that the concentration of CO_2 in the room will not exceed 3000 mg/m^3 ($CO_{2,max} = MPC_{CO_2}$), which, according to the recommendations of WHO, does not affect human health.

The volume fraction of CO_2 in the outdoor air must be converted into the mass fraction by considering the molar mass CO_2 ($M = 44$ kg/kmol) and the volume of one kmol of gas (22.4 m^3/kmol):

$$C_e = \frac{C_{e,ppm} \cdot M}{22.4} = \frac{550 \times 10^{-6} \cdot 44}{22.4} \times 10^6 = 1080 \left[\frac{m^3 \cdot kg \cdot kmol \cdot mg}{m^3 \cdot kmol \cdot m^3 \cdot kg} = \frac{mg}{m^3}\right]$$

$$\dot{V} = \frac{G}{C_{i,CO_2,max} - C_e} = \frac{25 \times 10^3}{3000 - 1080} = 13.0 \left[\frac{mg \cdot m^3}{h \cdot mg} = \frac{m^3}{h}\right]$$

For the dilution of CO_2 emitted by a seated person, the room must be ventilated with a supply of at least 13 m^3 of outdoor air every hour. However, because an increased level of physical activity also means an increased emission of emitted pollutants by humans, and as there are also other sources of pollution in the building, the fresh-air supply by ventilation is designed using the empirical values given in Table 2.3 (right).

- we detect poor air quality predominantly by smell. Therefore, in the 1990s, the method of ventilation design based on perceived air quality by smell was established. Odours are emitted by people, building materials and devices. The olf was adopted as the unit for the rate at which odours are emitted to the indoor air (from Latin word *olfactus*, meaning "sense of smell"). One olf represents the amount of perceived odours emitted by an average seated person doing light work (1–1.2 met, see Sect. 1.8). The total odour load is defined as the sum of the emissions from all sources. The odour expressed by olfs emitted by person or 1 m^2 of surfaces area of some building materials, are listed in Table 2.4. Buildings with odour loads lower than 0.1 olf per m^2 of room surface area are considered low-pollution buildings. The unit decipol (dp) was defined to describe the relation between the ventilation rate and the total amount of odours, expressed as the sum of olfs. By definition, the room is ventilated with one decipol (*dp*) if for each olf emitted into the indoor air, 10 L of outdoor air per second is supplied to the room (or 36 m^3/h). The volume air flow rate $\dot{V}$ for ventilation is defined by taking into account the sum of emitted olfs and the designed number of decipols *dp*:

$$\dot{V} = \frac{\sum olf \cdot dp \cdot 3600}{1000} \left[\frac{m^3}{h}\right]$$

The design number of decipols also determines the category of indoor air quality (IAQ). It is recommended that residential buildings are ventilated with one decipol of fresh outdoor air.

Table 2.4 Number of olfs emitted by people and some building materials. Substances emitting less than 0.1 olf/m^2 are considered to be low-polluting materials [7]

odour source	olfs emitted by the source
sitting person (1-1.2 met)	(olf/person) 1
smoker, lasting	25
athlete	30
	(olf/m^2)
wool carpet	30
synthetic floor covering	0.4
PVC covering	0.2
rubber	0.6
marble	0.01

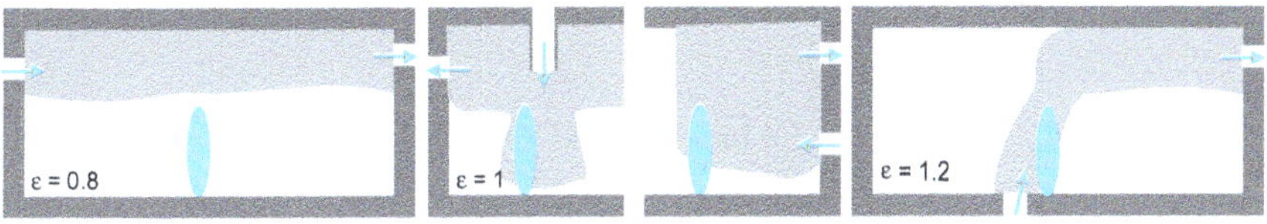

Fig. 2.2 Besides the amount of emitted pollutants, the IAQ depends on the efficiency of the dilution of pollutants, which is measured by ventilation effectiveness ε_υ. The most effective is displacement ventilation

Explanation Regardless of the method used to determine the volume flow rate $\dot{V}$ of the ventilation air, the dilution of the pollutants depends on the technique and the effectiveness ε_v of the ventilation (Fig. 2.2). Spaces can be ventilated with a short path ($\varepsilon_\upsilon = 0.8$), by mixing ($\varepsilon_\upsilon = 1$) or by displacement ($\varepsilon_\upsilon = 1.2$) ventilation. Consequently, the required volume air flow rate of the supply air is determined by the expression:

$$\dot{V}_\varepsilon = \frac{\dot{V}}{\varepsilon_v} \left[\frac{m^3}{h}\right]$$

2.2.3 *Modelling the Time-Dependent Concentrations of Indoor Air Pollutants*

Pollutants are generally classified into conservative and non-conservative pollutants. The first are inert substances, which do not participate in chemical reactions, nor do they decay, which would cause a change of their mass in the indoor air. Otherwise, the pollutants are classified as non-conservative. In modelling the time-dependent concentrations of non-conservative pollutants in the indoor air, we assume that this is a problem of the first order, which means that the change in the content of the pollutant with the time d*C*/d*t* depends on the current concentration and the decay rate of the observed pollutant:

$$\frac{dC}{dt} = -k \cdot C \left[\frac{mg}{h}\right]$$

where *k* (h^{-1}) is the decay-rate (Table 2.5). The negative sign in the expression accounts for the pollutant's natural decay and therefore the content of the pollutant in the indoor air decreases. The differential equation can be solved as follows:

$$\frac{dC}{dt} = -k \cdot C \rightarrow \frac{1}{C}\frac{dC}{dt} = -k \rightarrow \int_0 \frac{1}{C}\frac{dC}{dt}$$

$$= -\int_0^t k \rightarrow \ln\frac{C}{C_o} = -k \cdot t \rightarrow C = C_0 \cdot e^{-k.t}$$

where, C_o (g/m^3, mg/m^3 or μg/m^3) is the concentration of the pollutant in the air at time $t = 0$, and C (g/m^3, mg/m^3 or μg/m^3) is the concentration of pollutant over time t. Considering the law of the conservation of mass, the time-dependent concentration C of the pollutant in the indoor air is determine using the expression:

$$V\frac{dC}{dt} = \underbrace{G}_{\substack{\text{in building}\\ \text{source}}} + \underbrace{C_e \cdot \dot{V}}_{\substack{\text{entering}\\ \text{the building}}} - \underbrace{C_i \cdot \dot{V}}_{\substack{\text{exiting}\\ \text{the building}}} - \underbrace{k \cdot C_i \cdot V}_{\substack{\text{decaying inside}\\ \text{the building}}}$$

where V is the building nett volume (m^3), G is the amount of observed pollutant released into the indoor air (mg/h), C_e and C_i (mg/m^3) are the concentrations of the same pollutant in the supply (outdoor) air and in the indoor air, $\dot{V}$ (m^3/h) is the volume flow rate of the supply air for the ventilation and k (h^{-1}) is the decay-rate of the pollutant. In the steady-state condition, the change of the pollutant concentration dC_i/dt is equal to 0, and the concentration of the pollutant in the indoor air $C_{i,oo}$ is determined by:

Table 2.5 Decay-rate *k* for some building air pollutants

pollutant	decay-rate k (h^{-1})
Rn_{222}	0.0076
CO	0.0
NO_2	0.15
SO_2	0.23

$$C_{i,\infty} = \frac{G + C_e \cdot \dot{V}}{\dot{V} - k \cdot V} \left[\frac{\frac{mg}{h} + \frac{mg \cdot m^3}{m^3 \cdot h}}{\frac{m^3}{h} - \frac{m^3}{h}} = \frac{mg}{m^3} \right]$$

A typical non-conservative air pollutant in buildings is radon. Radon is a radioactive gas. Thus, it naturally decomposes into an isotope of polonium, and when these isotopes decay, the ionizing radiation in the form of α particles forms. When radon is inhaled, these particles cause damage to the cells in the respiratory tract. Most radon passes into buildings through leaks in the envelope of the building that is in contact with the earth and rocks, and is also emitted by natural building materials, such as bricks and concrete.

Case Study A room with a volume of 52 m^3 is ventilated with an air-change rate n equal to 0.7 h^{-1}. The breakthrough of the pipe to the outer wall of the room, which is dug into the ground, is not sealed and 234 mg of non-conservative pollutant enters into the room every hour. The decay-rate constant of the pollutant k is 0.0076 h^{-1}. In the outdoor air, the pollutant content is 0.003 mg/m^3. The maximum permitted content of observed pollutant in buildings is 12 mg/m^3. Determine whether the room is adequately ventilated to maintain the radon level below MPC.

$$C_{i,\infty} = \frac{G + C_e \cdot \dot{V}}{\dot{V} - k \cdot V} = \frac{G + C_e \cdot n \cdot V}{n \cdot V - k \cdot V} = \frac{234 + 0.003 \cdot 0.7 \cdot 52}{0.7 \cdot 52 - 0.0076 \cdot 52} = 6.5 \, \frac{mg}{m^3}$$

Solution: The ventilation rate is adequate in case of permanently ventilation.

In the presented case study, the calculated concentration of radon is established in the room after an infinite time. In order to determine the concentration of pollutant, at any time after a sudden change in the source of pollutant in the building (for example, when classes begin in a classroom), we could also use a differential equation derived using the law of conservation of mass. Since most of the air pollutants in buildings are conservative ($k = 0$), the expression for determining $C_{i,t}$ is simplified as:

$$C_{i,t} = C_{i,\infty} + \left(C_{i,0} - C_{i,\infty}\right) \cdot e^{-\frac{\dot{V}}{V} \cdot t} \left[\frac{mg}{m^3}\right]$$

where $C_{i,0}$ (mg/m^3) is the concentration of the pollutant in air in the unoccupied building, meaning at time 0.

Case Study Five people enter an office at 8:15 and start to work. The concentration of CO_2 measured in the room is shown in Fig. 2.3 and in the exterior before the workday it was 460 ppm (parts per million, there are 460 volume parts of CO_2 in the total of one million parts of the air volume). The office is

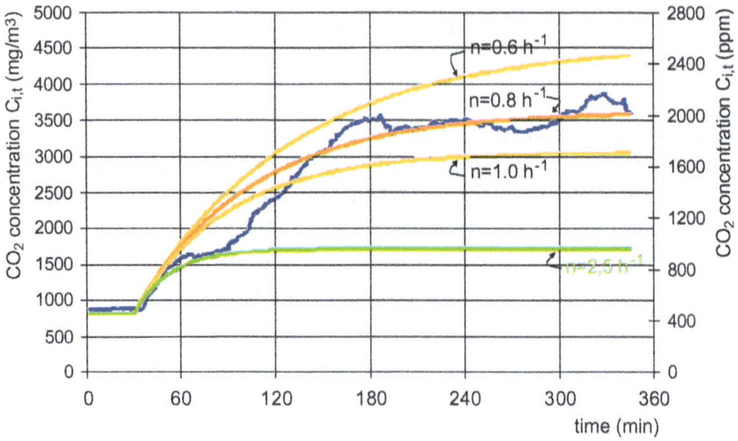

Fig. 2.3 Experimental and modelled concentrations of CO_2 in a naturally ventilated office occupied by five people at different air-exchange rates n (h^{-1})

naturally ventilated. Each person exhales 15 L of CO_2 per hour. The density ρ of CO_2 is 1.98 kg/m^3. The concentration of CO_2 in the office was then measured during the following hours. Based on the experimentally determined concentration of CO_2 in the indoor air, determine the air exchanges rate n (h^{-1}). The temperature θ_i in an office with a volume of 40 m^3 is 23 °C.

The concentration of CO_2 in the indoor air expressed as a mass fraction is equal to:

$$C_{i,CO_2} = C_{i,CO_2,ppm} \cdot 10^{-6} \cdot \frac{M}{22.4} \cdot \frac{273}{273 + \theta_i} 1000 \left[\frac{10^{-6} \cdot m^3 \cdot kg \cdot kmol \cdot K \cdot g}{m^3 \cdot kmol \cdot m^3 \cdot K \cdot kg} = \frac{g}{m^3} \right]$$

where the M corresponds to the molar mass of CO_2 (44 kg/kmol), θ_i is the office indoor-air temperature (°C) and constant 22.4 (m^3/kmol) is the volume of one kmol of gas in standard conditions (temperature 273 K and pressure 1 bar). The initial concentration $C_{i,CO_2,t=0}$ in the indoor is equal to the concentration of CO_2 in the outdoor air:

$$C_{i,CO_2,0} = C_{e,CO_2} = 460 \cdot 10^{-6} \frac{44}{22.4} \cdot \frac{273}{273 + 23} 1000 = 0.83 \frac{g}{m^3}$$

Since we do not know the air-exchange rate n (h^{-1}), the volume air-flow rate $\dot{V}$ is obtained by comparing the measured and the calculated concentrations of CO_2 for different assumed air-exchange rates. The total amount of CO_2 emitted by five people each hour G_{CO_2} (g/h) is equal to:

$$G_{CO_2} = N \cdot \frac{\dot{V}_{CO_2,person}}{1000} \cdot \rho_{CO_2} \cdot 1000 = 5 \cdot \frac{15}{1000} \cdot 1.98 \cdot 1000$$
$$= 148.5 \left[\frac{dm^3 \cdot m^3 \cdot kg \cdot g}{h \cdot dm^3 \cdot m^3 \cdot kg} = \frac{g}{h} \right]$$

The concentration of CO_2 in the air after an infinite time would be equal to:

$$C_{i,CO_2,\infty} = \frac{G_{CO_2} + C_{i,CO_2,0} \cdot n \cdot V}{n \cdot V - k.V} = \frac{G_{CO_2} + C_{i,CO_2,0} \cdot n \cdot V}{n \cdot V}$$
$$= \frac{148.5 + 0.83 \cdot n \cdot 40}{n \cdot 40} \left[\frac{g}{m^3} \right]$$

The concentration of CO_2 from the time when the people enter the room and until the establishment of the steady state, when the concentration of CO_2 reaches $C_{CO_2,oo}$, is given by the expression:

$$C_{CO_2,t} = C_{CO_2,\infty} + \left(C_{CO_2,t=o} - C_{CO_2,\infty} \right) \cdot e^{-\frac{\dot{V}}{V} \cdot t}$$
$$= C_{CO_2,\infty} + \left(0.83 - C_{CO_2,\infty} \right) \cdot e^{-n \cdot t}$$

Solution: The measured and modelled concentrations of $C_{CO_2,t}$ in the case of n 0.6, 0.8 and 1.0 h^{-1} are shown in Fig. 2.3. The best matching between the measured and modelled values can be seen for the air-exchanges rate n equals 0.8 h^{-1} and it can be assumed that office is ventilated with those air-exchange rates. The model can also be used to determine the required air-exchange rate in order not to exceed the concentration of the pollutant that ensures a high indoor air quality—below 1000 ppm. To reach that condition, an air-exchange rate $n = 2.5$ h^{-1} must be provided.

2.2.4 Ventilation and Indoor Environment Comfort

Similar to thermal comfort, empirical models have also been developed for a correlation between ventilation flow rate and the subjective perceived indoor-air quality (IAQ). Most models are based on the assumption that in buildings the most important source of air pollutants is people. The indoor air-quality classes A, B, C and D in [8] or classes I, II, III and IV in [9] were set based on a survey of the percentage of people who were not satisfied with the perceived air quality (PD). The classes can be determined with various indicators as a benchmark, for example, by the concentration of CO_2 in the indoor air above the concentration in outdoor air, or by the number of decipols of fresh air providing one olf of pollutants, as it is shown in Table 2.6.

Table 2.6 The IAQ quality classes and the predicted percentage of people dissatisfied with the indoor air quality *PD*. According to [9] the classes are defined by the concentration of CO_2 in indoor air over the concentrations in the outdoor air, or alternatively, with the designed number of decipol for ventilation of the room/building

perceived air quality class		percentage of dissatisfied with the perceived air quality PD	Co_2 concentration in the indoor air above outdoor ppm	vol%	number of decipols
A	I	< 15 %	350	0.035	1
B	II	< 20 %	500	0.050	0.7
C	III	< 30 %	800	0.080	0.4
	IV	> 30 %	> 800		< 0.4

Adequate ventilation of buildings is essential to achieve the pleasant conditions, but even more importantly, because a healthy and high productivity environment must be provided to the occupants. Figure 2.4 shows how the productivity in an office decreases with respect to the specific volume air-flow rate (in m^3/h per m^2 of room floor area) and specific emissions of odours (in olf per m^2 of room floor area) [10].

Case Study Determine the annual cost of productivity loss for an employee working in an office with an area A_{floor} 16 m^2 covered with synthetic floor cover emitting 0.34 olf/m^2 and ventilated with an air exchange rate n equals 0.5 h^{-1}, or in office floor painted with a low emission coating emitting 0.06 olf/m^2 and ventilated with an air-exchange rate n 1.0 h^{-1}. The height of the office h is 2.4 m. The annual income per company employee is €35,000. Assume that other structures do not emit odours.

The specific pollutants load in the office $c_{olf,1}$ and $c_{olf,2}$ are:

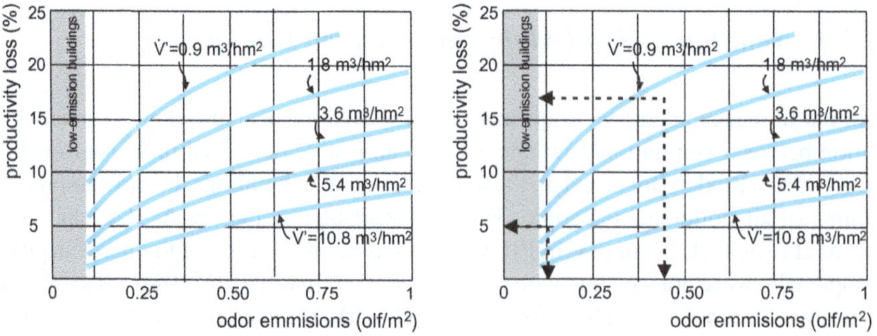

Fig. 2.4 Productivity loss with respect to the specific volume air-flow rate for ventilation (in m^3/h per m^2 of space floor area) and specific emissions of odours (in olf per m^2 of space floor area) as proposed by the Kosonen and Tan [10]

$$c_{olf,1} = \frac{A_{floor} \cdot c_{olf,floor} + N \cdot c_{olf,people}}{A_{floor}} = \frac{16 \cdot 0.34 + 1 \cdot 1}{16} = 0.40 \frac{olf}{m^2}$$

$$c_{olf,2} = \frac{A_{floor} \cdot c_{olf,floor} + N \cdot c_{olf,people}}{A_{floor}} = \frac{16 \cdot 0.06 + 1 \cdot 1}{A_{floor}} = 0.12 \frac{olf}{m^2}$$

and a specific volume air flow rate for ventilation $\dot{V}_1'$ and $\dot{V}_2'$ is:

$$\dot{V}_1' = \frac{\dot{V}}{A_{floor}} = \frac{\overbrace{A_{floor} \cdot h \cdot n_1}^{V_n}}{A_{floor}} = \frac{16 \cdot 2.4 \cdot 0.5}{16} = 1.20 \frac{m^3}{h \cdot m^2}$$

$$\dot{V}_2' = \frac{A_{floor} \cdot h \cdot n_2}{A_{floor}} = \frac{16 \cdot 2.4 \cdot 1.0}{16} = 2.4 \frac{m^3}{h \cdot m^2}$$

From Fig. 2.4 (right) [10] it is evident that the employee's performance in the second case is increased by ~ 12% (productivity loss will decrease from 17 to 5%), which would mean that the employer's annual income will increase by €4200. This indicates that an investment in low-emission building materials would be cost effective for employers.

2.2.5 How Does Building Ventilation Work

Air pressure is a consequence of the force exerted on the Earth's surface by the weight of the column of atmospheric gases. It is also known as hydrostatic pressure and is calculated as:

$$p_{atm} = \frac{V \cdot \rho \cdot g}{A} = \frac{A \cdot h \cdot \rho \cdot g}{A} = h \cdot \rho \cdot g \left[\frac{m^2 \cdot m \cdot kg \cdot m}{m^2 \cdot m^3 \cdot s^2} = \frac{N}{m^2} = Pa \right]$$

where V is the volume of the air column with base A (m^2) and height h (m), ρ is the density of air (kg/m^3) and g is the acceleration due to gravity (m/s^2). Although air pressure is constantly changing, it is often assumed in engineering practice to be a constant quantity with the reference value of 101.3×10^3 Pa at an elevation of 0 m. As the height increases, the air pressure decreases (approx. 10 Pa/m, a little less at low temperatures). Air mass movement in the atmosphere is a consequence of global pressure differences between the areas of high (cyclones) and low (anticyclones) air pressure. These fields arise due to the warming of the surface by solar radiation. Mass-flow quantity and direction in the atmosphere depend on the pressure-difference gradient, gravity, friction between atmosphere particles and the Coriolis effect that occurs because of the rotation of the Earth (Fig. 2.5). Nevertheless, we will assume

Fig. 2.5 Isobars, lines of the same atmospheric pressure, shown on the maps are recalculated to sea level, because the air pressure depends on the altitude of the location

that air flows between the interior of the building and its surroundings, as well as air flows inside the building, are driven by the differences in local air pressures due to:

- differences in air densities at different outdoor and indoor air temperatures;
- differences in air pressure on the building envelope due to the wind pressure acting on the façade surface;
- mechanical devices used for the mechanical ventilation and air-conditioning of buildings.

The air mass flow into the building can be random, as in the case of air infiltration through the gaps in the building envelope, or designed by purpose, as in the case of building ventilation. The pressure differences and the amount of air exchanged between the building and the external environment are determined using the law of the conservation of energy.

2.2.5.1 Bernoulli's Equation

The quantities of state of the air mass travelling through space (temperature, pressure, velocity and position) are constantly changing, but their total energy is always conserved. Assuming the air mass flow is isothermal, meaning that the observed volume of air does not change the temperature, the relations between the quantities of state can be expressed using the law of the conservation of energy in the form of Bernoulli's equation:

$$p_1 + \frac{1}{2} \cdot \rho \cdot v_1^2 + h_1 \cdot \rho \cdot g = p_2 + \frac{1}{2} \cdot \rho \cdot v_2^2 + h_2 \cdot \rho \cdot g \text{ [Pa]}$$

where, p is the pressure (Pa), v is the velocity (m/s), ρ is the air density (kg/m^3), g is the acceleration due to gravity (m/s^2) and h (m) is the elevation of the observed air

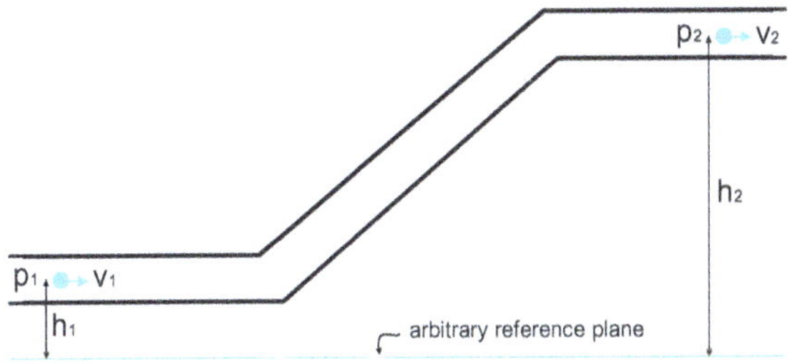

Fig. 2.6 Bernoulli's equation is one form of the law of the conservation of energy developed for isothermal fluid flow. The pressure, velocity and position of air mass particle determine its pressure, kinetic and potential energy, the sum of which remains the same all the time

mass volume with respect to a reference plane (Fig. 2.6). The Bernoulli equation can be written with "energy heights", by dividing all the terms in the equation by the air density ρ and the acceleration due to gravity g:

$$\frac{p_1}{\rho \cdot g} + \frac{1}{2 \cdot g} \cdot v_1^2 + h_1 = \frac{p_2}{\rho \cdot g} + \frac{1}{2 \cdot g} \cdot v_2^2 + h_2 \ [\text{m}]$$

Using Bernoulli's equation, we can explain and evaluate the natural ventilation of buildings and how the air infiltration into buildings is affected by the wind.

2.2.5.2 Principles of the Natural Ventilation of Buildings

According to the Ideal Gas Law, the pressure of air depends on its density, temperature and the gas constant (287 J/kg K for the air). The property of the air that its density depends on the temperature as well as on the humidity (Table 2.7). Moist air is lighter than dry, which affects the circulation of the air mass in the atmosphere. Nevertheless, this is usually not taken into account when the ventilation of the buildings is studied.

Assuming that air has the properties of an ideal gas, the density at an arbitrary absolute temperature (T_2) can be calculated based on the reference temperature and the density. For example, at the absolute temperature $T_1 = 273$ K, the density of air is $\rho_{T1} = 1.29$ kg/m^3:

Table 2.7 Air density at different air temperatures and relative humidities

θ (°C)	air density ρ (kg/m³)		
	φ 0%	φ 40%	φ 60%
-10	1.341	1.341	1.340
0	1.293	1.292	1.291
10	1.248	1.247	1.246
20	1.205	1.200	1.198

$$\frac{\rho_{(T2)}}{\rho_{(T1)}} = \frac{T_1}{T_2} \rightarrow \rho_{(T2)} = \rho_{(T1)}\frac{T_1}{T_2} \rightarrow \rho_{(T2)} = \frac{353}{T_2}\left[\frac{kg}{m^3}\right]$$

When the air temperature in the building and in the exterior is not the same, the hydrostatic pressure of the air in the building and the exterior also differs. As a consequence the gradients of outdoor $p_{e(z)}$ and indoor air pressure $p_{i(z)}$ with respect to the height z ($\Delta p/\Delta z$) are different, as shown in Fig. 2.7. A zone is formed where the air pressure is larger on the external side of the envelope, and the zone where the air pressure is larger on the inner side of the building envelope. The zones are separated by a neutral plane, where the pressure $p_{e(z)}$ and $p_{i(z)}$ are the same. Therefore, there is a difference in pressure on the building envelope, which causes an air mass flow from the exterior into the building and vice versa. This type of air exchange is called natural buoyancy ventilation or airing. By assuming a neutral plane as the reference, the Bernoulli equation for the ventilation opening below the neutral plane is written as (Fig. 2.7, right):

$$p_{1i} + \frac{1}{2} \cdot \rho_{(\theta e)} \cdot v_{1i}^2 + h_i \cdot \rho_{(\theta e)} \cdot g = p_{2i} + \frac{1}{2} \cdot \rho_{(\theta i)} \cdot v_{2i}^2 + h_i \cdot \rho_{(\theta i)} \cdot g$$

Since the volume of the outdoor environment as well as the interior of the building is very large, the air velocity v at points 1i and 2i is equal to 0 m/s. Considering that h_i is measured from the neutral plane and is therefore negative, the difference in the air pressure at level i is equal to:

$$p_{1i} - p_{2i} = (-h_i) \cdot g \cdot \left(\rho_{(\theta i)} - \rho_{(\theta e)}\right) = h_i \cdot g \cdot \left(\rho_{(\theta e)} - \rho_{(\theta i)}\right) \text{ [Pa]}$$

In the wintertime, when $\theta_{ii} > \theta_e$, the difference in pressures ($p_{1i} - p_{2i}$) is positive, which means that air flows through opening from the exterior into the building. Using the Bernoulli equation, the velocity of the air that flows into the building can be determined:

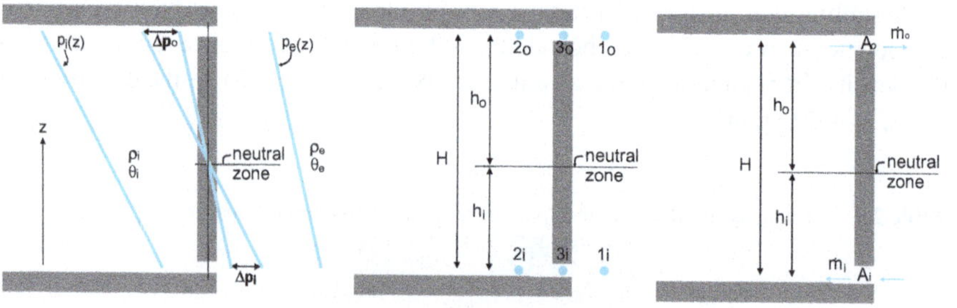

Fig. 2.7 Gradient of the air pressure on the outer and inner sides of the building envelope in the case of $\theta_i > \theta_e$ (left). The air mass-flow directions in case of $\theta_i > \theta_e$ (middle), reference points at level of air inlet (i) and outlet (e) openings (right)

$$p_{2i} - p_{3i} = p_{2i} - p_{1i} = h_i \cdot g \cdot \left(\rho_{(\theta e)} - \rho_{(\theta i)}\right) = \frac{1}{2} \cdot \rho_{(\theta e)} \cdot v_{3i}^2 \rightarrow$$

$$\rightarrow v_{3i} = \sqrt{\frac{2 \cdot h_i \cdot g \cdot \left(\rho_{(\theta e)} - \rho_{(\theta i)}\right)}{\rho_{(\theta e)}}} \left[\sqrt{\frac{m \cdot m \cdot kg \cdot m^3}{s^2 \cdot m^3 \cdot kg}} = \frac{m}{s}\right]$$

With a continuity equation that correlates the mass- and volume-flow rates of fluids, the air mass-flow rate that enters the building m_i is determined using the expression:

$$\dot{m}_i = C_d \cdot \rho_{(\theta e)} \cdot \dot{V}_i = C_d \cdot \rho_{(\theta e)} \cdot A_i \cdot v_{3i} \left[\frac{m^2 \cdot kg \cdot m}{m^3 \cdot s} = \frac{kg}{s}\right]$$

where A_i (m^2) is the cross-section area of the inlet opening and V_i (m^3/s) is the volume air-flow rate. Due to hydraulic losses, only part of the pressure energy changes into the kinetic energy of air particles and because of that the actual air mass flow is lower for the hydraulic loss coefficient C_d. The typical values of the coefficient C_d are between 0.6 and 0.7. The Bernoulli equation can again be used to determine the difference in the air pressure at level o, which is h_o above the neutral plane (Fig. 2.7, right):

$$p_{1o} + \frac{1}{2} \cdot \rho_{(\theta e)} \cdot v_{1o}^2 + h_o \cdot \rho_{(\theta e)} \cdot g = p_{2o} + \frac{1}{2} \cdot \rho_{(\theta i)} \cdot v_{2o}^2 + h_o \cdot \rho_{(\theta i)} \cdot g$$

$$p_{2o} - p_{1o} = h_o \cdot g \cdot \left(\rho_{(\theta e)} - \rho_{(\theta i)}\right) \text{ [Pa]}$$

The difference in the air pressure is positive if $\theta_I > \theta_e$, which means that the air flows from the building through the opening with a cross-sectional area of A_o (m^2) towards the exterior. The air mass-flow rate $\dot{m}_o$ is equal to:

$$p_{2o} - p_{3o} = p_{2o} - p_{1o} = h_o \cdot g \cdot \left(\rho_{(\theta e)} - \rho_{(\theta i)}\right) = \frac{1}{2} \cdot \rho_{(\theta i)} \cdot v_{3o}^2 \rightarrow$$

$$v_{3o} = \sqrt{\frac{2 \cdot h_o \cdot g \cdot \left(\rho_{(\theta e)} - \rho_{(\theta i)}\right)}{\rho_{(\theta i)}}} \rightarrow \dot{m}_o = C_d \cdot A_o \cdot \dot{V}_o = C_d \cdot A_o \cdot \rho_{(\theta i)} \cdot v_{3o} \left[\frac{kg}{s}\right]$$

The unknowns in the above equations are the distance between the neutral plane and the inlet/outlet openings level, i.e., h_i and h_o, which are determined by the law of the conservation of mass. Considering the relationship between density and air temperature, the following expression is derived:

$$\dot{m}_i = \dot{m}_o \rightarrow \frac{h_i}{h_o} = \left(\frac{A_o}{A_i}\right)^2 \frac{\rho_{(\theta i)}}{\rho_{(\theta e)}} = \left(\frac{A_o}{A_i}\right)^2 \cdot \frac{353}{T_i} \cdot \frac{T_e}{353} = \left(\frac{A_o}{A_i}\right)^2 \frac{\theta_e + 273.15}{\theta_i + 273.15}$$

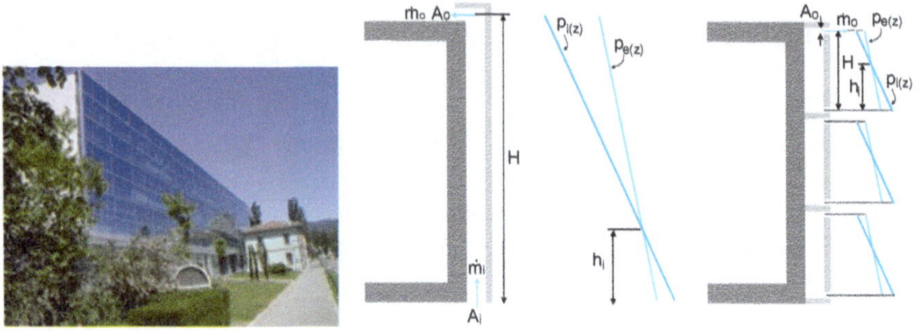

Fig. 2.8 Business building with a ventilated façade. The air-pressure gradients inside a multi-storey and a corridor façade ($p_{i(z)}$), and outside ($p_{e(z)}$)

The height difference between the openings is equal to H and equal to the sum of h_i and h_o.

Case Study Determine the specific volume air-flow rate in a 1-m-wide multi-storey double-skin façade (Fig. 2.8, left and middle) with height H 10.5 m. The multi-storey double-skin façade has an inlet opening with a surface A_i 0.6 m^2 and outlet opening with the surface A_o 0.3 m^2. The second building has corridor façade with opening in each floor and the distance between the openings that are 0.08 m wide ($A_i = A_o = 0.08$ m^2) H is 3.4 m. The outdoor temperature θ_e is 28 °C, and the air temperature in the double and corridor façade zone θ_i is 42 °C. The hydraulic loss coefficient of the openings C_d is 0.6

For the case of a double-skin façade, the distance between the inlet opening and the equilibrium plane h_i and the inlet air velocity v_i are equal to:

$$\frac{h_i}{h_o} = \frac{h_i}{H - h_i} = \left(\frac{A_o}{A_i}\right)^2 \frac{\theta_e + 273.15}{\theta_f + 273.15} \rightarrow h_i = \frac{H \cdot \left(\left(\frac{0.3}{0.6}\right)^2 \frac{301}{315}\right)}{1 + \left(\left(\frac{0.3}{0.6}\right)^2 \frac{301}{315}\right)} = 2.0 \text{ m}$$

$$v_i = \sqrt{\frac{2 \cdot h_i \cdot g \cdot \left(\rho_{(\theta e)} - \rho_{(\theta i)}\right)}{\rho_{(\theta e)}}} = \sqrt{\frac{2 \cdot 2.0 \cdot 9.81 \cdot \left(\frac{353}{301} - \frac{353}{315}\right)}{\frac{353}{301}}} = 1.3 \ \frac{\text{m}}{\text{s}}$$

and the specific air volume flow rate V'_i (per 1 m of the façade width) is:

$$\dot{V}_i' = C_d \cdot A_i \cdot v_i = 0.6 \cdot 0.6 \cdot 1.3 = 0.47 \ \frac{\text{m}^3}{\text{s} \cdot \text{m}}$$

In the case of a corridor façade the height h_i is approx. $H/2$ because the areas of the inlet and outlet openings are the same ($A_i = A_o = 0.08$ m^2/m per m of the façade width). The velocity of the air v_i entering the corridor façade

Fig. 2.9 Natural ventilation and building cooling by ventilation during the summertime can be enhanced with ventilation shafts (left) [Building Research Establishment, UK]. In contrast, to avoid large heat losses and discomfort, double slide or revolving doors are installed in tall buildings and staircases. In this way the inlet air-flow rate is reduced by up to 60% (for double slide doors) and by 98% in the case of revolving doors (right)

through the inlet opening and the specific volumetric flow rate V'_i are:

$$v_i = \sqrt{\frac{2 \cdot h_i \cdot g \cdot \left(\rho_{(\theta e)} - \rho_{(\theta i)}\right)}{\rho_{(\theta e)}}} = \sqrt{\frac{2 \cdot 1.6 \cdot 9.81 \cdot \left(\frac{353}{301} - \frac{353}{315}\right)}{\frac{353}{301}}} = 1.18\ \frac{m}{s}$$

$$\dot{V}'_i = C_d \cdot A_i \cdot v_i = 0.6 \cdot 0.08 \cdot 1.18 = 0.06\ \frac{m^3}{s \cdot m}$$

This case study shows that with so-called shaft ventilation (shaft has enlarged vertical distance between inlet and outlet ventilation openings) the air flow rate is significantly higher and such a technique of buoyancy-driven natural ventilation of the buildings is much more effective.

However, stack ventilation is not always desirable. For example, staircases in tall buildings are closed with double slide or revolving doors in order to reduce the unwanted infiltration of air. This reduces the heat losses and improves the indoor thermal comfort (Fig. 2.9).

2.2.6 *The Impact of the Wind on the Ventilation of Buildings*

Air is a viscous fluid. Due to the friction between the particles of air mass at different velocities and vortexes, part of the kinetic energy of the air mass dissipates into the heat. In nature, it is assumed that air mass particles stick to the solid surfaces, and their air velocity is 0 m/s. When increasing the distance from the ground, the frictional forces decrease and the wind speed increases in the Ekman layer. The wind velocity with respect to the elevation can be modelled taking into account the surface-roughness coefficient $\alpha(1)$ also known as the Hellmann exponent:

$$\left(\frac{v_{(z)}}{v_{10}}\right) = \left(\frac{z}{10}\right)^{\alpha}$$

where v_{10} (m/s) is the wind velocity measured by meteorologists at a height of 10 m, $v_{(z)}$ (m/s) is the wind speed at an arbitrary height z (m) above the ground (Fig. 2.10). The Hellmann exponent αdepends on the roughness of the terrain, as shown in Table 2.8.

As the air flows around the building, the zones of high (over-pressure) and low (under-pressure) air pressure are formed in the vicinity of the building. The largest pressure occurs along the building-envelope surfaces, where the wind speed decrease to 0 m/s and the whole kinetic energy of air particles is converted into pressure energy. A stagnation pressure Δp_t (Pa) at the building envelope is formed in such conditions. It can be determined by Bernoulli's equation:

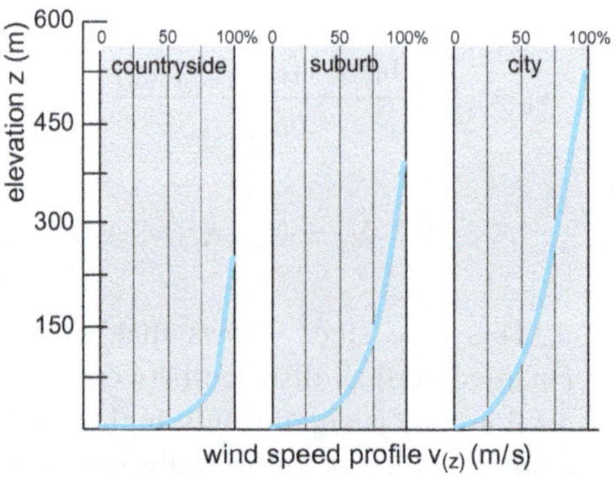

Fig. 2.10 Wind speed $v_{(z)}$ increase with respect to the height z depends on the surface-roughness coefficient α, known as the Hellmann exponent [3]

Table 2.8 Helmann's surface-roughness coefficient a for different types of surfaces found in nature and the built environment. Note that the values of α are informative and depend on several parameters, such as topology and atmosphere stratification stability [3]

surface type	surface roughness (1)
water body	0.10
meadow	0.13
forest	0.20
settlement	0.25
city with tall buildings	0.31

Table 2.9 Stagnation pressure Δp_t on the building envelope at different wind speeds in open space in the surrounding of the building $v_{(z),ref}$ (left). Stagnation pressure Δp_t on the building envelope at different heights z at a reference wind speed $v_{(10),ref}$ 25 m/s and surface roughness coefficient α 0.25 (right)

wind speed (m/s)	wind speed (km/h)	stagnation pressure Δp_t (Pa)
0.5	1.8	0.4
1	3.6	0.7
2	7.2	2.6
5	18	16
10	36	65
20	72	258
30	108	580

building height (m)	wind speed (m/s)	stagnation pressure Δp_t (Pa)
5	21	65
10	25	100
20	29.7	156
50	37.4	241
100	44.5	375

$$p_1 + \frac{1}{2} \cdot \rho \cdot v_{(z),1}^2 + z_1 \cdot \rho \cdot g = p_2 + \frac{1}{2} \cdot \rho \cdot v_{(z),2}^2 + z_2 \cdot \rho \cdot g \text{ [Pa]}$$

$$p_2 - p_1 = \Delta p_t = +\frac{1}{2} \cdot \rho \cdot v_{(z),1}^2 + \underbrace{\frac{1}{2} \cdot \rho \cdot v_{(z),2}^2}_{0} + \underbrace{z_2 \cdot \rho \cdot g - z_1 \cdot \rho \cdot g}_{0} \text{ [Pa]}$$

The stagnation pressures Δp_t at different open space air speeds $v_{(z),ref}$ are given in Table 2.9 (left). At wind speeds over 1 m/s, the stagnation pressure is greater than the pressure due to the thermal buoyancy (at common difference between indoor and outdoor air temperatures during the winter), and the wind has a dominant impact on the infiltration of the air into the building. The stagnation pressure also increases with the building height as wind speed increases by elevation (Table 2.9, right).

Looking at the closer surroundings of the building, the high-pressure zones around the building envelope are formed on the wind-exposed upwind side and the low pressure on the lee side of the building envelope. The actual pressures on the individual part of the building envelope are determined by the coefficient of wind pressure C_p. C_p and has a value between −1 and + 1. The value of the C_p is positive in high-pressure areas, where the wind causes intensive air flow into the building and equal to + 1 if the actual pressure is equal to the stagnant pressure Δp_t. In contrast, the C_p is negative if the wind causes the suction of air from the building.

The pressure distribution around the buildings is determine by experiments and numerical computing methods. Experiments are performed in wind tunnels, where the air is moved by fans and the pressures and speeds are measured at different point around a scale model of building or settlement. Tracer gas is often added into the air flow in order to form visible streamlines (Fig. 2.11). The areas of dense streamlines are high-pressure areas, while thinner streamlines are characteristic of pressure lows.

The wind-driven infiltration of air into the building is undesirable because air-exchange rates cannot be controlled. This leads to higher energy demands of the building and can cause living discomfort. The exception is natural cooling with night-time ventilation, which is an efficient technique for decreasing the energy demand for cooling. To avoid problems caused by high wind-driven infiltration, the building

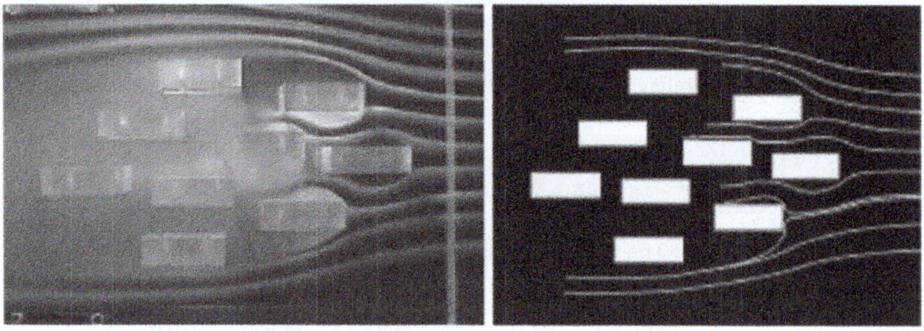

Fig. 2.11 Visualisation of streamlines in a model of a settlement of nine buildings using tracer gas. The areas of dense streamlines are high-pressure areas, and the areas with thinner streamlines are areas of low pressure (left). The streamlines are determined by computational fluid dynamics (CFD) simulations. The match is evident and the accuracy of the CFD tools can be seen

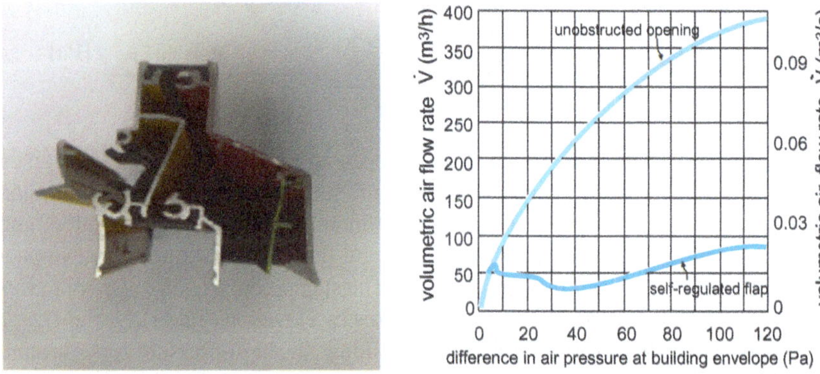

Fig. 2.12 "Aerodynamic flap" built in a ventilation grill (left). The volume air-flow rate that enters the building by infiltration through a ventilation grill with or without a control element according to the difference in the air pressure between the external environment and the building (right)

envelope must be designed with a wind barrier and built tight, while ventilation openings should be equipped with self-regulated elements, as shown in Fig. 2.12.

2.2.7 Complex Methods for the Prediction of Ventilation of Buildings

In the case study of a double-skin façade shown in Fig. 2.8, the air temperature in the façade was assumed to be a constant and at a pre-known value. However, the air temperature in the ventilated façade and the volume air flow rate in the façade are mutually dependent, because the ventilation is driven by the thermal buoyancy. Therefore, computer tools must be used for a detailed assessment of the natural

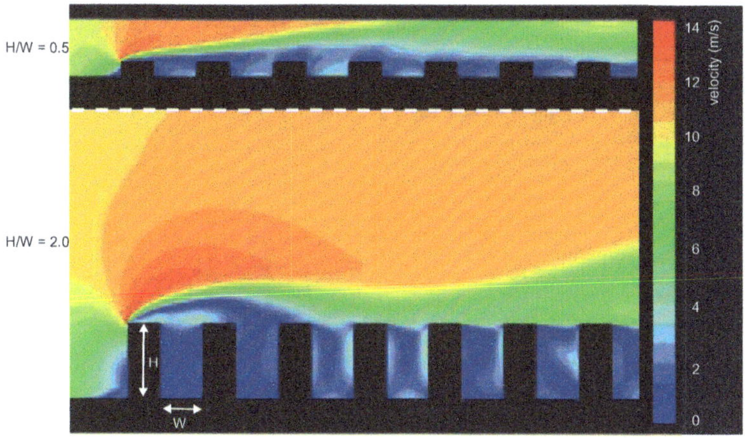

Fig. 2.13 CFD simulation of air flow in settelments consist of six street canyons with building height/street width ratio (H/W) of 0.5 and 2.0, at the reference wind speed (before the settlement) $v_{(10)}$ 10 m/s; velocity field in street canyons are shown [11]

ventilation of the buildings and, in particular, the cooling of complex buildings with natural ventilation. Such computer tools are based on the numerical solution of differential equations, based on the laws of the conservation of mass, energy and momentum. The solution procedure begins by creating a geometrical model, e.g., of a ventilated façade, building or settlement, and the steady or transient boundary conditions, for example, the profile of the wind speed in front of the building or settlement or the time-changed temperatures on the insulated surfaces of the building envelope. The problem can be considered as planar (2D) or spatial (3D). In both cases, the geometrical model is divided into a large number of small control volumes. For each control volume, the computer tool numerically determines the temperature, velocity, pressure and the amount of kinetic energy converted into heat in the case of the turbulent flow, or just the temperature if the control volume represents a solid material (Fig. 2.13).

This method of solving is called computational fluid dynamics (CFD). The development of computer hardware and software over the past decade made it possible to use these tools with basic engineering knowledge.

2.2.8 *Air Permeability of Building Structures*

The building envelope is never absolutely tight, and because of that air-mass infiltration is always present to some extent. One of the reasons is that most building materials are porous (Fig. 2.14). The porosity of building materials is measured by the permeability k (m^2). The permeability of some construction materials is presented

Fig. 2.14 Ventilated façade with a wooden cladding. A wind barrier is laid on top of the thermal insulation. The barrier must be water-vapour permeable to enable the transfer of water vapour without interstitial condensation (left). To ensure adequate air tightness of the building, all the joints between the doors or windows and the walls must be weatherstripped (right)

Table 2.10 Air permeability k of some building materials

construction material	air permeability k (m²)
brick	$6.8 \cdot 10^{-14}$
aerated concrete	$1.4 \cdot 10^{-12}$
concrete	$4.4 \cdot 10^{-15}$
lime mortar	$2.8 \cdot 10^{-14}$
fiber thermal insulation	$1.3 \cdot 10^{-10}$

in Table 2.10. The volume air flow rate $\dot{V}$ (m³/h) that flows through the layer of porous materials is determined by the expression:

$$\dot{V} = \frac{k}{\mu \cdot d} \cdot A \cdot \Delta p_{e-i} \cdot 3600 \left[\frac{m^2 \cdot m \cdot s \cdot m^2 \cdot Pa \cdot s}{kg \cdot m \cdot h} = \frac{m^3 \cdot m^2 \cdot Pa}{N \cdot h} = \frac{m^3}{h} \right]$$

where d (m) is the thickness of the building material layer, A (m²) its surface area and $\Delta p_{e\text{-}i}$ (Pa) is the difference in the air pressure on the opposite surfaces of the layer. The quantity μ (kg/ms) is a property of air, known as the dynamic viscosity. It depends on the temperature and is equal to 17×10^{-6} kg/ms at an average winter ambient temperature.

Case Study Determine the amount of air that passes through a layer of 150-mm-thick fibrous heat insulation with a surface of 100 m² in one hour. The difference in air pressure between the exterior and interior of the building.

$\Delta p_{e\text{-}i}$ is 10 Pa.

$$\dot{V} = \frac{k}{\mu \cdot d} \cdot A \cdot \Delta p_{e-i} = \frac{1.3 \cdot 10^{-10}}{17 \cdot 10^{-6} \cdot 0.15} \cdot 100 \cdot 10 \cdot 3600 \frac{s}{h} = 183 \left[\frac{m^3}{h} \right]$$

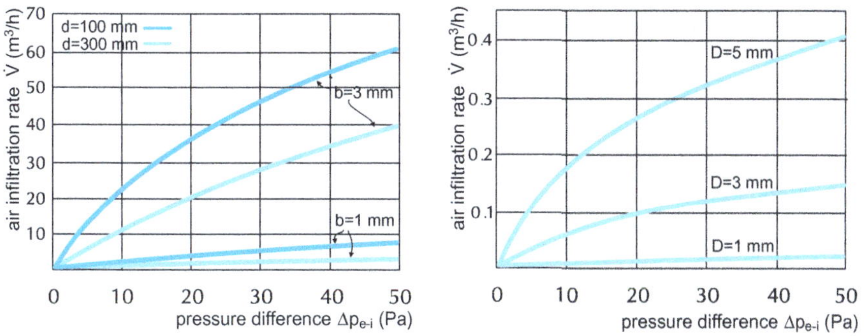

Fig. 2.15 Specific volume air flow through a 1 m lengh narrow crack with width b (mm) in otherwise air tight layer of thickness *d* (cm) (left); volume air flow rate through whole with diameter *D* (mm) in air tight wind barrier foil (right) [12]

The case study shows that a building structure consisting of porous materials must have a wind barrier to minimize the air infiltration. The wind barrier is a thin foil made from a material with a low air permeability. Air may also infiltrate the building through the joints between the doors/windows and the building structure, or through cracks in the wind barrier. The volume air flow rate $\dot{V}'$ (m^3/hm) that passes otherwise in an air-tight building structure with thickness *d* (mm) through a 1-m-long narrow crack of width *b* (mm) at difference in air pressure $\Delta p_{e\text{-}i}$ (Pa) is shown in Fig. 2.15 (left), while Fig. 2.15 (right) shows the volume air flow rate *V* (m^3/h) that passes the hole with diameter *D* (mm) in a damaged airtight wind barrier.

In the past, most residential buildings were ventilated naturally through joints between the window frame and the window sash, which were deliberately not made absolutely airtight. The air leakage through the windows (and doors) is evaluated by the airtightness factor *a* (m^3/hmPa$^{2/3}$), indicating the volume air flow rate $\dot{V}$ (m^3/h) passing through a 1-m-long joint at a predefined pressure difference. The air flow rate also depends on the mode of flow through the joint, which can be either laminar or turbulent. Therefore, the pressure difference $\Delta p_{e\text{-}i}$ is weighted with the exponent *n*. Typical values of the exponent *n* are between 0.4 and 1, and the default value of 2/3 is used in the calculations. The amount of air that passes into the building through all the window and door joints is determined by the expression:

$$\dot{V} = \sum_{j=1}^{m} a_j \cdot l_j \cdot \Delta p_{e-i}^{n} = \sum_{j=1}^{m} a_j \cdot l_j \cdot \Delta p_{e-i}^{2/3} \left[\frac{m^3}{m \cdot h \cdot Pa^{2/3}} m \cdot Pa^{2/3} = \frac{m^3}{h} \right]$$

where *j* is the numerator of the elements with joints in the building envelope, l_j is the length of the joint in the *j*-th element and m is total number of joints.

2.2.9 *Requirements on the Airtightness of Buildings*

The requirements concerning the airtightness of buildings cover the requirements of airtightness of windows and doors (Fig. 2.16), which are tested in laboratory conditions as part of the product's certification, and the requirements regarding to the airtightness of the building as a whole, which can only be tested on completed buildings.

2.2.9.1 Airtightness of Windows and Doors

Windows and doors are classified into air-permeability classes. These classes are determined by experiments performed in the laboratory measuring the volume air flow rate $\dot{V}$ that passes a closed window or a door built in an air-tight wall at different pressure differences $\Delta p_{e\text{-}i}$ on the opposite side of the tested element. The maximum pressure difference established during the test is 600 Pa, which corresponds to the stagnation pressure caused by the wind at a speed of ~ 32 m/s (~114 km/h) in the surrounding of the building. The air permeability class is determined at the reference pressure difference $\Delta p_{e\text{-}i}$ 100 Pa according to the specific quantity of air passing through the window or the door, expressed per 1 m^2 of element surface area ($V_{100,A,\max}$) or per 1 m of joint length ($V_{100,l,\max}$) (Table 2.12).

Explanation Subject to the regulations on the energy efficiency of buildings in Slovenia, only windows and doors with class 2 (for single- and two-storey buildings) or class 3 (taller buildings) airtightness must be installed because the wind speed (and stagnation air pressure) increases with the height above the terrain (Fig. 2.17).

Fig. 2.16 Window with three planes of sealing. Such windows should be built in energy-efficient buildings

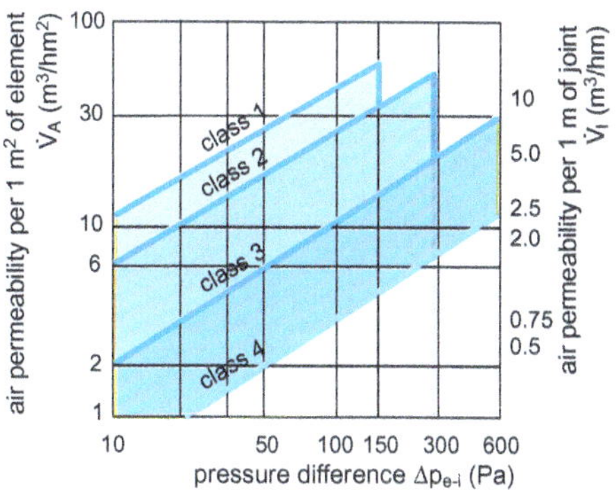

Fig. 2.17 Amount of air passing through the joints of the door or window at pressure differences $\Delta p_{e\text{-}i}$ expresed per 1 m^2 of element area or per 1 m of joint between the frame and sash lengh in one hour. Values define the airtightness class of the window or door at a pre-defined pressure difference (100 Pa) [13]

For any given difference in the air pressure between the exterior and interior of the building, the amount of air that passes into the building through a closed window or door is determined by the expressions (Table 2.11).

$$\dot{V} = \dot{V}_{100,l} \cdot l \cdot \left(\frac{\Delta p_{e-i}}{100}\right)^{2/3} \left[\frac{m^3}{h}\right]$$

$$\dot{V} = \dot{V}_{100,A} \cdot A \cdot \left(\frac{\Delta p_{e-i}}{100}\right)^{2/3} \left[\frac{m^3}{h}\right]$$

Table 2.11 Airtightness factor *a* for different windows and door sealing technologies

window or door design	a ($m^3/h \cdot m \cdot Pa^{2/3}$)
unsealed windows unsealed doors	0.6
windows with a single seal	0.3
windows with multiple seals	0.2
limit values achievable with current joint sealing technologies	0.01–0.05

Table 2.12 Airtightness classes for doors and windows, maximum allowable air permeability of windows $V_{100,A,\max}$ and $V_{100,l,\max}$ for different air-tightness classes [13]

airtightness class	maximum test pressure difference $\Delta p_{e\text{-}i,max}$ (Pa)	maximum air permeability $V_{100,l,max}$ at Δp 100 Pa ($m^3/h{\cdot}m{\cdot}Pa^{2/3}$) (expressed per 1 m of joint length)	maximum air permeability $V_{100,A,max}$ at Δp 100 Pa ($m^3/h{\cdot}m^2{\cdot}Pa^{2/3}$) (expressed per m^2 of building block)
1	150	12.5	50
2	300	6.75	27
3	600	2.25	9
4	600	0.75	3

Case Study The tenants of an apartment in an older building replaced their windows as part of an energy renovation. Old timber frame windows without gaskets were replaced with class-3 airtight windows. Evaluate whether the amount of the fresh air supplied in the case of naturally ventilated living room with two windows with an area 1 m^2 after renovation is still adequate. Assume that one person is occupying the room, so the amount of the fresh supply air of 30 m^3/h is needed to fulfil the IAQ requirements and that the pressure difference $\Delta p_{e\text{-}i}$ is 15 Pa.

Let us assume that the windows are square. Thus, the total length of the joints between the frame and the sash of the windows Σl is ~ 8 m. Old windows without gaskets had an air permeability of 0.6 $m^3/hmPa^{2/3}$ (Table 2.12). The airtightness $\dot{V}_{100,A}$ of new windows corresponds to the allowed minimum value for class 3. According to Fig. 2.17, 10.5 m^3/m^2 h of air will pass the window per 1 m^2 of area and per hour. Before renovation, the volume air flow rate was equal to:

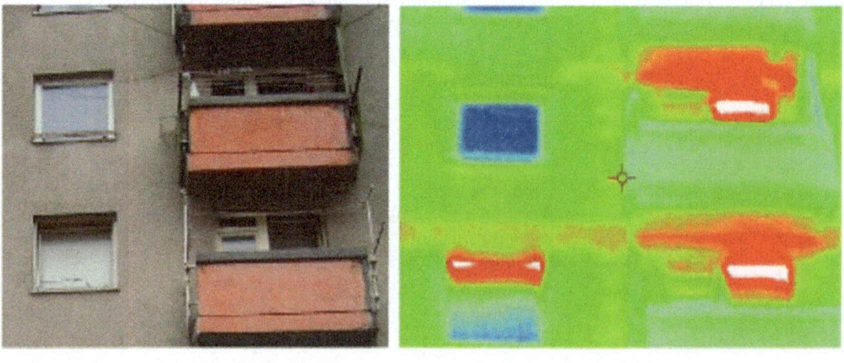

Fig. 2.18 Thermal image of the building confirms the findings from the case study. After the replacement of the windows, the tenants are very often leaving windows open in bathrooms over the night. IR image confirms high heat losses due ventilation in such case [14]

$$\dot{V} = \sum_{1}^{i} a_i \cdot l_i \cdot \Delta p_{e-i}^{n} = 0.6 \cdot 8 \cdot 15^{2/3} = 29.2 \frac{m^3}{h}$$

The amount of the supply air for ventilation in the case of old (closed) windows was approximately sufficient to maintain the IAQ. After replacing the windows, the volume air flow rate decrease to the amount:

$$\dot{V} = \dot{V}_{100,A} \cdot A \cdot \left(\frac{\Delta p_{e-i}}{100}\right)^{2/3} = 10.5 \cdot 2 \cdot \left(\frac{15}{100}\right)^{2/3} = 6.0 \frac{m^3}{h}$$

The amount of fresh air for the ventilation of living room after replacing the windows is no longer sufficient for adequate IAQ (Fig. 2.18). Because of that, the new windows must be opened frequently, additional ventilation openings must be built or a mechanical ventilation system must be installed.

2.2.9.2 Airtightness of Buildings

The airtightness of buildings has become a measure of the energy efficiency of buildings in recent years. The regulations most often define the maximum permissible air-change rate per hour for a 50-Pa pressure difference between the exterior and interior of the building, designated as $n_{50,max}$. For example, in naturally ventilated residential buildings, n_{50} should not be greater than 3 h^{-1} and for buildings with a mechanical ventilation system, $n_{50,max}$ must not exceed 2 h^{-1}. The requirements for passive houses are even more rigorous—$n_{50,max}$ may not exceed 0.6 h^{-1}.

Explanation ASHRAE—the American Society of Heating, Refrigerating and Air Conditioning Engineers demands mechanical ventilation for all buildings that have a n_{50} of less than 3 h^{-1}, as natural ventilation alone is unable to ensure the required indoor air quality.

In order to validate the airtightness of buildings, an experimental method has been established based on the measurement of the air flow rate between the exterior and interior of the building, with an increased difference between the indoor and outdoor air pressures. The difference is established by means of a fan installed in the building envelope, usually into the doorway, together with a device that seals the same opening. Hence, the technique was named the "blower-door" method. The conditions that must be fulfilled during the test are stated in [15]. By adjusting the operation of the fan, a pressure difference is established in the range of ~ 20 to 80 Pa. Because the permeability of the building envelope may depend on the direction of the air flow, the airtightness test is conducted twice, by establishing the over and under air pressure

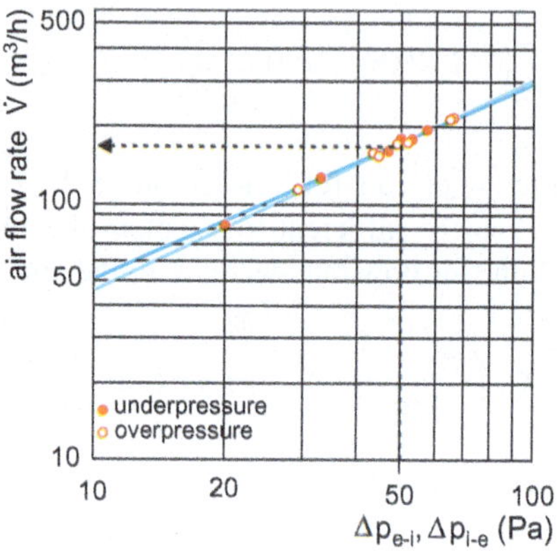

Fig. 2.19 Result of the airtightness test presented in a log–log diagram showing the measured volume air-flow rate V in the range of overpressure and underpressure air pressure differences 35 to 65 Pa. The reference air exchange rate n_{50} at 50 Pa pressure difference is determined from the approximation line. The diagram shows the airtightness test results for a passive building with a volume of 630 m^3. The reference air exchange rate n_{50} is 0.275 h^{-1} $-$170 (m^3/h)/630 (m^3/h). The building fulfils the requirements for passive buildings ($n_{50,max}$ 0.6 h^{-1})

in the interior of the building, relative to the external air pressure (Fig. 2.19). Once the pressure difference is steady, the air flow rate through the building envelope is the same as the air flow rate generated by fan. Taking into account the space volume of the building, the air-exchange rate is determined. By interpolation of the experimental data, the reference value n_{50} at the pressure difference of 50 Pa is determined. During the experiment, places of air leakage are located with smoke or by measuring the local air velocity, as shown in Fig. 2.20.

2.3 Ventilation Heat Losses and Energy Efficiency of Buildings

The energy consumption of the world is increasing, even though the current energy supply and demand are not sustainable, leading to climate change. In 2018, the EU countries converted 18,030 TWh of primary energy, mostly from fossil fuels (Fig. 2.21). Primary energy represents the amount of natural energy sources that were transformed to provide the needed amount of energy carriers to the users. The energy available to the users is classified as final energy. In the structure of the final energy (in total 13,072 TWh) in EU households (Fig. 2.22), heat demand represents almost 2/3 (64%) of all the final energy consumed, followed by domestic hot-water heating (14.8%) and electricity with a 14.1% share. In total, more than one-quarter of all final energy in the EU Member States is converted in households (27.2%). The development of new materials, advance construction technologies and highly

Fig. 2.20 The blower and the door panel system are sealed into the doorway, overpressure and underpressure airtightness test is done by rotation of the device (above); the untight point are detected by smoke generator or by velocity sensor; at pressure difference of 50 Pa, the air velocity on the site of the lequage can reach several m/s (below)

efficient smart building service systems, as well as a number of European directives on the energy performance of buildings, led to a significantly reduced consumption of heat and electricity in new and renovated buildings during the past decade.

Efficient energy use in buildings is enforced by implementing different measures, requirements and energy-efficiency indicators:

- with requirements related to the properties of buildings, such as the thermal transmittances of building structures U, the average thermal transmittance of the building envelope H'_t, a building's airtightness n_{50}, and an adequate shading of the transparent parts of the building envelope during the summer.

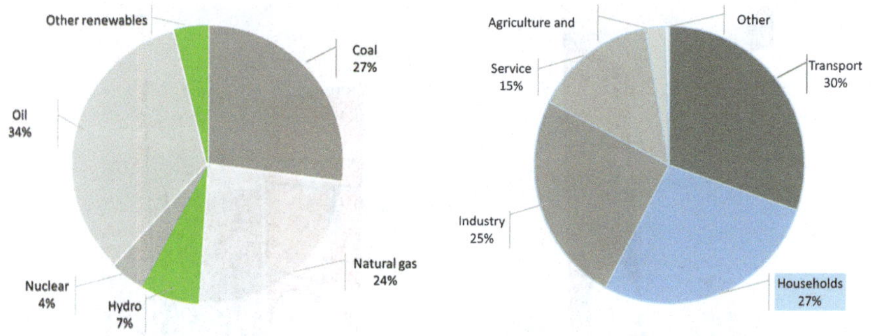

Fig. 2.21 World primary energy demand by energy sources in the year 2013 was 161,248 TWh (13,864 Mtoe), mostly from fossil fuels—gas, oil and coal (left); in the households, 27.2% of final energy was consumed in the year 2018 in EU28 Member states (right) [16]

Fig. 2.22 Final energy consumption in households in the EU Member state countries in the year 2016 by energy carriers [16]

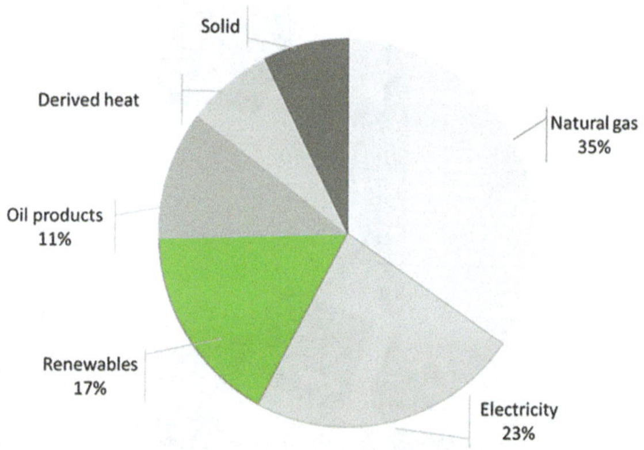

- efficient low-exergy building service systems, including natural and mechanical space heating and cooling systems, hybrid and mechanical ventilation with heat recovery and air-conditioning systems incorporating techniques of free cooling such as evaporative cooling, sanitary hot-water systems with heat recovery, electrical lighting adapted and controlled according to the daylight and smart control systems, including wireless communications and self-adjusted control algorithms based on the weather forecast and self-evaluated patterns of building use [17].
- by introducing technologies for the utilization of renewable energy sources, in particular solar energy, including the solar heating and cooling systems and the photovoltaic systems for on-site electricity production (Fig. 2.21).
- smart community energy-supply systems developed on the base of a sustainable municipality and federal energy policy that enable the integration of dispersed generators of heat, cold and electricity into smart, distant community systems and the optimized control of energy carriers' production and demand.

2.3.1 Energy-Efficiency Metric of Buildings

Awareness of the impact of energy use for the operation of building service systems on the environment, causing local problems related to human health, loss of biodiversity and global climate changes during the past 50 years, have led to increasing demands on the energy efficiency of buildings. Several indicators have been introduced for the validation of building design in engineering practice. Although the requirements differ among the countries, the most common implemented indicators are [18] (Fig. 2.23):

- the thermal transmittance of building structures U (W/m^{2}K) (Sect. 1.5.1) that consist of the building's thermal envelope, must be below the maximum permitted values (Fig. 2.24);
- the average thermal transmittance of building envelope H'_T (W/m^2 K) (Sect. 1.5.7) of a building that is thermally enveloped towards exterior or unheated spaces, must be lower than the permitted value (Fig. 2.24).

Fig. 2.23 Vacuum tube solar thermal collectors on the flat roof of a technology centre, flat solar thermal collectors on the roof of an inn (above), a solar thermal system for the heating of domestic hot water on the flat roof of a elderly-people's home, and a photovoltaic system on a two-family building. These technologies enable the on-site production of the heat and electricity needed for the operation of the buildings

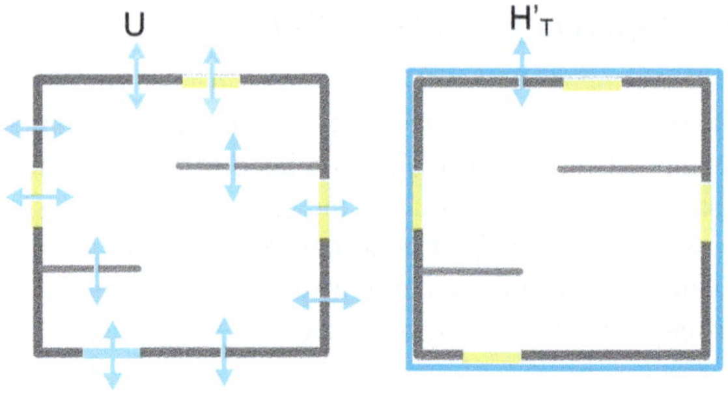

Fig. 2.24 Thermal transmittances U of building structures on the thermal envelope (left) and specific transmittance of building envelope H'_T (=area weighted average U of structures on the building envelope) (right) are the fundamental indicators of the energy efficiency of buildings

- the building envelope must be airtight. It is tested in-situ and expressed by the air-exchange rate for the air-pressure difference between the building and the exterior of 50 Pa n_{50} (h^{-1}) (Sect. 2.2.9.2);
- the transmission heat-transfer coefficient H_T (W/K). The transmission heat-transfer coefficient is determined using the expression:

$$H_T = \sum_{i=1}^{l} \overbrace{b_i \cdot U_{env,i} \cdot A_{env,i}}^{\text{envelope}} + \overbrace{H_{gr}}^{\text{ground}} + \sum_{j=1}^{m} \overbrace{\Psi_j \cdot l_j}^{\text{2D thermal bridges}} + \sum_{k=1}^{n} \overbrace{\chi_k \cdot n_k}^{\text{3D thermal bridges}} \quad [\text{W/K}]$$

where b_i (-) is the heat-loss correction factor for building structures equal to 1, except for the structures towards unheated spaces ($b < 1$) or in the case of a heated building structure (e.g., floor heating, $b > 1$), $U_{env,i}$ (W/m^2 K) and $A_{env,i}$ (m^2) are the thermal transmittance and surface area of the ith building envelope structure, H_{gr} (W/K) is the heat-transfer coefficient of the structures in contact with the ground (Sect. 1.5.5.1), Ψ_j (W/mK) and l_j (m) are the thermal transmittance of the j-th linear (2D) thermal bridge and its length and χ_k (W/K) and n_k (−) are the thermal transmittance of the kth point (3D) thermal bridge and the number of heat bridges (Sect. 1.5.6.1);

- ventilation heat-transfer coefficient H_V (W/K). *This* is measure of THE heat losses due to the air exchange between the exterior and interior of the building as a consequence of the air infiltration through leaking parts of the envelope and the designed natural or mechanical space ventilation. In the case of a naturally ventilated building, $H_{V,n}$ is determined by the expression:

$$H_{V,n} = \overbrace{0.34}^{\rho_{air} \cdot c_{p,air}} \cdot \left(\overbrace{n_{inf} \cdot V_n}^{\text{infiltration part}} + \overbrace{\underbrace{n \cdot V_n}_{\dot{V}}}^{\text{ventilation part}} \right) \left[\frac{W}{K} \right]$$

Table 2.13 Building airtightness expressed by the air-change rate n_{50} (h^{-1}) depends on the airtightness of the façade structures, the joints between structures and openings, and the air tightness of the windows and doors [15]

building airtightness category	n_{50} (h^{-1})
passive buildings	0.6
very airthight buildings with mechanical ventilation	1.5
airtight buildings with natural ventilation	3
older buildings with airthight facade and less airtigth windows without sealing	4
old buildings with moderate airtight facade and windows	6
old buildings with untight facade and windows	10

where V_n (m^3) is the net building volume (approx. $0.8 \times V_e$), n_{inf} (h^{-1}) is the air-change rate due to infiltration and $\dot{V}$(m^3/h) is ventilation air-flow rate. In the early design phase n_{inf} can be approximated as (1/10 to 1/20) × n_{50} (h^{-1}) (Table 2.13), while n (h^{-1}) is determined by the IAQ requirements.

If a heat exchanger is built into the mechanical ventilation system, the heat flux is transferred from the discharged to the fresh supply air (during winter) and the ventilation heat-loss coefficient $H_{V,m}$ is reduced in proportion to the temperature efficiency of the heat exchanger η_{rec} (Fig. 2.25):

$$H_{V,m} = 0.34 \cdot \left(\overbrace{n_{inf} \cdot V_n}^{\text{infiltration part}} + \overbrace{(1 - \eta_{rek}) \cdot \dot{V}}^{\text{mechanical ventilation part}} \right) \left[\frac{W}{K}\right]$$

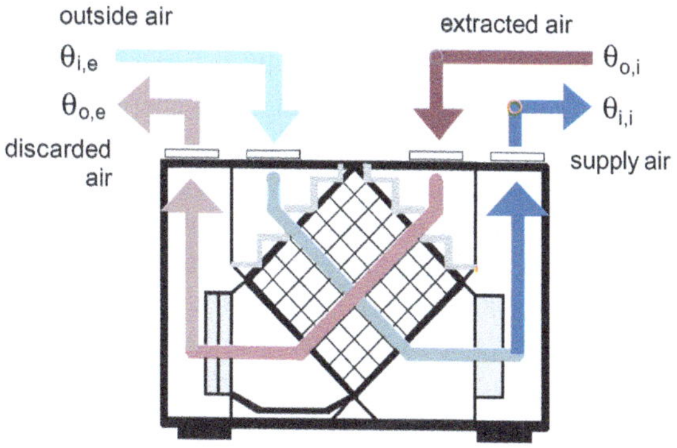

$$\eta_{rec} = \frac{\theta_{i,i} - \theta_{i,e}}{\theta_{o,i} - \theta_{i,e}} [1] \times 100 [\%]$$

Fig. 2.25 Temperature efficiency of heat recovery unit in ventilation system is defined by inlet and outlet air temperatures

- the heat load of the space heating system Φ_{NH} (W) is used for the design of the thermal power of the heat generator:

$$\Phi_{\mathrm{NH}} = (\mathrm{H_T} + \mathrm{H_v}) \cdot \left(\theta_\mathrm{i} - \theta_{\mathrm{e,p}}\right) [\mathrm{W}]$$

the heat load normalized per unit of temperate space useful area A_u (m^2) Φ'_{NH} (W/m^2) is a commonly used indicator of heat losses, e.g., for passive buildings [52] Φ'_{NH} must not exceed 10 W/m^2. The normalized cooling load Φ'_{NC} (W/m^2) can be determined and used for the design of the cooling power of the cold generator.

Explanation A space heating system with radiators can cover a heat load between 80 and 100 W per m^2 of A_u, the floor space heating 50 W/m^2, the thermal activated structures 30 W/m^2, while the cooling capacity of the air-conditioning system is 80–120 W per m^2 of A_u, the ceiling cooling 50–80 W/m^2 and the floor-space cooling system 30 W/m^2.

- energy needs for space heating, space cooling, domestic hot-water heating, ventilation, lighting and if applied, for air conditioning as well, must be determined according to the Energy Performance Building Directive EPBD [19–21]. Energy needs are calculated by a pre-selected category of indoor comfort (see Sect. 1.8.2) and with the assumption that the energy efficiency of all the appliances and technical building systems installed is assumed to be 100%, which means no heat losses are generated and no auxiliary energy is needed for the operation of the systems. The energy needs for heating ($Q_{H,nd}$), cooling ($Q_{C,nd}$) and ventilation ($Q_{V,nd}$) depend on the architecture of the building, the thermal quality and the air tightness of the building envelope, the type of ventilation and the solar and internal heat gains. The energy needs for domestic water heating ($Q_{W,nd}$) depend on the usage profile according to the type of building; meanwhile the energy needs for lighting (W_L) depend on the required illumination level and the quality of the daylighting. The energy needs for humidification ($Q_{HU,nd}$) and de-humidification ($Q_{DHU,nd}$) depend on the climate conditions and internal water-vapour sources. The energy needs are determined by taking into account the indoor-comfort requirements according to the type of the building. The energy needs are presented as specific values expressed per 1 m^2 of a building's useful floor area (kWh/m^{2}a). In countries where the energy demand for heating prevails, a specific value of the energy needs for heating ($Q'_{H,nd}$ in kWh/m^{2}a) (Fig. 2.26) is used for defining the building's energy-efficiency class (A to G) (Table 2.14), which is in fact the most recognizable indicator of the building's energy efficiency in general.
- the delivered (final) energy is expressed with the amount and the type of consumed energy carriers needed for the operation of the building's technical service systems

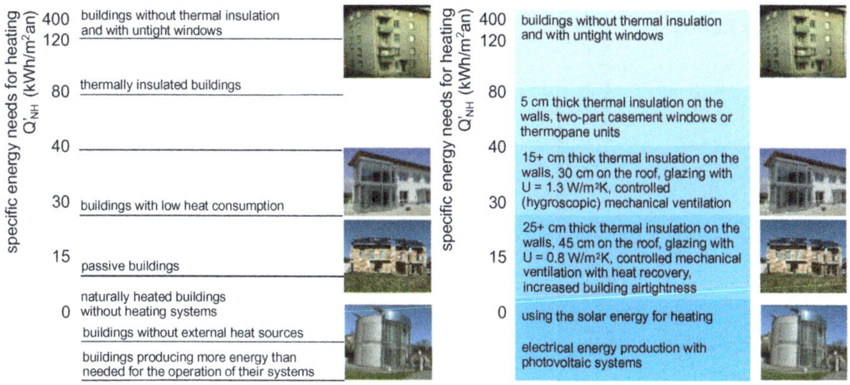

Fig. 2.26 Following the requirements of the EPBD, the specific energy needs for heating were adopted as the indicator for the public presentation of the energy efficiency of buildings. This indicator is also employed in the definitions of the various building categories, such as low-energy buildings ($Q'_{H,nd}$ ~ 25 kWh/m^{2}a) and passive buildings ($Q'_{H,nd}$ < 15 kWh/m^{2}a). The indicator is defined as the ratio of the energy needs for heating $Q_{H,nd}$ to the building's useful surface area A_u (kWh/m^{2}a)

Table 2.14 Building energy efficiency classes determined on the basis of energy needs for heating $Q_{H,nd}$ [19, 22]

energy efficiency class	delivered energy for heating Q'_{NH} (kWh/m^2 a)
A1	up to 10
A2	between 10 and 15
B1	between 15 and 25
B2	between 20 and 35
C	between 35 and 60
D	between 60 and 105
E	between 105 and 150
F	between 150 and 210
G	over 210

over the period of 1 year. The delivered energy is established separately for each of the installed building technical service systems, taking into account the energy needs, the energy losses of the generation and the distribution of the heat, cold or electricity, controlling of the systems as well as the use of auxiliary energy ($W_{AUX,f}$) needed for the operation of the building service systems. In this stage the efficiencies of the building service system's components, as stated by the manufacturer on the basis of standardized tests, are taken into account. The delivered energy for space heating ($Q_{H,f}$), space cooling ($Q_{C,f}$), ventilation ($Q_{V,f}$), preparation of domestic hot water ($Q_{W,f}$), lighting ($W_{L,f}$), air-conditioning ($Q_{DHU,f}$, $Q_{HU,f}$) must be assessed. Most often there is no limitation on the delivered energy demand in national legislation, but the components of systems must fulfil the minimum efficiency criteria. The requirements of energy efficiency for most commonly used appliances are presented in the EcoDesign Directives [23].

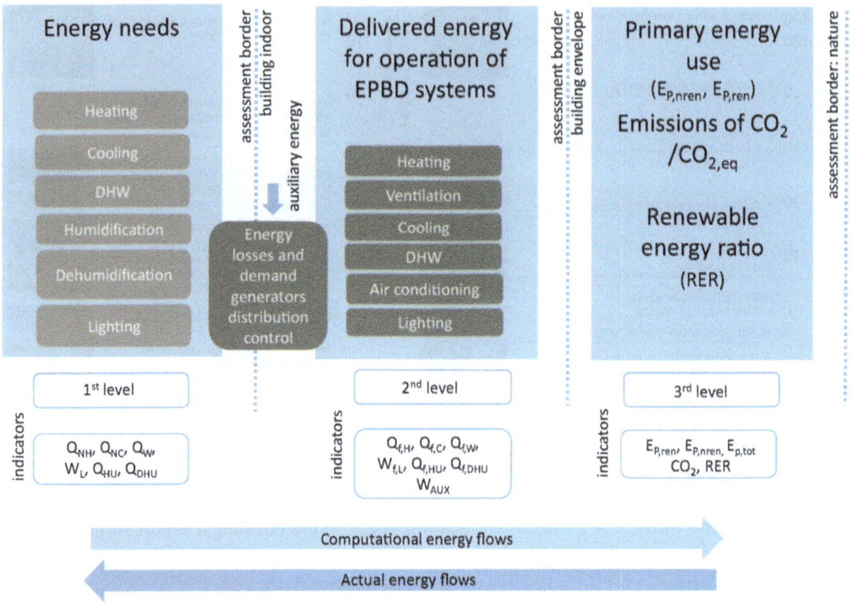

Fig. 2.27 Energy performance of the building is determined on three levels by energy balance carried out in the opposite direction according to actual energy flows

- once the type and amount of each energy carrier used for a building's technical service system's operation is known, the primary energy needed for the operation of the building and the CO_2 emissions can be calculated using pre-defined factors and emission coefficients (Fig. 2.27). The specific values of these two indicators are calculated per 1 m^2 of the building's useful floor area (in kWh/m^2a and kg/m^2a) and tested according to the maximum-allowed values stated in the national regulation on energy-efficient buildings. The primary energy needed for the operation of the building is expressed as the renewable ($E_{P,ren}$), non-renewable ($E_{p,nren}$) and total ($E_{P,tot} = E_{P,ren} + E_{P,nren}$) primary energy. At this stage, the share of the renewable energy sources (*RER* Renewable Energy Ratio) for the operation of the building is determined as the ratio between the renewable and the total primary energy required. This set of indicators is used for the characterization of nearly Zero Energy Buildings (nZEBs) [18].
- energy-performance certificates were introduced to acquaint the building's owners and occupants with the data on the energy efficiency of the building. An energy certificate can be issued on the basis of the theoretical calculations or based on the actually measured consumption of energy carriers over a whole year.

More details on nearly Zero Energy Buildings' requirements, technical service systems and assessment methods can be found in [24].

References

1. Masters, G.M., Ela, W.P.: Introduction to Environmental Engineering and Science. Prentice-Hall International Editions, New York (1991)
2. Keller B.: Bautechnologie III: Bauphysik, ETH, CH (2002)
3. Santamouris, M. (ed.): Environment Design of Urban Buildings: An Integrated Approach. Earthscan, London (2006)
4. Recknagel, H., et al.: Grejanje i klimatizacija: sa pripremom tople vode i rashladnom tehnikom, Interklima, CRO (1995)
5. ASHRAE Standard 55: Thermal environmental conditions for human occupancy; ASHRAE 2017
6. EN 16798-3:2017: Energy performance of buildings. Ventilation of buildings. For non-residental buildings. Performance requirements for ventilation and room conditioning system; CEN 2017
7. CEPHEUS, cost efficient passive houses as European standards: Mehrfamilienhaus Salzburg – Gnigl; THERMIE, Project nr. BU/0127/97, Energieinstitut Vorarlberg, A (1999)
8. CR 1752:1998: Ventilation of buildings—design criteria for the indoor environment; CEN 1998
9. EN 16798–1:2019: Energy performance of buildings—ventilation of buildings—part 1: indoor environmental input parameters for design and assessment of energy performance of buildings addressing indoor air quality, thermal environment, lighting and acoustics, CEN 2019
10. Kosonena, R., Tanb, F.: Assessment of productivity loss in air-conditioned buildings using PMV index. Energy Build. **36**, 10 (2004)
11. Vidrih B.; Impact of global climate changes and urbanization on micro-climate in urban environment. Ph.D. thesis (supervisor assoc. prof. Sašo Medved); University of Ljubljana (2010) (in Slovenian language)
12. Hagentoft C.E.: Introduction to Building Physics; Studentliteratur, Lund, S (2001)
13. EN 12114: 2000: Thermal performance of buildings. Air permeability of building components and building elements. Laboratory test methods, CEN 2000
14. EN 13187:2000: Thermal performance of buildings—qualitative detection of thermal irregularities in building envelopes—infrared method; ISO 2000.
15. EN ISO 9972:2015: Thermal performance of buildings—determination of air permeability of buildings—Fan pressurization method; ISO 2015
16. Final energy consumption by sectors and fuels in Europe. www.eea.europa.eu/data-and-mapsi/ndicators. 30 Oct 2019
17. Vidrih B., Arkar C., Medved S.: Generalized model-based predictive weather control for the control of free cooling by enhanced night-time ventilation. Appl. Energy **168**, 482–492 (2016)
18. EN ISO 52018-1:2017; Energy performance of buildings—Indicators for partial EPD requirements related to thermal energy balance and fabric features – Part 1: Overview of options, CEN 2017
19. The Energy Performance Building Directive (EU 2018/844); European Commission (2018)
20. EN ISO 52016-1:2017, Energy performance of Buildings – Calculation of energy needs for heating and cooling, internal temperature and heating and cooling load in a building or building zone. Part 1: Calculation procedures
21. The energy balance and design tool for efficient buildings and retrofit, Passive House Institute (2015)
22. Rules on efficient use of energy in buildings with a technical guideline; Ur.l. RS 52/2010 (in Slovenian language)
23. Directive 2009/125/EC of the European Parliament and of the Council of 21 October 2009 establishing a framework for the setting of ecodesign requirements for energy-related products (2009)
24. Medved S., Domjan, S., Arkar C.: Sustainable Technologies for Nearly Zero Energy Buildings: Design and Evaluation Methods. Springer Nature, Switzerland AG (2019)

Chapter 3
Moisture Uptake in Building Structures

Abstract Building structures are in constant contact with water in at least one of its states of matter. Water or water vapour enter and exit building structures as a consequence of precipitation, hydrostatic pressure, capillary suction and the water-vapour pressure difference between the indoor and outdoor air. The analysis of water and water-vapour transport in building structures is based on mass-transfer balance. Any moisture uptake in building structures is undesirable for a number of reasons. Moisture in construction materials increases the thermal conductivity and, consequently, the heat conduction. Moisture also deteriorates the mechanical properties of building structures, causing damage to the material and higher maintenance costs. Dampness is also closely connected with microbial growth, posing a health risk to the building's occupants and reducing the quality of life in affected buildings. We must, therefore, be acquainted with the moisture-transfer mechanisms, as well as the methods and criteria to verify that the building structures were designed in such a way that moisture accumulation would not impair the thermal and mechanical properties of building structures.

Learning objectives In this chapter you will learn about:

- why moisture uptake in building structures is undesirable;
- the thermodynamic properties of moist air;
- the mechanisms of water and water-vapour transfer in building structures;
- the mathematical models for evaluating the moisture transfer in building structures;
- the requirements relating to moisture uptake that have to be fulfilled in order to preserve the properties of building structures;
- the measures to prevent moisture uptake in building structures.

S. Medved, *Building Physics*, Springer Tracts in Civil Engineering,
https://doi.org/10.1007/978-3-030-74390-1_3

3.1 Symbols for Quantities and Units

a_v	Water-vapour diffusivity (m^2/s)
A	Water-sorption coefficient ($kg/m^2s^{0.5}$)
h_c	Convective surface heat-transfer coefficient (W/m^2K)
B	Water-penetration coefficient ($m/s^{0.5}$)
β_v, β_p	Surface mass-transfer coefficients expressed by moisture content of air (v), expressed by water-vapour pressure in the air (p) (m/s), ($kg/m^2sPa = s/m$)
$c_{p,a}$, $c_{p,v}$	Specific heat capacity of air, of water vapour (kJ/k_gK)
d, d_{vb}	Thickness, thickness of water-vapour retender or barrier (m)
$d_{p,v}$	Periodic moisture-penetration depth (m)
d_r	Thickness of the layer in which the interstitial condensed water is distributed (m)
D_v	Diffusion coefficient of water vapour in the air (m^2/s)
δ_v	Diffusion coefficient of water vapour of the building material (m^2/s)
f_{Rsi}, $f_{Rsi,\min}$	Indoor surface temperature factor, minimum required indoor surface temperature factor (1)
φ, φ_i, φ_e	Relative humidity, of indoor air (i), of outdoor air (e) (1, %)
g	Gravity acceleration (m/s^2)
$\dot{g}$, $\overline{g}$, $\dot{g}_{\max}$	Specific mass flux of water vapour, on average, the maximum (kg/m^2s, kg/m^2h, g/m^2s)
$\dot{g}_c$	Water vapour mass flux per 1 m^2 of building structure area (g/m^2h)
$\dot{g}_{ev}$	Water-vapour flux during drying per 1m^2 of building structure area (g/m^2h)
$\dot{g}'_c$	Quantity of liquid water condensing in a building structure per 1 m^2 of building structure area (g/m^2h)
$\dot{G}_v$	Source of water vapour (g/h)
h, h_a	Specific enthalpy, of moist air (a) (J/kg)
h_{cap}	Height of water column in the pore (m)
H_a, $H_{a,dry}$, H_v, H_w	Enthalpy of moist air (a), of dry air (a, dry), of water vapour (v), of water (w) (J)
λ	Thermal conductivity (W/mK)
m_a, $m_{a,\,dry}$, m_v, m_w	Mass of air (a), of dry air (a, dry), of water vapour (v), of water (w) (kg)
m'_v	Specific mass of water vapour (kg/m^2, g/m^2)
m_{dry}, m_{wet}	Mass of dry sample, of wet sample (kg)
M_c, $M_{c,\max}$, M_a	Amount of condensed water, maximum amount of condensed water, maximum allowed amount of condensed water, per 1m^2 of building structure area (kg/m^2)

(continued)

(continued)

$M_{m,v}$, $M_{m,a}$	Molar mass of the water vapour, of dry air (kg/mol)
n, n_{50}	Air exchange rate, air exchange rate of pressure difference 50 Pa (h-1)
N	Number of days in the month (day/m)
μ, μ_{vb}	Water vapour diffusion resistance, of water vapour retender or barrier (1)
Ψ	Volumetric moisture content (m^3/m^3, %); Linear thermal bridge transmittance (W/m K)
p_a, $p_{a,dry}$	Air pressure, of dry air (Pa)
p_v, $p_{v,i}$, $p_{v,e}$, $p_{v,k}{}'(p_{v,k}{}'')$	Water vapour pressure, in indoor air (v, i), in outdoor air (v,e), saturated at condensation interface (Pa)
$p_{v,si}$, $p_{v,se}$	Water vapour pressure at temperature of internal surface of building structure (v, si), of external surface of building structure (v, se) (Pa)
$p_{v,sat}$, p_{v,sat,θ_i}, $p_{v,sat,\theta_{si}}$	Water vapour pressure in saturated air, in saturated air at indoor air temperature θ_i, in saturated air at interior surface temperature θ_{si} (Pa)
p_w, p_l	Pressure of water, in capillary column (Pa)
r	Pore diameter (m)
$r_{s,d}$	Water-vapour diffusion resistance (m)
r_{vb}	Water-vapour resistance of retarder or barrier (m)
r_w	Heat of evaporation of water at 0 °C (kJ/kg)
R_v	Gas constant of water vapour (J/kgK)
ρ,ρ_a, $\rho_{a,dry}$, ρ_w, ρ_{dry}	Density, of air (a), of dry air (a, dry), of water (w), of dry sample (dry) (kg/m^3)
S_a, S_v, $S_{v,sat}$	Saturation of a building material with air, of building material with water vapour, maximum saturation of building material with water vapour (%)
σ	Surface-tension coefficient (N/m)
t	Time (s)
t_p	Length of period (s, h)
θ,$\overline{\theta}$,$\hat{\theta}$	Temperature (°C), contact angle (°), average temperature (°C), amplitude of temperature (°C)
θ_i, θ_e	Temperature of indoor air (i), of outdoor air (e) (°C)
θ_{si},θ_{se},$\theta_{se,\min}$	Temperature of internal surface of building structure (si), of external surface of building structure (se), minimum to prevent surface condensation (se, min) (°C)
$\theta_{e,p}$	Design heating period outdoor air temperature (°C)
θ_{dev}	Dew point temperature (°C)

(continued)

(continued)

$u, u_a, u_{bas}, u_{con}, u_{dif}, u_{tot}, u_{\max}$	Moisture ratio (humidity by mass), of the air (o), the basic of material (bas), increased by condensation (con), increased by interstitial condensation (dif), total, maximum allowable (max) (kg/kg, %)
U, U_w	Thermal (tml) transmittance, of window (w) (W/m^{2}K)
$v, v_a, v_e, v_{a,sat}, v_{si}$	Moisture content (humidity by volume), of air (a), of outdoor air (e), of saturated air (a, sat), of air at temperature of internal surface of building structure (si) (kg/m^3, g/m^3)
$\overline{v}$, $\hat{v}$	Daily average, daily amplitude of air moister content (g/m^3)
V, V_w	Volume, volume of the water in the matter (vs) (m^3)
$\dot{V}$	Ventilation air flow rate (m^3/h)
x, x_i, x_e	Absolute humidity of air, of indoor air (i), of outdoor air (e) (kg/kg, g/kg)

3.2 Causes and Effects of Moisture Uptake in Building Structures

Building structures accumulate moisture when exposed to water or water vapour. The moisture is accumulated in the materials that contain little pores in their structures. The pores in porous building materials contain a mixture of air and water and/or water-vapour molecules. The moisture content depends on the material properties and on the content of the water vapour in the air that surrounds the material, which can be expressed as the air's humidity. When a building material is in contact in steady-state conditions with moist air at a certain temperature and humidity, the hygroscopic equilibrium is reached. This is a state of matter where the water-vapour molecules inside the building material's pores are in an energy balance with their surrounding environment. Without technical measures, moisture cannot be below the hygroscopic equilibrium and building materials are said to be damp when the state of the hygroscopic equilibrium (under reference conditions) is exceeded. At the hygroscopic equilibrium, only the water vapour is present in the pores, while if the pores are filled with water, the material is saturated.

The possible causes for moisture uptake in building structures are:

- water transfer into the pores from the ground or precipitation;
- hydrostatic pressure of the water in the case that the building structure's surfaces are submerged in water; for example swimming-pool walls or water-tank walls;
- damaged plumbing or sewers;
- the high humidity of the indoor air surrounding the building structure because the moisture is emitted by people and pets, human activities such as cooking or bathing. The molecules of water vapour are transferred from the air into the

Table 3.1 The quantity of water-vapour released from different sources into the indoor air each

sources of water vapor in buildings	$\dot{G}_v$ (g/h)
physically inactive person	30 – 60
person doing hard work	do 300
cooking	100
laundry hung to dry	do 500
bathroom, bathing	70
bathroom, showering	2600
house plant	15
still water surface	40

building structure by a process known as sorption. The quantity of water vapour emitted each hour by different sources inside a building is shown in Table 3.1;

- condensation of water vapour from the air on the structure's surface in the case that the temperature of the surface of the structure is considerably lower than the air temperature, and the room with considerable water-vapour sources is not adequately ventilated;
- interstitial condensation of water vapour inside the building structure, which can occur during the moisture transfer by the diffusion of water-vapour molecules through the structure;
- the "embedded" moisture that remains in the structure after the construction.

All these causes of moisture uptake in building structures must be treated with equal care. The direct transfer of water into the building structure is prevented by water- and damp-proofing systems (for example, a coating of synthetic resin on the foundations or roof tiles), through the efficient draining of rainwater and by protecting the external surfaces with overhangs. Several mechanical, physical, chemical and electrical processes are used to modify the properties of the pores in building structures, so that the water transfer from the ground or exposed surfaces by capillary suction is prevented. Moisture accumulation due to the condensation of water vapour can be avoided with the proper thermal insulation of building structures (to raise the temperature on the surface of the wall) and by selecting materials that reduce the ingress of water-vapour molecules into the structure.

3.2.1 *Moisture Uptake in Building Structures Must Be Prevented*

Moisture uptake affects both the mechanical and thermal properties of building structures, as well as their appearance. The most important consequences of moisture uptake are:

Table 3.2 Influence of volumetric moisture content ψ on the thermal conductivity of generic materials with different densities. Thermal conductivity and density are mutually related [2]

ρ (kg/m³)	λ (Ψ 1%) (W/mK)	λ (Ψ 25%) (W/mK)
200	0.09	0.19
400	0.12	0.25
1000	0.24	0.51
2200	0.92	1.95

- damp structures increase the humidity of the indoor air and stimulate microbial growth, in turn detrimentally affecting the living comfort. The most severe pollutants are mould, mites, fungi, bacteria and viruses. Most of them survive room conditions where the relative air humidity is either lower than 30% or higher than 70%, while relative air humidity above 80% is especially critical. As discussed in Sect. 2.1, the risk of microbial growth can be mitigated by proper building ventilation and by increasing the building structure's indoor surface temperature with additional thermal insulation to avoid the problem of sick-building syndrome;
- as the moisture content increases, the thermal conductivity of the building materials increases, because the thermal conductivity of water ($\lambda \sim 0.15$ W/mK) that displaces the air in the pores is considerably higher than the thermal conductivity of the stationary air ($\lambda \sim 0.025$ W/mK) (Table 3.2). This is especially critical if the thermal insulation is moist. The influence of moisture uptake on the heat transfer by sensible heat flux is greater in materials that consist of open pores like aerated concrete or mineral wool;
- in porous building structures with water vapour tight surfaces, a periodic water transfer between the structure's external and internal surfaces occurs and a process similar to the heat transfer in "heat pipes": during a cold night, the water evaporates at the internal surface and water-vapour molecules migrate towards the external surface, where they condensate due to the lower temperatures. During the day, the absorbed solar radiation increases the temperatures in the outer part of the building structure and the process repeats in the opposite direction. During this, typically 1-day-long periodical process, the heat-transfer rate is much higher due to the additional latent heat flux, compared to the sensible heat transfer in dry building structures, causing significantly higher building heat losses;
- if a building structure contains water, the heat losses of the structure are also greater due to the latent heat flux needed to evaporate the moisture during the drying process of the building materials;
- most building materials swell when damp, causing additional mechanical loads. The expansions due to varying moisture can even exceed the expansions in building materials caused by a variation of its temperature (Fig. 3.1);
- maintenance costs rise as building materials that are too damp or are constantly in contact with water deteriorate, corrode, or in the worst scenario even fail.

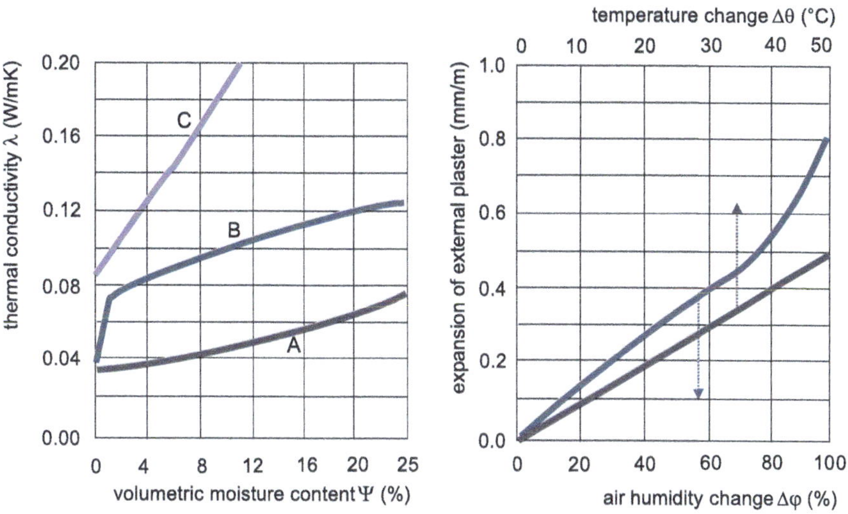

Fig. 3.1 Increasing thermal conductivity λ of thermally insulating materials (A—polystyrol, B—mineral wool, C—foamed concrete) [1] (left); the expansion of external plaster versus air temperature (thermal expansion) and humidity (hygroscopic expansion) [1] (right)

3.3 Psychrometric Properties of Moist Air

Air is a mixture of gases, predominantly N_2, O_2, Ar and CO_2. Air always contains water vapour, too. As is the case with mixtures of ideal gases, moist air behaves according to the Dalton's law, which states that the total air pressure p_a is equal to the sum of the partial pressures p_i of ith gas present in the air (also known as the dry air pressure $p_{a,dry}$), and the partial pressure of water vapour p_v:

$$p_a = \sum_i p_i + p_v = p_{a,dry} + p_v$$

Standard air or barometric pressure p_a is defined at sea level and is 1013 mbar or 101.3 kPa, and the partial pressure of water vapour p_v in the air is between 0.1 and 12 kPa according to the air temperature and humidity. It is a characteristic of moist air that at a certain temperature it can only contain a finite quantity of water vapour. In the state of saturation, the partial pressure of water vapour p_v in the air equals the water-vapour saturation pressure $p_{v,sat}$, which depends on the air temperature. This characteristic is one of the psychrometric properties of air at a temperature θ (°C) and can be determined by the expressions [3]:

$$\theta \leq 0^\circ C \quad p_{v,sat} = 611 \cdot e^{\left(\frac{22.44 \cdot \theta}{272.44 + \theta}\right)} \quad [Pa]$$

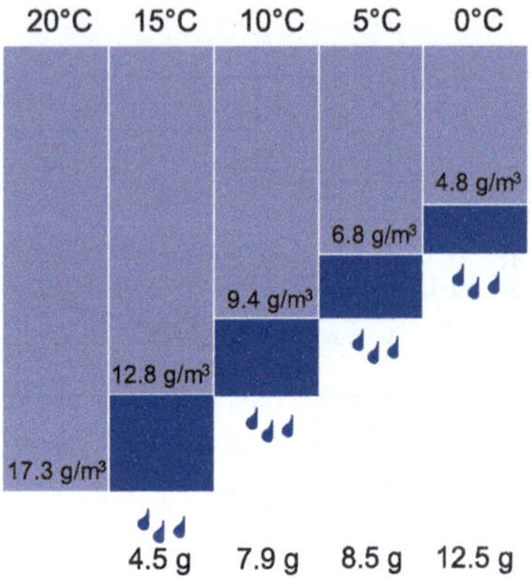

Fig. 3.2 Quantity of liquid water condensing from 1 m^3 of saturated air as the air is cooled from 20 °C to different dew point temperatures θ_{dew}

$$\theta > 0^{\circ}\mathrm{C}\quad p_{v,sat} = 611 \cdot e^{\left(\frac{17.08 \cdot \theta}{234.18 + \theta}\right)} \ [\mathrm{Pa}]$$

A state of saturation is achieved by moistening the air (i.e., by evaporation from the water surfaces or by spraying the water droplets through nozzles) or by cooling it to the saturation temperature, known as the dew-point temperature θ_{dew}. The water-vapour saturation pressure $p_{v,sat}$ increases with an increasing air temperature θ, which means that hot air can hold more water-vapour molecules than cold air. Figure 3.2 shows the quantity of (liquid) water that condenses from 1 m^3 of saturated air as the air is cooled from the initial temperature of 20 °C to lower temperatures. Any temperature shown is the "new" dew point temperature θ_{dew} at which air is saturated with water vapour and the water-vapour pressure is equal to saturated partial water-vapour pressure at that state of the air.

The quantity of water vapour in the air can be defined in several ways. The absolute humidity x is defined as the mass of water vapour (water) per 1 kg of dry air:

$$x = \frac{m_v}{m_{a,dry}} \left[\frac{kg}{kg}\right]$$

Using the equation of state for an ideal gas, the mass of the water vapour and the air can be substituted by the product of the water-vapour pressure p_w and the molar mass of the water vapour ($M_{m,v}$ 18.0 kg/kmol) and by the product of the dry-air pressure $p_{a,dry}$ and its molar mass ($M_{m,a}$ 28.9 kg/kmol):

$$x = \frac{m_V}{m_{a,dry}} = \frac{p_V \cdot \overbrace{M_V}^{18.0}}{p_{a,dry} \cdot \underbrace{M_{a,dry}}_{28.9}} = 0.622 \frac{p_v}{p_a - p_v} \left[\frac{kg}{kg}\right]$$

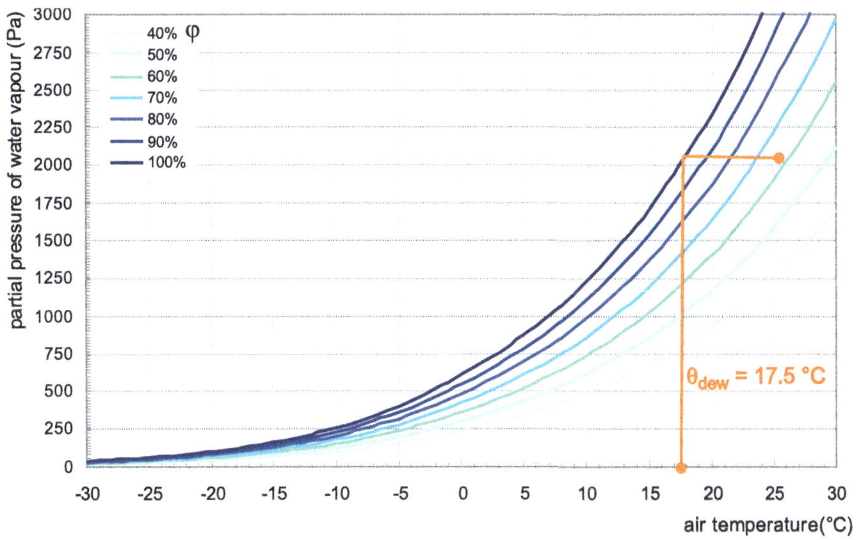

Fig. 3.3 Relation between the temperature θ, relative humidity φ and water-vapour pressure p_v of moist air

Since the mass of water vapour in 1 kg of dry air is relatively small, the absolute humidity x is often declared in grams per 1 kg of dry air. The absolute humidity of the air in nature is between 2 and 20 g/kg, and the absolute humidity of the indoor air is usually between 4 and 12 g/kg.

The humidity of the air can also be expressed as the relative air humidity φ, defined as the ratio of the actual partial pressure of the water vapour p_v to the water-vapour saturation pressure $p_{v,sat}$ at a specific air temperature. A relative quantity is limited to a range between 0 and 1, or between 0 and 100%. This means that the relative humidity of saturated air is 1 (100%).

$$\varphi = \frac{p_v}{p_{v,sat}}\left[\frac{Pa}{Pa} \equiv 1\right];\ \varphi = \frac{p_v}{p_{v,sat}} \cdot 100[\%]$$

The relative humidity of air in the atmosphere ranging from a couple of percent to 100%, which occurs just before precipitation. A diagram of the water-vapour partial pressure p_v vs. air temperature θ for different values of the relative humidity φ is presented in Fig. 3.3.

Case Study 1 A bottle of drink is taken from a fridge and placed in a room with 62% relative air humidity and a temperature of 24 °C. At what bottle surface temperature will a water droplet form? Find the solution using the diagram from Fig. 3.3.

Solution: water vapour will condense from the air on the bottle if the bottle surface temperature is lower than 17.5 °C.

Case Study 2 The temperature of outdoor air θ_e is 5 °C and the air pressure p_a is 101.3 kPa. The partial pressure of water vapour p_v in the outdoor air is 795 Pa. Calculate the air's relative and absolute humidity.

The water-vapour saturation pressure $p_{v,sat}$, the relative humidity φ, and the absolute humidity x of outdoor air in the air at 5 °C are:

$$p_{v,sat} = 611 \cdot e^{\left(\frac{17.08 \cdot \theta}{234.18 + \theta}\right)} = 611 \cdot e^{\left(\frac{17.08 \cdot 5}{234.18 + 5}\right)} = 873.2 \text{ Pa}$$

$$\varphi = \frac{p_v}{p_{v,sat}} \cdot 100 = \frac{795}{873.2} \cdot 100 = 91.0\%$$

$$x = 0.622 \frac{p_v}{p_a - p_v} = 0.622 \frac{795}{101300 - 795} = 0.0049 \frac{kg}{kg} = 4.9 \frac{g}{kg}$$

The quantity of water vapour in the air can also be described as the humidity by volume v with the unit g/m^3 or kg/m^3, defined by the water vapour mass per unit volume of air. It is calculated as:

$$v = \frac{p_v}{R_v \cdot (\theta + 273.15)} = \frac{p_{v,sat} \cdot \varphi}{462 \cdot (\theta + 273.15)} \left[\frac{\overbrace{Pa}^{\frac{N}{m^2}} \cdot kg \cdot K}{\underbrace{J}_{N \cdot m} \cdot K} = \frac{kg}{m^3} \right]$$

where R_v is the gas constant of water vapour (462 J/kg K) and θ is the air temperature (°C). The humidity by volume of air in the atmosphere is between 1.5 and 15 g/m^3. The humidity by volume v is the greatest in the case of saturated air. $V_{a,sat}$ is calculated using the equation above, taking the value of the relative air humidity φ to be 1.

Another property of moist air is the enthalpy H_a. Enthalpy is the total internal energy of moist air, measured relative to a reference temperature—the freezing point of water (0 °C). The enthalpy of dry air $H_{a,dry}$ with the temperature θ and the mass $m_{a,dry}$ is:

$$H_{a,dry} = m_{a,dry} \cdot c_{p,a} \cdot \theta \left[kg \cdot \frac{J}{kg \cdot K} \cdot \overbrace{{}^{\circ}C}^{=K} = J \right]$$

where $m_{a,dry}$ is the mass of dry air (kg), and $c_{p,a,dry}$ is the specific heat capacity of dry air at constant pressure. It is the quantity of heat in J that increases the temperature of 1 kg of air by 1 K. The specific heat capacity of air at 0 °C is 1005 J/kg. As the air in nature always contains water vapour and the water-vapour temperature is the same as the air temperature, the enthalpy of the water vapour H_v is:

$$\mathrm{H_v = m_v \cdot c_{p,v} \cdot \theta[J]}$$

where m_v is the mass of water vapour (kg), and $c_{p,v}$ is the specific heat capacity of water vapour. At 0 °C, $c_{p,v}$ is 1926 J/kg. If 1 kg of water is to be evaporated and form 1 kg of water vapour, its internal energy must be increased by the heat of vaporization r_w. The heat of vaporization at 0 °C r_w is 2.5 10^6 J/kg. The total enthalpy of moist air is the sum of the enthalpy of dry air, the enthalpy of water vapour, and the heat of vaporization necessary to create a given mass of water vapour in the air:

$$\mathrm{H_a = H_{a,dry} + H_v + m_v \cdot r_w = m_{a,dry} \cdot c_{p,a} \cdot \theta + m_v \cdot c_{p,v} \cdot \theta + m_v \cdot r_w[J]}$$

The unit for the enthalpy H_a is the unit for energy, i.e., the joule. Dividing the above expression by the mass of dry air $m_{a,dry}$ and remembering that the absolute air humidity x describes the mass of water vapour m_v in 1 kg of dry air, we obtain an expression for the specific enthalpy h_a:

$$\begin{aligned}\mathrm{\frac{H_a}{m_{a,dry}}} &= \mathrm{h_a = \frac{m_{a,dry}}{m_{a,dry}} \cdot c_{p,a} \cdot \theta + \frac{m_v}{m_{a,dry}} \cdot c_{p,v} \cdot \theta + \frac{m_v}{m_{a,dry}} \cdot r_w} \\ &= \mathrm{\left(c_{p,a} + x \cdot c_{p,v}\right) \cdot \theta + x \cdot r_w[J/kg]}\end{aligned}$$

The enthalpy H_a and the specific enthalpy h_a of air are defined relative to a selected reference temperature, i.e., the freezing point of water (0 °C). At that temperature, the absolute humidity of saturated air x is 4 g/kg, or 0.004 kg/kg. The specific enthalpy in this state is 0 J/kg for dry air and 10 kJ/kg for saturated air (h_a, or shortly $h =$ 0.004 kg/kg × 2500 kJ/kg). Enthalpy is an important property of air, as it can be used to evaluate the changing states of indoor air as it undergoes heating, cooling, humidification or dehumidification, according to the requirements of thermal indoor comfort. Psychrometry is a branch of thermodynamics that deals with the evolution of transformations in the state of moist air. The transformations in the states of air can also be plotted in psychrometric charts. One of the best-known psychrometric charts is the Mollier h–x diagram for moist air shown in Fig. 3.4.

Case Study 1 Air from a room with the indoor air temperature $\theta_1 = 20$ °C and the relative humidity $\varphi_1 = 50\%$ is heated with a hairdryer to a temperature $\theta_2 = 38$ °C. Find the absolute and the relative humidity of the hot air flowing out of the hairdryer.

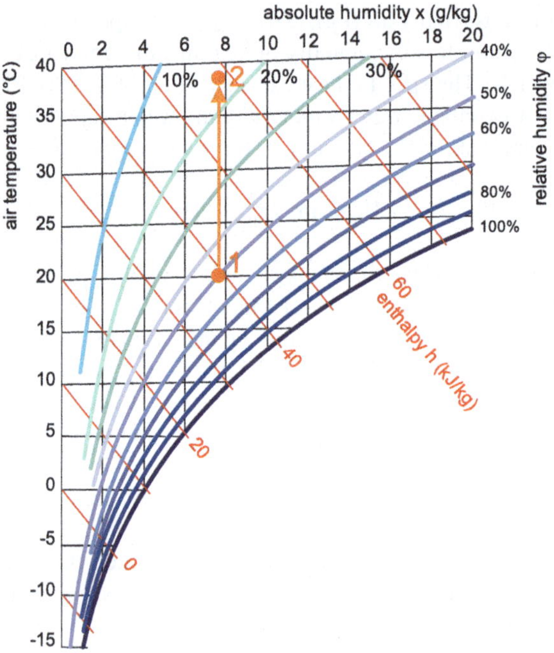

Fig. 3.4 Mollier h–x diagram for moist air shows the relations between the quantities of state—temperature θ, relative humidity φ, absolute humidity x and specific enthalpy h [3]

Solution: As the hairdryer heats the air, the state of the air changes from point 1 to point 2. The absolute humidity of the heated air remains unchanged ($x_1 = x_2 = 7.6$ g/kg), but the relative humidity of the air decrease to $\varphi_2 = 18\%$.

Case Study 2 Air with mass $m_a = 2$ kg, temperature $\theta_1 = 21$ °C and relative humidity $\varphi_1 = 60\%$ is cooled to $\theta_2 = 5$ °C. How much heat must be removed from the air and how much water condenses?

Solution: The absolute humidity of the air in state 1 is $x_1 = 9.5$ g/kg. The mass of dry air $m_{a,dry}$ is therefore 2 kg—2 kg × 0.0095 kg/kg = 1.981 kg. As the air is cooled, it becomes saturated with water vapour at 13 °C (Fig. 3.5, top, left). The water vapour is condensed from the air as the air continues to cool. The quantity of heat removed depends on the mass of dry air and the difference in the specific enthalpies:

$$Q = m_{a,dry} \cdot (h_1 - h_2) = 1.981 \cdot (45 - 20) = 49.5\, kg \cdot \frac{kJ}{kg} = kJ$$

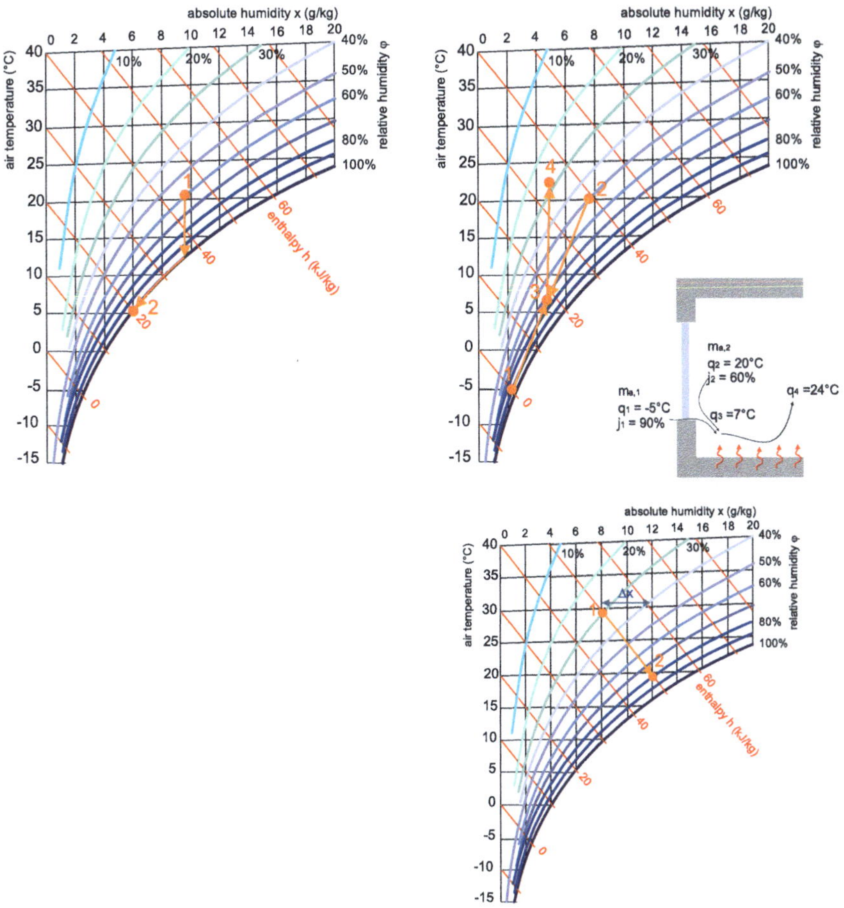

Fig. 3.5 Case studies' solutions: case study 2 (top, left), case study 3 (top, right), case study 4 (bottom)

and the amount of condensed water depends on the difference of the absolute humidity of the air in state 1 and end state 2:

$$\Delta m_w = m_{a,dry} \cdot (x_1 - x_2) = 1.981 \cdot (9.5 - 6) = 6.9\ \text{kg} \cdot \frac{g}{kg} = g$$

Case Study 3 Explain why indoor air may be too dry in wintertime when buildings are heated and ventilated with outdoor air.

Solution: Fresh air of state 1 ($\theta_1 = -5$ °C, $\varphi_1 = 90\%$), enters a room through a ventilation opening, and mixes with the indoor air of state 2 ($\theta_2 = 20$ °C, $\varphi_2 = 60\%$). Assuming the mass of air $m_{a,1}$ and $m_{a,2}$ are equal, the temperature of the mixture is found at the centre of the line between the points 1 and 2 (shown as state 3) ($\theta_3 = 7$ °C, $\varphi_3 = 75\%$). Once the mixture of fresh air and indoor air is heated with a radiator, the state of the air changes to point 4 ($\theta_4 = 22$ °C, $\varphi_4 = 28\%$) (Fig. 3.5, top, right). If there are no sources of water vapour in the room (such as people or plants), or the air is not treated with a humidifier, the humidity of the air will be lower than the value required by the thermal comfort criteria ($\varphi_i > 30\%$).

Case Study 4 Why is air near fountains cool during the summertime?

Solution: The air surrounding the fountains is cooled due to the evaporation of water droplets. The energy needed for the water-vapour molecules to escape a water droplet is drawn from the internal energy of the air. The sensible heat of the air is transformed into the latent heat of the water-vapour molecules. Since the enthalpy of the air remains unchanged, the air becomes cooler and moister. This is a natural process, which also takes place in the leaves of a plant, and is also mimicked by air-conditioning systems based on evaporative cooling. If 4 g of water evaporates into 1 kg of air of state 1 ($\theta_1 = 29$ °C, $\varphi_1 = 40\%$), the absolute humidity will increase by Δx 4g. Meanwhile, the air will cool down (as the enthalpy remaining constant) to $\theta_2 = 18$ °C (Fig. 3.5, bottom).

3.4 Dampness and the Uptake of Moisture in Building Materials

Most building materials are porous; some examples are concrete, brick, wood, mineral wool, and even stone. The pores vary in size, and the volume of pores ranges from 5% in granite to 95% in mineral wool. Pores can contain dry air only (which is s theoretical case only), moist air, water molecules, or molecules of some other liquid. In theory, when the pores only contain dry air, the degree of saturation of a building material with air S_a is equal to 100%. In addition to air, the pores in building materials also contain water molecules, which means the materials are damp. The moisture content of saturated materials is defined by $S_{w,sat}$ (%). In real conditions, the moisture content in building materials is in the range between S_a and $S_{w,sat}$ and can be described as:

- The moisture ratio or humidity by mass u is defined by the mass of water molecules (water vapour and/or water) per 1 kg of dry material, measured in kg/kg or %;
- the moisture content or humidity by volume v is defined by the mass of water molecules) per 1 m^3 of material, measured in kg/m^3;
- the volumetric moisture content ψ, defined as the ratio of the water-molecules volume to 1 m^3 of material, measured in m^3/m^3 or %.

The quantities used to describe the moisture content are mutually related. The moisture ratio is determined by weighing the original wet sample of building material with a volume V and mass m_{wet}, drying it in an oven at 105 °C for a day or two. The dry sample has lower mass m_{dry} and density of the sample ρ_{dry} can be determined from its volume. The moisture ratio or humidity by mass is:

$$u = \frac{(m_{wet} - m_{dry})}{m_{dry}} 100 \ [\%]$$

The moisture content or humidity by volume v is:

$$v = \frac{(m_{wet} - m_{dry})}{V} \left[\frac{kg}{m^3}\right]$$

The volumetric moisture content Ψ equals:

$$\Psi = \frac{V_w}{V} = \frac{\frac{(m_{wet} - m_{dry})}{\rho_w}}{\frac{m_{dry}}{\rho_{dry}}} = u\frac{\rho_{dry}}{\rho_w}\left[\frac{m^3}{m^3} = 1\right] \equiv u\frac{\rho_{dry}}{\rho_w} 100 \ [\%]$$

Explanation In practice, the high moisture content in building materials is treated as an "undesirable" property, so declaring values that are as low as possible is the goal. For high-density materials ($\rho > 500$ kg/m^3), the moisture content is the lowest if expressed as the moisture ratio u (kg/kg), and for porous materials, it is the lowest if expressed as the volumetric moisture content ψ (m^3/m^3). Additional explanations are presented in [4].

Case Study A wooden sample measuring 500 × 100 × 10 mm^3 is weighed. The wooden sample's mass is 0.35 kg. The sample is then put into an oven to dry at 105 °C for 40 h. The mass of dry sample is 0.30 kg. The measured mass is used to determine the density of dry wood $\rho_{dry}(\rho_{dry} = m_{dry}/V)$, which is 600 kg/$m^3$. Determine the moisture ratio u, the moisture content v and the volumetric moisture content ψ of the original sample:

$$u = \frac{(m_{wet} - m_{dry})}{m_{dry}} \cdot 100 = \frac{0.35 - 0.30}{0.30} \cdot 100 = 16.7\%$$

$$v = \frac{(m_{wet} - m_{dry})}{V} = \frac{0.35 - 0.30}{0.5 \cdot 0.1 \cdot 0.01} = 100 \ \frac{kg}{m^3}$$

$$\Psi = \frac{V_w}{V} = \frac{\frac{(m_{wet} - m_{dry})}{\rho_w}}{\frac{m_{dry}}{\rho_{dry}}} = \frac{\frac{(0.35-0.30)}{1000}}{\frac{0.30}{600}} \cdot 100 = 10\%$$

The moisture uptake in building structures is attributed to different phenomena and mechanisms, which are described below.

3.4.1 Sorption Water-Vapour Uptake in Building Materials

The water-vapour molecules are distributed through the air and find their way into the open pores in the building material, where they adhere to the walls as individual particles or form a thin film. This phenomenon is called adsorption, and such materials are referred to as hygroscopic. A few examples of hygroscopic materials are concrete, wood and plaster. In contrast, water-vapour molecules do not enter non-hygroscopic materials. Some examples of non-hygroscopic materials include brick, plastics and the majority of insulating materials.

Explanation Sorption is a general principle related with how the molecules of gas adhere to the surfaces of pores in solid mater. We distinguish between adsorption, or the process of depositing molecules to the surface of the pores, and the opposite process of desorption, when molecules release from the surface of the pores, which happens during the drying of the solid mater.

The amount of water molecules contained in the pores of hygroscopic materials in contact with moist air depends on the temperature θ and the relative humidity φ of the air defined by the partial pressure of the water vapour in the air. However, a hygroscopic material surrounded by moist air at steady state will only take up moisture by absorption to a certain final moisture content v. This state is known as the hygroscopic equilibrium.

The point of hygroscopic equilibrium at constant temperature and different values of relative air humidity φ in contact with the material is described by the sorption isotherm or sorption curve.

An example of a sorption isotherm is shown in Fig. 3.6 (left). For any building material, the shape of the absorption curve (moisture-uptake process) is slightly

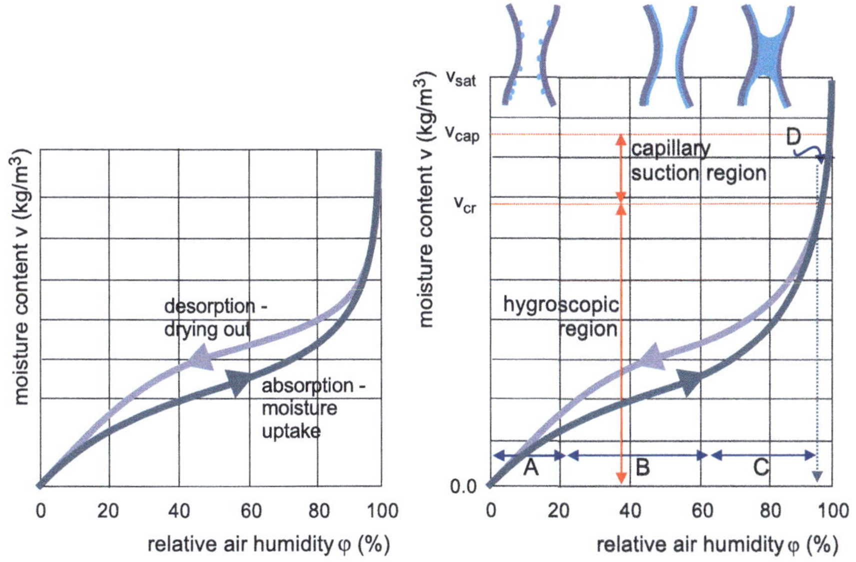

Fig. 3.6 Sorption isotherm and hysteresis between the processes of absorption and desorption (left); regions/processes taking place in a hygroscopic material during moisture uptake caused by increase of the air humidity at the surface of the sample (right)

different from the shape of the desorption curve (drying process) and this is also referred to as the sorption hysteresis. This difference can be explained by the different bell-shaped surfaces of the water column in a saturated pore taking up water or drying out, as well as by the fact that the saturation pressure of the water vapour in the air over a concave water surface is greater than over a level surface. This property depends on the material properties of the water and on the diameter of the pores (capillaries), and is described analytically as the Gibbs–Thomson effect [2].

Explanation An isothermal process is a thermodynamic change of the state of matter that takes place at a constant temperature (Table 3.3). The term sorption isotherm originates from the fact that the equilibrium between the relative air humidity φ in the contact with the surface of the sample and the sample moisture

Table 3.3 Volumetric moisture content of wood (cellulose) Ψ (%) in state of the hygroscopic equilibrium at different air temperatures θ and relative air humidity φ [2]

	Ψ	
θ (°C)	φ 30%	φ 90%
0	6%	22%
30	6%	20%
40	6%	19%

content v or moisture ratio by mass u of hygroscopic materials depends above all on the relative humidity of the air, and much less on the air temperature (Table 3.3). The influence of the temperature on the sorption process is usually neglected, so the properties of the materials can be simplified:

$$v(\varphi, \theta) \doteq v(\varphi) \quad u(\varphi, \theta) \doteq u(\varphi)$$

Different conditions occur in hygroscopic materials during hygroscopic water intake:

- $v = 0$ — the matter is dry matter; therefore, the moisture content v is 0 kg/m^3. This state can only be achieved if the material sample is heated at 105 °C for a day or two;
- $0 < v < v_{cr}$ — hygroscopic water-vapour uptake (identified by areas A, B and C in Fig. 3.6 (right)). At values of the relative air humidity under 20% (area A), individual water-vapour molecules are deposited on the surfaces of the pores. A at relative air humidity over 60% (area B) the molecular layer grows and forms a thin film of liquid water. If the air humidity is even greater (area C), the pressure drop in the pores causes capillary condensation and the pores start to fill with liquid water. The critical moisture content v_{cr} (state at which majority of pores are saturated with water) is reached when the relative air humidity is approximately 98%. The moisture content in the material rises rapidly once vcr is exceeded;
- v_{cap} — the pores are completely filled with liquid water (area D). Once the moisture content reaches the critical capillary moisture content v_{cap}, liquid water is transported into the material by capillary suction;
- v_{sat} — all the pores in the material are completely filled with water and the air has been completely expelled from the pores. Such a condition cannot be established solely by natural processes.

The hygroscopic water uptake in building materials is influenced by salts. For example, magnesium, calcium or sodium chloride reduce the critical moisture content v_{cr} (the critical moisture content v_{cr} for $MgCl_2$, CaCl and NaCl is 33%, 75% and 86%, respectively). This means that walls containing CaCl cannot dry out if the relative air humidity is greater than 75%.

Explanation Salty clothes take a long time to air dry, so they must be washed in fresh water before they are hung out to dry.

3.4.2 Capillary Water Uptake in Building Materials

Capillary water uptake in building structures is a phenomenon involving the transport of liquid water into the pores of the building material. Water is most often transported into the pores of the building structure from the soil or from water droplets during precipitation by capillary suction. Capillary suction is a natural process, explained by the attractive forces between the water molecules and the surface of the pore wall. Surface tension occurs as a result of the uneven surface of a water column in the pore, which is also known as the meniscus. This causes a drop in the pressure p_l on the wetted side of the meniscus compared to the air pressure over the column p (Fig. 3.7). Capillary forces occur on the water-column envelope, acting in the direction opposite to gravity. The suction potential is defined by the difference between p_a*and* p_l on the wetted side of the meniscus and is determined by:

$$p_a - p_l = \frac{2 \cdot \sigma}{r} \cdot \cos \theta \text{ [Pa]}$$

where σ is the surface-tension coefficient (σ for water at 20 °C is 0.073 N/m), θ is the contact angle (°), and r is the pore diameter (m) [5, 7].

Explanation The contact angle θ (Fig. 3.7) can be positive or negative. In the first case, the angle θ is greater than 0° and the liquid is referred to as a wetting liquid (one example is water). A consequence of this property is that the water in the pore rises, due to the lower air pressure in the pore above the water column. In non-wetting liquids, the angle θ is negative and the capillary

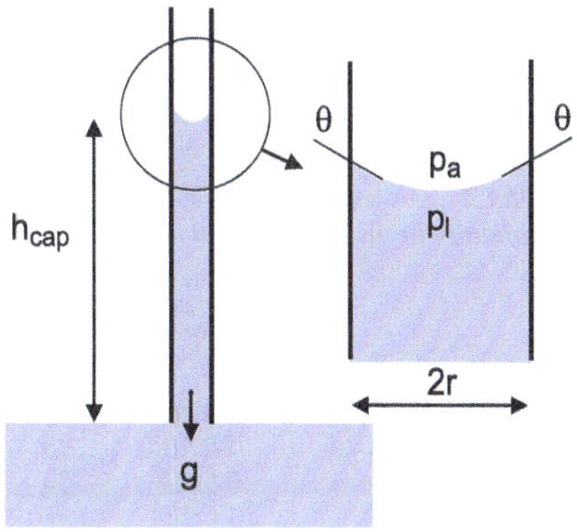

Fig. 3.7 Water column of height h_{cap} in a pore, and the wetting angle θ; in the case shown, the wetting angle θ is positive, as characteristic for the wetting liquids; p_a is pressure of moist air and p_l is pressure of water in pore proportional to ~ ln (φ_a); the difference in p_a–p_l define the suction potential

forces push the liquid column downwards, compared the the liquid level in the surroundings. Mercury is an example of a non-wetting liquid. For most building materials, the wetting angle θ is greater than 0° and capillary suction occurs.

The water column rises in the pores of a building material up to the height h_{cap}, where the capillary forces acting on the water column are balanced with the force of gravity acting upon the water column in the opposite direction. The suction height h_{cap} is determined as follows:

$$h_{cap} = \frac{2 \cdot \sigma}{r \cdot g \cdot \rho_W} \left[\frac{\overbrace{N}^{kg\frac{m}{s^2}} \cdot s^2 \cdot m^3}{m \cdot m \cdot m \cdot kg} = m \right]$$

where g is the acceleration due to gravity (m/s^2) and ρ_w is the density of water (kg/m^3). By assumption, θ *is* approximately 0° and cos(θ) is approximately 1. Table 3.4 shows the theoretical water-column height h_{cap} in a vertical cylindrical capillary with radius r. Capillary suction is, therefore, only relevant when the diameter of the pores in the material is small enough.

Capillary suction can magnify the moisture uptake in a building structure because moist air in a pore at a lower pressure above the meniscus is saturated with water vapour at a lower relative air humidity. When the pressure of the air over the surface of the liquid water column in a pore drops, water vapour can start condensing from the air over the wetted area of the pore. This phenomenon is known as capillary condensation. Since capillary suction depends on the diameter of the capillary, so does the relative humidity of the air when the capillary condensation starts taking place in the pore. The critical relative humidity φ_{cr} is determined with Thomson's equation:

Table 3.4 Water-column height h_{cap} in a cylindrical capillary tube vs. the capillary tube radius r (left); pore radius and the humidity of air in the pore over the meniscus, which gives rise to the capillary condensation (right)

radius r (m)	h_{cap} (m)
10^{-3}	0.015
10^{-4}	0.15
10^{-5}	1.5
10^{-6}	15
10^{-7}	150

radius r (m)	φ (%)
$0.5 \cdot 10^{-3}$	99.8
$0.5 \cdot 10^{-4}$	98
$1.5 \cdot 10^{-5}$	93
$0.5 \cdot 10^{-5}$	80
$1.5 \cdot 10^{-6}$	47
$0.5 \cdot 10^{-6}$	10

$$\varphi = e^{\left(-\frac{2\cdot\sigma}{r}\cdot\frac{1}{\rho_W\cdot R_V\cdot(\theta+273.15)}\right)} [1]$$

where σ is the surface-tension coefficient (N/m), r is the pore radius (m), ρ_w is the density of water (kg/m^3), R_v is the gas constant for water vapour (462 J/kg K) and θ is the temperature of the air in the pore (°C).

Case Study If the pore radius is greater than 0.5×10^{-5} m, capillary condensation occurs when the relative humidity of the air in the pore exceeds 80% (and not 100% as in the atmosphere).

Most building materials are porous, but not all pores are of regular cylindrical shape or oriented vertically. This influences the transport of water by capillary suction in the pores. The height h_{cap}, up to which a building structure in contact with water becomes saturated due to capillary suction, is determined empirically from the building materials' water-penetration coefficient B, which is measured in m/s$^{0.5}$ (Table 3.5). The height of the water due to the capillary suction h_{cap} also depends on the duration of the process t (s):

$$h_{cap} = B \cdot \sqrt{t}[m]$$

The specific mass flow rate of water $\dot{g}$ entering a building structure by capillary suction varies with the duration of the moisture uptake. It is determined from the building material's water-sorption coefficient A (kg/m^2 s$^{0.5}$) (Table 3.5) and the duration of the moisture-uptake process $t(s)$. The amount of liquid water m_w accumulated in a building structure during the capillary-suction process is determined by integrating the time-dependent specific mass flow rate of water $\dot{g}(t)$ over the entire period of the moisture uptake. The quantity is determined per m^2 of the building

Table 3.5 Water sorption coefficient A and the water- penetration coefficient B for some building materials [5]

building material	A (kg/m^2 s$^{0.5}$)	B (m/s$^{0.5}$)
brick	0.4	$1.4 \cdot 10^{-3}$
aerated concrete	0.1	$0.4 \cdot 10^{-3}$
concrete (w/c 0.3)	0.01	$0.1 \cdot 10^{-3}$
concrete (w/c 0.7)	0.03	$0.3 \cdot 10^{-3}$
lime mortar	0.3	$1.0 \cdot 10^{-3}$

structure's cross-section area, through which the capillary suction is taking place:

$$\dot{g}(t) = \frac{A}{2 \cdot \sqrt{t}} \left[\frac{kg}{m^2 s}\right] \rightarrow m_w = \int_0^t \dot{g}(t) \cdot dt = A \cdot \sqrt{t} \left[\frac{kg}{m^2}\right]$$

Case Study Determine the moisture content v and moisture ratio u for a concrete wall with density $\rho = 2000$ kg/m^3 (V/C = 0.7) after 6 months of moisture uptake by capillary suction of the groundwater, while neglecting the simultaneously drying of the wall due to the desorption and the water-vapour diffusion:

$$h_{cap} = B \cdot \sqrt{t} = 0.3 \cdot 10^{-3} \sqrt{3600 \cdot 24 \cdot \frac{8760}{2}} = 5.8 \text{ m}$$

$$m_w = A \cdot \sqrt{t} = 0.03 \sqrt{3600 \cdot 24 \cdot \frac{8760}{2}} = 584 \frac{kg}{m^2}$$

The moisture content v in the wall equals the ratio of the suction water mass to the volume of the wall (V equals to 1 m^2 × h_{cap}) in which the suction water is distributed:

$$v = \frac{m_w}{1 \cdot h_{cap}} = \frac{584}{1 \cdot 5.8} = 101 \frac{kg}{m^3} \quad u = \frac{v}{\rho} = \frac{101}{2000} = 0.051 \frac{kg}{kg} = 5.1\%$$

3.4.3 *Diffusion Moisture Transfer in Building Structures*

We have learned that the quantity of water vapour in the air can be quantified in different ways, for example, as the absolute x (kg/kg) or relative humidity φ (%) or with the partial pressure of the water vapour p_v (Pa) in the air. In the case that two regions of stagnant air with different levels of humidity make up an observed system, a diffusion mass flux of water-vapour molecules is established between these two regions. The water-vapour mass flow stops once the internal energy is equal in both regions, which means that the partial pressures of the water vapours in both regions are equal. The diffusion mass flux of the water vapour $\dot{g}$ between the two regions of stagnant air with different humidity levels is determined by Fick's law:

$$\dot{g} = -D_v \frac{dv}{dx} \left[\frac{kg}{m^2 s}\right]$$

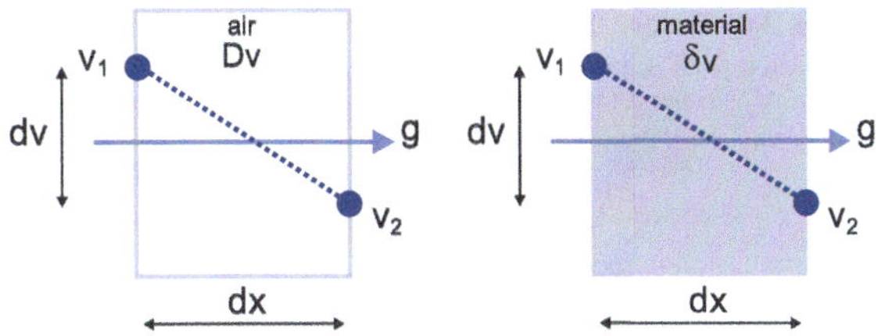

Fig. 3.8 Quantities determining the diffusion flux of water vapour $\dot{g}$ in the air and in a construction material

The diffusion mass flux is a specific quantity, measured in kg/s per m^2 of the building structure's surface area. The humidity by volume changes (dv/dx) inside the observed system is known as the humidity gradient. In the case that the observed system is filled with air, the constant D_v in the expression is the diffusion coefficient of water vapour in the air. The negative sign means the diffusion mass flux is directed in the opposite direction to the gradient of the humidity (Fig. 3.8).

Explanation Diffusion mass flux $\dot{g}$ as presented in the equation is one dimensional, meaning the humidity by volume only varies in one axis of the spatial coordinate system (x). The material is also assumed to be isotropic, so its material properties are constant throughout the volume.

The diffusion coefficient D_v for stagnant air is:

$$D_v = (22.2 + 0.14 \cdot \theta) \cdot 10^{-6} \left[\frac{m^2}{s} \right]$$

where θ is the air temperature (°C). At 10 °C and 1013 mbar, the diffusion coefficient of water vapour in the air D_v is 2.360 $\times$ 10^{-5} m^2/s.

A diffusion mass flux of water vapour also occurs in porous materials that separate two regions of air with different values of the humidity by volume v. The diffusion mass flux of the water vapour $\dot{g}$ is once again determined using Fick's law:

$$\dot{g} = -\delta_v \frac{dv}{dx} \left[\frac{kg}{m^2 s} \right]$$

The diffusion flux of the water vapour through a layer of material exists as long as the air humidity is different on the two sides of the layer. As presented, Fick's law defines the steady diffusion mass flux, assuming that the boundary conditions do not change over time. In the expression, δ_v (m^2/s) is the diffusion coefficient of water vapour of the building material. The diffusion coefficient δ_v depends on the moisture content in the material v, which means we are dealing with a nonlinear system of the first order. The calculation of the water-vapour diffusion mass flux can be simplified by substituting the material's diffusion coefficient δ_v with the water-vapour diffusion resistance μ, and evaluating the diffusion potential by the difference in the partial

Table 3.6 Water-vapour diffusion coefficient for some construction materials (left); diffusion resistance μ of some construction materials (right) [4, 6, 7]

material	δ_v (m^2/s)
PE film	$0.02 \cdot 10^{-6}$
wood	$0.5 \cdot 10^{-6}$
concrete	$1.0 \cdot 10^{-6}$
gypsum	$2.0 \cdot 10^{-6}$
brick	$3.0 \cdot 10^{-6}$
cellulose thermal insulation	$12.9 \cdot 10^{-6}$
mineral wool	$17.5 \cdot 10^{-6}$

material	μ (1)
PE film	800.000
wood	70
concrete	30
gypsum	4
brick	7
cellulose thermal insulation	1
mineral wool	1

pressures of the water vapour p_v in the air on the boundaries of the material layer instead of the difference in the moister content v (Table 3.6). The diffusion resistance μ is a dimensionless number, defined as the ratio of the diffusion coefficient of the water vapour in the air D_v to the diffusion coefficient of the water vapour in the particular material δ_v:

$$\mu = \frac{D_v}{\delta_v}\left[\frac{m^2 s}{m^2 s} = 1\right]$$

The expression used to determine the moisture content v of the air in differential notation is:

$$dv = \frac{dp_v}{R_v \cdot (\theta + 273.15)}$$

The diffusion mass flux of the water vapour is therefore:

$$\dot{g} = -\frac{D_v}{\mu}\frac{dv}{dx} = -\frac{1}{\mu} \cdot \frac{D_v}{R_v \cdot (\theta + 273.15)} \cdot \frac{dp_v}{dx}\left[\frac{kg}{m^2 s}\right]$$

The negative sign in the expression means the diffusion flux is directed in the opposite direction to the gradient of the water vapour's partial pressure.

Case Study Prove that the diffusion mass flux of water vapour through a porous material, determined using the expression derived from Fick's law, is measured in kg per second and m^2. Use the basic units of measurements from the International System of Units (kg, m and s). The gas constant R_v of water vapour is 462 J/kgK. A joule can be expressed as Nm. A substitute unit for N can be determined from Newton's first law: a force of 1 N accelerates a 1 kg body to 1 m/s^2. It follows that J = Nm = kgm^2/s^2. The pressure difference is measured in Pascals. A Pascal is N/m^2, and thus also kgm/s^2/m^2 or kg/ms^2.

$$\dot{g} = -\frac{1}{\mu} \cdot \frac{D_V}{R_v \cdot (\theta + 273.15)} \cdot \frac{dp_v}{dx}$$
$$\equiv \frac{1}{1} \frac{m^2 \cdot kg \cdot K}{s \cdot J \cdot K} \frac{Pa}{m} \equiv \frac{1}{1} \frac{m^2 \cdot kg \cdot K \cdot s^2}{s \cdot kg \cdot m^2 \cdot K} \frac{kg}{m \cdot s^2 \cdot m} \equiv \left[\frac{kg}{m^2 s}\right]$$

Case Study Determine the value of the constant C in the expression for the diffusion mass flux of water vapour $\dot{g}$ at the temperature $\theta = 10\ °C$.

$$\dot{g} = -\frac{1}{\mu} \cdot \underbrace{\frac{D_v}{R_v \cdot (\theta + 273.15)}}_{C} \cdot \frac{dp_v}{dx} \rightarrow C$$
$$= \frac{2.260 \cdot 10^{-5}}{462 \cdot (10 + 273.15)} \cdot \frac{1000Pa}{kPa} \cdot \frac{3600s}{h} \cdot \frac{1000g}{kg} = 0.622 \frac{g}{kPa \cdot m \cdot h}$$

Water-vapour pressure gradient expressed in the differential form dp_v/dx can be substituted with the differences $\Delta p_v/\Delta x$. In this case, Δp_v is the partial water-vapour pressure difference at the boundaries of the building structure layer, and Δx is the layer thickness d. Accounting for the value of the constant C (from Case study 2), the diffusion mass flux of the water vapour $\dot{g}$ is:

$$\dot{g} = 0.622 \cdot \frac{p_{v,1} - p_{v,2}}{\mu \cdot d} \left[\frac{g}{m^2 \cdot h}\right]; \; p_{v,1} > p_{v,2}$$

where the product of the water-vapour diffusion resistance and the layer thickness ($\mu \times d$) represents the water-vapour diffusion resistance r of the building structure layer. Instead of r, the designation s_d is used to describe the layer's resistance to water-vapour diffusion.

Explanation Fick's law defines the diffusion mass flux of water vapour $\dot{g}$ through the layer of a substance corresponding to Fourier's law defining the heat flux $\dot{q}$ conducted through a layer of a building material. The resistance to heat conduction d/λ is in the case of water-vapour diffusion substituted with the resistance to water-vapour transport $\mu \cdot d$, and the temperature difference as the "driving potential" for heat conduction is substituted by the difference in the partial water-vapour pressures in the air at the boundary of the layer. The differences in the temperatures and in the partial water-vapour pressures are arranged so that the values of $\dot{g}$ and $\dot{q}$ in the mathematical expressions are positive.

Explanation The constant C depends on the temperature and the air pressure. For the average ambient conditions [8], define the value of the constant C equal to 0.72 g/(kPa m h).

3.5 Modelling the Uptake of Moisture in Building Structures

So far, we have become familiar with the basic mechanisms of water and water-vapour transfer in building structures, i.e., hygroscopic moistening, capillary suction and diffusive water-vapour transport, and we have learned why the accumulation of moisture in building structures must be prevented. However, indoor comfort can be improved as the internal parts of the building structure absorb the water vapour in a period when the humidity of the indoor air is high, and release it once the humidity falls under the conditions of indoor comfort. The periodic exchange between the absorption and desorption processes improves the indoor microclimate. Nevertheless, air-conditioning will still be needed in many cases. In contrast, capillary suction is always unwanted and must be prevented in any case. Therefore, the diffusion of water-vapour molecules in the building structure is the most common reason that building structures moisten.

3.5.1 Indoor Air Moisture Control by the Sorption of Water Vapour

The sorption process whereby water-vapour molecules from the indoor air pass into and out of the building structure, affect the microclimate conditions in the building. The process is transient because the humidity of the indoor air varies due to the changing occupancy and the activities performed in the building and are periodic with the period $t_p = 24$ h (Fig. 3.9). It is assumed that the changing of the moisture content in the indoor air $v(t)$ during the observed period (a day) is described with a sine function:

$$\mathrm{v(t)} = \overline{\mathrm{v}} + \hat{\mathrm{v}} \cdot \sin\left(\frac{2 \cdot \pi \cdot \mathrm{t}}{\mathrm{t_p}}\right)\left[\frac{\mathrm{g}}{\mathrm{m}^3}\right]$$

where $\overline{v}$ is the average air-moisture content (g/m^3) and $\hat{v}$ is the amplitude (the greatest difference between the instantaneous value of v and the average daily moisture

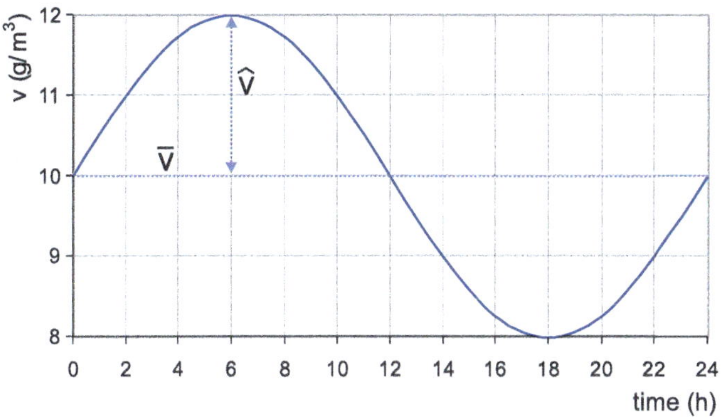

Fig. 3.9 Periodic variation of indoor-air moisture content $v(t)$ during the day. For the presented case the average air humidity $\overline{v}$ is 10 g/m^3 and the amplitude $\hat{v}$ =2 g/m^3

content $\overline{v}$) in a period t_p. In a residential building with an average indoor-air temperature of 20 °C and a relative humidity of 60%, the typical value of the average air humidity $\overline{v}$ is 10 g/m^3 and the typical amplitude $\hat{v}$ induced by the sources of water vapour inside the building is 1–2 g/m^3. To model the transient sorption water-vapour mass flow rate a new material property, referred to as the water-vapour diffusivity a_v (m^2/s), must be introduced. The water-vapour diffusivity defines the changing moisture content inside a substance as a result of the changing air humidity close to the surface in transient conditions. The water-vapour diffusivities a_v for some building materials are listed in Table 3.7. Because it was assumed that the variation of the indoor air humidity is periodic, the building structure's moisture content only changes down to a certain depth, as it takes up water for half of the time period and dries out during the other half. This depth is known as the periodic moisture-penetration depth $d_{p,v}$, calculated as [5, 6, 9]:

Table 3.7 Water-vapour diffusivity a_v and periodic moisture penetration depth $d_{p,v}$ for some construction materials (t_p = 24 h). The depth of moisture penetration in wood is considerably smaller compared to the cellulose insulation or mineral wool insulation

material	a_v (m^2/s)	$d_{p,v}$ (mm)
wood	$0.76 \cdot 10^{-10}$	1.4
concrete	$2.4 \cdot 10^{-10}$	2.6
gypsum	$9.2 \cdot 10^{-10}$	5.0
brick	$51 \cdot 10^{-10}$	11.8
cellulose thermal insulation	$370 \cdot 10^{10}$	31.9
mineral wool	$6480 \cdot 10^{10}$	133.5

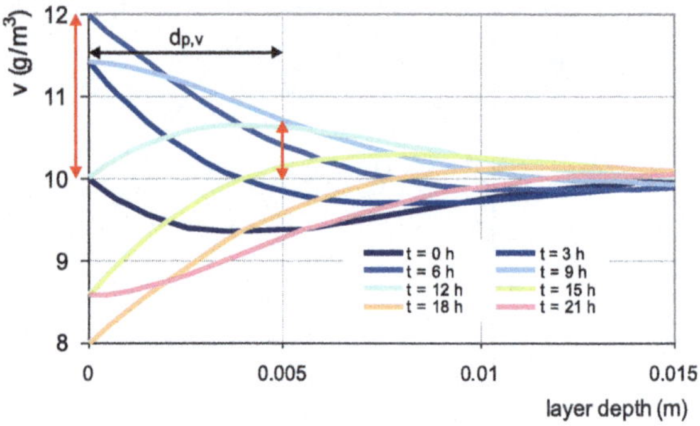

Fig. 3.10 Periodic daily variation of moisture content in a drywall due to the sorption of water vapour from the indoor air at $\overline{v}$ 10 g/m^3 and $\hat{v}$ 2 g/m^3. The periodic moisture

$$d_{p,v} = \sqrt{\frac{a_v \cdot t_p}{\pi}} = \sqrt{\frac{a_v \cdot 24 \cdot 3600}{\pi}} = 165.8 \cdot \sqrt{a_v}[\text{m}]$$

The moisture-content amplitude in the building structure at depth $d_{p,v}$ is 36.8% of the air moisture content amplitude $\hat{v}$, at the depth $2 \cdot d_{p,v}$ is reduced to 13.5% of $\hat{v}$, and at $3 \cdot d_{p,v}$ it falls to 5% of $\hat{v}$. Therefore, it can be assumed that the moisture content in the homogeneous single-layer building structure will only be changing down to a depth of approximately $3 \cdot d_{p,v}$ and that the thickness of the building structure beyond this depth will not be utilized for the indoor-air humidity control during the day. An example of periodically changing the moisture content in the building structure is shown in Fig. 3.10. The water-vapour mass flux rate that is entering and exiting the building structure is a time-dependent quantity and the maximum mass flux $\dot{g}_{max}$ in a period t_p is given by:

$$\dot{g}_{max} = \sqrt{2} \cdot \delta_v \cdot \frac{\hat{v}}{d_{p,v}} \cdot 3600 \frac{s}{h} \cdot 1000 \frac{g}{kg} \left[\frac{g}{m^2 \cdot h}\right]$$

where δ_v is the diffusion coefficient of water vapour in the building material (m^2/s, Table 3.7). Assuming that the variation of the indoor-air humidity is defined with a sine function, the average sorption flux in a 12-h half-period of moisture uptake is determined with:

$$\overline{g} = \dot{g}_{max} \cdot \frac{\int_0^{\pi} \sin v(t) \cdot dv(t)}{\pi} = \dot{g}_{max} \cdot \frac{\overbrace{-\cos\pi - (-\cos 0)}^{2}}{\pi} \dot{g}_{max} \cdot \frac{2}{\pi} \left[\frac{g}{m^2 \cdot h}\right]$$

and the specific water vapour m'_v absorbed by the building structure per 1 m^2 of surface area during the sorption process is:

$$m_v = \overline{g} \cdot \frac{t_p}{2} = \dot{g}_{max} \cdot \frac{2}{\pi} \cdot \frac{t_p}{2} = \dot{g}_{max} \cdot \frac{t_p}{\pi} \left[\frac{g}{m^2}\right]$$

The same quantity of water vapour that entered the building structure is then released back into the room during the desorption process in the next 12 h ($t_p/2$) of the sorption process.

Case Study Determine the periodic moisture-penetration depth $d_{p,v}$ and the quantity of water vapour stored (and released) each day in a 15-mm-thick gypsum board, if the daily average air humidity $\overline{v}$ is 10 g/m^3 and the air humidity amplitude $\hat{v}$ is 2 g/m^3.

$$d_{p,v} = 165.8 \cdot \sqrt{a_v} = 165.8 \cdot \sqrt{9.2 \cdot 10^{-10}} = 0.005 = 5 \text{ mm}$$

$$\dot{g}_{max} = \sqrt{2} \cdot \delta_v \cdot \frac{\hat{v}}{d_{p,v}} \cdot 3600 \frac{s}{h} = \sqrt{2} \cdot 2.0 \cdot 10^{-6} \cdot \frac{2}{0.005} \cdot 3600$$

$$\dot{g}_{max} = 4.07 \frac{g}{m^2 \cdot h}$$

$$m_v = \dot{g}_{max} \cdot \frac{t_p}{\pi} = 4.07 \cdot \frac{24}{\pi} = 31 \frac{g}{m^2}$$

Figure 3.10 shows how the moisture content inside a gypsum board is changing during the time period t_p equal to 24 h. It can be concluded that the gypsum board is well chosen because the thickness of the board is equal to 3·$d_{p,v}$. Comparing the amount of stored water vapour over the adsorption phase to the quantity of water vapour emitted by an average sitting person (approx. 50 g/h), it can be concluded that an area of ~ 14.5 m^2 of building structure will be needed to absorb the water vapour emitted by a person during a working day. The case study confirms that the sorption process in building structures has an impact on the indoor air humidity and therefore on the indoor comfort. Concrete, wood and cellulose plaster behave in a similar way to gypsum.

3.5.2 Modelling the Diffusive Water-Vapour Transfer in Building Structures

As heat transfer is similar to mass transfer, there is a similarity between convective heat and convective mass transfer. Convective mass transfer combines the diffusion of water-vapour molecules in the air and the motion of moist air molecules in a space. Let us assume that the indoor air at temperature θ_i and moisture content v_i is heated or cooled to the temperature θ_{si} at the surface of the building structure. This causes a change in the air humidity v_{si} for the air in contact with the structure's surface.

The direction of the mass transfer is determined by the difference in the humidity of the indoor air v_i and the air in contact with structure surface v_{si}, and accordingly by the difference in the partial pressures of the water vapour in the air $p_{v,i}$ and $p_{v,si}$. The specific mass flux of the water vapour $\dot{g}$ transferred at the surface of the building structure is [9]:

$$\dot{g} = \beta_v \cdot (v_i - v_{\theta si}) \left[\frac{kg}{m^2 \cdot s}\right] = \beta_p \cdot \left(p_{v,\theta i} - p_{v,\theta si}\right) \left[\frac{kg}{m^2 \cdot s}\right]$$

The surface mass-transfer coefficients β_v and β_p are determined from the similarity between heat transfer and mass transfer using Lewis' relation:

$$\beta_v = \frac{h_c}{\rho_a \cdot c_{p,a}} \left[\frac{m}{s}\right]; \quad \beta_p = \frac{\beta_v}{R_v \cdot (T + 273.15)} \left[\frac{\overbrace{kg}^{\frac{Ns^2}{m}}}{m^2 \cdot s \cdot \underbrace{Pa}_{\frac{N}{m^2}}} = \frac{s}{m}\right]$$

where α_c is the convective surface heat-transfer coefficient (W/m^{2}K), ρ_a is the density of the air (kg/m^3) and $c_{p,a}$ is the specific heat capacity of the air (J/kgK). The convective heat-transfer coefficient h_c is determined from the criteria equations according to the orientation of the building structure (vertical, horizontal) and the velocity of the air close to the surface of the structure. The typical values are between 1 and 4 W/m^2 K on the interior surface of the building structure, and from 5 to 10 W/m^2 K on the external side, owing to the influence of the wind speed. Accounting for these values, the resistance to water-vapour transfer ($1/\beta_v$) between the surface of the structure and the surround air is in the range 0.002–0.03 s/m [10].

The diffusion mass flux of water vapour through the structure depends on the actual water-vapour pressure in the building structure and the water-vapour transfer resistance r, also designated as s_d, determined as:

$$r = \mu \cdot d[m]$$

For a multilayer structure, the total water-vapour transfer resistance r_{tot} is the sum of the individual resistances r_i of the i-th layer of the structure:

$$\sum r_{tot} = \sum_i r_i = \sum_i \mu_i \cdot d_i \; [m]$$

The diffusion mass flux of the water vapour $\dot{g}$ through a multilayer building structure is determined by taking into account the constant C (Sect. 3.4.3) and the difference in the partial pressures of the water vapour in the air at the building-structure boundaries $(p_{v,i} - p_{v,e})$ with the extension:

$$\dot{g} = 0.622 \cdot \frac{p_{v,i} - p_{v,e}}{\sum_i r_i} \left[\frac{g}{m^2 \cdot h}\right]; \quad p_{v,i} > p_{v,e}$$

Explanation The calculations of diffusion water-vapour transfer through building structures usually neglect the resistance to water-vapour transfer at the surface of the building structure β_p, and only account for the resistances to diffusion of the water vapour r_i of the building structure layers.

3.6 Requirements for the Uptake of Water in Building Structures

The requirements that must be fulfilled by a building structure to ensure that the uptake of water does not significantly alter the material properties and affect the heat transfer, can be classified as follows:

- capillary suction of the water from the ground into the structure must be prevented;
- capillary suction of precipitation water into the structure at the building envelope in contact with the outdoor air must be limited with a suitably small water-sorption coefficient A of the outer façade layer;
- the conditions for microbial growth on the building construction surfaces must be avoided;
- the condensation of water vapour from the indoor air on the interior surfaces of the building structures must be avoided;
- interstitial condensation of water vapour in the building structure due to the diffusion of water-vapour molecules that could increase the moisture content to a level that affects the durability and the load capacity of building structures must be prevented.

The following sections describe in detail the measures necessary to fulfil these requirements.

3.6.1 Requirements for the Prevention of the Capillary Suction of Water into the Structure Having Contact with the Ground

The capillary suction of water into the building structure most often occurs at the foundations, flat roofs, or other structures in contact with the ground. The capillary suction can be prevented by placing damp proofing between the water source and the porous layer of the building structure, by sealing the pores or by neglecting capillary

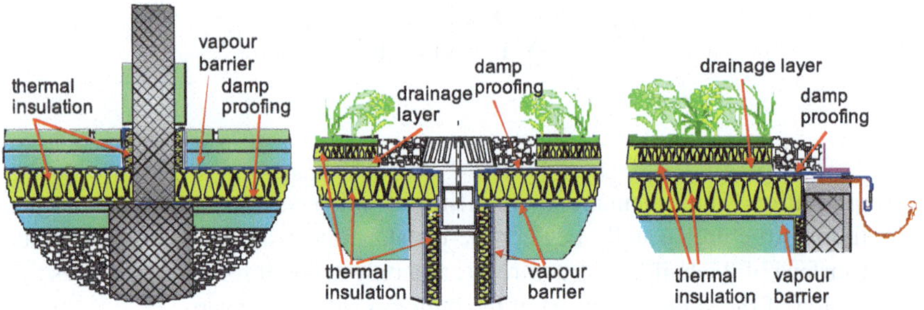

Fig. 3.11 Some examples of damp proofing solutions; a layer of damp proofing is laid on the external side of thermal insulation, while a vapour barrier finds its place on the interior side of the thermal insulation, where it prevents moisture uptake in thermal insulation by diffusion; the damp proofing of building foundations (left), the damp proofing around storm-water runoff and water gutter on a flat roof (middle, right) are shown [11]

suction due to chemical or electrical processes (Fig. 3.11). Damp proofing is realized with materials that are impermeable to water, containing no pores or only closed pores. Metal sheets, e.g., stainless-steel sheet, or natural or artificial pitch-based materials, are often used for damp proofing. Consistent sealing of the damp proofing is indispensable, as even a small hole can cause a considerable ingress of water. To reduce the capillary suction, it is also necessary to drain the water from the ground in contact with the building structures. This involves laying perforated pipes lined with filter fabric in a gravel backfill, and by installing suitable rainwater-drainage systems on the roof. Structures in contact with the ground must also be thermally insulated with materials that are impermeable to water, e.g., extruded polystyrene or foamed glass. Capillary water transport in new buildings can be efficiently prevented with intelligently designed building structures and by using well-proven materials. Existing buildings without a proper damp proofing pose more of a problem.

There are multiple remedies available for the protection of building structures against capillary suction in the scope of building renovation. They can be divided into mechanical, chemical and electrical techniques:

- mechanical solutions: the capillaries in the vertical walls are intersect by inserting a damp-proofing layer transversely into the structure. The wall is cut and a bitumen strip or sheet metal is inserted, or a corrugated stainless-steel sheet is driven into the wall by vibration (Fig. 3.12); this is the most effective way of preventing capillary suction. However, this approach is not possible in buildings that have thick walls, if installations are routed inside the walls, if the building has no foundations, or if the mechanical loads could cause the structures to crack.
- chemical processes: the capillary suction is preventing by solutions injected into drilled holes. The solution reacts with the building material and changes the properties of the pores (for example, by modifying the contact angle θ). One example is injecting a silicone emulsion. The emulsion is injected into holes (spaced at 100–200 mm, Fig. 3.13), and after 1 or 2 months the wall is rendered with a special

Fig. 3.12 Cutting the wall and inserting a bitumen strip (left) [12], inserted corrugated steel sheets (right) [12]

Fig. 3.13 Injection holes are visible in the wall of an older building before it is rendered with the final drying plaster

drying plaster. The large pore size in such drying plaster prevents capillary suction, but enables drying of the water vapour.

- electro-chemical processes: such process cause a change in the properties of the water molecules with an electrical voltage or with electromagnetic waves, and can effectively offset the force of the capillary suction.

3.6.2 Requirements for the Prevention of the Capillary Suction of Water from the Surface into the Façade of Building Structures

Façades exposed to precipitation can take up moisture by capillary suction, because the water droplets stick to the external surface of the structure. An exterior building-structure layer must be made of a material with a sufficiently low water-sorption coefficient *A*. The value of the water-sorption coefficient must not exceed 2 kg/m^2 h$^{0.5}$ or 0.033 kg/m^2 s$^{0.5}$. The values for some façade layers are given in Table 3.5.

3.6.3 Requirements for the Prevention of the Condensation of Water Vapour on the Interior Surfaces of Building Structures and the Conditions Stimulating Microbial Growth on the Building Structure's Surface

Condensation on the interior surface of a building structure can occur due to the following conditions:

- if the temperature of the interior surface of the structure is too low, then indoor air with temperature θ_i and a relative humidity φ_i circulating in the indoor space might come into contact with the cool surface of an external building structure. The temperature of the air in contact with the surface in this case falls and its relative humidity increases. If the air is cooled down to the dew point θ_{dew}, it becomes saturated, and if it is cooled even further, it stays saturated, but the excess water vapour condenses on the surface of the structure as liquid water.
- if the indoor air humidity is high due to the various sources of water vapour in the room. In this case the partial pressure of the water vapour in the air is high and might be close to the point of saturation and the air can cool below the dew-point temperature, even on the surfaces of thermally insulated building structures, especially the one with heat bridges. If the indoor water-vapour sources cannot be eliminated, the humidity of the indoor air can be decreased by room air-conditioning and/or adequate room ventilation.
- if stagnant air zones between the furniture and the external wall are formed; the convective and radiative heat transfer because of the blockage is less intense, and the temperature of the wall surface is lower and the condensation of water vapour from stagnant air is more likely.
- in the case of rapid heating of a cold room after being occupied. As the room is heated, the temperature of the air rises quickly, but the building structures takes considerably more time to heat up. Warmer air absorbs more water vapour emitted by the occupants, which might than condense on the surface of the structures until their temperature rises above the dew point, which might take several hours.

Occasional condensation may be allowed on building structures that do not absorb water (glass, ceramic tiles on bathroom walls), but microbial growth is possible on the surface of such structures, if the relative humidity of the air φ_i exceeds 80% for a couple of days. The microbial growth can be avoided by raising the temperature of the interior surface of the structure. The suitability of building structures with respect to the surface condensation and protection against microbial growth are verified with the methods discussed below.

3.6.3.1 Method of Minimum Partial Water-Vapour Pressure in the Air in the Vicinity of the Surface of the Building Structure

- The calculation begins with the determination of the structure's thermal transmittance U and the temperature θ_{si} (see Sect. 1.5.1) on the interior surface of the building structure, taking into account the predefined indoor θ_i and outdoor θ_e air temperatures.

Explanation According to the methodology in [8], the temperature θ_{si} is determined for each month of the year, taking into account the mean monthly outdoor air temperature θ_e. As an example, the values of θ_e over the year for the city of Ljubljana are given in Table 3.8.

- The actual pressure of the water vapour in the air p_v is determined from the indoor air temperature θ_i and the relative humidity φ_i. Their values are predetermined by the conditions of thermal indoor comfort (e.g., for living rooms θi is 20 °C and φ_i is 65%; see Sect. 1.8); the actual water-vapour pressure is determined with the expression:

$$p_{v,\theta i} = p_{v,sat,\theta i} \cdot \varphi_i = 611 \cdot e^{\left(\frac{17.08 \cdot \theta_i}{234.18 + \theta_i}\right)} \cdot \varphi_i [Pa]$$

- The saturation water-vapour pressure $p_{v,sat}, _{\theta si}$ at the temperature θ_{si} is determined. If the actual water-vapour pressure p_v is lower than the saturation pressure $p_{v,sat,\theta si}$, there will be no condensation on the surface of the building structure.

$$p_{v,\theta i} \leq p_{v,sat,\theta si}$$

- To prevent microbial growth, the maximum actual water-vapour pressure p_v at the temperature of the air in contact with building structure's surface must be limited as follows:

$$p_{v,\theta i} \leq 0.8 \cdot p_{v,sat,\theta si}$$

Table 3.8 Average monthly θ_e and φ_e in Ljubljana, with Cfb climate zone according to the Koppen-Geiger classification [13]

	θ_e (°C)	φ_e (%)
Jan	-1.6	90
Feb	0.5	92
Mar	4.5	86
Apr	11.5	88
May	15.5	84
Jun	20.5	77
Jul	23.6	78
Aug	17.7	91
Sep	17.7	93
Oct	11.8	94
Nov	5.0	93
Dec	0.2	90

Case Study The temperature of the indoor air θ_i is 20 °C and its relative humidity φ_i is 65%. Determine whether there is a risk of water-vapour condensation on the interior surface of a thermally insulated external wall, if the wall surface temperature θ_{si} is 18.3 °C?

The actual pressure of the water vapour in the indoor air p_v is:

$$p_{v,\,sat\,,\theta i} = 611 \cdot e^{\left(\frac{17.08 \cdot \theta}{234.18+\theta}\right)} = 611 \cdot e^{\left(\frac{17.08 \cdot 20}{234.18+20}\right)} = 2342.6\text{Pa}$$

$$p_{v,\theta i} = p_{v,\,sat\,,\theta i} \cdot \varphi_i = 2342.6 \cdot 0.65 = 1522.7\text{Pa}$$

The water-vapour saturation pressure $p_{v,sat,\theta si}$ for the wall's surface temperature is:

$$p_{v,sat,\theta si} = 611 \cdot e^{\left(\frac{17.08 \cdot 18.3}{234.18+18.3}\right)} = 2107.1\text{Pa}$$

The actual water-vapour pressure in the air ($p_v = 1522.7$ Pa) is lower than the water-vapour saturation pressure at the wall's surface temperature ($p_{v,sat,}$ $\theta_{si} = 2107.1$ Pa), so water vapour does not condense on the building structure's interior surfaces. The following condition is fulfilled as well:

$$p_{v,\theta i} = 1522.7 \leq 0.8 \cdot p_{v,sat,\theta si} = 0.8 \cdot 2107.1 = 1686.5\text{Pa}$$

Therefore, we do not expect microbial growth on the surface of the building structure.

3.6.3.2 Determining the Dew-Point Temperature on the Surface of a Building Structure

The dew-point temperature can be determined from the psychrometric diagram of moist air shown in Fig. 3.14. The risk of condensation on the surface of the building structure is determined as follows:

- like in the previous method, the procedure begins with a calculation of the building structure's thermal transmittance U and the temperature θ_{si} on the interior surface of the structure, taking into account the month with the lowest average daily temperature θ_e;
- the state of the indoor air, defined by the air temperature $\theta_I$and relative humidity φ_i, is plotted in the diagram (point 1 in Fig. 3.14);
- starting from the point of the indoor air state, a horizontal line is drawn to the saturation curve ($\varphi = 100\%$) and the dew-point temperature θ_{dew} is read on the x axis;
- the dew point is compared to the calculated building structure's surface temperature θ_{si}; the condition for non-condensation on the interior surfaces of the structure is fulfilled if θ_{dew} is lower than θ_{si}:

$$\theta_{\mathrm{dew}} < \theta_{\mathrm{si}}$$

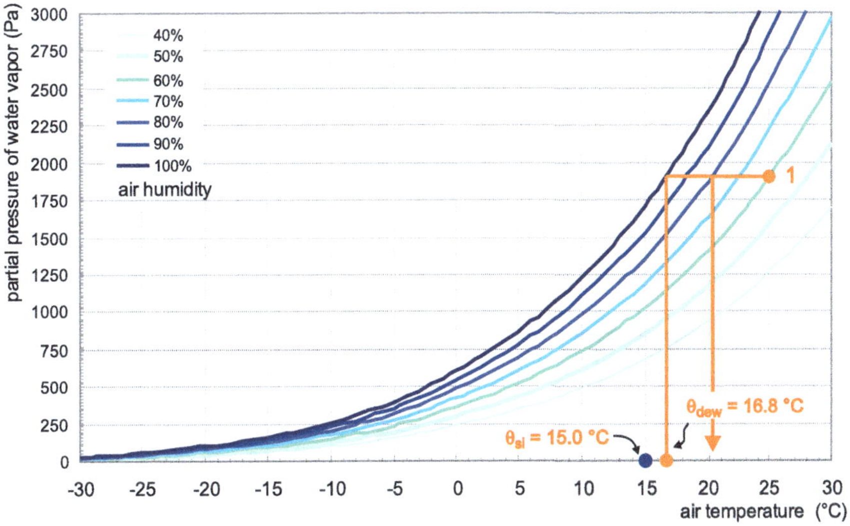

Fig. 3.14 Psychrometric diagram of moist air, the case study states are plotted

Case Study The temperature of the indoor air θ_i is 25 °C and its relative humidity φ_i is 60%. Check whether water-vapour condensation will occur on the interior surface of a wall with a temperature θ_{si} 15 °C. What is the minimum temperature on the surface of the structure that prevents microbial growth?

The dew-point temperature of air at the defined state θ_{dew} is 16.8 °C (Fig. 3.14). As the dew-point temperature is higher than the building structure's surface temperature ($\theta_{si} = 15$ °C), there is a risk of water-vapour condensation. Such a condition is not allowed and the wall must be thermally insulated in order to ensure a higher surface temperature. The lowest temperature on the interior surface of the structure θ_{si} that disables microbial growth is 20.2 °C.

3.6.3.3 Method of Minimum Surface-Temperature Factor

This method involves calculating the indoor-surface-temperature factor f_{Rsi} for the building structure and comparing it to the minimum acceptable temperature factor $f_{Rsi,\min}$. $f_{Rsi,\min}$ must be chosen so that water-vapour condensation and hence mould growth on the building structure's surface is prevented (Fig. 3.15). The surface temperature factor f_{Rsi} is given by:

$$f_{Rsi} = \frac{\theta_{si} - \theta_e}{\theta_i - \theta_e} > f_{Rsi,\min} = \frac{\theta_{si,\min} - \theta_e}{\theta_i - \theta_e} [1]$$

where θ_i and θ_e (°C) are the indoor and outdoor air temperatures, θ_{si} is the temperature of the interior surface of the building structure, and $\theta_{si,\min}$ (°C) is the minimum

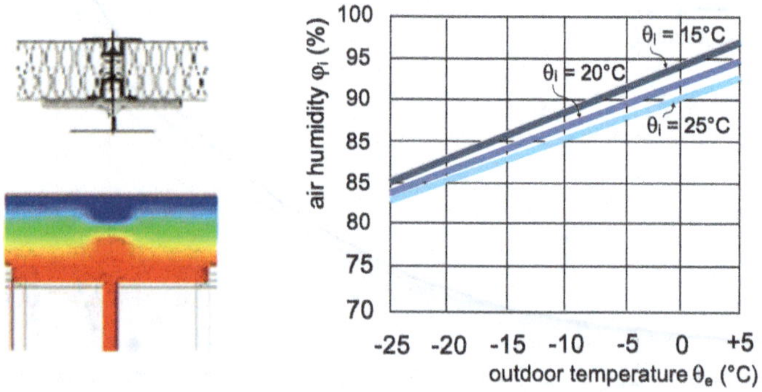

Fig. 3.15 Detail shows how a 100-mm-thick building envelope panel is fixed to a steel bearing structure. The linear thermal transmittance ψ of the joint is 0.218 W/m K. The diagram shows the highest φ_i at which there will be no risk of the condensation of water vapour on the thermal bridge interior area at different indoor θ_i, and outdoor air temperatures θ_e. The designed minimum temperature factor $f_{Rsi,\min}$ is taken to be 0.8 [14, 15]

required temperature for the interior surface of the building structure. $\theta_{si,\min}$ is given by the expression:

$$\theta_{\text{si,min}} = \frac{237.3 \cdot \ln\left(\frac{p_{v,\theta i}}{0.8 \cdot 610.5}\right)}{17.269 - \ln\left(\frac{p_{v,\theta i}}{0.8 \cdot 610.5}\right)} \; [^\circ\text{C}]$$

Case Study Check if there is a risk of mould growth on the interior surface of an external wall with the temperature $\theta_{si} = 18.3$ °C. The indoor-air temperature θ_i is 20 °C and the relative humidity of the indoor air φ_i is 65%. The water-vapour saturation pressure under such conditions for the indoor air is 1522.7 Pa. The temperature of the outdoor air θ_e is −1 °C.

$$\theta_{\text{si,min}} = \frac{237.3 \cdot \ln\left(\frac{p_{v,\theta i}}{0.8 \cdot 610.5}\right)}{17.269 - \ln\left(\frac{p_{v,\theta i}}{0.8 \cdot 610.5}\right)} = \frac{237.3 \cdot \ln\left(\frac{1522.7}{0.8 \cdot 610.5}\right)}{17.269 - \ln\left(\frac{1522.7}{0.8 \cdot 610.5}\right)} = 16.7\ ^\circ\text{C}$$

$$f_{\text{Rsi}} = \frac{18.3 - (-1)}{20 - (-1)} = 0.919 \quad f_{\text{Rsi,min}} = \frac{16.7 - (-1)}{20 - (-1)} = 0.843$$

Since f_{Rsi} is greater than $f_{Rsi,\min}$, microbial growth is not expected on the surface of the building structure.

The method of a minimum-surface-temperature factor is also used to analyse the condensation arising on the thermal bridging area. Using software tools (see Sect. 1.5.6) for the numerical modelling of the 3D heat transfer in building structures, the temperature factor f_{Rsi} is determined for the point of the lowest interior surface temperature and compared to the minimum value $f_{Rsi,\min}$.

The condensation of water vapour on the surface of a building structure can be avoided by one of the following measures:

- Better thermal insulation of structures. The most important means to prevent the condensation of water vapour on a building structure's interior surfaces is an adequate thermal transmittance U. Figure 3.16 shows the temperature of the building structure's interior surface θ_{si} versus the structure's thermal transmittance U. The values are determine for the outdoor air temperature θ_e—16 °C.
- More intense room ventilation. People, devices, plants and various processes release water vapour into the indoor air, increasing the air's humidity. This increases the risk of the water vapour condensing on the cold surfaces of a building structure. As this problem occurs particularly during cold periods of the year, and the risk of condensation could be minimized by appropriate space ventilation, with outdoor air that has low humidity at that time. The temporal variation of the indoor air's humidity $v_i(t)$ is determined from the mass balance of the water vapour in the building. The process is treated as a first-order problem, because the

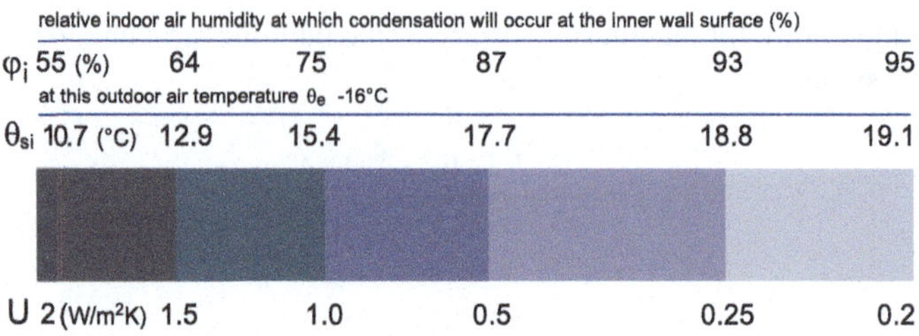

Fig. 3.16 Temperature at the building structure's interior surface θ_{si} and the relative humidity of air φ_i giving rise to the condensation of water vapour on the interior surface vs. the building structure's thermal transmittance U. The indoor-air temperature $\theta_i = 20$ °C

change in the air's humidity with respect to time (dv_i/dt) depends on the current air humidity $v_i(t)$. The mass balance of the water vapour accounts for the internal sources releasing water vapour at a constant mass flow rate $\dot{G}$ (kg/h), the humidity of the outdoor air v_e (kg/m^3) that flows into the building, and the humidity of the air removed from the space v_i (kg/m^3) in the same time. The building volume is V (m^3). The quantity of air that flows into the room $\dot{V}$(m^3/h) equals the quantity of air removed from the building and is given by the air exchange rate n (equal to $\dot{V}/V$, h^{-1}). The differential equation describing the rate of change of the air humidity v_i with respect to time t is:

$$\frac{\mathrm{dv_i}}{\mathrm{dt}} \cdot \mathrm{V} = \mathrm{n} \cdot \mathrm{V} \cdot \mathrm{v_e} - \mathrm{n} \cdot \mathrm{V} \cdot \mathrm{v_i(t)} + \dot{\mathrm{G}} = \mathrm{n} \cdot \mathrm{V} \cdot (\mathrm{v_e} - \mathrm{v_i(t)}) + \dot{\mathrm{G}} \quad [\mathrm{kg/h}]$$

The solution of the differential equation is:

$$\mathrm{v_i(t)} = \mathrm{v_e} + \frac{\dot{\mathrm{G}}}{\mathrm{n} \cdot \mathrm{V}}\left(1 - \mathrm{e}^{-\mathrm{n} \cdot \mathrm{t}}\right) \quad \left[\frac{\mathrm{kg}}{\mathrm{m}^3}\right]$$

where t (h) is the time elapsed from the moment the sources of water vapour are introduced into the building. Before that moment, the humidity of the indoor air v_i equals the humidity of the outdoor air v_e, because $\dot{G}$ is zero.

Case Study A building with an internal volume V 400 m^3 is ventilated with air-exchange rates (a) $n = 0.3$ h^{-1} and (b) $n = 1.2$ h^{-1}. The temperature of the outdoor air is θ_e is 25 °C and the relative humidity φ_e is 65%. Fourteen kg of water vapour is released into the indoor air each day. The indoor air temperature θ_i is 26 °C. Determine the relative humidity of the indoor air φ_i at steady state assuming a constant ventilation rate and a constant water-vapour emission throughout the day. The differential equation is solved by taking the time t to be infinite. The water-vapour saturation pressure and the actual pressure of the water vapour in the outdoor air are equal to $p_{v,sat,e}$ 3173.7 Pa and $p_{v,e}$ 2062.9 Pa ($\theta_e = 25$ °C, $\varphi_e = 65\%$.). The moisture content of the outdoor air v_e is [3]:

$$v_e = \frac{p_{v,e}}{R_v \cdot (\theta_i + 273.15)} = \frac{2062.9}{462 \cdot (25 + 273.15)} = 0.015 \ \frac{kg}{m^3}$$

In a steady state ($t \to \infty$), the expression for the temporal variation of the air humidity $v_{a,}$ (t) simplifies to:

$$v_i(t \to \infty) = v_e + \frac{\dot{G}}{n \cdot V}\left(1 - e^{-n \cdot \infty}\right) = v_e + \frac{\dot{G}}{n \cdot V}$$
$$= 0.015 + \left(\frac{14}{24} \cdot \frac{1}{0.3 \cdot 400}\left[\frac{kg \cdot day \cdot h}{day \cdot h \cdot m^3}\right]\right) = 0.0199 \ \frac{kg}{m^3}$$

The pressure of water vapour in the indoor air $p_{v,1}$, the saturation pressure $p_{v,sat,i}$ and the relative humidity of indoor air φ_i in the steady state are:

$$v_i(t \to \infty) = \frac{p_{v,i}}{R_v \cdot (\theta_i + 273.15)} = \frac{p_{w,i}}{462 \cdot (\theta_i + 273.15)}$$
$$\to p_{v,i} = v_i(t \to \infty) \cdot 462 \cdot (\theta_i + 273.15) = 2750.3 Pa$$

$$p_{v,sat,\theta i} = 611 \cdot e^{\left(\frac{17.08 \cdot \theta_i}{234.18 + \theta_i}\right)} = 611 \cdot e^{\left(\frac{17.08 \cdot 26}{234.18 + 26}\right)} = 3367.5 Pa$$
$$\varphi_i = \frac{p_{v,\theta i}}{p_{v,sat,\theta i}} \cdot 100 = \frac{2750.3}{3367.5} \cdot 100 = 81.7\%$$

The relative air humidity φ_i in case (a) when the building is ventilated with an air exchange rate of $n = 0.3$ h^{-1} is 81.7%; and in case (b) with the building ventilated with an air exchange rate $n = 2$ h^{-1} is 64.6%. This example underlines the significance of adequate ventilation for indoor comfort.

Case Study Figure 3.17 shows the temperature θ_i and the relative humidity φ_i of the indoor air for two apartments in an older apartment block, as well as the external temperature θ_e for a 10-day winter period. The first apartment (diagram on the left in Fig. 3.17) has old wooden casement windows, and the second apartment (diagram on the right) has been retrofitted with new windows with plastic frames and glazing with thermal transmittance U_w 1.3 W/m^{2}K. The airtightness of both apartments was measured using the blower-door method (see Sect. 2.2.9). The air-exchange rate n_{50} at a 50 Pa pressure difference was

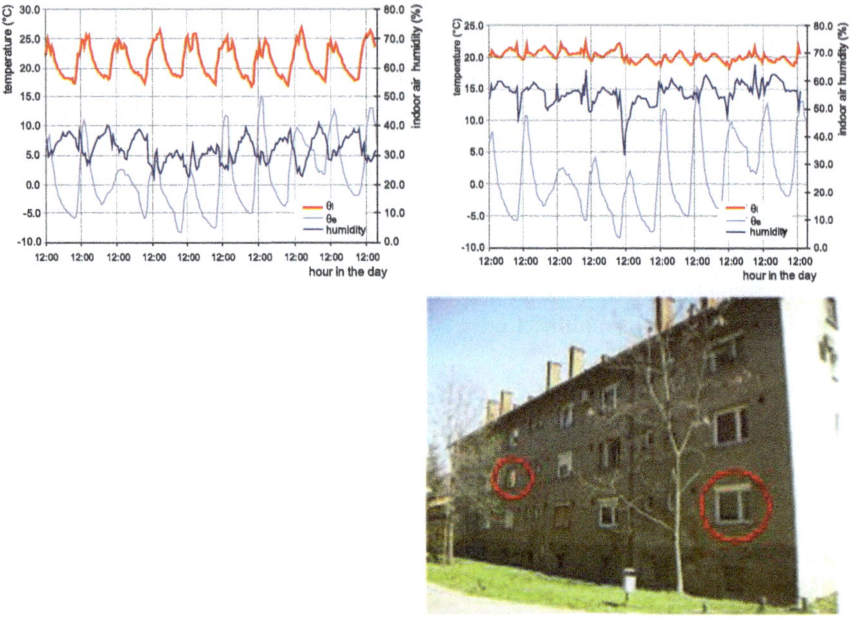

Fig. 3.17 Two apartments in an apartment block (right). The apartment on the left has old windows ($n_{50} = 14\ h^{-1}$(!) and the apartment to the right has new windows ($n_{50} = 2.4\ h^{-1}$). The indoor and outdoor air temperatures and the indoor air humidity in old (left) and retrofitted apartment (right) are shown on the graphs above. The variation of the indoor-air temperature in the old apartment is much larger, which is the consequence of larger heat losses [16]

14 h^{-1} for the apartment with old windows, and 2.4 h^{-1} for the apartment with new windows. In the renovated apartment, the measurements show that the relative humidity of the indoor air is constantly above 60%, which is inadequate. In contrast, the apartment with old windows is better ventilated (by infiltration) and the average relative air humidity φ_i assumes a more appropriate value of 35%. This example illustrates the role of adequate space ventilation in maintaining the air humidity in the indoor-environment comfort range. However, the ventilation must be done in an energy-efficient manner, which is clearly not the case in the example discussed here.

- One possible way to prevent surface condensation is by tempering the building structures all year round (Fig. 3.18). The heating pipes embedded in the walls up to 1 m above the ground, heat the structure to a temperature between 15 and 20 °C. In this way, the surface temperature of structure is higher and the moisture content in the walls is lower. This method is often employed for buildings listed as historic buildings or monuments. The users report better air quality, and the wall paintings and relics are more durable.

Fig. 3.18 Heating pipes were laid in two levels in the walls at the St. Peter and Paul church in Zagorje ob Savi, Slovenia. Water with temperature between 25 °C and 30 °C circulates the pipes all year long. The church has been fitted with a solar heating system to make the heating both energy and cost efficient [16]

3.6.4 Requirement for the Prevention of Interstitial Condensation of Water Vapour

The formation of liquid water inside a building structure can occur as a consequence of the diffusion of water-vapour molecules from the areas of higher partial pressure to the areas of lower partial pressure of the water vapour. In the winter period, it follows that the diffusion flux of water-vapour molecules in the air inside the pores of building material is directed from the building structure's interior outwards. In this case, the temperature inside the building structure decreases in the direction of water-vapour transfer, possibly leading to a state of air saturation. If the structure is cooled below the dew-point temperature, liquid water condenses from the water-vapour mass flux. This results in an increased thermal conductivity of the building materials and possible damages, especially in a common situation when the interstitial condensation occurs in a part of the structure where the temperatures are below the freezing temperature of water. It is, therefore, the best to avoid condensation with an appropriate design of the building structure. Nevertheless, there are certain situations where interstitial condensation is unavoidable. In such cases, interstitial condensation is only allowed if the following conditions are met:

- the quantity of liquid water accumulated by the end of the moisture uptake period per 1 m^2 of building structure M_c (kg/m^2) must not exceed the allowable value. The value of M_c is limited by the moisture content in the pores that can cause the transfer of liquid water by capillary suction through the building material;

Explanation The method adopted from [8] stipulates that the quantity of liquid water discharged into the building structure is determined for each month of the year, taking into account the average monthly temperatures and the humidity of the outdoor and indoor air. The example of meteorological conditions for the City of Ljubljana is presented in Table 3.8.

- the moisture ratio (humidity by mass) u (%) in a material exposed to interstitial condensation must not exceed the moisture level that enables capillary water transfer. As the building materials are never completely dry, the value u_{bas} at hygroscopic equilibrium is taken as their natural moisture ratio and the total moisture ratio u_{tot} in a material with interstitial condensation is the sum of the hygroscopic u_{bas} and excess moisture ratio u_{con}, determined from the quantity of condensed water. The total moisture ratio in any layer of a building structure must not exceed the maximum moisture ratio $u_{\max}$ (%), which is one of the hygroscopic properties of building materials (Table 3.9). The maximum allowable moisture content (humidity by volume) $v_{\max}$ (kg/m^3) is sometimes defined in place of the maximum moisture ratio.

Table 3.9 Hygroscopic moisture content u_{bas} and allowable moisture content $u_{\max}$ for some building materials. The total moisture ratio u_{tot}, which is the sum of $u_{bas} + u_{con}$,*must be below* $u_{\max}$ at all times [8, 17]

	u_{bas} (%)	u_{max} (%)
solid brick	1.5	4.0
concrete	2.1	3.8
mortar	2.0	6.0
soft timber	10.0	25.0
stone/glass wool	9.0	17.0

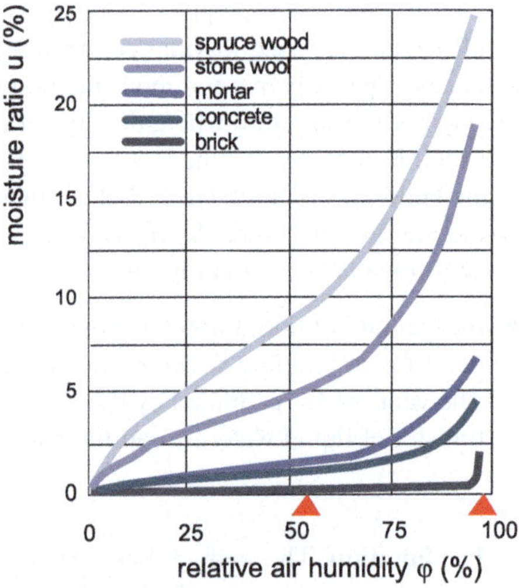

Fig. 3.19 Sorption curves for different building materials [12]

Explanation The values of u_{bas} and u_{max} for a building material are determined as follows from the Fig. 3.19, which shows the sorption curves for different building materials. The curves represent the relation between the moisture ratio of the material u (%) and the relative humidity φ_i (%) of the air surrounding the material. A comparison of the values u_{bas} and u_{max} from Table 3.9 shows that the basic moisture ratio u_{bas} is declared for the typical values of indoor-air humidity (φ_i ~ 50%), while the maximum allowable values of the moisture ratio u_{max} are declared for air humidities above ~ 96%, since higher values of the moisture ratio in the building material would lead to unacceptable capillary suction.

- all the materials that make up a building structure must dry out during the drying period (during the summer months) to the hygroscopic moisture ratio u_{bas}. It is necessary that a building structure is liquid–water free for at least of one month in a year.

3.7 Modelling the Transmission of Water Vapour by Diffusion in Building Structures

In engineering practice we use the Glaser method to determine if interstitial condensation occurs as consequence of the water-vapour diffusion in a building structure. It is an analytical method, but the solutions can also be found graphically. The problem is treated as a steady-state process, which means that:

- the boundary conditions do not change with respect to the time and the air temperature and the humidity on the interior and exterior side of the building structure (θ_i, θ_e, φ_i and φ_e, respectively) are constant during the observation period (which will be one month);
- the material properties are isotropic and do not change with respect to the material's temperature and humidity.

The calculation consists of the following steps:

❶ the indoor air temperature θ_i and the relative humidity φ_i are selected first. The values depend on the requirements for thermal comfort. Typical values are θ_I = 20 °C and φ_i = 65%, but other sets of values can be used when dealing, e.g., with building structures in bathrooms (θ_I = 22 °C, φ_i = 70%) or if the building is air-conditioned. According to [8], different values for each month can be assumed.
❷ the outdoor air temperature θ_e and humidity φ_e are selected. Example values are given in Table 3.8;
❸ the building structure's thermal transmittance U and the temperature on the surfaces and at the interfaces between the building structure layers are calculated: θ_{si},θ_1,θ_2,θ_3, ...θ_{se} (see Sect. 1.5.1);

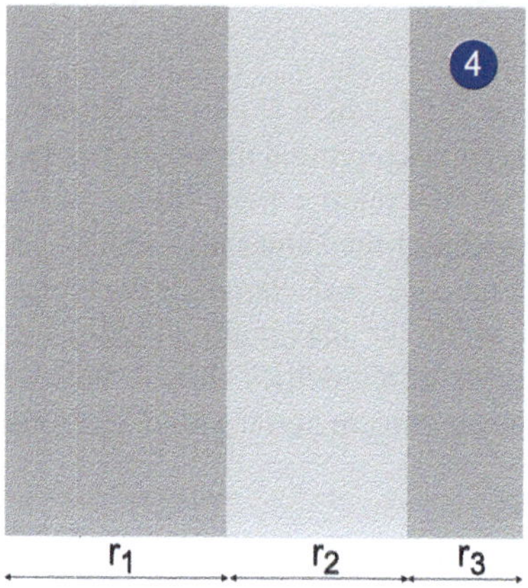

Fig. 3.20 Rather than using the layer thicknesses, on the Glaser diagram the layer's resistances to water-vapour transfer r are shown on the x axis (step 4)

❹ the Glaser diagram with the structure's layers is then drawn. The layer thickness d is replaced by the water-vapour diffusion resistance r (Fig. 3.20) r (the *sd* tag can be used instead of r) is the product of the layer thickness d and the corresponding water-vapour diffusion resistance μ;

$$r = \mu \cdot d[m]$$

❺ actual partial pressures of the water vapour in the indoor $p_{v,i}$ and outdoor air $p_{v,e}$ are calculated, and the values are plotted on the Glaser diagram. The resistance to water-vapour transfer on the surfaces of the building structure is neglected, so $p_{v,i}$ and $p_{v,e}$ are the actual partial pressures of the water vapour at the building structure's interior and exterior surface.

$$p_{v,i} = p_{v,sat,\theta i} \cdot \varphi_i \quad p_{v,e} = p_{v,sat,\theta e} \cdot \varphi_e \text{ [kPa]}$$

❻ the points shown $p_{v,i}$ and $p_{v,e}$ are connected by a line representing the distribution of the actual water-vapour partial pressure across the structure.

❼ the water-vapour saturation pressure $p_{v,sat}$ is calculated for the temperatures of the surfaces and at the interfaces between the building structure layers (θ_{si},θ_1,θ_2,θ_3, ... θ_{se}), and the values $p_{v,sat}$ are plotted on the Glaser diagram (Fig. 3.22). The points are connected with straight lines.

If the actual pressure of the water vapour p_v is smaller than the saturation pressure $p_{v,sat}$, elsewhere, on the surfaces of the structure and inside of the structure, there will be no condensation of the water vapour. Such an example is shown in Fig. 3.21. The design of such a building structure can be treated as optimal.

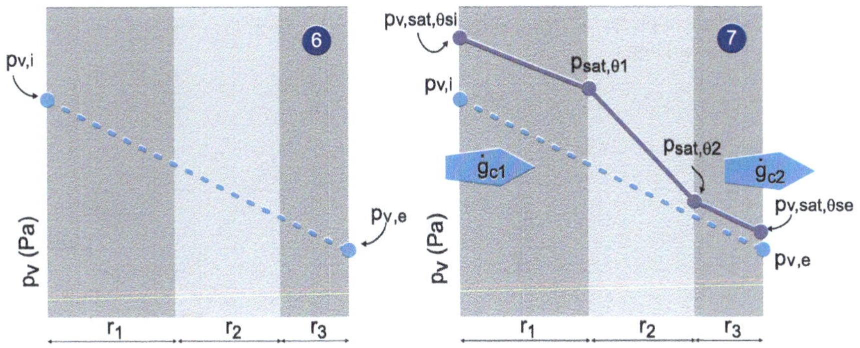

Fig. 3.21 Profile of actual water-vapour pressure (left) and profile of saturation water-vapour pressure (right) in a building structure. Throughout the building structure, the water-vapour saturation pressure is greater than the actual water-vapour pressure, so there will be no water-vapour condensation

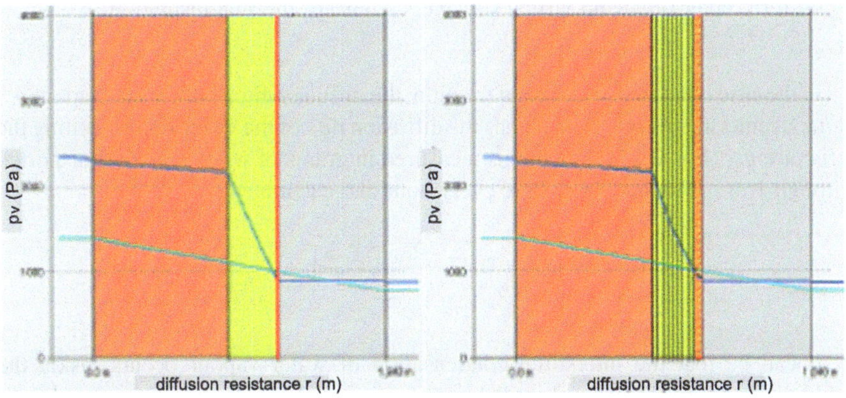

Fig. 3.22 Calculated profiles of actual water-vapour pressure (light blue) and saturation pressure (dark blue) in two building structures: one modelled with a uniform layer of thermal insulation (left), and the second modelled with a layer of thermal insulation divided so the resistance to heat transfer of individual sublayers does not exceed 0.25 m^2K/W (right). It can be seen that in the case of a large thermal resistance, the course of the water-vapour pressure in such a layer is linear [18]

Explanation The Glaser diagrams are drawn based on the assumption that the actual pressure and the saturation pressure of the water vapour changes linearly across a single layer of a structure. However, the water-vapour transfer is influenced both by the pressure gradient and the temperature gradient, resulting in a non-linear course of pressure. The error can be reduced by dividing the layers with a high thermal resistance to the heat conduction R_λ into any number of virtual sublayers, each of which has a thermal resistance lower than 0.25 m^2K/W. The example in Fig. 3.22 shows the profile of the actual pressure

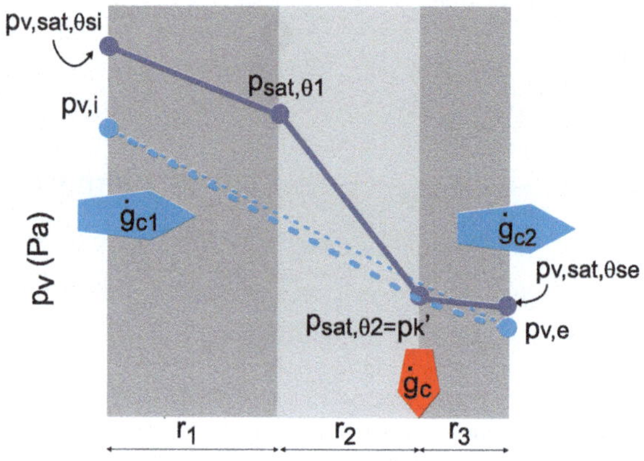

Fig. 3.23 Example of planar condensation in a building structure

and the saturation pressure of the water vapour for a building structure with a uniform layer (left) and virtual sublayers (right) of thermal insulation.

In the case of no interstitial condensation, the diffusion flux of the water vapour $\dot{g}_{c1}$ (g/m^{2}h) entering the structure equals the diffusion flux of the water vapour exiting the structure $\dot{g}_{c2}$ (Fig. 3.21). The flux is measured in grams of water vapour transported through 1 m^2 of building structure every hour and calculated as:

$$\dot{g}_{c1} = \dot{g}_{c2} = 0.72 \cdot \frac{p_{v,i} - p_{v,e}}{\sum_i r_j} \left[\frac{g}{m^2 \cdot h}\right]$$

It can be that the interstitial condensation of water vapour occurs inside the building structure. An example of the water-vapour condensation between two layers is shown in Fig. 3.23. This situation is known as planar condensation and takes place when the water-vapour saturation pressure $p_{v,sat}$ is lower than the actual pressure p_v at any interface between two adjacent layers of the structure. As such, a state ($p_v > p_{v,sat}$) would not be possible, the actual water-vapour pressure is reduced to the saturation pressure ($p_v = p_{v,sat}$) and designated as $p'_{v,k}$. A water vapour that condenses, increase the humidity of both building materials in contact with the plane of condensation. In this case the inlet water-vapour flux $\dot{g}_{c1}$ is greater than the outlet flux $\dot{g}_{c2}$ and the difference between the two is the amount of liquid water $\dot{g}'_c$ accumulated inside the structure in each hour for 1 m^2 of the structure area:

$$\dot{g}_{c1} = 0.72 \cdot \frac{p_{v,i} - p'_k}{r'}; \quad \dot{g}_{c2} = 0.72 \cdot \frac{p''_k - p_{v,e}}{r''}; \quad \dot{g}'_c = \dot{g}_{c1} - \dot{g}_{c2} \left[\frac{g}{m^2 \cdot h}\right]$$

where r' is the sum of the resistances to the water-vapour diffusion r of all the layers between the interior surface and the plane of condensation (in Fig. 3.23, $r' = r_1 + r_2$), and r'' is the sum of the resistances to water-vapour

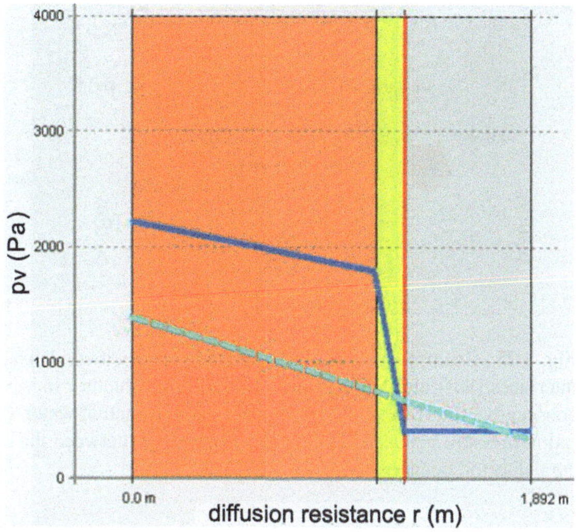

Fig. 3.24 Glaser diagram of water-vapour transfer for the structure from a calculation example [7].

Table 3.10 Study case data on materials

temperatures and pressures of water vapour						
	layers (inside-out)	d (cm)	r (m)	θ (°C)	p_v (Pa)	$p_{v,sat}$ (Pa)
	indoor air			20		
	inner surface			19.1	1403.4	2213.6
1	hollow brick	29	1.16	15.3	764.6	1740.3
2	mineral wool	12	0.132	-4.6	691.5	414.0
3	mortar	2	0.6	-4.7	361.0	410.5
	outer surface			-4.7	361.0	410.5
	outdoor air			-5		

transfer for all layers between the plane of condensation and the exterior surface of the building structure (in Fig. 3.23, $r'' = r_3$).

Study Case Find whether there is a risk of water-vapour condensation in an external wall that consists of three layers. The temperature and relative humidity of the outdoor air are θ_e = -5 °C and φ_e = 90%, respectively. See Table 3.10 for a list of materials, layer thicknesses d and layer resistance to water-vapour diffusion r, temperatures at the surface and at the interfaces between the layers, actual water-vapour pressures p_v and the saturation water-vapour pressures $p_{v,sat}$.

The Glaser diagram presenting the distribution of the saturation and actual pressures of the water vapour is shown in Fig. 3.24. The structure is subject to planar condensation. The quantity of liquid water $\dot{g}'_c$ condensing each hour per 1 m^2 of building structure is:

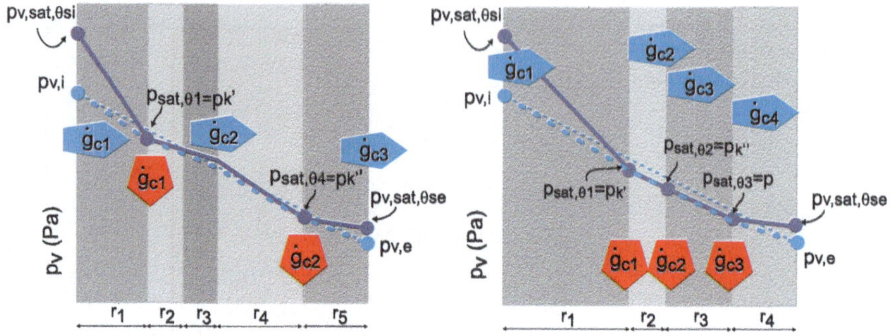

Fig. 3.25 Example of a building structure in which condensation occurs in two non-consecutive interfaces (left) and an example of a building structure in which condensation occurs in several consecutive interfaces (right). In these cases the actual water-vapour pressure is equal to the saturation pressure between points of $p'_{v,k}$ and $p''_{v,k}$ (between the first and the last interface where the condensation occurs)

$$\dot{g}_{c1} = 0.72 \cdot \frac{1.4034 - 0.414}{1.16 + 0.132} = 0.551\ \frac{\mathrm{g}}{\mathrm{m^2 \cdot h}}$$

$$\dot{g}_{c2} = 0.72 \cdot \frac{0.414 - 0.361}{0.6} = 0.064\ \frac{\mathrm{g}}{\mathrm{m^2 \cdot h}}$$

$$\dot{g}'_{c} = \dot{g}_{c1} - \dot{g}_{c2} = 0.551 - 0.064 = 0.49\ \frac{\mathrm{g}}{\mathrm{m^2 \cdot h}}$$

Water-vapour condensation can also take place in several non-consecutive interfaces of condensation. In such a case the actual water-vapour pressure p_v is equal to the saturation pressures $p_{v,sat}$ at each of the interfaces of condensation, as shown in Fig. 3.25 (left). The amount of the liquid water condensing per 1 m^2 of structure surface area in such a case must be determined for each of the interfaces of condensation separately as $\dot{g}'_{c,1}$ and $\dot{g}'_{c,2}$ in the following way:

$$\dot{g}'_{c,1} = \dot{g}_{c1} - \dot{g}_{c2}\left[\frac{\mathrm{g}}{\mathrm{m^2 \cdot h}}\right];\quad \dot{g}'_{c,2} = \dot{g}_{c2} - \dot{g}_{c3}\left[\frac{\mathrm{g}}{\mathrm{m^2 \cdot h}}\right]$$

$$\dot{g}'_{c,1} = 0.72 \cdot \left(\frac{p_{v,i} - p'_k}{r_1} - \frac{p'_k - p''_k}{r_2 + r_3 + r_4}\right)\left[\frac{\mathrm{g}}{\mathrm{m^2 \cdot h}}\right]$$

$$\dot{g}'_{c,2} = 0.72 \cdot \left(\frac{p'_k - p''_k}{r_2 + r_3 + r_4} - \frac{p''_k - p_{v,e}}{r_5}\right)\left[\frac{\mathrm{g}}{\mathrm{m^2 \cdot h}}\right]$$

where $r_{i,}$ is the resistance to the water-vapour diffusion of the ith layer of the building structure, as shown in Fig. 3.25 (left). Water-vapour condensation can also take place in several consecutive interfaces of condensation. Such a case is presented in Fig. 3.25 (right). If this is the case, the amount of liquid water condensing per 1 m^2 of structure

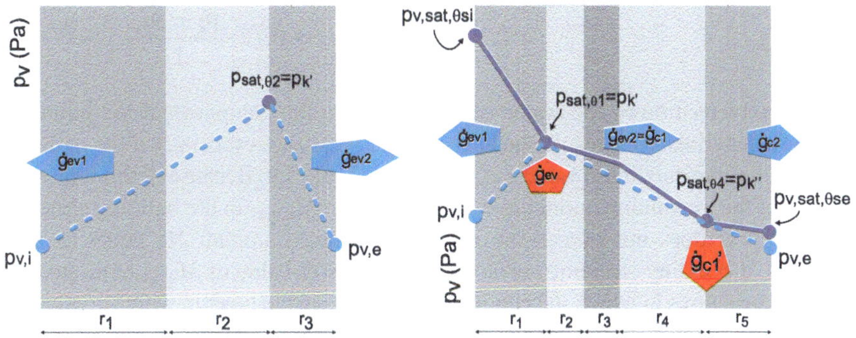

Fig. 3.26 Drying of a building structure can occur in any moment if the actual water-vapour pressure in the vicinity of the interface of condensation is lower than the saturation water-vapour pressure (left). The drying and condensation processes can occur in building structures at the same time (right)

surface area $\dot{g}'_{c,y}$ at the yth interface of condensation is determined like in the previous case, while the outlet water-vapour flux $\dot{g}_{c,y}$ is equal to the inlet water-vapour flux $\dot{g}_{c,y+1}$ to the next interface of condensation.

If the building structure contains condensed water and the actual water-vapour pressure p_v at the interface(s) of the condensation is higher than the actual water-vapour pressure at any surface of the structure ($p_{v,i}$ or/and $p_{v,e}$), the construction begins to dry. Such a case is shown in Fig. 3.26 (left). The water-vapour flux $\dot{g}_{ev}$ during drying by evaporation is directed towards the building structure surface at which the actual water-vapour pressure is lower; therefore, the building structure could dry from the inside and the outside at the same time. Total water-vapour flux $\dot{g}_{ev}$ in such a case is the sum of the indoor $\dot{g}_{ev1}$ and the outdoor $\dot{g}_{ev2}$ mass fluxes:

$$\dot{g}_{ev1} = 0.72 \cdot \frac{p'_k - p_{v,i}}{r'}; \quad \dot{g}_{ev2} = 0.72 \cdot \frac{p'_k - p_{v,e}}{r''}; \quad \dot{g}_{ev} = \dot{g}_{ev1} + \dot{g}_{ev2} \left[\frac{g}{m^2 \cdot h}\right]$$

where r' and r'' are the sum of resistances to the water-vapour diffusion r of all the layers between the interface of condensation and the interior or exterior surfaces of building structure, respectively (according to Fig. 3.26 (left) r' is equal to $r_1 + r_2$ and r'' is to equal r_3). In special cases, both the drying and the condensation of water vapour can occur in the building structure and amount of the water decrease ($\dot{g}_{ev}$) in one part, and increase ($\dot{g}'_c$) in the other part of the building structure. An example is shown in Fig. 3.26 (right). The sum of the resistances to the water-vapour diffusion r'' in this case is determined by the sum of the resistances to the nearest plane of condensation ($r'' = r_2 + r_3 + r_4$) and the drying diffusion water-vapour flow $\dot{g}_{ev,2}$ is equal to the water-vapour flow entering the plane of condensation. The drying mass flow and the amount of condensing water are defined using the following expressions:

$$\dot{g}_{ev} = \dot{g}_{ev1} + \dot{g}_{ev2} = 0.72 \cdot \frac{p'_k - p_{i,v}}{r_1} + 0.72 \cdot \frac{p'_k - p''_k}{r_2 + r_3 + r_4} \left[\frac{g}{m^2 \cdot h}\right]$$

$$\dot{g}'_c = \dot{g}_{c1} - \dot{g}_{c2} = \dot{g}_{ev2} - \dot{g}_{c2} = 0.72 \cdot \frac{p'_k - p''_k}{r_2 + r_3 + r_4} - 0.72 \cdot \frac{p''_k - p_{v,e}}{r_5} \left[\frac{g}{m^2 \cdot h}\right]$$

The condensation of the water vapour increases the dampness of the building-structure materials. As a result, the heat-transfer increases, especially if the moisture ratio reaches the maximum value u_{max} (Fig. 3.19, Table 3.9) Because of this it must be verified if the maximum amount of condensed water $M_{c,max}$ in the building structure over the year does not exceed the maximum allowed amount M_a equal to 1 kg per 1 m^2 of building-structure surface. An even lower value of M_a is requested for building structures in which the condensation occurs in the layer in which condensed water cannot be spread into the whole volume of the layer by local capillary suction. This is the property of most fibrous thermal insulation materials. For such building structures M_a is limited to 0.5 kg per 1 m^2 of building structure surface over the year. For certain materials, the allowable increase in the moisture ratio is also restricted. For example, for solid wood products the increase in the moisture ratio u_{dif} due to the interstitial condensation of water vapour must not exceed $u_{dif,max}$ 5% above the hygroscopic moisture ratio u_{bas} and for other wood products $u_{dif,max}$ must not be more than 3%. The increase of the moisture ratio is determined using the expression:

$$u_{dif} = \frac{M_{c,max}}{d_r \cdot \rho} \cdot 100 < u_{max}[\%]$$

where d_r (m) and ρ (kg/m^3) are the virtual layer thickness, in which the condensed water is distributed, and the density of this layer. The typical value of d_r is in the range 0.02–0.07 m, but not more than the actual layer thickness d.

Study Case It was found that the maximum amount of condensed water $M_{c,max}$ in the building structure over the year is 0.25 kg/m^2. Condensation occurs in a 2-cm-thick layer of the wooden board. Determine if the increase of the moisture ratio u_{dif} is acceptable. The density of the wooden board in which capillary suction cannot occur is 600 kg/m^3 and $u_{dif,max}$ for such a product is 3%.

$$u_{dif} = \frac{M_{c,max}}{d_r \cdot \rho} \cdot 100 = \frac{0.25}{0.02 \cdot 600} \cdot 100 = 2.08\%$$

Solution: the interstitial condensation of water vapour is acceptable because (a) the maximum amount of condensed water $M_{c,max}$ in the building structure is below 0.5 kg/m^2 and (b) because the increase of the moisture ratio $u_{dif,max}$ is below 3%. *Note*: check whether M_a is a state with a different value in the national regulations.

3.7.1 Modelling of Water-Vapour Mass Transfer by Diffusion in Engineering Practice

The thermodynamic state of the outdoor and the indoor air is changing all the time. The influence of transient boundary conditions can be simplified by assuming that the water-vapour diffusion process is steady in each of the months during the year. This calculation procedure is defined in [8]. It is based on the Glaser's diagram, where, as we already know, the building-structure layers' thicknesses d are substituted by their resistances to water-vapour transfer r. The state of the indoor air is described based on thermal comfort criteria (e.g., $\theta_i = 20$ °C during the heating period, and 25 °C in the summer time). The state of the outdoor air is described using the mean monthly meteorological data for a particular site. As an example, the data for Ljubljana is given in Table 3.8. The procedure includes the following steps:

❶ the thermal transmittance of the building structure U is determined, and then the temperatures at the surface and at the interfaces between the building structure layers are calculated for each month in a year (θ_{si}, θ_1, θ_2, θ_3, …θ_{se}).
❷ these temperatures are then used to determine the water-vapour saturation pressure and the actual pressure on the interior and the exterior surfaces of the building structure ($p_{v,sat,\theta si}, p_{v,sat,\theta 1}, p_{v,sat,\theta 2}, p_{v,sat,\theta 3}$ … and $p_{v,\theta si} = p_{v,i}, p_{v,\theta se} = p_{v,e}$).
❸ twelve Glaser diagrams are drawn. The layers are represented with their resistances to water-vapour transfer r, which is equal to the product of the layer thickness d and the corresponding water-vapour diffusion resistance μ of the layer material.
❹ the points $p_{v,i}$ and $p_{v,e}$, representing the actual water-vapour pressure in the indoor and outdoor air, are drawn on the Glaser diagram and are connected by a straight line showing the actual water-vapour pressure course inside the structure. The saturation water-vapour pressure for the corresponding temperatures (θ_{si}, θ_1, θ_2, θ_3,… θ_{se}) at the interfaces of layers are also plotted and connected with straight lines. The first Glaser diagram is created for the last month when there is no condensation in the building structure. For example, if the condensation does not occur in the period from June to October, the first diagram will show the state of building structure in October.
❺ the analysis continues with the next month (e.g., November).
❻ ① if it turns out that the saturation pressure in the building structure is greater than the actual water-vapour pressure through the next month, there will be no condensation and the analysis moves on to the next month as well.
❻ ② if the saturation pressure is lower than the actual pressure at one or several interfaces between the building-structure layers, there is going be condensation of water vapour there. The monthly inlet and outlet water-vapour mass flux is determined (the inlet mass flux is always greater than the outlet mass flux on water vapour in this cases). The monthly amount of condensed water is determined as well. Here is an example for December (Fig. 3.27, left, N = 31):

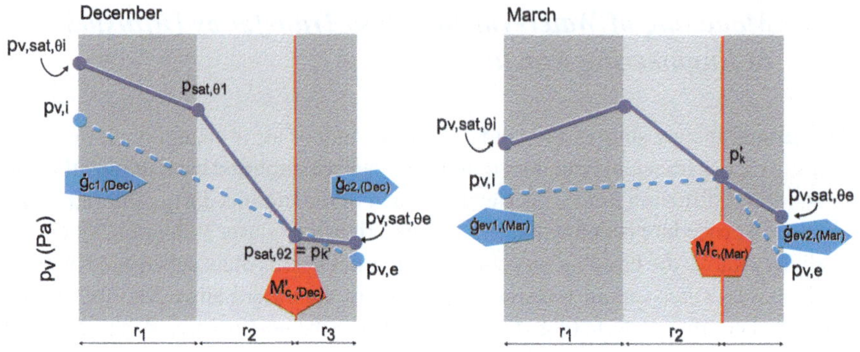

Fig. 3.27 An example of water-vapour condensation in the one interface in the December, quantity of the condensed water $M'_{c,(dec)}$ is equal to the difference between inlet and outlet mass flux of the water vapour $(g_{c1,(Dec)} - g_{c2,(Dec)})$ (left); an example of drying process during the March (right)

$$\dot{g}_{cl,(Dec)} = 0.72 \cdot \frac{p_{v,i,(Dec)} - p'_{k,(Dec)}}{\underbrace{r'}_{r_1+r_2}} \cdot 24 \cdot N\left[\frac{g}{m^2 \cdot h} \cdot \frac{h}{day} \cdot \frac{day}{month} = \frac{g}{m^2 \cdot month}\right]$$

$$\dot{g}_{c2,(Dec)} = 0.72 \frac{p'_{k,(Dec)} - p_{v,e,(Dec)}}{\underbrace{r''}_{r_3}} \cdot 24 \cdot N\left[\frac{g}{m^2 \cdot h} \cdot \frac{h}{day} \cdot \frac{day}{month} = \frac{g}{m^2 \cdot month}\right]$$

$$M'_{c,(Dec)} = \dot{g}_{cl,(Dec)} - \dot{g}_{c2,(Dec)}\left[\frac{g}{m^2 \cdot month}\right]$$

❼① the analysis continues with the next month. If the condensation of water vapour reappears, the monthly quantity of condensed water $M_{(i)}$ is calculated and added to the amount of the water $M_{c,(i-1)}$ condensate in the preceding month(s), yielding the total quantity of condensate $M_{c,(i)}$:

$$M_{c,(i)} = M_{c,(i-1)} + M'_{c,(i)}$$

❼② if there is no condensation ($M'_{c(i)} = 0$), the structure will be drying during this month. The monthly quantity of evaporated water is (example for March, Fig. 3.27, right, N = 31):

$$\dot{g}_{evl,(Mar)} = 0.72 \cdot \frac{p'_{k,(Mar)} - p_{v,i,(Mar)}}{\underbrace{r'}_{r_1+r_2}} \cdot 24 \cdot N\left[\frac{g}{m^2 \cdot h} \cdot \frac{h}{day} \cdot \frac{day}{month} = \frac{g}{m^2 \cdot month}\right]$$

$$\dot{g}_{ev2,(Mar)} = 0.72 \cdot \underbrace{\frac{p'_{k,(Mar)} - p_{v,e,(Mar)}}{r''}}_{r_3} \cdot 24 \cdot N \left[\frac{g}{m^2 \cdot month}\right]$$

The total quantity of the previous amount of the condensate water at the end of the February $M_{c.(Feb)}$ is reduced by the quantity evaporated during March:

$$M_{c,(Mar)} = M_{c,(Feb)} - \left(\dot{g}_{evl,(Mar)} + \dot{g}_{ev2,(Mar)}\right)\left[\frac{g}{m^2 \cdot month}\right]$$

❽ the procedure is repeated for all 12 months of the year.

The design of the building structure is appropriate with respect to the water-vapour transfer by diffusion if the following conditions are met:

- there is no condensation of water vapour in the building structure for any month of a year;
- all the layers where the condensation of water vapour appears are completely dried out during the summer months. Therefore, before the first month when the condensation reappears. In this case the value M_c equals 0 kg/m^2 in the last month of the calculation cycle.
- the maximum amount of condensed water $M_{c,max}$ in the building structure in any month during the year is less than the allowable value M_a per unit of the building structure surface area.

❾ if any of the above-mentioned conditions are not fulfilled, the building structure must be redesigned. The design is modified based on the guidelines given in Sect. 3.5.2.

Case Study An analysis of water-vapour transfer by diffusion in a building structure with the layers indicated in Table 3.11 will be carried out. The structure has a thermal transmittance of 0.317 W/m^{2}K. The water-vapour pressure on the interior $p_{v,i}$ and exterior $p_{v,e}$ sides of the structure is determined for the boundary conditions shown in Table 3.12, and values are shown in Table 3.13. Twelve Glaser's diagrams are constructed and shown

Table 3.11 Composition of the virtual building structure analysed in the case study

	d (cm)	λ (W/mK)	μ (1)
plaster	2.5	0.81	10
brick	29	0.61	6
thermal insulation	8	0.034	1.1
mortar	2	0.99	25
silicate facade brick	12	1.1	20

Table 3.12 Monthly mean values θ_i and φ_i (left) and monthly mean values θ_e and φ_e (right) for the building discussed in the case study

	θi (°C)	φi (%)	θe (°C)	φe (%)
Jan	20	59	-1.0	90
Feb	20	57	1.0	92
Mar	20	59	5.0	86
Apr	20	57	9.0	88
May	20	58	14.0	84
Jun	20	54	18.0	77
Jul	20	51	20.0	78
Aug	20	51	19.0	91
Sep	20	51	15.0	93
Oct	20	56	10.0	94
Nov	20	52	4.0	93
Dec	20	56	1.0	90

Table 3.13 Water-vapour pressure, water-vapour mass flux and amount of condensate water in case study building structure.

	pi (kPa)	pk' (kPa)	pe (kPa)	ġc1(+),ġev1(-) (kg/m² month)	ġc2(+),ġev2(-) (kg/m² month)	M'c,(i) (kg/m² month)	Mc,(i) (kg/m²)
Oct	1.332		1.300			0	0
Nov	1.332	0.923	0.810	0.088	0.017	0.071	0.071
Dec	1.378	0.669	0.557	0.158	0.017	0.141	0.212
Jan	1.332	0.589	0.481	0.165	0.017	0.148	0.360
Feb	1.355	0.683	0.582	0.135	0.015	0.120	0.480
Mar	1.261	0.893	0.724	0.082	0.027	0.055	0.535
Apr	1.191	1.398	1.193	-0.044	-0.032	-0.076	0.459
May	1.191	1.788	1.478	-0.133	-0.049	-0.182	0.277
Jun	1,191	2.406	1.855	-0.261	-0.085	-0.346	0
Jul	1.308		2.270			0	0
Aug	1.215		1.842			0	0
Sep	1.308		1.882			0	0

in Fig. 3.28, starting with October, the month in which there is no condensed water in the structure ($M_{c,(Oct)} = 0$ kg/m^2). In the next month, November, we can see that condensation will occur at the interface between the mortar and the thermal insulation layer (for this case, $r' = 2.078$ m and $r'' = 2.9$ m). Water-vapour condensation also takes place in the building structure in December, January, February and March. The total quantity of condensate reaches a maximum value at the end of March ($M_{c,(Mar)} = M_{c,\max}$). The structure begins to dry in April and is completely dried out during the next three months. The monthly values of the water-vapour mass fluxes and the amount of condensate water month-by-month is shown in Fig. 3.28. The building structure is appropriate because it dries out during the summer months and because the maximum amount of condensate water is 0.53 kg/m^2 (in March). This is below the allowed maximum value for M_a equal to 1 kg/m^2.

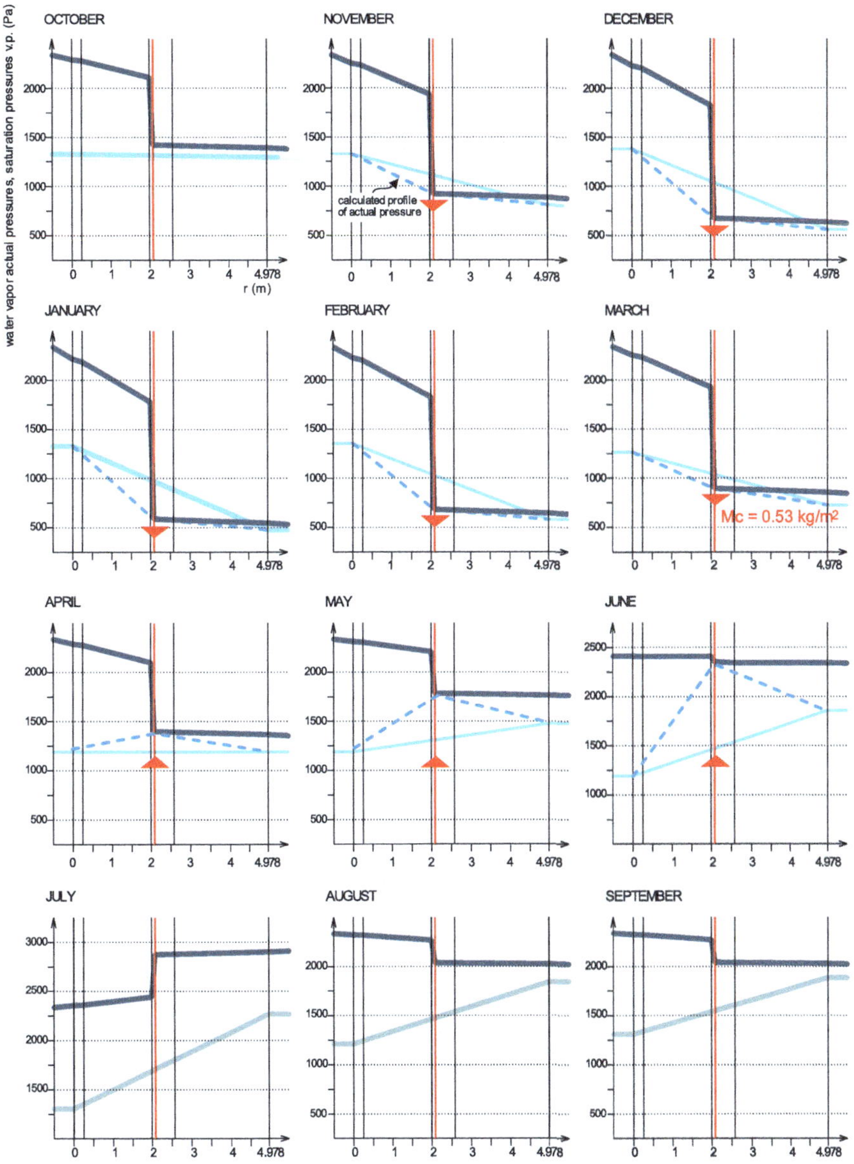

Fig. 3.28 Glaser's diagrams constructed for twelve mouths, starting with October, the month when there is no condensed water

3.7.2 *What to Do if the Condensation Criteria Are not Fulfilled?*

In case that either the maximum moisture ratio u_{max} or the maximum amount of condensate water $M_{c,max}$ that appears in the structure over the year is above the allowable values or the building structure is not completely dry in at least one month of the year, the building structure must be modified. Two options are common:

- by reducing the mass flux of water vapour from the indoor air into the building structure, or from the internal (warmer) part of the structure into its external (colder) part;
- by enabling the water vapour that enters the building structure to be transferred through the structure and pass into the outdoor air as unimpeded as possible.

The first measure calls for the installation of a vapour barrier or vapour retarder, a thin layer made of a material with a high water-vapour diffusion resistance μ and therefore a high resistance to water-vapour transfer r. Such a layer prevented water-vapour condensation by the reduction of the inlet water-vapour mass flux into the structure. To ensure this the vapour barrier or retarder must be installed on or close to the interior surface of the building structure.

Explanation Water-vapour retarders or water-vapour barriers are classified according to the resistance to water-vapour transfer r_{vd} (also designated as $s_{d,vd}$); layers with $r_{vb} > 10^+$ m are classified as vapour retardants, while layers with $r_{vb} \sim 100^+$ m are classified as vapour barriers. Some examples are given in Table 3.14.

The required water-vapour resistance r_{vb} and thickness of the water-vapour barrier or retarder d_{vb} can be determined graphically using the Glaser diagram. The procedure is as follows (Fig. 3.29, left):

Table 3.14 Thickness d and resistance to water-vapour transfer r of some materials used as vapour barriers or retarders [6]

	d (mm)	$r = s_d$ (m)
polyethylene sheet	0.15	50
polyethylene sheet	0.25	100
PVC sheet	0.20	30
aluminium foil	0.05	1500
bitumen building paper	0.1	2
vapor-permeable membrane	0.4	0.2

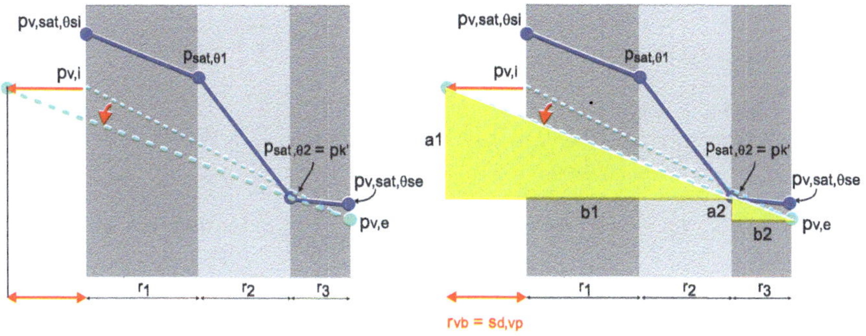

Fig. 3.29 Required resistance of water-vapour barrier r_{vb} ($s_{d,vp}$) is determined graphically in the Glaser diagram (left) for analytically using the rule of similar triangles (right)

❶ the line representing the water-vapour pressure distribution is rotated about the point $p_{v,e}$ downwards, until it touches the saturation-pressure point p'_k;
❷ the point $p_{v,i}$ is translated horizontally to the new line representing the water-vapour pressure distribution;
❸ considering the scale, the required additional resistance of the vapour barrier r_{vb} ($s_{d,vp}$) is read from the diagram;
❹ the manufacturers' catalogues are then used to find the appropriate product.

The required vapour-barrier resistance r_{vb} can also be calculated using the similar-triangles method (note: the expression below is adapted to the example in Fig. 3.29, right):

$$\frac{a_1}{b_1} = \frac{a_2}{b_2} \rightarrow \frac{p_{v,i} - p'_k}{r_{vb} + r_1 + r_2} = \frac{p'_k - p_{v,e}}{r_3} \rightarrow r_{vb} = \frac{p_{v,i} - p'_k}{p'_k - p_{v,e}} \cdot r_3 - (r_1 + r_2)[\mathrm{m}]$$

Study Case Find the required thickness of a polyethylene-sheet vapour barrier ($\mu_{vb} = 80{,}000$) to avoid interstitial condensation in the structure defined in Table 3.15, and define the position where the barrier should be installed. The water-vapour pressure in the indoor air $p_{v,i}$ is 1403 Pa, and the water-vapour pressure in the outdoor air $p_{v,e}$ is 234 Pa. Condensation takes place at the interfaces between layers 2–3 and 5–6. In order to design a water-vapour barrier that will prevent interstitial condensation, the water-vapour pressure at condensation interfaces should be lower (at least equal) than the saturation pressure of water vapour $p_k{}' = 606.4$ Pa. The analytical procedure for the determination of the vapour barrier is as follows:

$$r_{vb}(\equiv s_{d,vb}) = \frac{p_{v,i} - p_k'}{p_k' - p_{v,e}} \cdot (r_3 + r_4 + r_5 + r_6) - (r_1 + r_2)$$

$$= \frac{1.403 - 0.606}{0.606 - 0.234} \cdot (0.340 + 0.114 + 0.132 + 0.072) - (0.144 + 0.336)$$

$$r_{vb} = 0.930\text{m} \rightarrow d_{vb} = \frac{r_{vb}}{\mu_{vb}} = \frac{0.93}{80000} \approx 0.1\text{mm}$$

The graphical solution, also shown in Fig. 3.30, gives the same result as the analytical solution. The polyethylene sheet must be installed on the interior (warm) side of the building structure, before the first interface of the water-vapour condensation.

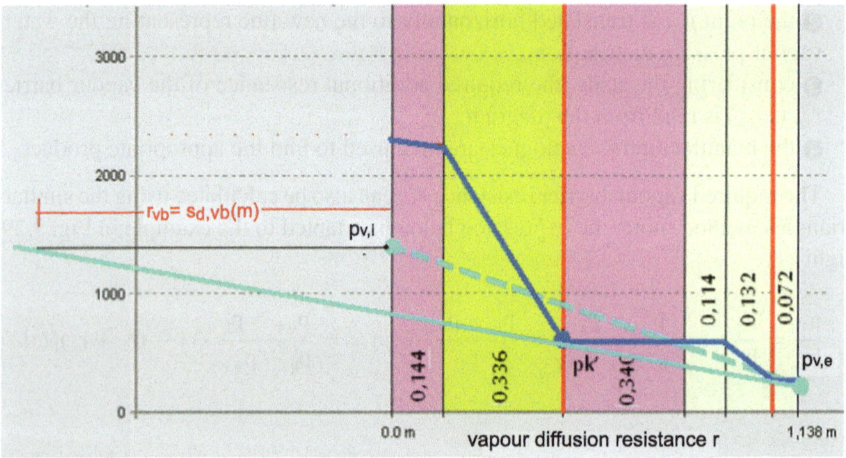

Fig. 3.30 Graphical solution of the case study

Table 3.15 Composition of building structure from the calculation example

	layers (inside-out)	d (cm)
	indoor air	
	inner surface	
1	plasterboard	2
2	mineral wool	24
3	chipboard-hard	2
4	cement mortar	0.3
5	mineral wool	12
6	facade mortar	0.3
	outer surface	
	outdoor air	

Warning The installation position of the water-vapour barrier is very important. If the barrier is installed on the warmer side of the building structure (e.g., before the thermal insulation), this reduces significantly the diffusive water-vapour mass flux that enters the structure and very likely interstitial condensation will not occur. If the water-vapour barrier is installed, as a result of designer error, in the exterior part of the building structure (e.g., on the outer side of the thermal insulation), the water-vapour mass flux towards the outdoor air will be significantly reduced and very likely interstitial condensation will occur. In such a case an additional problem occurs in this case when condensed water freezes inside the building structure, potentially causing the render to come off. Therefore, any layer installed on the colder, exterior part of the building structure must be vapour-permeable. Nevertheless, it can be watertight, such as the membranes for secondary roofing. The two properties (vapour permeability and water tightness) are not mutually exclusive, because the diameter of the water-vapour molecules is orders of magnitude smaller than the diameter of the water droplets.

The interstitial condensation can be prevented by arranging the material properties of the layers in building structures. The resistance to the water-vapour transfer r of building-structure layers should decrease in the direction of the colder part of the structure. Therefore, the final façade layer(s) must have a low resistance to water-vapour diffusion and the capillary-water absorption coefficient A must be smaller than 0.5 (kg/ms$^{0.5}$) to prevent moistening of the structure by capillary suction. An efficient way to prevent interstitial condensation is to design the building structure with a ventilated air gap (Fig. 3.31). In this way the water-vapour molecules that pass the building structure can be transported by the buoyancy-driven air flow inside

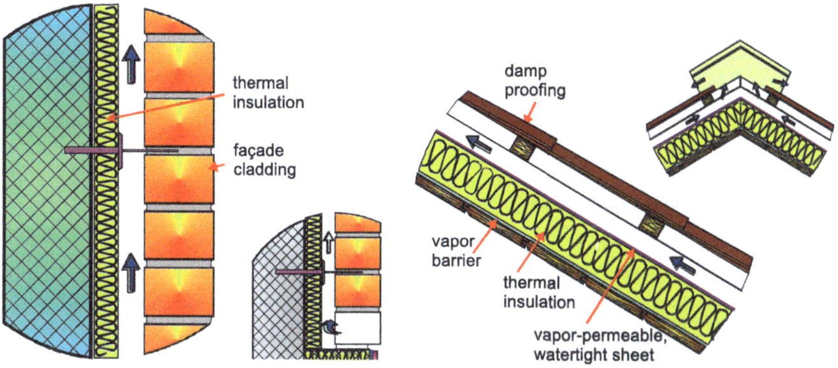

Fig. 3.31 The air circulating through the buoyancy ventilated gap transports the water vapour in the direction of the outdoor environment. The gap thickness must be between 5 and 8 cm with an appropriate free cross-section of the inlet and outlet openings [11]

the gap towards the outdoor air. The inlet and outlet openings of the ventilated air gap must be designed by following the guidelines presented in Sect. 1.5.3. Typical examples of such solutions are ventilated façades and roofs.

3.7.3 *Transient Water-Vapour Diffusion in Building Structures*

The water-vapour transfer models based on the Glaser method rest on several assumptions: the building material's thermal conductivity λ and diffusion coefficients δ_v do not change with respect to the moisture content in the material, capillary suction and surface adsorption are omitted from the model of the moisture uptake, the structures are treated as perfectly tight, and the transfer of heat by latent heat flow in the processes of condensation and drying is disregarded (Fig. 3.32).

Those simplifications in the stationary models of diffusive water-vapour transfer can only be to a large extent avoided by employing transient numerical methods solved by numerical tools. The transient numerical approach involves solving differential thermodynamic equilibrium equations for infinitesimally small elements that constitute a building structure. Methods such as the finite-difference method, the

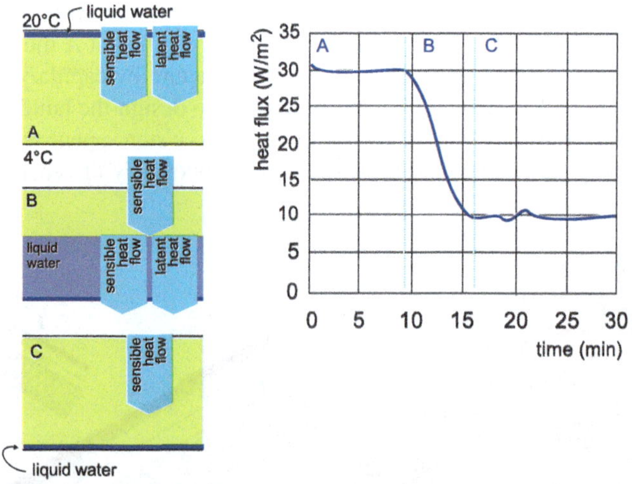

Fig. 3.32 An example of transient water-vapour transfer in building structure. The water-vapour transfer in an insulated panel containing liquid water at warmer side of the panel. Water start to evaporate and the heat is transferred by sensible as well by latent heat flux, increasing heat loses (A); water vapour that flows towards colder side of the panel, condensed and latent heat is realised, causing decrease of the heat flux that transfer through the structure (B), after all the water-vapour condensate, heat is only transferred as sensible heat and heat flux through the panel is much lower now (C) [19].

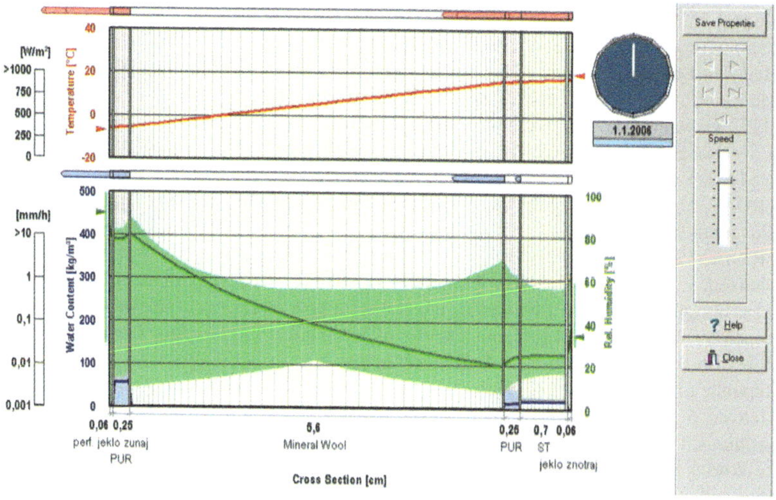

Fig. 3.33 Transient heat and water-vapour transfer in a façade panel consist of two metal sheets with polyurethane lining (PUR) and fibrous thermal insulation core. The temperature (red curve) and the moisture content (blue curve) profile at a particular hour during the year, and the moisture content range that occurs in the panel during the year (light-green area), as determined with the WUFI software tool [9, 20]

finite-volume method or the finite-element method are used to transform the differential equations into difference equations, which can be solved as set of algebraic equations. The building-material properties can be changed in each time step in order to encompass the non-stationary nature of the water-vapour diffusion process. One software tool for the evaluation of non-stationary water-vapour transfer by diffusion in building structures is WUFI [20] (Fig. 3.33).

References

1. Bauphysik Kalender 2001, Ernest & Sohn, D, 2001.
2. Keller B.; Bautechnologie IV, Teil 3. Grundlagen für die Konstruktion, ETH, CH, 2002.
3. Oprešnik, M.: Thermodynamics of moist air. University of Ljubljana, Faculty of Mechanical Engineering, Ljubljana (2000).(in Slovenian language)
4. ISO 10456:2007; Building material and products – Hygrothermal properties – Tabulated design values and procedures for determining declared and design thermal values.
5. Hagentoft C.E.; Introduction to Building Physics, Studentlitteratur, Lund, S, 2001
6. Medved, S.: Thermal techniques in building. University of Ljubljana, Faculty of Mechanical Engineering, Ljubljana (1997).(in Slovenian language)
7. Kladnik R.; Physics for university students – Acoustic and optic, DZS, Ljubljana, 1989 (in Slovenian language).
8. EN ISO 13788:2012: Hygrothermal performance of building components and building elements – Internal surface temperature to avoid critical surface humidity and interstitial condensation – Calculation method; ISO 2007.

9. Künzel Hartwig M.; Simultaneous Heat and Moisture Transport in Building Components, Fraunhofer-Institut für Bauphysik, Stuttgart, D, 1995.
10. Hens H.; Building Physics - Heat, Air and Moisture, Ernest & Sohn, D, 2007.
11. Medved S., et al.; Termo encyclopaedia, Termo d.o.o., Trebnje, 2003.
12. , [13] www.hidrosanir.si; Technical data (1.1.2020).
13. Slovenian Environment Agency (digital data base meteo.arso.gov.si (2019).
14. Medved, S., Černe, B.: Heat transfer on heat bridges in large area thermo-insulating panels TRIMO. University of Ljubljana, Faculty of Mechanical Engineering, Ljubljana (2002).(in Slovenian language)
15. TRISCO 12.0w, Physibel, 2018.
16. Medved, S., et al.: FP6-SUSDEV REMINING LOWEX – Redevelopment of European mining areas into sustainable communities by integrating supply and demand side based on low exergy principles. University of Ljubljana, Faculty of Mechanical Engineering, Ljubljana (2008)
17. Rules on thermal insulation and efficient energy use in buildings, Official Gazette of the Republic of Slovenia, No. 42/02.
18. Medved, S., et al.: KI Energy 2010: EPBD tool for evaluation of energy performance of the buildings. University of Ljubljana, Faculty of Mechanical Engineering, Ljubljana (2010)
19. Leskovšek U. et al.; Heat and moisture transfer in fibrous thermal insulation with tight boundaries and a dynamic temperature; International journal of heat and mass transfer, Sept. 2011, Vol. 54, Iss. 19/20; 2011.
20. WUFI 2D, Fraunhofer-Institut für Bauphysik, Holzkirchen, D, 2008.

Chapter 4
Visual Comfort and Architectural Lighting

Abstract Vision is the sense that provides us with the majority of information about the natural and built environments. The information received depends on the characteristics of the light source, the optical properties of the objects reflecting the incoming light, and the way we perceive light with our ability to see. Besides that, light is involved in the chemical and biological processes taking place in our bodies, so proper lighting is indispensable to our well-being and health. In this chapter, light will be treated as electromagnetic radiation over a range of wavelengths that can be detected by our vision. The most important source of natural light is the Sun, from which the Earth receives light in form of direct radiation emitted from the Sun's photosphere and as the diffuse radiation from the sky. Natural light in the outdoor and indoor environments is complemented by artificial light sources, predominantly electric lamps. The technological development of electric light sources over the past two decades has given us products that are capable of emitting light that is very similar to natural light, yet with a significantly reduced consumption of electrical energy.

Designing the visual comfort in buildings is a process of optimizing the illumination of the spaces as well as light perception in the way that building occupants are provided with a comfortable, highly productive and healthy living environment.

Learning objectives In this chapter you will learn about:

- the physical and physiological aspects of the perception of light;
- the characteristics of daylight and the modelling of natural light sources;
- the sources of artificial lighting;
- the optical properties of building materials that influence the transmission of light;
- the criteria and techniques for the improvement of visual comfort; the significance of energy-efficient lighting for the efficient use of energy in buildings.

S. Medved, *Building Physics*, Springer Tracts in Civil Engineering,
https://doi.org/10.1007/978-3-030-74390-1_4

4.1 Symbols for Quantities and Units

A, A_s	Area, of sun disk (s) (m^2)
A_w,A_f, A_g, $A_{envelope}$	Area of window (w), floor (f), glazing (g), of room envelope (envelope) (m^2)
A_u	Useful area of building (m^2)
a, b	Depth of the room, width of the room (m)
α, α_v, $\alpha_{v,\lambda}$	Absorptivity, of light (v), of light wavelength λ(v, λ) (–, %)
α_s, $\alpha_{s,noon}$, α_P	Sun's altitude, at solar noon (noon), of sky patch (p) (°)
β	Angle between normal to the surface and view of observer (°)
CCT	Correlated colour temperature (K)
CRI	Colour-rendering index (–)
C_s, C_g, C_o	Quotient of illumination of vertical plane, by sky (s), reflected on ground (g), reflected on nearby object (o) (–)
DF, DF_{av}, $DF_{\min}$	Daylight factor, average (av), recommended (min) (%)
d	Thickness (m)
DGP	Daylight glare probability (–)
δ	Declination (°)
γ_s	Sun's azimuth (°)
γ_s, γ_g, γ_o	Obstruction angle, sky (s), ground (g), objects (o) (°)
E, Ep, EH	Illuminance, of observed point (p), of horizontal plane (H) (lm/m^2, lx)
E_V, $E_{V,s}$, $E_{V,g}$, $E_{V,o}$	Illuminance of vertical plane (V), with sky luminance flux (s), with luminance flux reflected on the ground (g), with luminance flux reflected on nearby objects (o) (lm/m^2, lx)
E_l	Luminous efficiency of luminaire (%)
ET	Equation of time (min)
F_D	Daylight dependency factor (–)
$F_{D,s}$	Daylight supply factor (–)
$F_{D,c}$	Electric lighting control factor (–)
F_o	Occupancy factor (–)
Φ_v, $\Phi_{v,\lambda}$	Luminous flux (lm/μm)
$\Phi_{v,\lambda,\rho}$, $\Phi_{v,\lambda,\alpha}$, $\Phi_{v,\lambda,\tau}$	Spectral luminous flux reflected (ρ), absorbed (α), transmitted (τ) (lm)
$G_{glob,0}$, $G_{dir,0}$, $G_{dif,0}$	Solar irradiation on the horizontal plane, global (glob), direct (dir), diffuse (dif) (W/m^2)
h, h_w	Height, height of window-head (w) (m)
K_l, $K_{l,dir}$, $K_{l,dif}$	Luminous efficacy of solar irradiation, of direct (dir), of diffuse (dif) (lm/W)
I, I_o, I_β	Luminous intensity, in direction of normal to the surface (o), in direction of observation (β) (cd)

(continued)

(continued)

i, i_1, i_2	Incident angle between vector of the luminous flux and normal to the surface, elevation angle of the sky patch (1, 2) (°, rad)
L, L_β	Luminance, in the direction of view (β) (cd/m^2)
L	Latitude (°)
L_P, L_Z	luminance of the sky patch (P), of zenith (z) (cd/m^2)
L_s, L_b	Luminance of observed object (s), of surroundings of the object (b) (cd/m^2)
LENI	Light energy numerical indicator (kWh/m^{2}a)
l	length (m)
λ	Wavelength (nm, μm)
λ, λ_{st}	Longitude, longitude of standard meridian (st) (°)
μ	Attenuation coefficient (m^{-1})
N	Running day in the year (–)
n, n_λ	Refractive index, refractive index of light with wavelength λ(λ) (–)
P	Position index of observer (–)
P_n, P_p	Electric power of installed luminaires (n), emergency light electric power (p) (W)
R_a	Value of CRI (–)
r, r_s	Distance from the luminaire centre, radius of the sphere, of Sun (s) (m)
ρ,ρ_o,ρ_v, $\rho_{v,\lambda}$	Reflectance, albedo, of object (o), of light (v), of light of wavelength λ(v, λ) (-, %)
ρ_{av}, ρ_{bottom}, ρ_{top}	Average reflectance of surfaces in the room (av), of surfaces below the work plane in the room (bottom), of surfaces above work plane in the room (top) (–)
UGR	Unified glare rating (–)
V_λ	Relative spectral sensitivity of the eye (–)
Ω	3D solid angle (sr)
ω	2D solid angle (rad), Sun's hour angle (°)
TC	Colour temperature (K)
t_{civil}, t_{sun}	Civil time (civil), solar time (sun) (h:m:s)
t_d, t_n, t_p	Duration of electric lighting operation during daytime (d), during night-time (n), duration of emergency light operation (p) (h/an)
τ, τ_v, $\tau_{v,\lambda}$	Transmittance, of light (v), of light of wavelength λ(v,λ) (-, %)
$\dot{Q}_{r,\lambda}$	Spectral radiative heat flux (W/nm)

4.2 Architectural Lighting Throughout History and the Importance of High-Quality Lighting

The purpose of the first dwellings, initially sought in the natural environment and later built by the people themselves, was to shelter their occupants from the harsh climate. The entrance was the only "window" into such a dwelling. However, intuition led mankind to explore the rules of daylight and to start using it, beginning with sacral buildings. People began to use the combination of light and shadows to create a feeling of holiness in the room (Fig. 4.1).

Although glass was discovered thousands of years BCE, the first glass-forming equipment was used no sooner than the 17th Century. However, the costs of glass manufacture only started to fall sharply after glass-pressing machines were invented at the beginning of the 19th Century. The technology used to manufacture window glass, i.e., floating molten glass on the surface of molten tin, has been in use since 1950. The improved thermal characteristics of glazing have led to buildings with increasingly large glass surfaces.

For thousands of years, people supplemented the natural light by burning animal fat and vegetable oils. Oil lamps made of clay were characteristic of early civilisations. Candles were in use by the time of the Roman Empire. Much later, gas lighting was widespread until the electrification of cities began. The first gas-lighting systems for public streets were built at the beginning of the 19th Century in the United States and Europe. At the end of the 19th Century, the first electric lightbulb was produced

Fig. 4.1 A beam of light entering the Pantheon through the oculus in the dome creates a heavenly light (left); the natural illumination in modern buildings is controlled using several mechanical, optical and electronic systems (right) [1]

and since then the efficiency of electrical light sources improved continuously until today.

Humans gather more than 80% of all the information about their surroundings by visual perception. From an engineer's point of view, visual comfort has been established if a person situated in a room is not hindered when performing tasks where he or she detects the environment mostly by vision. The performance of visual work is associated with:

- the size of an observed detail that is determined by the solid angle (ω, measured in minutes of arc), which is defined by the size of the observed detail h and its distance from the observer l (Fig. 4.2). The solid angle is used to determine the visual acuity, for example, with a Snellen chart in an ophthalmologist's office (Fig. 4.3, left);
- the contrast between the observed detail and its immediate surroundings. To be able to see a detail, its brightness must differ from the brightness of its surroundings or background (Fig. 4.3, right).

Fig. 4.4 provides the basic guidelines for designing the lighting of residential and work spaces:

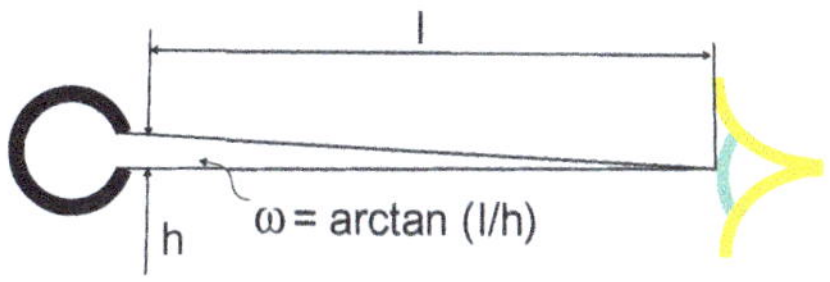

Fig. 4.2 Solid angle ω defines the size of a detail that can still be discerned at a certain illuminance. This defines the visual acuity

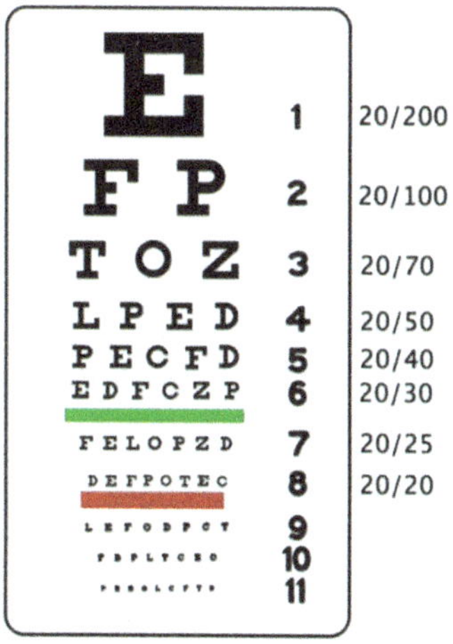

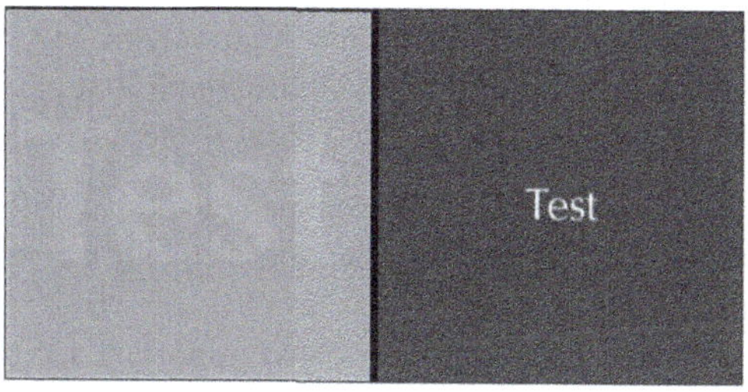

Fig. 4.3 Ophthalmologists use the Snellen chart for visual acuity testing [2] (left). A higher contrast makes it easier to discern the details and to perform more demanding tasks (right)

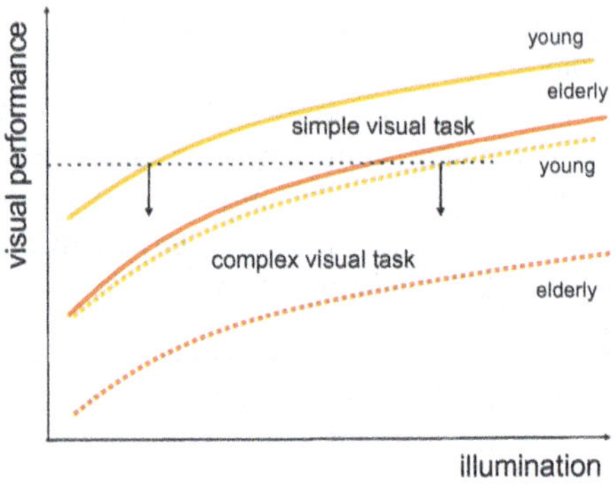

Fig. 4.4 Visual complexity of a task affects the work efficiency more than the illumination [3, 40]

- in the case of simple visual tasks (observation of the details with large solid angle ω), the efficiency of the work increases rapidly with the increasing illumination, but only to a certain limit: above that limit, increased illumination only causes an increased energy consumption;
- in the case of complex visual tasks, the efficiency of the work is lower and increases the performance to a lesser extent, despite an increasing illumination of the work-plane, the efficiency of discerning details and the efficiency of work for complex visual tasks can never be as efficient as when performing simple visual tasks.

It is, therefore, more important to ensure that the residential and working spaces in a building are designed so it is easier for the occupants to detect their environment, than to just keep increasing the illumination of the rooms.

In addition to the performance efficiency, indoor visual comfort also incorporates the psychological and health aspects of proper illumination. Considering the psychological aspects, daylight and artificial lighting should be designed to ensure that occupants:

- will be directed and be able to orientate in the room;
- will focus their attention and be enable to discern the significance of information;
- will sense the individuality in different illuminated areas (e.g., a desk) in larger office rooms;
- will have an internal sense of time that will be stimulated by daylight;
- will have no fear in an environment that may otherwise appear to be dangerous, for example, in long hallways, if the other end is not illuminated.

This is in addition to increasing the economic value of the building, because sociological studies have shown that insolation is one of the key factors in the purchase of a new apartment.

Appropriate daylight is also important for the health of a building's occupants because:

- it contributes to the well-being, mental health and vitality;
- it fills us with contentment, optimism and confidence;
- it enables the synthesis of vitamins in our bodies;

- it has been proven that a lack of daylight causes psychosomatic disorders such as depression, lack of concentration, tiredness and insomnia.

4.3 Physical and Physiological Fundamentals of Light Perception

4.3.1 Light Is Electromagnetic Radiation

Physics treats light in general as a form of electromagnetic radiation. Electromagnetic radiation propagates through vacuum at a speed of ~ 300,000 km/s, also known as the speed of light. The distance between two consecutive points with the same amplitude of oscillation is known as the wavelength λ. It is measured in micrometres ($\mu m = 10^{-6}$ m) or nanometres ($nm = 10^{-9}$ m). Electromagnetic radiation with wavelengths between 380 and 780 nm is detected by the eyes and is referred to as light in the field of building physics (Fig. 4.5).

There are several natural sources of light that the Earth receives from space: solar radiation, which includes the wavelengths of light, the reflected sunlight from the Moon's surface, and any other star in space that emits light as well. Natural sources of light on the Earth are phenomena such as fires, volcanic eruptions and other natural processes, such as lightning or bio-luminescence. Modern life demands that the natural sources of light are supplemented with artificial sources of light. For more than a century, buildings and the external environment have been illuminated with electrical energy, converted into light by lamps.

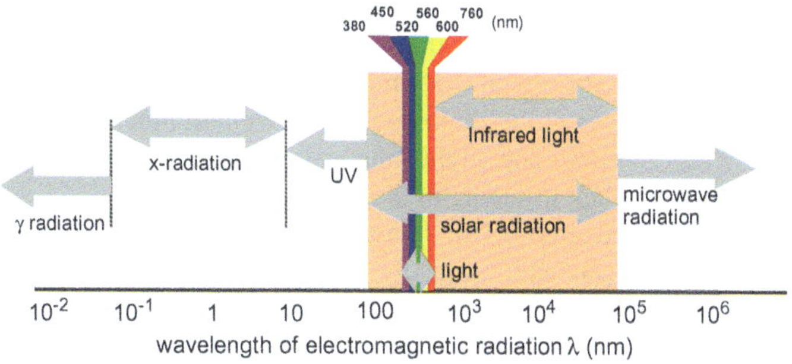

Fig. 4.5 Wavelengths of different types of electromagnetic waves; radiation in the spectral range between 380 and 760 nm (0.38–0.76 μm) is known as light and can be detected with our eyes [5]

4.3.2 Perceiving the Light

Our vision organ has evolved to adapt to the intensity of light and to the distance to the observed objects, as well as to discern the details and the colour of the objects in our surroundings. These are the tasks for the elements of eye: the cornea, iris, pupil, lens and retina, where the light rays are converted into neural impulses and then visualised in the brain (Fig. 4.6).

The iris is a circular muscle, expanding and constricting to open or close the pupil, and in this way control the amount of light entering the eye. Different pigments give the iris its distinctive colour. The eye's pupil is seen as a dark surface and behind it lies the eye socket. The width of the pupil is controlled, depending on the brightness of the observed objects. In the case of excessive irradiance, the eye protects itself by closing the eyelids. The lens is the size of a coffee bean and consists of approximately 2200 tiny layers called the laminae. In the process of accommodation (adaptation for the perception of depth), the ciliary muscles contract or relax to shape the lens, changing its optical properties so that the eye focuses on the observed object. The eye is protected on the external side by the cornea. The cornea directs the light striking the surface of the eye inside. The lens forms an image of the object on the retina, where photoreceptors convert the light into an electrical signal. There are two types of photoreceptors, known as the rods and the cones due to their characteristic shape. Both contain chemicals that are altered by the light in chemical and electrophysiological processes. The cones also contain light-sensitive pigments, which enable us to see colour, while the rods only give us the ability to detect shades of grey. The cone cells enable photopic vision in daylight conditions, and the rods are responsible for the night-time scotopic vision (Fig. 4.7). In both types of vision, the detection of light depends on the wavelength of light. The ganglion cells transmit electrical impulses into the brain over the optic nerves. Three kinds of cones are responsible for distinguishing colour, reacting to the wavelengths of blue (with the peak of the response at 430 nm), yellow-green (535 nm) and red light (590 nm). Human vision also adapts to the intensity of the illumination in the environment. Day-time vision is considerably less sensitive than night-time vision. As the sun sets slowly, the eye has evolved to slowly adapt to seeing in the dark too [41].

The field of vision covered by the human eyes is large: approximately 180° in the horizontal direction and approximately 130° in the vertical direction, measured from the centreline of the view. The retina has a specific structure and the sharpness of the vision varies over the field of view. Our vision is very sharp only within the

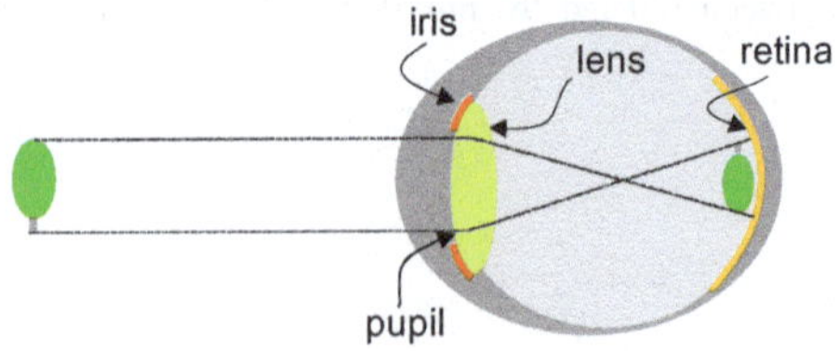

Fig. 4.6 Most important elements of the eye

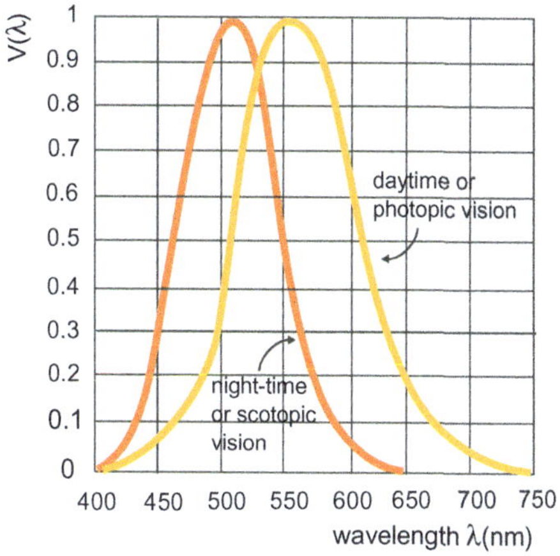

Fig. 4.7 A human eye detects differently light sources of equal intensity, emitting at different wavelengths/colours. The detection of colours also depends on the time of the day (during the daytime or during the night-time). The eye's relative ability to detect the wavelengths is known as the relative spectral sensitivity of the eye $V(\lambda)$. It is defined for a standard observer in the CIE methodology [6]

central 2° of the field of vision. Objects also appear sharp in the central visual field within a 30° cone around the centre of vision. The brightness of objects is extremely important for detection in this field. In the peripheral field of vision (within 90° of the centre line), the objects can be discerned, but are not seen as being sharp (Fig. 4.8).

The detection of the luminous flux depends on both the intensity and on the wavelengths of the light. Because of that the lighting is evaluated with the physical quantities (e.g., emitted flux by luminaire can be expressed as its power in W and the density of the flux received by illuminated surface in W/m^2) as well as with

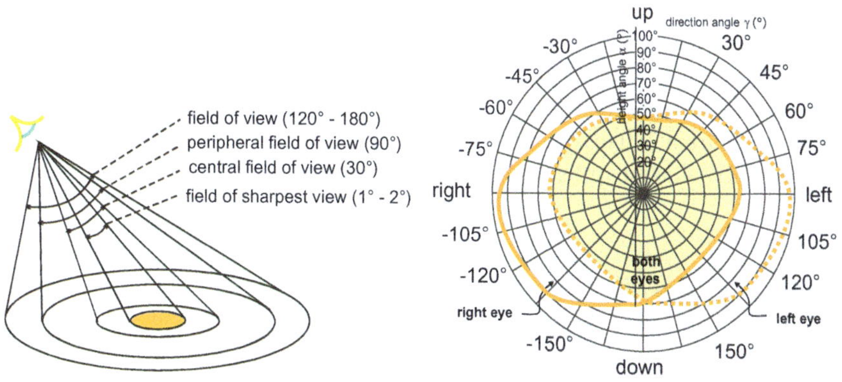

Fig. 4.8 The sharpness of human sight depends on the angle measured from the centre line of the view [7] (left); human field of view [42] (right)

Table 4.1 Radiometric (physical) and photometric (physiological) quantities

radiometric quantities	photometric quantities
radiant flux density (W/m^2)	luminous flux (lm)
radiant flux (W)	luminous intensity (cd)
thermal emittance, the thermal power emitted by a radiator per unit area (W/m^2)	luminance (cd/m^2)
irradiance, the thermal power incident on a surface per unit area (W/m^2)	illuminance (lm/m^2 = lx)

physiological quantities (e.g., the luminous intensity of a light source is measured in candelas and the luminous flux per m^2 of illuminated surface is measured in lux). In the case of physiological quantities, the sensitivity of the eye to different wavelengths of light is accounting. The detection of light and the perceived visual comfort are also affected by vision defects. A decreased sharpness of vision is a consequence of cornea defects. In some people, the focal point of the eye's lens lies in front of the retina, causing near-sightedness (myopia). A near-sighted eye is not capable of producing an accurate image of distant objects because it cannot accommodate to view closer objects. Far-sightedness is a defect caused by the inadequate refraction of light rays. It is common in older people, because the eye's lens gradually loses its flexibility. In the case of astigmatism, light rays in the eye are focused at different points and some people are unable to see individual colours or colours in general. This defect is known as colour blindness or daltonism. Adequate visual comfort is very important for older people, who spend most of their time indoors. A 60-year-old person requires two- to three- times the illumination to perceive the same visual comfort as a 20-year-old. The requirements and the design for indoor visual comfort will be discussed in detail in Sect. 4.7.

4.4 Photometry and the Photometric Quantities

Photometry is the science of light processes and the measurement of light. In contrast to radiometry, which deals with the whole spectral range of electromagnetic radiation (Table 4.1), photometry is only interested in electromagnetic radiation with wavelengths from 380 to 760 nm. Therefore, in photometry physiological units are used for the evaluation of light sources and lighting comfort.

4.4.1 Photometric Quantities

To be able to describe the basic photometric quantities, we must first define the differential solid angle of space as the ratio of the surface-area differential of a sphere *dA* to the square of its radius *r* (Fig. 4.9). The solid angle is measured in steradians. One steradian is the solid angle subtended at the centre of a sphere with

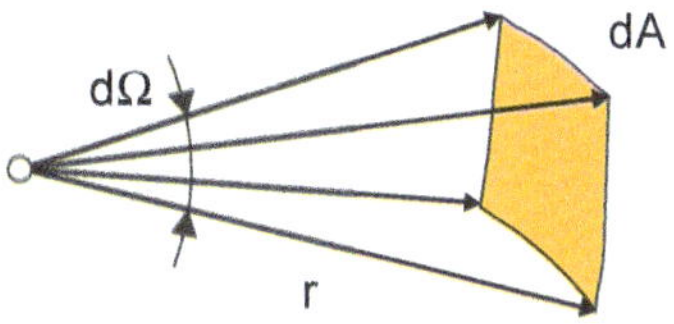

Fig. 4.9 The differential solid angle of space $d\Omega$ defines the differential of a sphere area dA with radius r

a radius $r = 1$ m by a 1 m^2 patch on the surface of the sphere. The total solid angle in three-dimensional space is [4]:

$$d\Omega = \frac{dA}{r^2} \quad \rightarrow \quad \Omega = \frac{A}{r^2} = \frac{4 \cdot \pi}{1^2} = 4\pi \ [sr]$$

An isotropic point-light source emits an equal luminous flux into any differential solid angle of space $d\Omega$. The Sun is an example of an isotropic point source. A flat light-emitting surface only illuminates half the total solid angle (2π), as it can only be "seen" from the space in front of it.

The luminous flux Φ_v is a photometric quantity, representing the quantity of energy emitted by a light source in a unit of time. It is determined based on the spectral radiant heat flux $\dot{Q}_{r,\lambda}$ (W/nm) emitted by a light source over visible wavelengths. The luminous flux is measured in lumens (lm). The lumen is defined as follows:

- the radiant flux emitted by a thermal emitter at a certain temperature is determined by Planck's law (when the individual wavelengths λ of the radiation is observed) and the Stefan-Boltzmann law (when the total radiant flux at all wavelengths to be determined), as explained in Sect. 1.3.3. For example, the radiant flux emitted by an optically black emitter with 1 m^2 of surface area at 2043 K is ~ 2780 W. By convention, a radiant flux of 1 W emitted at a wavelength of 555 nm (this is the wavelength where the human eye is the most sensitive under daylight conditions—adapted to photopic vision and at which spectral sensitivity V_λ of the eye is equal to 1), equals a luminous flux of 683 lm. This constant is known as the theoretical maximum of luminous efficacy K_l (lm/W) for monochromatic light ($\lambda = 555$ nm). The total luminous flux Φ_v emitted by a light source is calculated as the integral over visible wavelengths, considering the sensitivity of the human eye V_λ to different wavelength, is equal:

$$\Phi_v = K_l \int_{\lambda=380nm}^{\lambda=760nm} \dot{Q}_{r,\lambda} \cdot V_\lambda \cdot d\lambda \ [lm]$$

Explanation The value of the luminous efficacy K_l 683 lm/W is adopted for daytime conditions. As a human eye is more sensitive at night-time vision, a radiant flux of 1 W emitted at a wavelength of 507 nm (this is the wavelength where the human eye is the most sensitive under night-time conditions adapted to scotopic vision), equals 1700 lm/W.

The ratio of the luminous flux to the radiant flux gives the luminous efficacy of a light source. The luminous efficacy of real light sources is considerably smaller than the theoretical maximum luminous efficacy—for an incandescent light bulb it is 10 lm/W, for a fluorescent tube it is 90 lm/W, even the Sun is far less efficient than an ideal light source and its luminous efficacy K_l is between 100 and 110 lm/W. Among all light sources, light-emitting diodes (LEDs) have the greatest luminous efficacy, up to 160 lm/W, which is close to the theoretical maximum luminous efficacy of a white-light emitter (~ 200 lm/W). The luminous efficacy of light sources is directly related to efficient energy use.

The Sun is an example of an isotropic point-light source, while it emits light uniformly into three-dimensional space. However, not all light sources are isotropic and the luminous flux can vary with the direction. The luminous flux emitted by a light source in a specified direction is defined as the luminous intensity I. The luminous intensity is measured in candela (cd = lm/sr). For an isotropic point-light source, the luminous intensity equals the luminous flux divided by the total solid angle of space $\Omega = 4\pi$. This means that such source with a luminous intensity of 1 cd emits 4π lumens:

$$I = \frac{d\Phi_v}{d\Omega} = \frac{\Phi_v}{4 \cdot \pi}\left[\frac{lm}{sr} = cd\right]$$

Case Study A 40-W incandescent light bulb made of milky glass is an almost isotropic light source. Considering the luminous efficacy ($K_l = 10$ lm/W), the luminous flux emitted by the light bulb is 40 W × 10 lm/W = 400 lm. The luminous intensity I equals 400 lm/4 × π ~ 32 cd.

A planar light source emits light in a half solid angle of space. Such a source has the property of a Lambertian emitter if the luminous intensity I_β depends on the cosine of the angle β measured from the normal to the surface of the light emitter (Fig. 4.10):

$$I_\beta = I_o \cdot \cos\beta \text{ [cd]}$$

where I_o is the luminous intensity in the direction of the normal of the emitters surface. Most diffuse surfaces in the buildings can be assumed to be Lambertian surfaces.

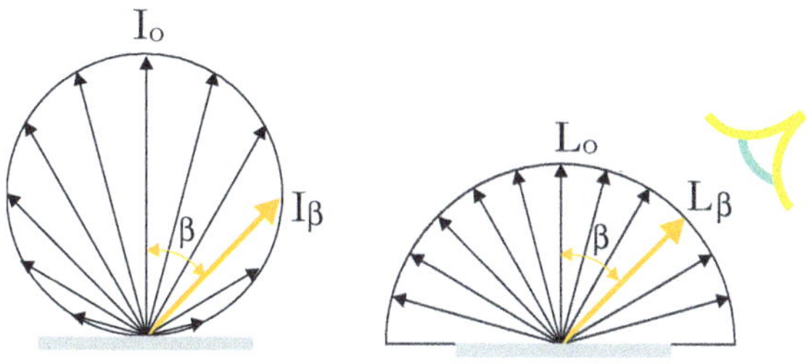

Fig. 4.10 Properties of Lambertian light emitter: luminous intensity (left) and its luminance (right)

The light emitted by most luminaires is not isotropic, as they are fitted with mirrors that direct the light in a specified direction. Because of that, the manufacturers show the luminous intensity distribution diagram, known as the "candela diagram", for their products. The candela diagram could be in 2D form (for a long, line-type luminaire) or in 3D form (for point-type luminaires). An example is shown in Fig. 4.11.

The only photometric quantity that we can perceive is the luminance of the light source or the surface that reflects the light. The luminance is defined as the ratio of the luminous intensity I to the surface area of the light emitted or the reflected surface:

$$L = \frac{I}{A} \left[\frac{cd}{m^2}\right]$$

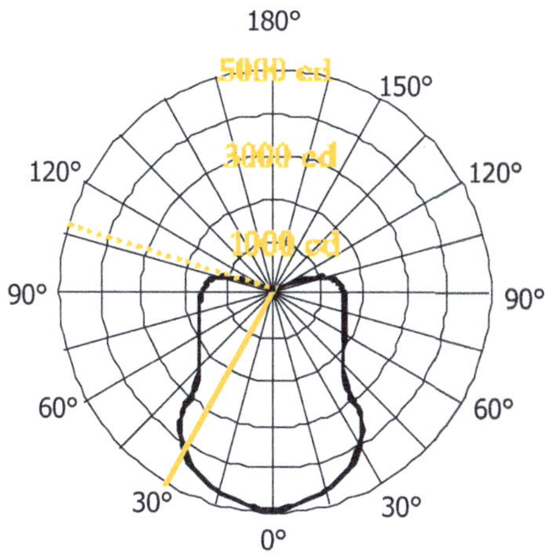

Fig. 4.11 The candela diagram for the point-type luminaire. The luminous intensity of the luminaire in the direction towards down (angle 0°) is 5000 cd, while in the direction 30° from the vertical it is 4000 cd. If there will be no reflection of the light on the surfaces in the room, the ceiling of the room will be unlighted because the luminaire does not emit light in the space at an angle greater than 107° from the vertical direction

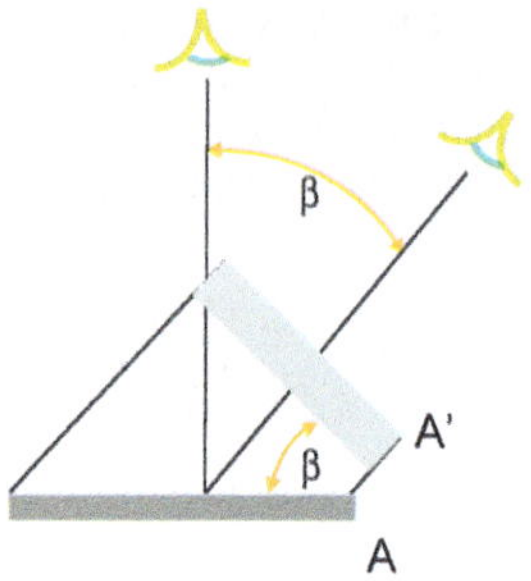

Fig. 4.12 The luminance of the Lambertian light emitter (source or reflector) will be equal regardless the angle of the observation and observer will for example, perceived diffuse sheet of paper equally bright regardless from where those objects will be viewed

where A is the apparent area of the light source defined according to the direction of the observer's view towards the light source. Some examples: the luminance of the Sun is 1.9×10^9 cd/m^2, of an incandescent light bulb is 6×10^6 cd/m^2, of a candle is 5×10^3 cd/m^2 and of the Moon is 3×10^3 cd/m^2. The definition of luminance leads to the conclusion that a smaller light emitter with the same luminous intensity as a larger one will appear brighter for the observer. And if the luminance of the observed subject is above ~ 1000 cd/m^2 (~ 2000 cd/m^2 if it is lighted by daylight), the object will be to "bright" and will cause a sense of glade. The Lambertian light source will have the same luminance regardless of the direction of observation, despite the fact that the luminance intensity depends on the cosine of the observation angle (Fig. 4.10). The reason is that the size of the light emitter depends on the cosine of the observation angle as well (Fig. 4.12):

$$L_\beta = \frac{I_\beta}{A'} = \frac{I_o \cdot \cos\beta}{A \cdot \cos\beta} = L_o \quad \left[\frac{cd}{m^2}\right]$$

In engineering practice the light comfort is designed on the basis of the illuminance of the objects. Illuminance is defined by the luminous flux Φ_v on the unit of the object's surface area *A*. In the case of an isotropic point-type light source with luminous intensity *I* (lm), the illuminance of the surface *A* (m^2) of the sphere around the isotropic source is:

$$E = \frac{\Phi v}{A} = \frac{4 \cdot \pi \cdot I}{4 \cdot \pi \cdot r^2} = \frac{I}{r^2} \quad \left[\frac{lm}{m^2} = lx\right]$$

Illuminance is measured in lumens per m^2 or in lux (lx). From the above equation, it can be concluded that illuminance decreases by the square of the distance to the light source. In the presented case, the light-flux vectors are perpendicular to the surface of the receiver (sphere). If this is not the case, the illuminance of the surface depends on the cosine of the incident angle:

$$E = \frac{I}{r^2} \cdot \cos\beta \quad \left[\frac{lm}{m^2} = lx\right]$$

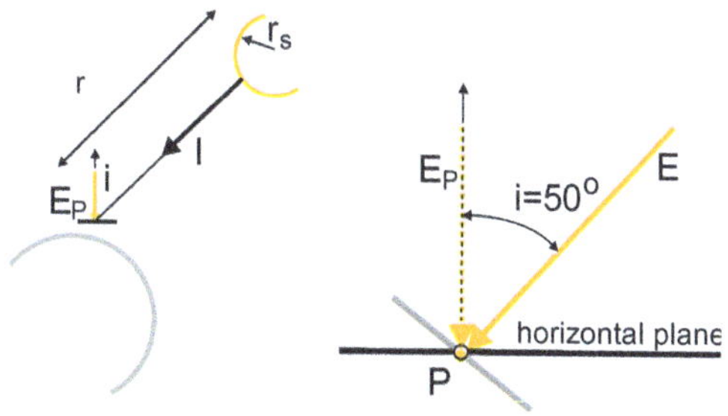

Fig. 4.13 The quantities used in the calculation example (left); the Sun is an isotropic light source and its luminous intensity I is equal in any direction (right)

Case Study The Sun is an isotropic light source. The sun luminance L equals 1.9×10^9 cd/m^2. Determine the illuminance E_P of the patch P on a horizontal plane E_H at the outer edge of the atmosphere, if the angle between the sun's rays and the normal to the plane is 50° (Fig. 4.13). The radius of the Sun r_s is 6.96×10^8 m; therefore, the apparent surface of the Sun (Sun disk) A_s is:

$$A_s = \pi \cdot r_s^2 = \pi \cdot 6.96^2 \cdot 10^{16} = 1.52 \cdot 10^{18}\ \mathrm{m}^2$$

The luminous flux Φ_V emitted by the Sun and its luminous intensity I, assuming the Sun as isotropic point-type light source, are:

$$\Phi_V = 4 \cdot \pi \cdot L \cdot A_s = 4 \cdot \pi \cdot 1.9 \cdot 10^9 \cdot 1.52 \cdot 10^{18} = 3.63 \cdot 10^{28}\ \mathrm{lm}$$

$$I = \frac{\Phi_V}{4 \cdot \pi} = \frac{3.63 \cdot 10^{28}}{4 \cdot \pi} = 2.89 \cdot 10^{27}\ \mathrm{cd}$$

The distance between the Sun and a horizontal plane at the edge of the atmosphere is $r = 1.49 \times 10^{11}$ m. The illuminance E of a surface perpendicular to the direction of the luminous flux and the illuminance E_P of the horizontal plane are:

$$E = \frac{I}{r^2} = \frac{2.89 \cdot 10^{27}}{1.49^2 \cdot 10^{22}} = 130174\ \mathrm{lx} \quad \text{and} \quad E_P = E \cdot \cos 50° = 83670\ \mathrm{lx}$$

In the case that the luminance L of the light source is known, the illuminance of the patch P is defined by the expression:

$$dE = \frac{d\Phi v}{dA} = \frac{I \cdot \cos i \cdot d\omega}{dA} \rightarrow E = \frac{L_\beta \cdot A \cdot \cos i \cdot \Delta\omega}{A}$$
$$= L_\beta \cdot \cos i \cdot \Delta\omega \left[\frac{\mathrm{cd}}{\mathrm{m}^2} \cdot \mathrm{sr} = \frac{\mathrm{lm}}{\mathrm{m}^2} = \mathrm{lx}\right]$$

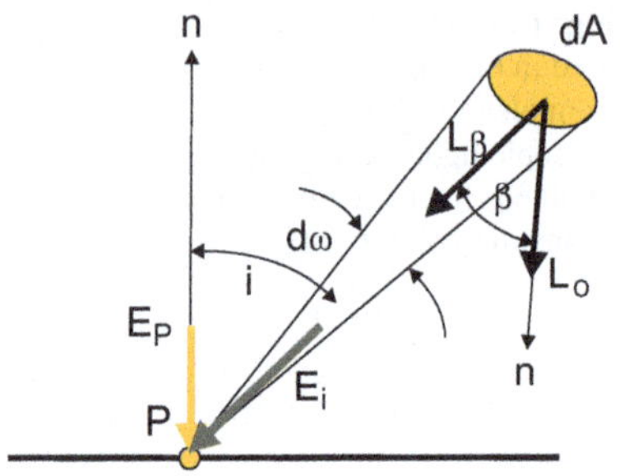

Fig. 4.14 Quantities used to determine the illuminance E_p of patch P on an arbitrary surface illuminated by a light source with luminous intensity L_β

where L_β is the luminance of the light source in the direction towards to the patch, i is the angle between the light source and the normal of the patch and dω is the size of the light source measured in a differential solid angle (Fig. 4.14).

Case Study The illuminance of the patch I from a previous case study can be determined alternatively using data on the Sun's luminance L. The Sun in the sky subtends a differential angle equal to the ratio of the Sun's disc surface A_s to the square of the distance R between the Sun and the patch P on the horizontal plane:

$$\Delta\omega = \frac{A_s}{r^2} = \frac{\pi \cdot r^2}{r^2} = \frac{\pi \cdot 6.96^2 \cdot 10^{16}}{1.49^2 \cdot 10^{22}} = 6.85 \cdot 10^{-5}\ \mathrm{sr}$$

The illuminance of the patch on the horizontal plane on the edge of the Earth atmosphere is:

$$E_P = L \cdot \Delta\omega \cdot \cos i = 1.9 \cdot 10^9 \cdot 6.85 \cdot 10^{-5} \cdot \cos 50° = 83710\ \mathrm{lx}$$

For Lambertian light sources L_β is equal to L_o and the illuminance of the patch P with a light source having a size of the solid angle ω (which is the integral of all $d\omega$) is equal to:

$$E_P = \int_A L_\beta \cdot \cos i \cdot d\omega = Lo \cdot \int_A \cos i \cdot d\omega$$

and in the case of non-Lambertian light sources:

$$E_P = \int_A L_\beta \cdot \cos i \cdot d\omega = L_o \cdot \int_A \cos\beta \cdot \cos i \cdot d\omega$$

Let us assume that the patch P is located on the horizontal plane illuminated by a hemispherical light source above it—for example, the sky. The surface integral can be

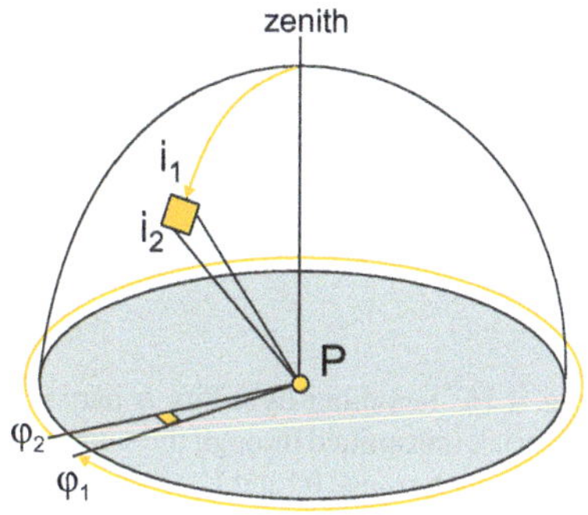

Fig. 4.15 Angles i and φ determining the surface area of a hemisphere that illuminates the patch P on a horizontal plane under the hemisphere

divided into two single integrals and the cosine β is equal to 1, because each sky patch is orientated towards the patch on the horizontal plane (Fig. 4.15). The differential solid angle dω can be replaced by $d\omega = \sin i \times di \times d\varphi$, where the product $di \times d\varphi$ represents the area of the differential patch on the hemisphere, as shown in Fig. 4.15. The total illuminance of the patch P can be determined by integrating the size of the unobstructed hemisphere (integration limits are $i_1 = 0$ and $i_2 = \pi/2$ radians and $\varphi_1 = 0$ and $\varphi_2 = 2\pi$ radians as shown in Fig. 4.15):

$$E_P = L_o \cdot \int_{\varphi=0}^{\varphi=2\pi} \int_{i=0}^{i=\pi/2} \cos i \cdot \sin i \cdot di \cdot d\varphi \rightarrow L_o \cdot \frac{\sin^2 i_2 - \sin^2 i_1}{2} \cdot (\varphi_2 - \varphi_1) \quad [\text{lx}]$$

Case Study Determine the illuminance E_P of a patch on a horizontal plane that is illuminated by the isotropic sky with the luminous intensity $L_{(o)} = 5000\ \text{cd/m}^2$. The boundary angles are $i_1 = 0° = 0$ rad; $i_2 = 90° = \pi/2$ rad; $\varphi_1 = 0° = 0$ rad and $\varphi_2 = 360° = 2 \cdot \pi$ rad.

$$E_P = 5000 \frac{\sin^2(\pi/2) - \sin^2(0)}{2} (2 \cdot \pi - 0) = 5000 \cdot \pi = 15707\ \text{lx}$$

4.4.2 *Radiative (Optical) Properties of Matter*

When a ray of light reaches the specimen, part of the light flux is reflected from the surface, absorbed into the material and transformed mostly into heat, and in the case of a transparent substance, transmits through it (Fig. 4.16).

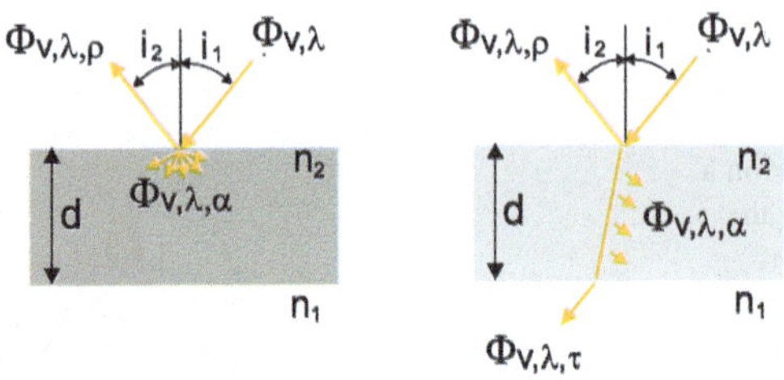

Fig. 4.16 Incoming light flux is partially reflected and absorbed, and in the case of transparent material, transmitted through the substance. The magnitude of each process is expressed by relative values (in the range 0^+ and 1^-) defined on the basis of the energy-conservation law as the reflectance ρ_V, absorptance α_V and transmittance τ_V of the light

Explanation Besides being converted into heat, the absorbed light flux can begin other process as well, like photoluminescence or photovoltaic phenomena in PV cells.

Taking into account the law of energy conservation, the sum of all three energy fluxes is equal to the incoming light flux Φ_V, while the magnitude of the energy fluxes is defined by the radiative properties of the substance, i.e., the reflectance, absorptance and transmittance of the light flux. Although the radiative properties depend on several impact factors, the most pronounced are the wavelength λ of light and the incident angle *I* of the light flux towards the surface of the sample:

$$\overbrace{\Phi_{V,\lambda}}^{\text{incoming}} = \overbrace{\Phi_{V,\lambda,\rho}}^{\text{reflected}} + \overbrace{\Phi_{V,\lambda,\alpha}}^{\text{absorbed}} + \overbrace{\Phi_{V,\lambda,\tau}}^{\text{transmited}} \quad [\text{lm}]$$

$$1 = \frac{\Phi_{V,\lambda,\rho}}{\Phi_{V,\lambda}} + \frac{\Phi_{V,\lambda,\alpha}}{\Phi_{V,\lambda}} + \frac{\Phi_{V,\lambda,\tau}}{\Phi_{V,\lambda}} = \overbrace{\rho_{V,\lambda}}^{\text{reflectance}} + \overbrace{\alpha_{V,\lambda}}^{\text{absorptance}} + \overbrace{\tau_{V,\lambda}}^{\text{transmitance}} \quad [1]$$

The reflectance of light on the surface of the specimen depends on the surface's reflectance, the surface texture and the incident angle of the light flux (beam). The surface texture defines that the reflection is specular, diffuse or directed diffuse (Fig. 4.17). Specular surfaces reflect the light flux at the angle of incidence and

Fig. 4.17 The reflection of light can be specular, diffuse or directed diffuse

Table 4.2 Reflectance of the light of common built materials [44]

building materials, finish layers	reflectance of luminous flux ρ_v(-)
brick, red	0.1 – 0.15
granite	0.2 – 0.25
glass (ordinary, reflective)	0.07; 0.3
plywood	0.25 – 0.40
concrete	0.20 – 0.30
timber, oak	0.1 – 0.15
timber, maple	0.6
polished marble	0.3 – 0.7
gypsum	0.8
dark red masonry paint	0.15 – 0.2
light green masonry paint	0.45 – 0.55
white masonry paint	0.75 – 0.85

such surfaces are used for optical elements that direct the light into the depth of the room, thus improving the uniformity of the daylight in the space. Diffuse surfaces are normally treated as Lambertian emitters. The average light reflectance ρ_v over the light wavelength ranges of some built materials are shown in Table 4.2 Such surfaces scatter the light without the risk of glare. It is common that in simplified lighting-evaluation methods the indoor surfaces are determined by assuming the indoor surfaces are diffuse. In reality, diffuse surfaces reflecting light more pronouncedly in the direction opposite to the incident angle, but can only be analysed using advanced computer tools [31].

If the specimen is opaque, the amount of the light flux that will be absorbed on the surface of the matter is defined by the absorptance $\alpha_{v,\lambda}$, which is equal to the difference between unity and the reflectance of the specimen surface $(1 - \rho_{v,\lambda})$. In the case of transparent matter, part of the light flux is transmitted through the specimen and partly absorbed on its way through the sample. The transmittance of a substance depends on the amount of the absorbed light flux on its way through the specimen (Fig. 4.18, left) and the reflection losses at the surface between two materials of different refractive index n (Fig. 4.18, right). The absorbed light flux in a single passage through the specimen is defined by the attenuation coefficient μ, which is a property of the material (window glass has an attenuation coefficient of 0.003 m^{-1},

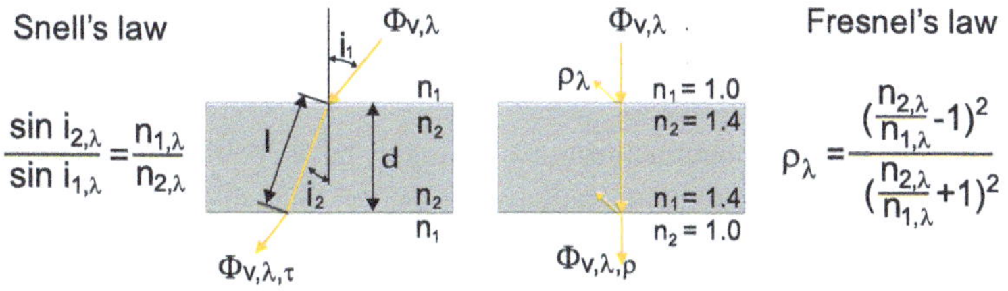

Fig. 4.18 Transmittance τ_v of transparent materials depends on the absorption of light flux on its propagation through material which is modelled by Beer-Lambert law (left) and on reflection of light flux on the boundary between two materials of different refractive indexes modelled by Fresnel law of reflection

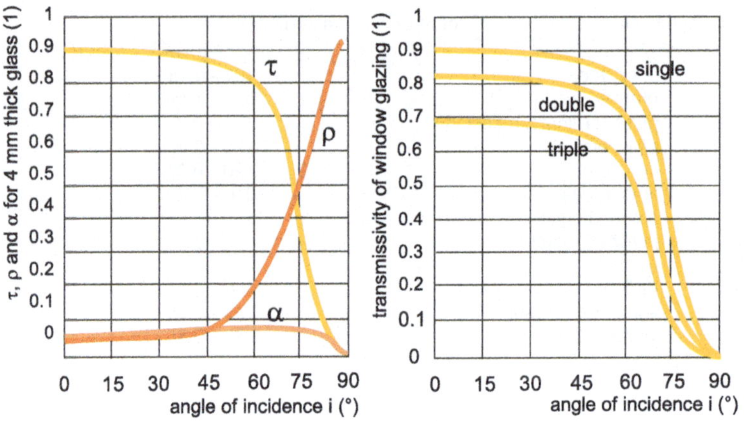

Fig. 4.19 Radiative properties ($\rho_{V,\lambda}$, $\alpha_{V,\lambda}$ and $\tau_{V,\lambda}$) of 4-mm-thick glass pane versus light-flux incident angle (left). An example of the transmittance of single-, double- and triple-pane glazing. Note that the values are informative and can significantly differ according to the type of the glass and the thin coating sputtered on the glass panes to decrease the heat losses and the risk of overheating

low-iron glass is 0.00004 m^{-1}, and polycarbonate is 0.0165 m^{-1}) and the length l of the light-flux path through the transparent material. The mathematical expression for this phenomenon is known as the Beer-Lambert law, while the reflectance losses are modelled by the Fresnel law (Fig. 4.18). Both phenomena caused a determination of the radiative properties of transparent materials complex, especially in the case of laminated glass, multiple glass-pane glazing or glazing with low-emissivity layers. Purpose-developed computer codes, such as Window 7.0 [45], are used to determine the radiative properties of modern multi-pane glazing (Fig. 4.19).

Explanation The refractive index n and the attenuation coefficient m depend on the wavelength of the light, which is not addressed here.

4.4.2.1 Selective and Dynamic Radiative Properties of Glazing

Advance materials and manufacturing technologies have enabled the production of glazing with selective and dynamic radiative properties. Selective glazing is designed in the way that the radiative properties of glazing are changing over time as response to a pre-selected impact parameter—outdoor temperature, density of light flux or incident angle (Fig. 4.20), while the dynamic radiative properties of (smart) glazing is adapted to the current needs by the user or a computer algorithm. Such types of glazing include:

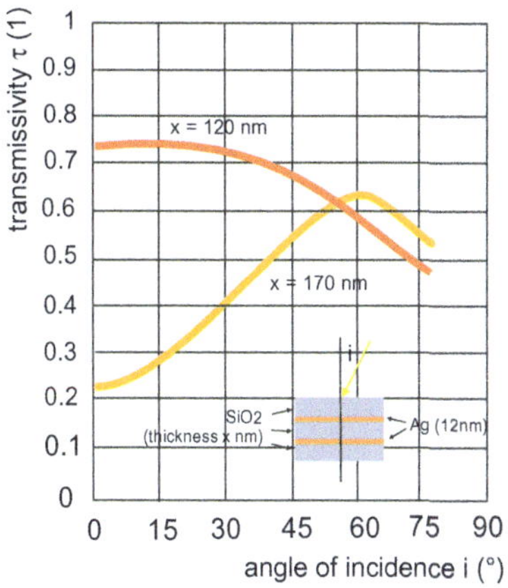

Fig. 4.20 Selective glazing with thin layers of silicon dioxide (SiO_2) and gold (Ag); transmittance of the glazing at different incidence angle of luminous flux can be adjusted by varying the thickness of the SiO_2 layer [10]; case (a) is more appropriate for south orientated windows, case (b) for east/west orientated windows because lower incident angle of sun rays

- photochromic glass, where the transmissivity changes according to the luminous flux;
- thermochromic glass, where the transmissivity is a function of temperature;
- electrochromic glass, where the transmissivity is changed by applying the voltage from an external source;
- gasochromic glass with a layer that interacts with a particular gas, changing the colour and thereby the transmissivity.

Despite intensive development, such glazing is still not commonly used in buildings.

4.5 Daylight

4.5.1 The Sun as a Source of Daylight

The temperature of the Sun's photosphere, the sphere seen as disk in the sky, is approximately 6000 K. As a thermal emitter the Sun emits electromagnetic radiation over a wide range of wavelengths, including the wavelengths of visible light (λ between 380 and 780 nm). The Sun's photosphere has a luminous efficacy, the ratio between emitted luminous flux Φ_v (lm) and emitted heat flux $\dot{Q}$ (W), 95 lm/W, which is the highest of the thermal emitters, even ones with a higher temperature.

The surface of the Earth is illuminated by the luminous flux coming directly from the Sun, and by diffuse luminous flux from the sky, because a part of the solar

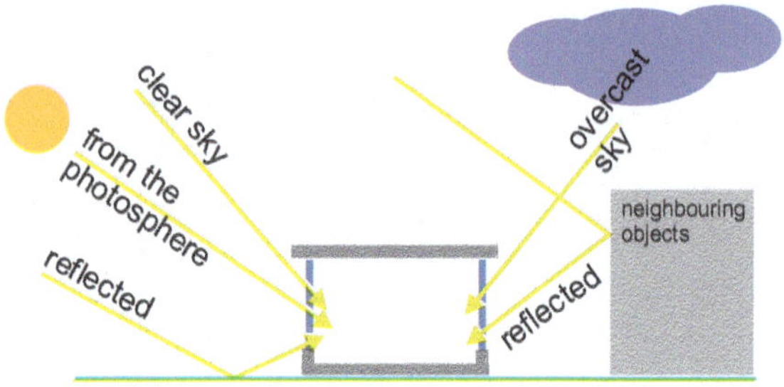

Fig. 4.21 The environment and the indoor spaces are illuminated by the direct, diffuse and reflected light

radiation is scattered as it travels through the Earth's atmosphere. A surface can also be illuminated by the luminous flux reflected from nearby natural or man-made objects (Fig. 4.21). The availability of daylight depends on the position of the Sun in the sky and whether the sky is clear or covered by clouds. The sky's cloud coverage is measured by the cloud index, the ratio of the solid angle subtended by the clouds in the sky to the total solid angle of the sky over a horizontal surface, which is equal to 2π steradians. By convention, the sky is clear if the cloud index is less than 0.15 and cloudy if the cloud index exceeds 0.6. In the past, the cloud index was determined empirically with a visual observation of the sky by meteorologists. Nowadays, the cloud index is constantly monitored using satellite images, comparing the albedo of the clouds and the Earth's surface. Albedo is a property of natural surfaces with respect to the reflection of light (Fig. 4.22).

The following sky characteristics are commonly assumed in the design of the daylight in buildings (Fig. 4.23):

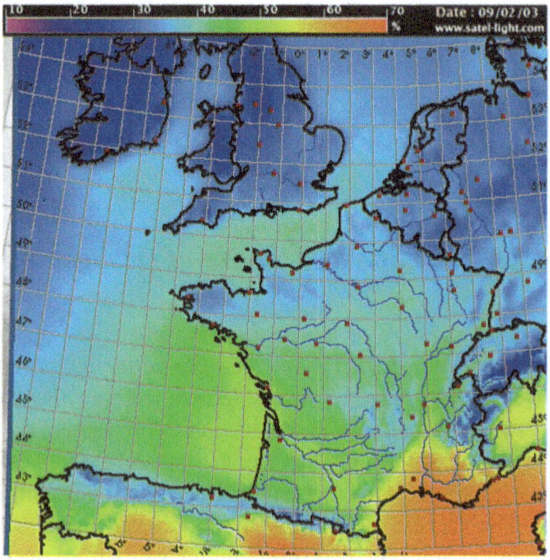

Fig. 4.22 The clear sky index, calculated from satellite images of the Earth's surface [11]

Fig. 4.23 Clear, intermediate and overcast sky conditions

- clear sky condition; the majority of the light is received directly from the Sun. As there are no clouds in the sky, only a small part of the direct luminous flux dissipates into the diffuse one. In a clear sky the solar corona patch is the brightest and the dark-blue patch opposite (90°) the Sun is the darkest;
- overcast sky condition; the sky is covered by a homogeneous cloud layer that is an isotropic source of diffuse daylight. It is assumed that any patch of the sky has the same luminance L (cd/m^2), regardless of the patch elevation above the horizon or the celestial orientation (azimuth). Therefore, the luminance of the zenith is equal to the luminance of the patches on the horizon;
- CIE overcast sky condition; the luminance of the patch in the sky depends on the elevation of the patch above the horizon, but is independent of its celestial orientation (azimuth). This means that patches 45° above the horizon orientated towards the south or north will have the same luminance at the same time during the day. Regardless of the time, the brightest point in the sky is the zenith and the darkest patches are on the horizon and the ratio of luminescence between them is 3:1. Such sky conditions are assumed when the quality of daylight of the buildings is evaluated;
- intermediate sky condition; is most realistic model of sky luminance, commonly used in computer simulations of daylighting and visual comfort indicators on the short time step. Such sky conditions are commonly approximated by using historical hourly meteorological data in form of test reference year or a local weather forecast.

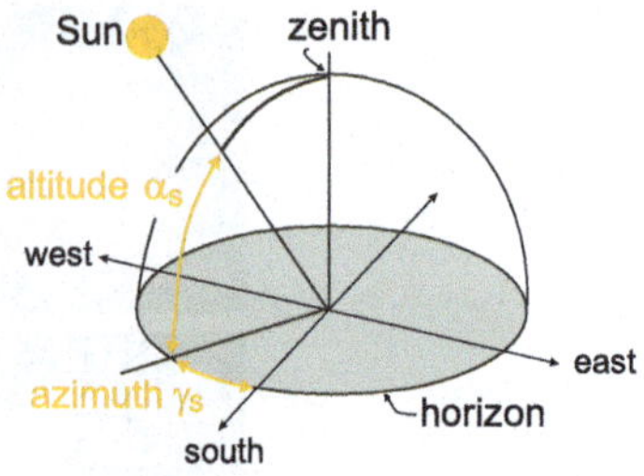

Fig. 4.24 Solar altitude α_s and azimuth γ_s

Explanation In 2003, the International Commission on Illumination (CIE) adopted a standard defining 15 different models of sky conditions, among them the CIE overcast sky model.

Regardless of which model of sky luminance is used, the data on the current position of the Sun in the sky must be defined by two angles (Fig. 4.24):

- the solar altitude angle α_s, the angle between the sun's rays and the horizontal plane;
- the solar azimuth angle γ_s, the angle between the projection of the sun's rays to the horizontal plane and the direction of South.

The current position of the Sun in the sky depends on the rotation of the Earth as well as on the relative tilt of the Earth's rotation axis on its orbit, described as the declination $\delta(°)$. While the position of the Earth's rotation axis in space remains constant (at ~ 23.45° according to the plane of orbit) throughout the year, the position of points on the Earth relative to the Sun changes all the time during the period of one year.

Declination δ is defined as angle between Sun's rays and Earth's equatorial plane. At equinox (spring and autumn) the declination is 0°. For the places in the northern hemisphere at the time of winter solstice the declination is −23.45° and at the summer solstice the declination is +23.45°. The declination on any given day of the year is determined as:

$$\delta = 23.45 \cdot \sin\left(\frac{360}{365}(284 + \mathrm{N})\right) \; [°]$$

where N is the running day in the year, for example 1 January is the 1st day and 15 February is the 46th day in the year. In contrast to the declination, which is the same at any location on the Earth, the solar altitude and solar azimuth depend on the location on the Earth from where the Sun is observed. A position of the observer is defined by its latitude L and longitude λ. The latitude is the angle between the equatorial plane and the observation point, and the longitude is the angle between the prime (Greenwich) meridian and local meridian towards West or East. The latitude L in the northern hemisphere is positive and in the range between 0° at the Equator and 90°at the North Pole.

The declination contains information about the time in the year, while daytime is taken into account by the hour angle ω(°). The hour angle changes by 15° each hour (360°/24 h). It has been defined that the hour angle is 0° at solar noon (12 o'clock by solar time), is −15° at 11 a.m., is −30° at 10 a.m. and so on, meanwhile, the hour angle in the afternoon is positive (e.g., +45° at 3 p.m.).

The solar altitude α_s and azimuth γ_s are given by [8]:

$$\alpha_s = \sin^{-1}(\sin L \cdot \sin \delta + \cos L \cdot \cos \delta \cdot \cos \omega) \ [^\circ]$$

$$\gamma_s = \sin^{-1} \frac{\cos \delta \cdot \sin \omega}{\cos \alpha_s} \ [^\circ]$$

The solar altitude α_s at sunrise and sunset equals 0°, and reaches the maximum value for a particular day at sun noon:

$$\alpha_{s,noon} = 90^\circ - L + \delta \ [^\circ]$$

Case Study Determine the direction and the length of the shadow cast by a 1.5-m vertical stick at a location with a geographical latitude L 46° on the 15th of April at half past 10 a.m. (Figs. 4.25 and 4.26).

The 15th of April is the 105th (31 + 28 + 31 + 15) day of the year, and the hour angle ω at 10:30 is:

$$\omega = (10.5 - 12) \cdot 15^\circ = -22.5^\circ$$

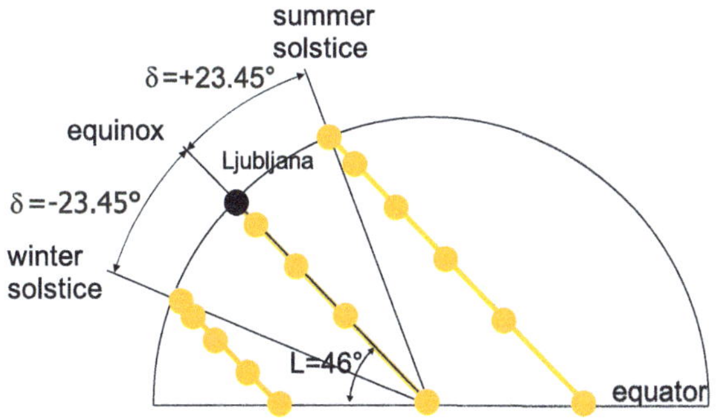

Fig. 4.25 The latitude of the observation point L defines the minimum and maximum solar altitude angles at the solar noon. For the site with latitude 46°, the minimum as at the solar noon at winter solstice is 20.6° and the maximum as at summer solstice is 67.5°

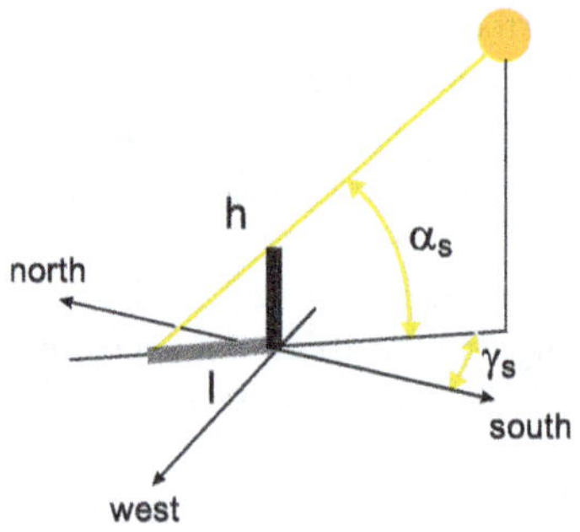

Fig. 4.26 Quantities used in the case study

The declination δ, solar altitude α_s and solar azimuth γ_s are:

$$\delta = 23.45 \cdot \sin\left(\frac{360}{365}(284 + n)\right) = 23.45 \cdot \sin\left(\frac{360}{365}(284 + 105)\right) = 9.4^\circ$$

$$\alpha_s = \sin^{-1}(\sin L \cdot \sin \delta + \cos L \cdot \cos \delta \cdot \cos \omega)$$

$$\alpha_s = \sin^{-1}(\sin 46 \cdot \sin 9.4 + \cos 46 \cdot \cos 9.4 \cdot \cos(-22.5)) = 48.6^\circ$$

$$\gamma_s = \sin^{-1}\left(\frac{\cos \delta \cdot \sin \omega}{\cos \alpha}\right) = \sin^{-1}\left(\frac{\cos 9.4 \cdot \sin(-22.5)}{\cos 48.6}\right) = -34.8^\circ$$

and the length of the shadow l_p is:

$$l = \frac{h}{\tan \alpha_s} = \frac{1.5}{\tan 48.6} = 1.32\ \text{m}$$

The direction of the shadow cast by a 1.5-m vertical stick is 34.8° northwest.

4.5.2 *Sun Path Diagram*

The solar altitude α_s and solar azimuth γ_s through the day can be presented in a solar-path diagram. This polar (Fig. 4.27) or cylindrical (Fig. 4.30) diagram is drawn for any location with a given latitude *L*. In the polar diagram, the diurnal part of the Sun is projected on the plane parallel to the equatorial plane (the observer is above the Earth's pole), while in the cylindrical diagram the path of the Sun is projected to the surface of a cylinder wrapped around the horizon, seen from the observation point.

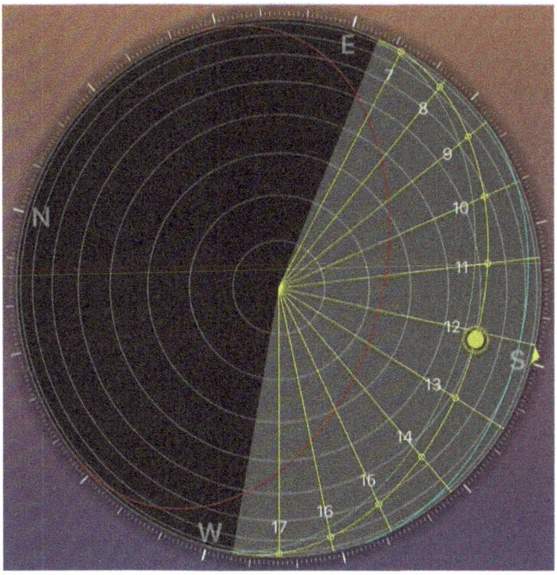

Fig. 4.27 Sun's path shown in a polar-type diagram [20]

In both types of diagrams, the solar altitude and azimuth are presented in terms of the solar time. This means that at the time of the solar noon, the solar altitude α_s reaches the maximum in that day, and the solar azimuth γ_s at that moment is 0°. Because the Earth's rotation around the Sun follows an elliptical orbit, the length of the arc on the orbit travelled by the Earth and the declination varies from day to day. Because of that the solar time and the civil time differs and the difference because of astronomical features (elliptic orbit and position of the Earth rotation axis in the space) is designated as the Equation of Time (*ET*) [9]:

$$\mathrm{ET} = 9.87 \cdot \sin(2 \cdot \mathrm{Y}) - 7.355 \cdot \cos(\mathrm{Y}) - 1.5 \cdot \sin(\mathrm{Y})[\mathrm{min}];\ \mathrm{Y} = \frac{360}{364} \cdot (\mathrm{N} - 81)$$

where *N* is the running day number in the year. *ET* is equal to 0 on only 4 days in a year. On other days it is between $-14'15''$ and $+16'30''$. The mathematical expression can be shown in a graph called an "analemma" (Fig. 4.28). It is common that "analemma" curves are plotted on solar path diagrams, as shown in Fig. 4.30.

Another reason for the difference between solar and civil time is that we set our watches to equal civil time inside the whole time zone, even though if it takes the Sun one hour to "pass" a time zone between two standard meridians. Because of that, the difference between the solar and civil time depends not only on *ET*, but on the difference between the longitude of the observation point λ and the longitude of the time zone's standard meridian λ_{st} as well (Fig. 4.29). The standard meridian of Central European Time Zone has a longitude of $-15°$.The civil time can be determined knowing the solar time (e.g., read from sun watch), and the solar time (which defines the true position of sun-made shade) can be determined by knowing

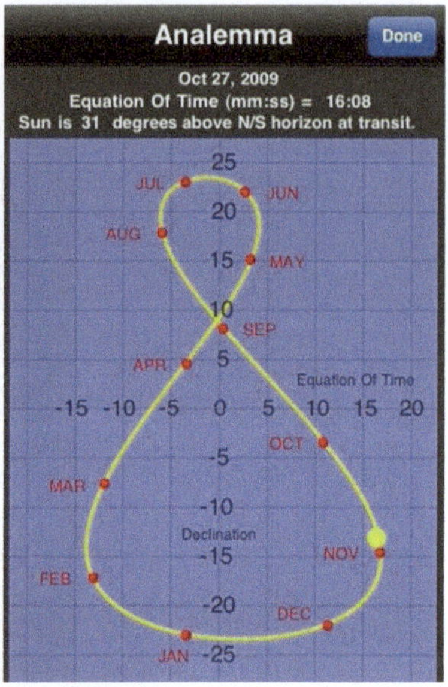

Fig. 4.28 The difference between solar and civil time because of astronomical features is shown in analemma diagram. The figure shows the Earth's declination δ (°) on the vertical axis; both features (ET and δ) are planetary features and do not depend on the geographical position of the observation point on the Earth's surface [12]

Fig. 4.29 Audemars Piguet Equation of Time wrist watch has a hand that shows the *ET* variation over the year and the solar noon time, the bezel is factory made for the geographical location of Geneva, with the longitude λ—6°24′. Because of that, the solar noon will be on 4 days during the year exactly at 12:35:24 civil time (because *ET* is 0, while $4 \times (\lambda - \lambda_{st}) = 4 \times (-6.15 - (-15.00))$, while on the rest of the days the solar noon will be a little bit earlier (but not before 12:21) or later (but not later than 12:52) [13]

the civil time using the expressions:

$$t_{civil} = t_{sun} - \overbrace{4 \cdot (\lambda - \lambda_{st})}^{minutes} - \overbrace{ET}^{minutes} \quad t_{sun} = t_{civil} + \overbrace{4 \cdot (\lambda - \lambda_{st})}^{minutes} + \overbrace{ET}^{minutes}$$

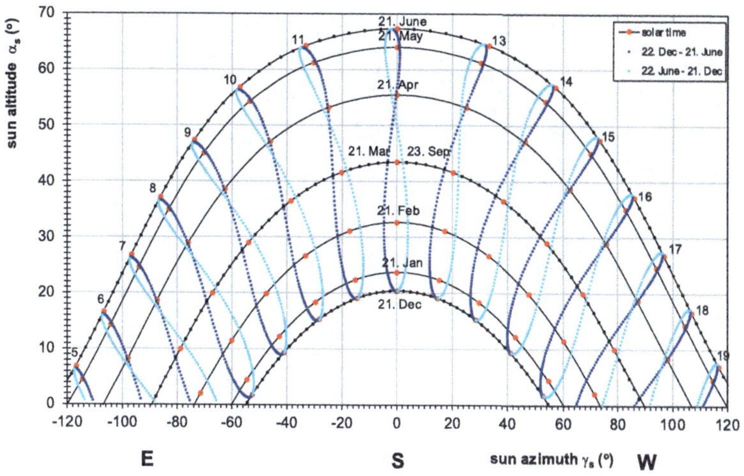

Fig. 4.30 Cylindrical type of solar-path diagram designed for Ljubljana (L 46.1°, λ–14.5°) [14]. The difference between solar and civil time when ET is 0 is only 2 min because Ljubljana lies close to the standard time-zone meridian. The sun path shown for the 21st February corresponds for 21st October; the analemmas are coloured with light blue colour for the period between 22nd June and 21st December and by dark blue colour line for the period between 22nd December and 21st June (solar noon will be at ~ 11:50 civil time on 21st October and at ~ 2:15 on 21st February. *Note* the summer time not taken into account)

In Fig. 4.30 the hours of solar time are shown with red points, while analemmas show the derivation of the civil time from the solar time.

4.5.3 *The Luminance of the Clear Sky*

The luminance of a clear sky depends on several geographical and meteorological parameters. The luminance of each patch L_P in the sky is different and depends on the instantaneous solar altitude and azimuth (α_s, γ_s), the altitude and azimuth of patch P (α_P, γ_P) in the sky and the optical properties of the atmosphere (the luminance of the observed patch in the sky will be different in an urban, countryside or industrial environment). The brightest area by far is the zone of the solar corona ($L_P > 50{,}000$ cd/m^2), while the darkest point in the sky ($L_P < 1{,}000$ cd/m^2) lies opposite (~90°) the Sun (Fig. 4.31). In engineering practice, software tools are used to determine the luminance of the clear sky patches [14]. This type of sky is used for glare-risk analyses.

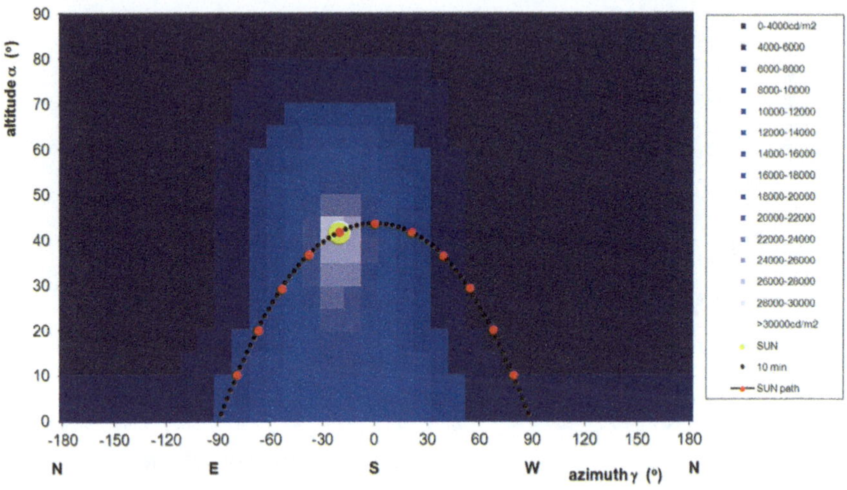

Fig. 4.31 Luminance of patches in the sky determined for clear sky conditions. The sun path is show for the 21st of March for a location with the geographical latitude L 46° and the illuminance of the sky is shown for 11 o'clock on the same day [14]

4.5.4 *The Luminance of the Overcast Sky and the CIE Overcast Sky*

Overcast sky conditions are commonly used to assess the quality of daylight in indoor spaces during a building's design process. Such sky conditions simplify the numerical evaluation, although quite faithfully representing the daylight potential in the case of the least favourable daylight conditions. In the case of overcast sky conditions, each of the patches in the sky has the same luminance L_P, including the highest point in the sky L_Z—the zenith. The luminance of the patch L_P changes during the day according to the instantaneous solar altitude α_s (°) (Fig. 4.31):

$$\mathrm{L_P} = \mathrm{L_Z} = \frac{9}{7\cdot\pi}\cdot(300 + 21000\cdot\sin\alpha_s)\quad\left[\frac{\mathrm{cd}}{\mathrm{m}^2}\right]$$

Unlike the overcast sky model where the position of the patch regarding its elevation and celestial direction has no impact on the patch's luminance, the CIE overcast sky model assumes that the zenith is the brightest point in the sky and the patches on the horizon are the darkest (Fig. 4.32). The luminance of an individual patch L_P depends on the instantaneous solar altitude α_s and the patch elevation angle above the horizon α_P:

$$\begin{aligned}\mathrm{L_P} &= \mathrm{L_Z}\cdot\left(\frac{1+2\cdot\sin\alpha_\mathrm{P}}{3}\right)\\ &= \frac{9}{7\cdot\pi}\cdot(300+21000\cdot\sin\alpha_s)\cdot\left(\frac{1+2\cdot\sin\alpha_\mathrm{P}}{3}\right)\quad\left[\frac{\mathrm{cd}}{\mathrm{m}^2}\right]\end{aligned}$$

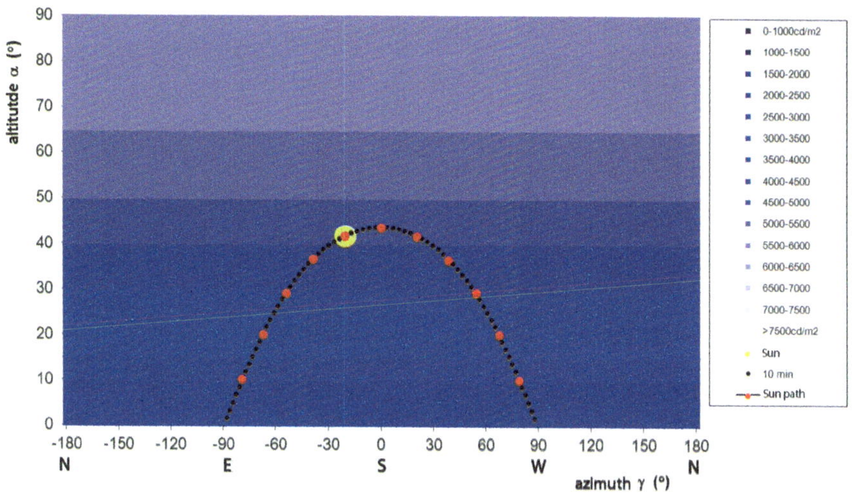

Fig. 4.32 Luminance of patches in the sky determined using the CIE overcast sky model. The virtual sun path is shown for the 21st of March for a location with the geographical latitude L 46° and the luminance of the sky is shown for 11 o'clock in the same day [14]

Explanation In the CIE overcast sky luminance model the luminance of the zenith is 3-times higher than the luminance of patches on the horizon, but the luminance of the patches does not depend on their celestial orientation. This means that the space daylight will be better if transparent openings are designed in such a way that the space interior receives light from the zone in the sky closer to the zenith rather than to the horizon. An example of such an opening is a skylight. At the same time, there will be no difference in the modelled daylight of spaces with south- or north-orientated openings.

4.5.5 *Illuminance of a Horizontal Plane Under an Overcast Sky*

The illuminance of an outdoor plane by luminous flux received from an overcast sky is determined as the integral of the differential luminous flux emitted by a patch with a differential area dA over the total area of the sky hemisphere (Fig. 4.33). The differential luminous flux is evaluated by the patch's luminance dL_P and the cosine of the angle i between the normal to the plane and the normal to the surface of the differential patch. The fact that in the model of an overcast sky, as well as in the model of the CIE overcast sky, the luminance of the patches is independent of the celestial orientation (azimuth) γ_P of the patch, simplifying the solution of the surface

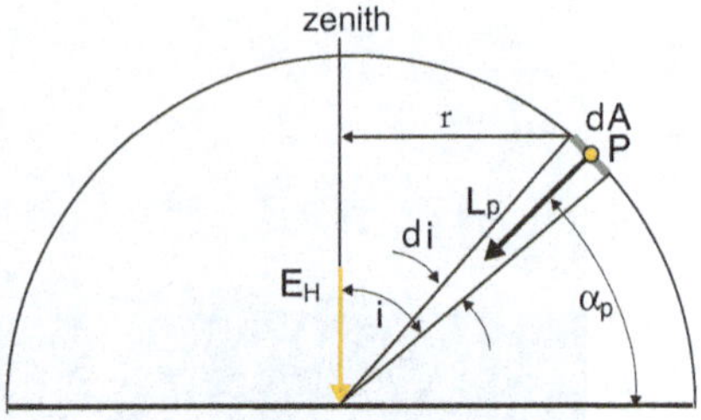

Fig. 4.33 The quantities used in the calculation of the illuminance of a horizontal plane by the light emitted from uniform and CIE overcast sky with a given luminance

integral. In the integral the differential size of the patch dA can be expressed by its 3D solid angle $d\omega$ (Fig. 4.33). Later, the 3D solid angel $d\omega$ can be replaced by the 2D angle di and the circumference of the ring r in the following way:

$$d\omega = \overbrace{2 \cdot \pi \cdot r}^{\text{ring circumreference}} \cdot di = 2 \cdot \pi \cdot \sin i \cdot di$$

taking into account that the distance between the patch and the horizontal plane is equal to 1. In the case of the CIE overcast-sky luminance model, the illumination of the horizontal plane E_H in integral form is:

$$E_H = \int_A L_p \cdot \cos i \cdot d\omega = \int_A L_z \left(\frac{1 + 2 \cdot \sin \alpha_p}{3} \right) \cdot \cos i \cdot d\omega$$

$$= 2 \cdot \pi \cdot \int_0^{\pi/2} L_z \left(\frac{1 + 2 \cdot \sin \alpha_p}{3} \right) \cdot \cos i \cdot \sin i \cdot di$$

$$E_H = \frac{2}{3} \cdot \pi \cdot L_z \cdot \int_0^{\pi/2} \left(1 + 2 \cdot \sin \alpha_p\right) \cdot \cos i \cdot \sin i \cdot di$$

because α_P can be expressed as ($\pi/2 - i$), sin a_P is equal to cos i, and the integral can be divided into two terms:

$$E_H = \frac{2}{3} \cdot \pi \cdot L_z \cdot \int_0^{\pi/2} \cos i \cdot \sin i \cdot di + \frac{4}{3} \cdot \pi \cdot L_z \cdot \int_0^{\pi/2} \cos^2 i \cdot \sin i \cdot di$$

By substituting the variables and changing the integral boundaries (instead of integrating between 0 and $\pi/2$, we will integrate from $-\pi/2$ to 0), we obtain the following solution:

$$E_H = \frac{2}{3} \cdot \pi \cdot Lz \cdot \left[\frac{\cos^2 i}{2} \right]_{i=-\pi/2}^{i=0} + \frac{4}{3} \cdot \pi \cdot Lz \cdot \left[\frac{\cos^3 i}{3} \right]_{i=-\pi/2}^{i=0}$$

$$E_H = \frac{2}{3} \cdot \pi \cdot L_z \cdot \left(\frac{1}{2} - 0\right) + \frac{4}{3} \cdot \pi \cdot L_z \cdot \left(\frac{1}{3} - 0\right)$$

Finally, the illuminance of an outdoor horizontal plane under the CIE overcast sky is:

$$E_H = \frac{7}{9} \cdot \pi \cdot L_z \left[\frac{lm}{m^2} = lx\right]$$

In the case of an overcast sky, the term $((1 + 2 \times \sin \alpha_p)/3)$ in the integral of E_H can be omitted because the luminance of the patches in the sky do not depend on the elevation of a patch in the sky above the horizon, the illumination of the horizontal plane is:

$$E_H = \pi \cdot L_z \left[\frac{lm}{m^2} = lx\right]$$

4.5.6 Illumination of a Vertical Plane

The illumination of vertical surfaces E_V consists of the contributions of the direct luminous flux from the Sun and the sky ($E_{V,s}$), the luminous flux reflected from the outdoor natural and built surfaces ($E_{V,g}$) and the luminous flux reflected from buildings ($E_{V,b}$). The total illumination of the vertical plane E_V is therefore:

$$E_V = E_{V,s} + E_{V,g} + E_{V,o} \quad [lx]$$

When the sky is overcast, the vertical surfaces only receive the luminous flux $E_{V,s}$ emitted by the half of the celestial dome between the zenith and the horizon. Besides that, the luminous flux received from the sky could be reduced by the nearby obstacles (e.g., hills and neighbourhood buildings) that mask the sky. By limiting the boundary angles in integrals developed to determine the illumination of the horizontal plane E_H, the illumination of the vertical surface on the buildings can be determined. The surrounding buildings that mask the part of the sky can be treated as 2-dimensional objects, with the height described with the obstruction angle γ_z(°) measured from the zenith (Fig. 4.34). If the CIE overcast-sky luminance model is used, the illumination of the vertical surface is equal to [15, 16]:

$$E_{V,s} = L_z \cdot \left(\frac{\pi}{6}(1 - \cos \gamma_z) + \frac{4 \cdot \pi}{9} \sin^2 \gamma_z\right) = \frac{3.5}{9} \cdot \pi \cdot L_z \; [lx]$$

The determination of the luminous flux on the vertical surfaces, for example, on the window surface, could be simplified by introducing quotients of the luminous

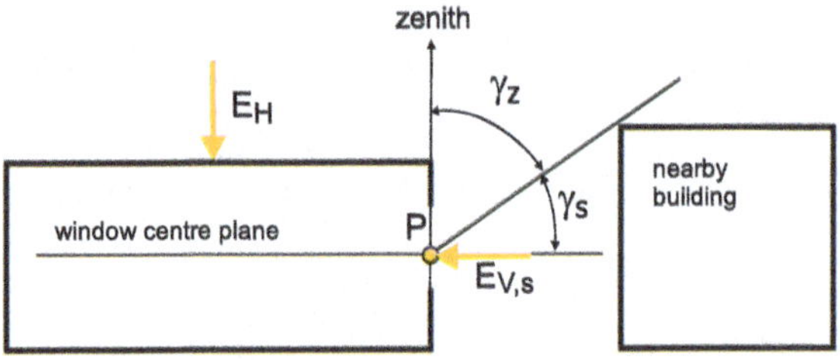

Fig. 4.34 Point *P* on the vertical façade receives the luminous flux from the sky. The uncovered part of the sky is limited by the angle between the zenith and the obstacle γ_s. This could be a hill in the vicinity of the building or adjacent structures

flux. The quotient of the illumination of a vertical surface caused by the luminous flux received from the sky C_s is the ratio of the illumination of a point on the vertical surface $E_{V,s}$ to the illumination of a point on the horizontal unshaded plane E_H:

$$C_s = \frac{E_{V,s}}{E_H}\ [l]$$

and the quotients of illumination of a vertical surface by the luminous fluxes reflected from the outdoor surfaces C_g and the neighbouring buildings C_b can be defined as:

$$C_g = \frac{E_{V,g}}{E_H}\ [1] \qquad C_o = \frac{E_{V,o}}{E_H}\ [1]$$

The total illumination E_V of a point on the vertical plane in the urban environment is determine as:

$$E_V = \left(C_{V,s} + C_g \cdot \rho_g + C_o \cdot \rho_o\right) \cdot E_H\ [\mathrm{lx}]$$

The quantities ρ_g and ρ_o are the albedo of the ground surfaces and the neighbouring buildings, respectively (Table 4.4). The values of the quotients C_s, C_g and C_o depend on the solid angles γ_s, γ_g, γ_o, as presented in Figs. 4.34, 4.35 and 4.36 and presented for the CIE overcast sky conditions in Table 4.3.

A part of the received luminous flux is reflected off the natural and built surfaces according to the surface's reflectance ρ_s. Except in special cases, such as the surface of water that exhibiting specular reflection, outdoor surfaces are commonly assumed to reflect the light according to the laws of Lambertian surfaces. Such surfaces reflect diffuse light and the brightness of those surfaces depends on the solid angle between the normal of the surface and the observer. The luminous flux reflected off the outdoor surfaces improves the daylight of the indoor spaces by 10–15% in a typical built environment and up to 50% if the outdoor surfaces are covered by snow. In the urban environment with narrow street canyons having brightly painted buildings, the daylight of the indoor spaces is improved because the luminous flux is reflected off

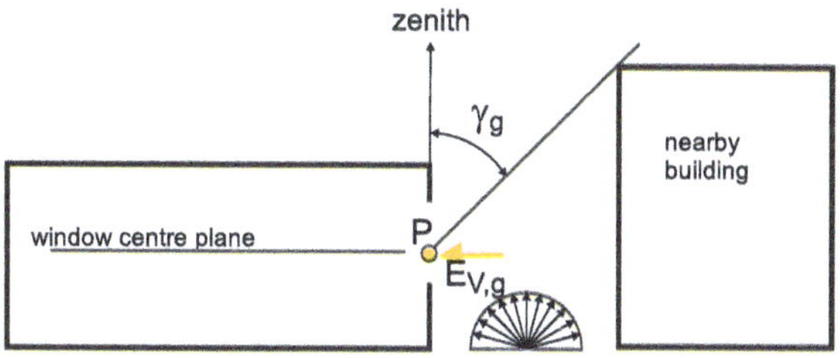

Fig. 4.35 Point *P* on the vertical surface receives the luminous flux that is reflected from the ground. The received luminous flux depends on the reflectivity of the ground (Table 4.4) and the size of the angle of the unobstructed sky γ_g (Fig. 4.3)

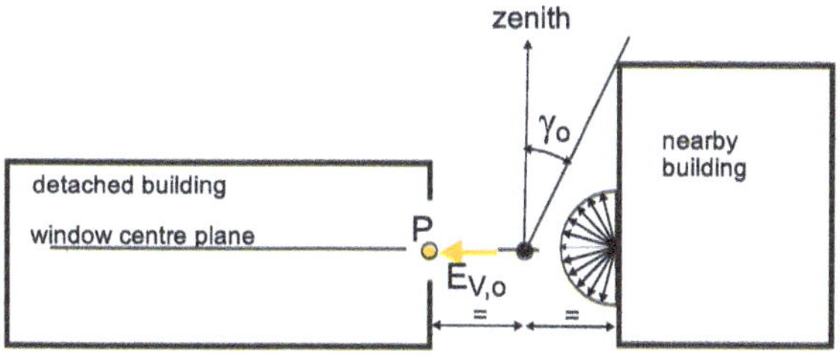

Fig. 4.36 Reflected luminous flux from nearby objects depends on the reflectivity ρ_o of the objects (Table 4.3) and the angle γ_o, measured from a point in the centre of the street canyon. The received luminance flux is proportional to the quotients C_o, which differ for detached buildings ($C_{o,d}$) and the buildings in the street canyon ($C_{o,c}$) (Fig. 4.3) [16, 17]

Table 4.3 Values of the quotients C_s, C_g and C_o in case that the sky luminance is determined using the CIE standard overcast-sky model. C_o differs depending on whether the surrounding building is detached ($C_{o,d}$) or the building is part of a street canyon ($C_{o,c}$)

γ_s γ_g γ_o [°]	C_s [1]	C_g [1]	$C_{o,d}$ detached building [1]	$C_{o,c}$ street canyon [1]
0	0.396	0	0	0
10	0.353	0.050	0	0.200
20	0.302	0.100	0.020	0.190
30	0.244	0.160	0.030	0.170
40	0.183	0.220	0.050	0.150
50	0.125	0.290	0.060	0.130
60	0.074	0.360	0.060	0.100
70	0.034	0.430	0.050	0.070
80	0.009	0.480	0.030	0.030

Table 4.4 Reflectivity or albedo of natural and built surfaces

surface	albedo (1)
grass (0.02 m, 1.0 m tall)	0.26; 0.16
soil (wet, dry)	0.05; 0.40
large water surfaces	0.9
snow (fresh, old)	0.95; 0.4
forest (deciduous, coniferous)	0.20; 0.15
building materials	reflectance (1)
walls (concrete, brick, stone, wood)	(0.20; 0.3; 0.3; 0.1–0.4)
roofs (bitumen, clay, sheet, straw, white reflective paint)	(0.10; 0.1; 0.2–0.5; 0.2; 0.8
painted surfaces (bright, dark)	0.50–0.75; 0.2–0.4
glass (normal, reflective)	0.07; 0.3
city (built-up areas, average)	0.10–0.27; 0.15

the facades of the surrounding buildings. However, this will probably not compensate for the reduced luminous flux received from the shaded sky.

4.5.7 Luminous Efficacy of the Sun and the Sky Irradiation

The models that can be used to determine the sky's luminance presented in Sects. 4.5.3 and 4.5.4 are simplified compared to the real conditions, because they do not account for the amount and distribution of the cloud cover and the local atmospheric pollution. More realistic data on the site daylight potential can be determined based on the solar irradiation data. Such databases are widely available and include data for hourly values of the global $G_{glob,o}$ (W/m^2), direct $G_{dir,o}$ (receiving directly from the sun) and diffuse $G_{dif,o}$ (receiving from the sky) solar irradiance. By employing experimentally determined approximation models, different forms of solar irradiation can be converted into the luminous flux Φv by the luminous efficacy factors K_l. Different luminous efficacy factors are developed for the global ($K_{l,glob}$), direct ($K_{l,dir}$) and diffuse solar irradiance $(K_{l,dif})$. The commonly used values are presented in Table 4.5. The luminous efficacies $K_{l,glob}$ and $K_{l,dir}$ are commonly defined, taking into

Table 4.5 Typical values of the luminous efficacy of the global solar irradiance $K_{l,glob}$, direct solar irradiation $K_{l,dir}$ and diffuse solar irradiance $K_{l,dif}$. When the sky is clear, the luminous efficacy depends on the instantaneous solar altitude α_s [18, 19]

	luminous efficacy of solar irradiation K_l (lm/W)
overcast sky	$K_l = 107$
clear sky, for global irradiance	$K_{l,glob} = 91.2 + 0.702\alpha_s - 0.00063\alpha_s^2$
clear sky, for direct irradiance	$K_{l,dir} = 51.8 + 1.646\,\alpha_s - 0.01513\,\alpha_s^2$
clear sky, for diffuse irradiance	$K_{l,dif} = 144$

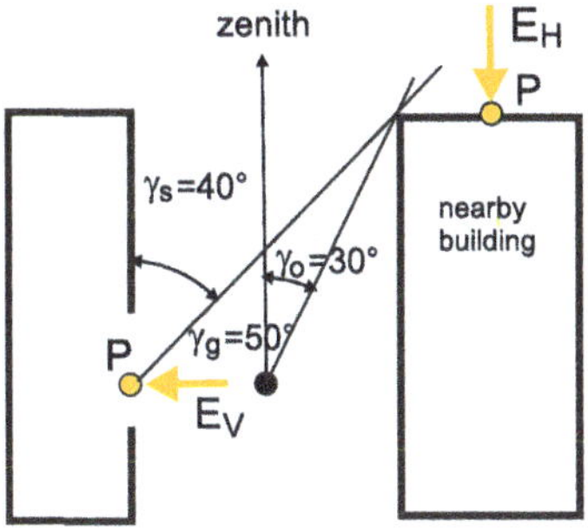

Fig. 4.37 Buildings and boundary angles from the case study

account the instantaneous solar altitude angle α_s. The illuminance of the horizontal plane E_H is:

$$E_H = K_l \cdot G_{glob,0}\left(\frac{lm}{W} \cdot \frac{W}{m^2} = \frac{lm}{m^2} = lx\right)$$

$$E_H = K_{l,dir} \cdot G_{dir,0} + K_{l,dif} \cdot G_{dif,0}\ (lx)$$

where $G_{glob,o}$, $G_{dir,o}$ and $G_{dif.o}$ are the global, direct and diffuse solar irradiation on the horizontal plane,

Case Study What is the illuminance of an open horizontal plane E_H and what is the illuminance of the point P on a vertical façade E_V of a street canyon building? The global irradiance of a horizontal plane $G_{glob,o}$ at the observed moment is 380 W/m^2 and consists of the direct irradiance $G_{dir,o}$ 210 W/m^2 and the diffuse irradiance $G_{dif,o}$ 170 W/m^2. The reflectivity of the ground surfaces ρ_g is 0.2, and the reflectivity of the nearby buildings ρ_o is 0.3. The corresponding obstruction angles are presented in Fig. 4.37.

From the solar-path diagram (Fig. 4.30) we can read that the solar altitude α_s on the day and the hour of the observation is equal to 30°. The luminous efficacy of the solar radiation is taken from Table 4.5:

$$K_{l,dir} = 51.8 + 1.646 \cdot 30° - 0.01513 \cdot 30°^2 = 87.6\left[\frac{lm}{W}\right]$$

$$K_{l,dif} = 144\left[\frac{lm}{W}\right]$$

and the illumintion of the horizontal plane is:

$$E_H = K_{l,dir} \cdot G_{dir,o} + K_{l,dif} \cdot G_{dif,o}$$

$$E_H = 87.6 \cdot 210 + 144 \cdot 170 = 42876\ \text{lx}\left[\frac{\text{lm} \cdot \text{W}}{\text{W} \cdot \text{m}^2} = \text{lx}\right]$$

The quotients $C_s = 0.183$ ($\gamma_s = 40°$), $C_g = 0.290$ ($\gamma_n = 0°$) and $C_o = 0.170$ ($\gamma_b = 30°$) can be read from Table 4.3. The illuminance of point P on the vertical façade is equal to:

$$\begin{aligned} E_V &= \left(C_s + C_g \cdot \rho_g + C_o \cdot \rho_o\right) \cdot E_H \\ &= (0.183 + 0.290 \cdot 0.2 + 0.170 \cdot 0.3) \cdot 42876 \doteq 12500\ \text{lx} \end{aligned}$$

4.5.8 Availability and Autonomy of Daylight

The availability of daylight is defined by the fraction of the occupancy period of the building over the year at which daylight alone provides sufficient illumination of the indoor spaces. It can be determined by hour-to-hour meteorological data of the illuminance of a horizontal plane E_H or when such data are unavailable, by hourly solar irradiation for a given hour in the year, and the luminous efficacies of solar irradiance K_s (Table 4.5). Figure 4.38 shows the hour-by-hour illuminance of an outdoor, unobstructed horizontal plane over the year determined with the solar irradiation data for the City of Ljubljana. The outdoor-illuminance data can be rearranged as cumulative values, showing the percentage of the time interval in which the illuminance E_H will be equal to a certain value. An example of such an evaluation is shown in Fig. 4.39, made for the daytime interval that corresponds to working hours in an office building (from 8 to 17 o'clock) over the whole year [23]. In this way the autonomy of daylight is defined. As an example, Fig. 4.39 shows that during 92% of the working hours the outdoor illuminance E_H will be at least 5000 lx. This means that if the office was designed in a way to fulfil daylight visual-comfort requirements at that particular outdoor illuminance, the artificial light in such an office will only be needed during 8% of the working hours. Regarding the example site daylight potential, it can also be seen that in the case of poor daylight design when an outdoor illumination of 20,000 lx is required for adequate daylighting, the daylight autonomy will be 68% and during the rest of the working hours, artificial lighting will be needed.

4.6 Artificial Electric Lighting

Daylight is the most pleasant and energy-efficient means of lighting. Nevertheless, it is not available all the time when buildings are in use and the daylight must be complemented with artificial lighting. The sources of artificial light are distinguished

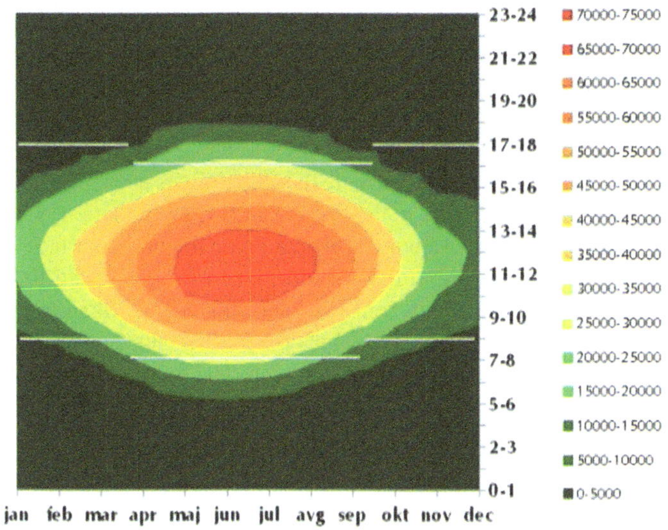

Fig. 4.38 Illuminance of an external horizontal surface E_H for the average day in a given month of the year for Ljubljana. The lines designate the beginning and the end of business hours. The values were determined based on the solar irradiance G_{glob} in a given hour in the year, and the corresponding luminous efficacy of the solar irradiance K_s. Whenever the illuminance E_H is in excess of a limit value, e.g., 5000 lx, the daylight should suffice to illuminate a properly designed room

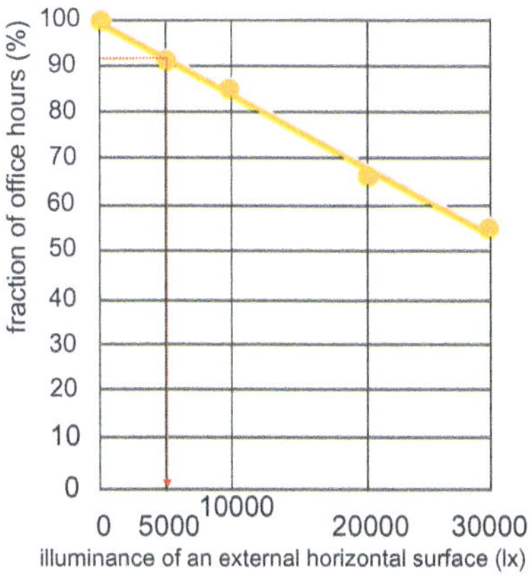

Fig. 4.39 Daylight availability curve for the office hours between 8 and 17 o'clock. The curve was derived from data on the hourly solar irradiance for a horizontal surface G_{glob} in Ljubljana. The daylight autonomy can be read from the diagram

with respect to the process used to transform the electrical energy into light. Incandescent light bulbs have an element that produces electromagnetic radiation in the visible part of the spectrum as the electric current heats a tungsten filament. The source of light can also be an arc that is formed between two electrodes in an arc lamp. Such lamps are known as high-intensity discharge (HID) lamps. Fluorescent lamps (FLs) are gas-discharge lamps where the electricity is used to excite molecules of gas, which in turn emit energy in the form of ultraviolet radiation. This radiation is then converted into light by a fluorescent coating on the inside surface of the lamp assembly. The latest technology in the field of electric lighting are light-emitting diodes (LEDs), which are made of semiconductor materials. All sorts of lamps are integrated into luminaires, in which various optical elements designed to direct the light into the space are united. The properties of electric light sources are defined by:

- the luminous efficacy K_l(lm/W) is the ratio of the emitted luminous flux Φ_v in the wavelength range of visible light (380 nm < λ < 780 nm) to the electrical power P_{el} of a lamp expressed in W. The emitted luminous flux Φ_v is weighted by the relative spectral sensitivity of the eye V_λ. A monochromatic light source emitting light with a 550-nm wavelength will have the greatest luminous efficacy of 683 lm/W. Real light sources have a lower luminous efficacy—incandescent light bulbs up to 20 lm/W, compact fluorescent lamp up to 70 lm/W, and LED lamps up to 160 lm/W (even above 200 lm/W in laboratory conditions). Luminous efficacy indicates the energy efficiency of a light source (Fig. 4.40);
- the luminous efficiency E_l (%) is the ratio of a light source's luminous efficacy K_l to the maximum luminous efficacy of a monochromatic light source (683 lm/W). E_l is given in percent. A thermal radiator at 3000 K has E_l 4%, a thermal radiator at temperature 4000 K has E_l 7%, and the Sun's photosphere at 6000 K has E_l 34%,

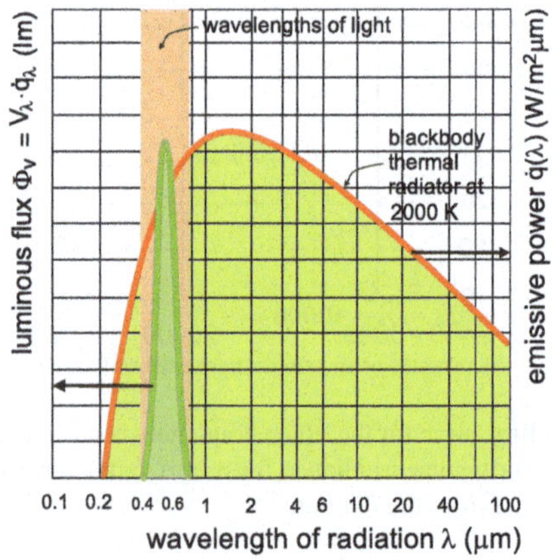

Fig. 4.40 The luminous efficacy of electric light source K_l is the ratio of luminous flux (in lumens) emitted by a light source to its electric power, which equals the total emitted heat flux (in W). Although the definition applies to the thermal emitters, it is used for all types of electric light sources

Table 4.6 Colour temperature T_c of thermal emitters. The values corresponds to the correlated colour temperature CCT of other electrical light sources

colour temperature	Tc, CCT (K)
warm	< 3300 K
neutral white	3300–5000 K
cold daylight	> 5000 K

which is the highest among all thermal emitters. Luminous efficiency indicates how close to the theoretical maximum of emitted light flux a technology of light generation is;

- the colour temperature T_c (K) is a characteristic of light sources that generate the light as thermal radiators, having a sufficiently high temperature. It is determined by comparing the spectrum of light emitted by a particular electrical light source and the spectrum of light emitted by an ideal thermal radiator—an optical black emitter at temperature T_c. The radiation spectrum of an optical black emitter is determined by Planck's law of the spectral distribution of radiation (Sect. 1.1.3). Hotter radiators emit light at higher wavelengths (more blue light), while bodies at a lower temperature emit longer wavelengths towards the red part of the spectrum (Table 4.6);

Explanation The colour temperature is determined on the basis of physical laws of thermal radiation. Taking the physiological point of view, the colour hues are described based on our experience of observing nature—the blue colour reminds us of ice and water and is perceived as cold, while the red colour is associated with fire and warmth.

- gas-discharge lamps and LEDs do not emit the light as thermal emitters, because of that, the colour of the emitted light is evaluated by the apparent colour temperature *CCT* (correlated colour temperature). It is determined by comparing the appearance of the colour of light emitted by such lamps to the appearance of the colour of light emitted by optically black thermal radiator at temperature T_C.

Figure 4.41 shows a colour space that was prepared in 1931 by the CIE (International Commission on Illumination) based on the three basic colours detected by sight—blue, green and red. The diagram also shows the Planck's locus, designating the colours of light emitted by an optically black thermal emitter at different temperatures. The colour temperature is a property of natural and artificial light sources—the colour temperature of the blue sky is 8000 K, of sunlight at solar noon 5500–6500 K, and of sunlight at sunset 3500 K. The colour temperature of a candle flame ~2000 K, of an incandescent light bulb somewhere between 2700 and 3000 K, and of a xenon arc lamp, 4000 K. LED light sources can produce a large range of CCTs—between 2700 K (warm white) and 5000 K (cool white).

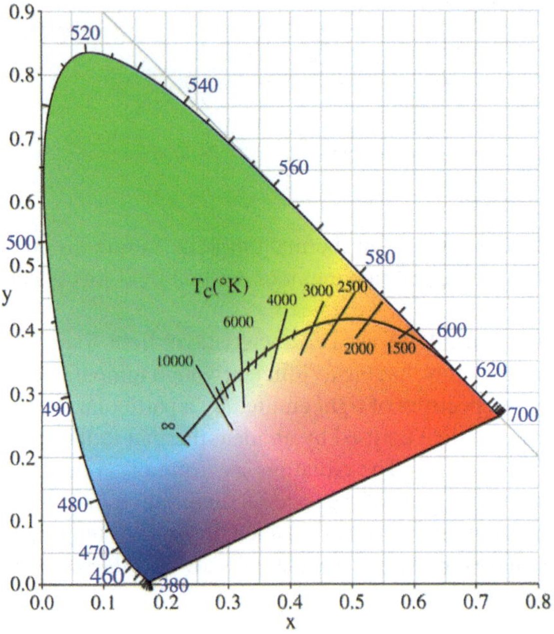

Fig. 4.41 The CIE colour space. The x, y and z (perpendicular to x and y) axes represent the fraction of red, green and blue colour, respectively. The white point is defined with the values $x = y = z = 0.333$. The external locus covers monochromatic light of different wavelengths, and the Planck's locus in the middle of the diagram represents the colour temperature of thermal radiators at different temperatures T_C. The white colour corresponds to the colour emitted by a thermal radiator at 6500 K, similar to sunlight at noon. The crossing lines designate the correlated colour temperatures *CCT* of non-radiating gas-discharge lamps [6]

- the colour-rendering index (*CRI*) is a measure of the faithfulness of colours of artificial illuminated objects, when compared to the perceived colour of an object illuminated with an ideal natural source of light—the sunlight at a solar noon. The colour rendering index is expressed as R_a with values between 0 and 100. An artificial light source with R_a equal to 100 produces the same detected colours of the body's surface as if it were illuminated by clear sky daylight at a solar noon. The artificial lighting's design must account for the required values of R_a with respect to the intended use of the space, as shown in Table 4.7.
- electric light sources are fitted into lighting fixtures, which must conform to several technical and safety requirements. Lighting fixtures are shaped in a way to distribute the light in a desired spatial pattern showing in polar photometric distribution 2D or 3D diagram ("the candela diagram"); according to the shape of the candela diagram, luminaires are divided into five categories. The boundary cases are the direct lighting fixtures, sending the light straight downward in a narrow angle (Type I), and indirect lighting fixtures directing the light up to the ceiling (Type V).

Table 4.7 Quality classes and colour rendering indexes R_a for the illumination of rooms with different intended uses

	R_a	intended use
1A	>90	galleries
1B	80 to 90	apartments, hotels, restaurants, hospitals, schools
2A	70 to 80	industrial buildings
2B	60 to 70	industrial buildings
3	40 to 60	rooms where the colour detection is less important
4	20 to 40	rooms where the colour detection is not important

The CRI values are also listed for transparent materials such as window glazing, determining the faithfulness of the colours of the environment when observed from inside through the glazing

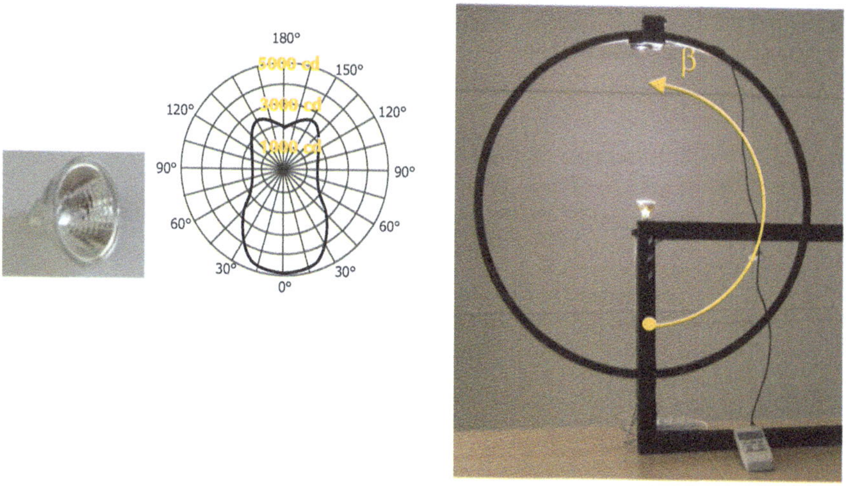

Fig. 4.42 Lamp installed in the centre of the ring with an illumination sensor used to determine the polar photometric distribution diagram of the lamp (left); polar 2D photometric distribution diagram for the lamp (bottom)

Case Study 1 A polar photometric distribution diagram can be determined by experiment (Fig. 4.42) measuring the illumination on the points equidistant from the source with an instrument that is positioned perpendicular with respect to the illuminance flux vector. If the emitted light flux is axisymmetric (2D light source), the measurement points are located on a single plane perpendicular to the axis of the luminaire. Otherwise, the experiment is repeated for several planes rotating around the central vector of the light flux (i.e., 3D luminaires). The measured illumination $E_{(\beta)}$ and the distance between the point and the light source r determine the luminous intensity with the expression:

$$E = \frac{I}{r^2} \cdot \cos i \quad \rightarrow \quad I_{(\beta)} = \frac{E_{(\beta)} \cdot r^2}{\underbrace{\cos i}_{1}} \; [\mathrm{cd}]$$

The sensor is moved around the ring at the same angular distances ($\Delta\beta$) and the calculated luminosity of the light source is that plotted in the polar photometric distribution diagram (Fig. 4.42, middle).

Case Study 2 Determine the direct illuminance E_P of a point P on a horizontal plane illuminated by two similar lamps installed 3.0 m ahead and at a distance of 2 m. The polar photometric distribution diagram of the lamp and the position of the observed point are shown in Fig. 4.43.

The luminous intensity $L_{(\beta)}$ in the direction of the observed point read from the polar photometric distribution diagrams are equal to:

$$L_{0^\circ} = 4900\ \mathrm{cd},\ L_{(25^\circ)} = 4200\ \mathrm{cd},\ L_{(30^\circ)} = 4000\ \mathrm{cd},\ L_{(46^\circ)} = 2500\ \mathrm{cd}$$

The distances r_1 and r_2 between the lamp centre and the illuminated point are:

$$r_1 = h = 3\mathrm{m},\ \ r_2 = \frac{h}{\cos 30^o} = 3.46\ \mathrm{m},\ \ r_3 = \frac{h}{\cos 46^o} = 4.32\ \mathrm{m},$$

$$r_4 = \frac{h}{\cos 25^o} = 3.31\ \mathrm{m}$$

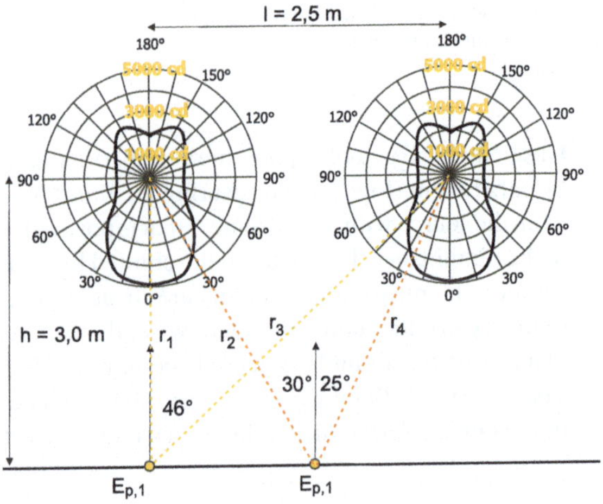

Fig. 4.43 Lamp's luminous-intensity-distribution diagram and the quantities used in the case study

The direct illuminance E_P at the observed point P is:

$$E_{P,1} = \frac{L_{(0^\circ)}}{r_1^2} \cdot \cos i_1 + \frac{L_{(46^\circ)}}{r_2^2} \cdot \cos i_2$$
$$= \frac{4900}{3^2} \cdot \cos 0^\circ + \frac{2500}{4.32^2} \cdot \cos 46 = 733 \text{ lx}$$

$$E_{P,2} = \frac{L_{(30^\circ)}}{r_2^2} \cdot \cos i_3 + \frac{L_{(25^\circ)}}{r_4^2} \cdot \cos i_4 = \frac{4000}{3.46^2}$$
$$\cdot \cos 30^\circ + \frac{4200}{3.31^2} \cdot \cos 25^\circ = 636 \text{ lx}$$

A fraction of direct luminous flux received by the illuminated patch from the lamp is reflected on the surfaces and scattered around the room. Therefore, complex ray-tracing computer tools have to be used for detailed illumination studies.

4.6.1 *Electric Light Sources*

Incandescent lamps replaced gas lamps at the end of the 19th Century. A coiled tungsten filament, which melts at 3680 °C, is inserted into a glass enclosure, filled with an inert gas or evacuated to prevent the filament from oxidising. As a result of its high electrical resistance, the filament heats up to 2500 °C. Tungsten slowly evaporation during operation, which limits the life of incandescent light bulbs to about 1000 h (Fig. 4.44).

Explanation The service life of incandescent lamps is defined as the period during which 50% of the light bulbs fail.

Fig. 4.44 An incandescent light bulb; these lamps have been banned due to their low luminous efficacy and resulting high electric energy consumption (left); halogen bulb with a filament (middle), a low-voltage halogen bulb

If the glass enclosure is filled with an inert gas (argon or krypton) with added atoms of halogen elements such as iodine or bromine, the tungsten evaporating from the surface of the heated filament is not deposited on the glass enclosure, instead it reacts with the halogen and later returns back to the filament. The service life of a light bulb is substantially extended in this way (2500–3500 h). Such lamps are known as halogen lamps. Because of the higher temperature of the filament, these lamps have a better illuminance efficacy, but the glass bulb must be made of highly heat-resistant quartz glass. The first light bulbs of this kind were made around 1960. They can operate at different supply voltages.

Fluorescent lamps have a considerably better luminous efficacy than incandescent light bulbs. They have been available on the market since the 1940s. The electrons emitted from a hot cathode collide with the atoms of mercury vapour, which are added to the argon, neon, xenon or krypton gas inside the lamp. The electrons pass their energy to the mercury atoms and excite them. The excited atoms are not stable, so they return to the "normal" energy state, producing ultraviolet light at 253 nm and 185 nm. The invisible ultraviolet radiation is converted into visible light inside a thin fluorescent coating on the internal surface of the lamp tube. Phosphor with special additives is used to regulate the colour of the emitted light (Fig. 4.45).

Explanation Fluorescence is a form of luminescence, a chemical, biological or other process that creates light. In the case of fluorescence, a substance absorbs short-wavelength UV-radiation and emits radiation at longer wavelengths—the visible light.

The fluorescent lamps are distinguished by their high luminous efficacy (up to 90 lm/W) and a long service life of between 12,000 and 15,000 h. The lamps are equipped with a ballast and a starter. The ballast regulates the voltage in the lamp during operation, while the starter provides the high voltage necessary for the initial ionisation of the mercury atoms. In older lamps, these two elements often produced an irritating hum and flicker. High-frequency electronic ballasts are integrated into modern lamps, also known as compact fluorescent lamps. This lamp variant has been on the market since the 1990s. Their luminous efficacy is comparable to the classic fluorescent lamps and their service life reaches up to 10,000 h. However, frequent starting cycles reduce the service life considerably, by about two-thirds after 20,000 ON/OFF cycles (Fig. 4.46).

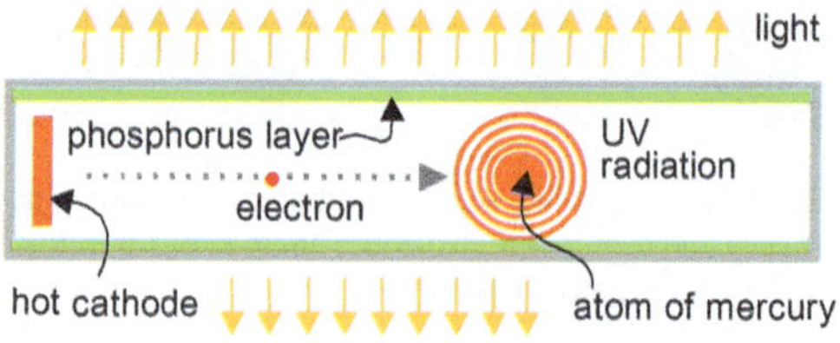

Fig. 4.45 Operating principle of fluorescent lamps

Fig. 4.46 Fluorescent lamps contain mercury vapour, which requires special disposal and recycling; the presence of mercury must be marked on the lamp (left); a compact fluorescent lamp is fitted with a silent electronic ballast and the Edison screw fitting to facilitate the replacement of old incandescent light bulbs (right)

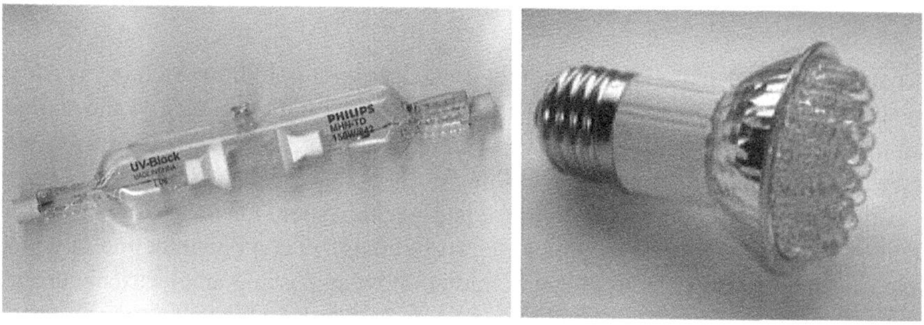

Fig. 4.47 A metal-halide discharge lamp (left); a LED lamp with the standard Edison screw fitting (right)

Explanation The service life of gas-discharge lamps is defined as the time period in which the emitted luminous flux is reduced to 80% of the initial value.

One source of light is the arc created between the electrodes inside an evacuated inner quartz-glass arc tube filled with an inert gas (such as xenon), and enclosed within an external glass bulb that stops the ultraviolet radiation. Mercury, metal halide (binary compounds of metal with halogen elements like fluorine, chlorine, bromine, iodine) or sodium can be added to the inner arc tube of so-called high-intensity discharge (HID) lamps. The arc is formed by a high-voltage current that causes the gas molecules to ionize and emits light. The service life of HID lamps is long, over the 100,000 h. Low- and high-pressure sodium lamps are a special type of gas-discharge lamps. Low-pressure sodium lamps have a double glass envelope, containing solid sodium, and neon and argon gas. Within a few minutes of turning on the lamp, the sodium is vaporized, the vapour atoms are ionized and start emitting photons. These lamps have a high luminous efficacy; nevertheless, such lamps emit almost monochromatic yellow colour. These lamps are often used in street lighting

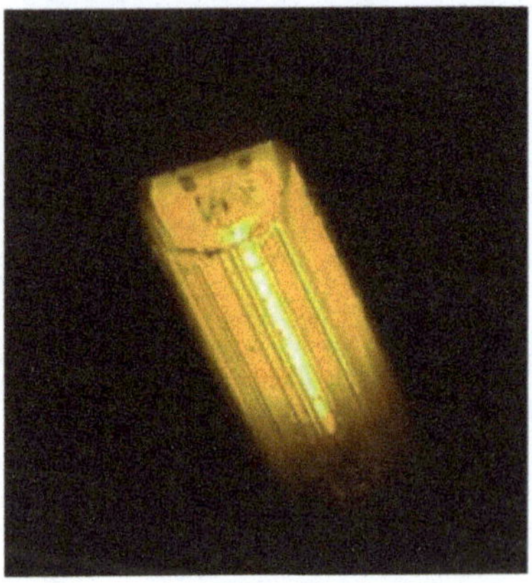

Fig. 4.48 Low-pressure sodium lamps emit characteristic yellow light; they are commonly used in street lighting

because the wavelengths of yellow light cause less light pollution. Mercury is added to sodium in high-pressure sodium lights. Due to the high pressure of the sodium vapour and the ionisation of mercury, such lamps emit light in a wider spectrum, and have an illuminance efficacy up to 100 lm/W (Figs. 4.47 left, 4.48).

Light-emitting diodes (LEDs) are the latest achievement in electric lighting technology. They utilize the principle of electro-luminescence, a phenomenon in which some substances emit electromagnetic radiation when passing an electrical current. They are made of semiconducting materials, in which the added electrons and gaps form a band gap. LEDs emit monochromatic light at wavelengths (colours) depending on the size of the band gap. They are manufactured from compounds of different inorganic chemical elements, such as gallium, arsenic, indium and aluminium. LEDs emitting white light can be made of three individual LEDs emitting the primary monochromatic colours (red, green and blue). Nowadays "white LED" (YAG) are high energy and cost efficient. Such LED emits blue light and use property of fluorescence of phosphor or other material coating. By adjusting the emission and fluorescence process, the colour temperature of the light can be adjusted from warn to cold. LEDs have a high luminous efficacy (up to 160 lm/W and in laboratory conditions up to 200 lm/W) and a long service life—from 75,000 to 100,000 h (Fig. 4.47, right; Table 4.8).

4.7 Requirements and Criteria of Visual Comfort

The main criteria that must be fulfilled with design of lighting and that are used for an assessment of the visual comfort in the indoor environment are:

- the minimum duration of exposure to sunlight. This criterion is defined by the number of hours for a given day(s) in the year, when a room receives direct (beam) solar radiation;
- an adequate minimum and average daylight factor;

Table 4.8 Characteristics of electric light sources

electric light source	K_l (lm/W)	E_l (%)	T_c, CCT (K)	R_a (1)
incandescent light bulb	<15	2.5	2700	95+
halogen lamp with a filament	<25	3.5	3000	95+
fluorescent lamp	<105	15	3000	95+
compact fluorescent lamp	60	10	3000	> 90
low-pressure Na lamps	<200	22	1800	5
high-pressure Na lamps	<140	35	2200	25
light-emitting diodes (LED)	<140	22	3000	90

- an adequate level of illumination provided by daylighting, electric lighting, or a combination of both;
- protection from glare that could occur when too bright spots or patches are observed in the central field of vision;
- adequate light colour temperature of artificial light that provides the most appropriate conditions for a comfortable stay and high productivity at work.

4.7.1 Duration of Exposure to Sunlight

The requirements regarding the minimum duration of exposure to sunlight originates in the well-being and healing effects of solar radiation. The duration is measured by the number of hours in a day when the room in a building can be illuminated by direct solar radiation, accounting for the nearby natural obstacles (e.g., hills) and surrounding buildings. The procedure takes into account the solar geometry and the clear sky conditions. The requirements vary according to the national legislation. For example, in Slovenia, the minimum required duration of exposure to sunlight of the rooms in residential buildings that are mainly used during the day-time is 1 h at the winter solstice, 3 h at the equinox and 5 h at the summer solstice, while 4 h of sunlight are recommended in at least one habitable room of a dwelling at equinox in Germany. In the UK it is requested that interiors are exposed to sunlight at least during 25% of the annual probable sunlight hours, of which at least 5% are between the 21st of September and 21st of March [46]. The duration of exposure to sunlight can be determined using photographic images, graphical methods or software tools. Panoramic or fish-eye photographs of the surroundings are taken, and the solar path is overlaid by drawing or AR applications (Fig. 4.49, left). The duration of exposure to sunlight is determined graphically using polar (Fig. 4.27) or cylindrical (Fig. 4.30) solar-path diagrams with plotted nearby objects that cast a shadow on the observed point.

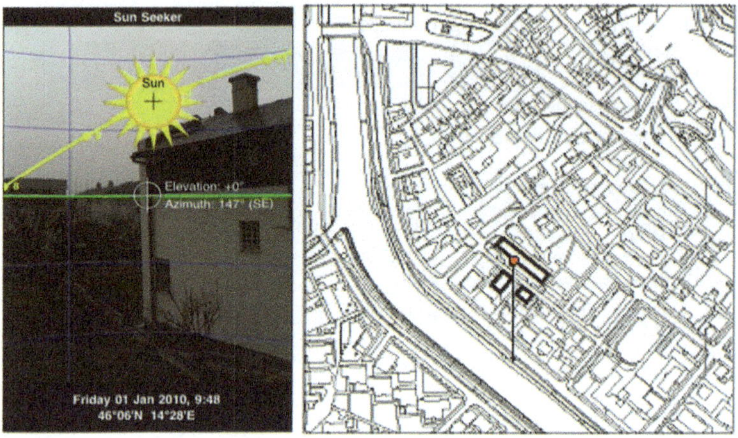

Fig. 4.49 Modern electronic devices equipped with an electronic compass, GPS and augmented reality (AR) apps overlay the solar path over the photo of the surroundings [20] (left); ground plan of the building where observed point P from case study is located, in case study shadow casting by buildings 1 and 2 will be included in the analyse [21] (right)

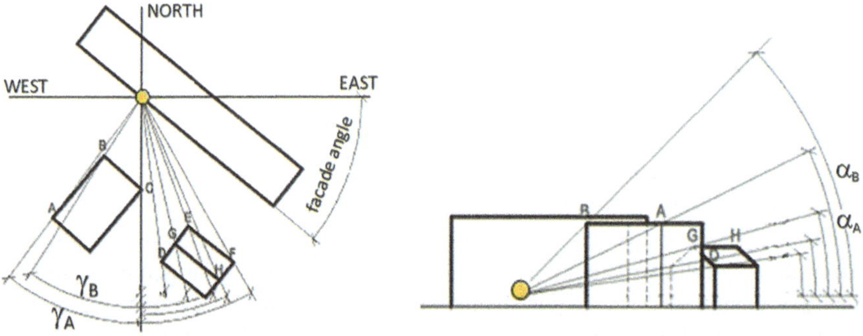

Fig. 4.50 Azimuth angles γ (left) and elevation angles α (right) of boundary points on buildings 1 and 2 that cast a shadow of observation point P

Case Study Determine the duration of exposure to sunlight of point P on the vertical façade of a building that is shaded by two surrounding buildings, as shown in Fig. 4.49 (right). The latitude L of the site is 46° North. The procedure is as follows:

- on the ground plan the boundary points A to C on building 1 and D to H on building 2 casting a shadow are marked (Fig. 4.50), the azimuth angles γ (the angular deviation from the South direction), of each boundary point are determined;

- the distances ΔL between observation point P and each of the boundary points is measured on the ground plan, and the difference in the height Δh between each of the boundary points and the observed point P is measured on the elevation view;
- the elevation angle of each boundary point relative to the point P is determined as:

$$\alpha = \tan\left(\frac{\Delta h}{\Delta L}\right) [^\circ]$$

For the case study, the values of the elevation angles of the boundary points are shown in Table 4.9;

- all the boundary points are plotted in the solar-path diagram and connected to each other to form the silhouette of both buildings; the two boundary azimuth angles of the façade where the observed point is located, are plotted as well as the vertical lines at the azimuth angles -51° (SE) and $+129^\circ$ (NW, not seen in the figure) respectively, Fig. 4.51;
- the duration of the sunlight for the selected day in a year is determined as the sum of all the partial time periods when the sun is "visible" from the observed point; the duration of exposure to sunlight for the observed point is given in Table 4.10.

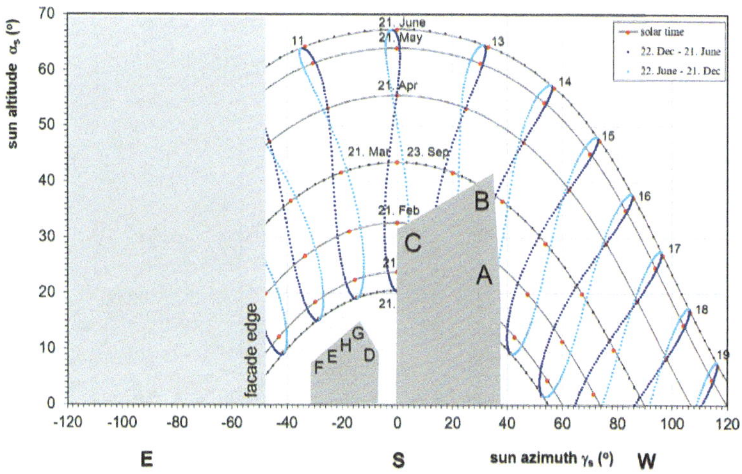

Fig. 4.51 Cylindrical solar-path diagram showing the silhouette of the buildings that cast shadow on the observed point P, the boundary angle of the SE building façade on which observation point is located is—51° towards the East

Table 4.9 Elevation angles α (°) and azimuth angles γ (°) of boundary points on building 1 and building 2 measured relative to observation point *P* (Fig. 4.50)

boundary point	elevation angle α (°)	azimuth angle of boundary point γ (°)
A	21	37
B	40	33
C	32	0
D	8	-7
E	10	-21
F	7	-30
G	15	-13
H	11	-24

Architectural CAD software tools enable an assessment of the duration of exposure to sunlight during the design phase (Fig. 4.52). The analyses can also incorporate a digital model of the environment, so the duration of exposure also accounts for the natural obstacles. Such tools can also show the animation of the building's shadow through the day, which determines the building's zone of influence.

Table 4.10 The duration of exposure to sunlight of the observation point *P* for characteristic days in a year (Fig. 4.51)

day	duration of exposure to sunlight	together daily (h:m)
21 Dec	8:10 - 12:00 and 14:40 - 16:10	5:10
21 Mar	9:20 - 13:30 and 13:50 - 18:10	8:30
21 Jun	10:20 - 19:30	9:20
21 Feb	9:00 - 12:20 and 14:20 - 17:30	6:10

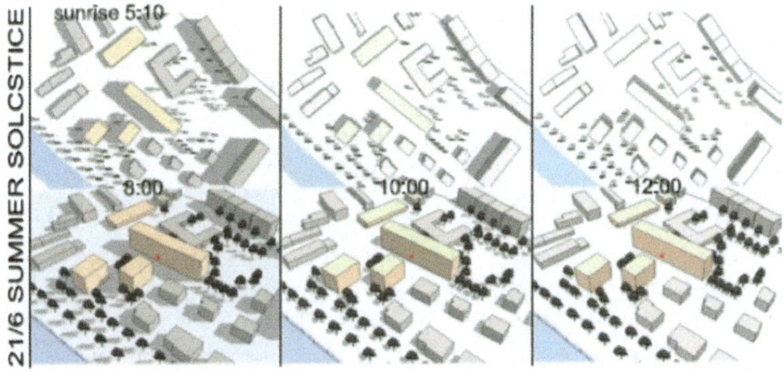

Fig. 4.52 Computer-generated visualization of shadows in the settlement from a case study at 8:00 a.m., 10:00 a.m. and at solar noon during the summer solstice

4.7.2 *Illuminance Level in the Indoor Environment*

An adequate illumination of objects and work surfaces is required to distinguish objects and perform various tasks. The illumination level E (lm/m^2 or lx) is a measure of the density of the luminous flux on the illuminated surface. To recognise an upright human face, it must be illuminated with approximately 10 lx, which corresponds to 20 lx for a horizontal surface. To distinguish facial details, the illuminance must be considerably higher, at least 200 lx. The required illuminance level depends on the difficulty of the visual tasks. The recommended illuminance for visual tasks of different complexity are given in Table 4.11. The adequate illumination level must be provided both for daylighting and electric lighting. The assessment object is usually a work surface situated 0.85 m above the floor and 0.5 m away from the internal walls or furniture.

The general illumination of the service areas in the room should be at least one-third of the required illuminance at the workplace, and in spaces that are not constantly occupied (such as corridors) at least one-sixth of the service areas illumination. For example, in the case that the recommended workplace illuminance is 500 lx, the general lighting should ensure at least 170 lx, and the corridors should be illuminated with 60 lx in this case at 0.20 m above the floor.

Software tools are used to calculate the illuminance of surfaces in the scope of the indoor visual comfort design. The results are represented graphically with isolux lines, the curves connecting all the points in space of equal illuminance.

Table 4.11 Recommended illuminance levels for different complexities of visual tasks

illuminance level E (lx)	complexity of visual taks
50	no detail perception required, safe walking enabled; warehouses, corridors
100	occasionally used spaces, correct identification of objects; sanitary facilities, corridors
200	long-term work, perception of larger details; service counters
300	long-term work, easier visual tasks, perception of details > 10 angular minutes at high contrast; conference rooms
500	moderate visual tasks, perception of details with size from 5 to 10 angular minutes at low contrast; conference rooms, dinning rooms, offices, classrooms
750	complex visual taska, perceptin of details with size from 3 to 5 angular minutes, good colour recognition; self-service stores, bureaus, drawing rooms
1000	very complex visual tasks, perception of details with size from 2 to 3 angular minutes, low contrast, detailed colour recognition; assembly rooms, colour control
> 1500	very complex visual tasks, perception of details < 1 angular minute, recommended use of optical aids; fine mechanical work, colour control

4.7.3 Daylight Factor

The Daylight Factor (DF) is an indicator of the quality of daylighting. It is the ratio of the illuminance at a selected point in the room $E_{P,H}$ with daylight, to the illuminance E_H of an outdoor horizontal, unshaded plane (Fig. 4.53), expressed as a percentage:

$$\mathrm{DF} = \frac{\mathrm{E_{P,H}}}{\mathrm{E_H}} \cdot 100\ [\%]$$

As the luminous intensity of natural sources (the Sun and the sky) varies, some design boundary conditions have to be defined for the determination of *DF.* The CIE overcast-sky model is commonly used. The daylight factor is determined for points situated on a work plane, usually 0.85 m above the floor, and with the illuminance of the outdoor horizontal unshaded plane $E_H = 5000$ lx (Fig. 4.54).

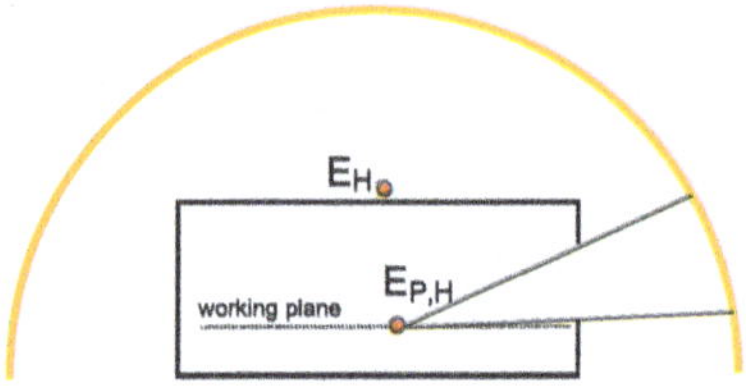

Fig. 4.53 The daylight factor DF is the ratio of the illuminance at a selected point on a working plane in the room $E_{p,H}$ to the illuminance at a horizontal unshaded point $E_H$under the CIE overcast sky, expressed as a percentage

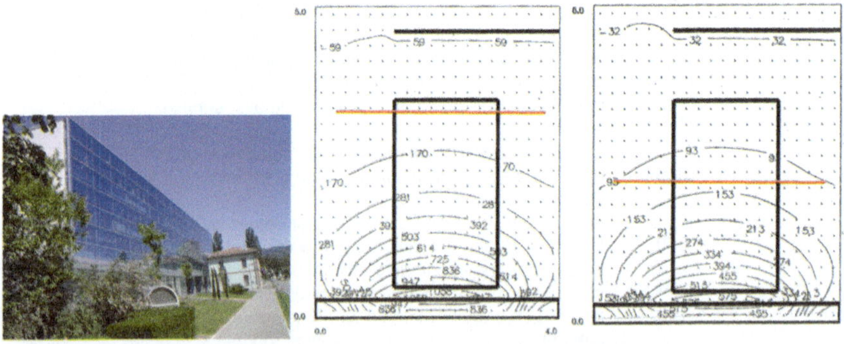

Fig. 4.54 An office building renovated with double glass façade (left); isoluxes in typical office determined on a working plane 0.85 m above the floor before (middle) and after (right) renovation; illuminance level was determined at standard CIE overcast shy and for outdoor horizontal illuminance 5000 lx; the lines set at 150 lx, represent the depth of space that is adequate daylight; attached glass façade decrease the daylight factor, although traffic noise level and energy demand for heating and cooling of the office are significantly decreased

Table 4.12 Recommended values of average (DF_{av}) and minimum daylight factors (DF_{min}) for different type of spaces in residential buildings

	DFav (%)	DFmin (%)
offices	5	2
classrooms	5	2
corridors	1 to 2	0.6
living rooms	1.5	0.5
bedrooms	1	0.3
kitchens	2	0.6

Table 4.13 Perception of room illumination with regard to the average daylight factor DF_{av}

DFav	perception of room illumination
< 2%	room is perceived as insufficiently lit
2% to 5%	room is perceived as sufficiently lit, from time to time natural illumination must be supplemented with electrical
> 5%	room is perceived as good naturally lit, possible overheating of a room due to large glass surfaces in building envelope should be carefully considered

Besides the sky component, illumination of the interior of the room depends on the internal reflected component and indirect light flux reflected on the outdoor surfaces.

The area of the patch of sky diminishes with an increasing distance from the window, and as consequence the sky component of daylight falls as we move deeper into the space. Because of that, the quality of the daylight is described by the average (DF_{av}) and the minimum daylight factor ($DF_{\min}$). The typical recommended values of DF_{av} and $DF_{\min}$ established according to the physiological needs of humans are listed in Table 4.12. According to the average daylighting factor, we can predict how bright the room will be perceived by the occupants (Table 4.13).

4.7.3.1 Determining the Daylight Factor

The daylight factor depends on the building's architecture, i.e., the geometric design of the rooms, and the size and layout of the windows. Different methods for the assessment of daylight were developed for architects and engineers, including empirical rules, engineering methods or computer-aided design tools.

Empirical rules can be used to determine the size of the windows and their position on the façade of the ceiling and the maximum room depth that still enables an adequate daylight factor. For example, the maximum room depth should not be deeper that approximately two times that of the window head height and the minimum glazed façade surface area should be one-seventh to one-fifth of the room's floor area. If the room is illuminated with skylights, distributed uniformly across the ceiling, a total net surface of the openings between 1/20 and 1/10 of the room's floor area will suffice. Besides that, the height of the window parapet and the minimum light transmittance of the glazing could be the subject of regulations. For example, in the Netherlands,

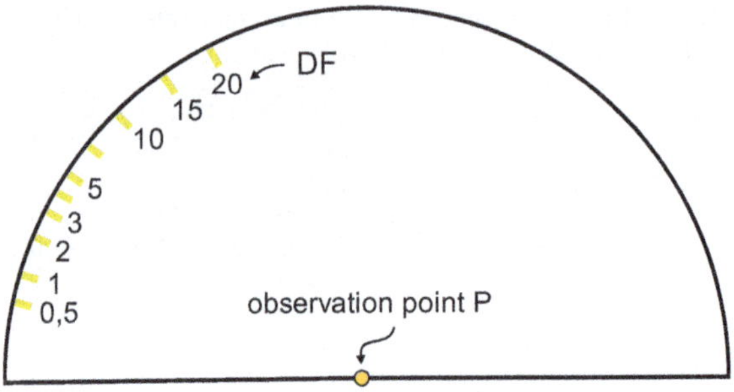

Fig. 4.55 The protractor method is based on a graphical tool on which the sky component *DF* can be read for any selected point in the room; the standard CIE overcast sky model is used [16]

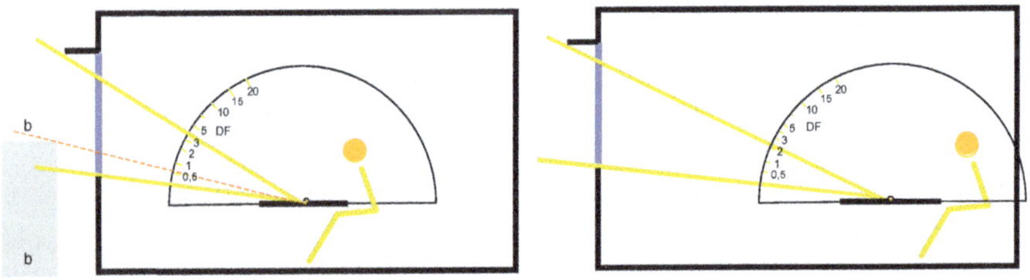

Fig. 4.56 A protractor is a simple graphical tool to determine the sky component daylight factor. The *DF* is determined as the difference between the DF read from tool for the angles subtending the point's view of the sky

the transmittance of the light of the glazing τ_{vis} of at least 65% is required, while it should be at least of 50% in Slovenia.

The daylighting of square-shaped rooms can be evaluated with nomograms, graphical tools or simplified calculation methods. The protractor method uses a graphical tool that is placed on the location of the observed point in the room. The protractor shown in Fig. 4.55 can be used in the case of an infinitely long window in the facade, while correction factors must be considered in the case of a different window design.

Case Study Determine the *DF* for two points on a work-plane at ½ (Fig. 4.56, left) and ¾ (Fig. 4.56, right) of the depth of the room, in an office with an infinitely long façade window. There is no surrounding object in front of the window. Assume that the internal reflected component increases the daylight factor on the working plane by 0.4%.

The sky component of $DF_{1/2}$ is 3.8–0.2 = 3.6% and total $DF_{1/2}$ is 4.0%

The sky component of $DF_{3/4}$ is 2.2–0.18 = 2.02% and total $DF_{3/4}$ is 2.4%

Let's assume that there is an object (b) with reflectance ρ_v 0.45 in front of the window that shades the window of the office. The sky component of daylight decreases because the size of the patch of the sky seen from observation point is smaller, while an additional external reflected component increases DF:

$$DF_{1/2} = \overbrace{DF_{1/2,sky}}^{\text{daylight from sky}} + \overbrace{DF_{1/2,erc}}^{\text{external reflected}} + \overbrace{DF_{1/2irc}}^{\text{internal reflected}} = [\%]$$

$$DF_{1/2} = 3.8 - 0.55 + (0.55 - 0.20) \cdot 0.45 + 0.4 = 3.7\ [\%]$$

A simple empirical model for determining the average DF_{av} was proposed by Tregenza and Loe [47]:

$$DF_{av} = \frac{A_g}{A_{envelope}} \cdot \frac{\gamma_z \cdot \tau_v}{\left(1 - \rho_{av}^2\right)} \cdot 100\ \ [\%]$$

where A_g is the area of glazing (m^2), $A_{envelope}$ is the total area of the internal envelope of the space (m^2), γ_z(°) is the obstruction angle shown in Fig. 4.34, τ_v (–) is the transmittance of the light of the glazing and ρ_{av} (–) is the average reflectance of the light of the structures (including glazing) that form the inner envelope of the space. Longmore's formula [24, 48] is another well-known empirical model for determining the average DF_{av} at the working plane in the simple formed spaces:

$$DF_{av} = A_g \cdot \tau_v \cdot k_1 \cdot k_2 \left(\frac{C_s}{A_{bottom}} + \frac{C_s \cdot \rho_{bottom} + 0.05 \cdot \rho_{top}}{A_{envelope}(1 - \rho_{av})} \right) \cdot 100\ \ [\%]$$

The terms in the brackets account for the sky component of daylight and the external and internal reflected components of daylight, the external component of the daylight reflected from the outside objects, and the internal reflected component of the daylight reflected from the surfaces inside the room. The room is virtually split between the bottom part, which includes surfaces below the plane of observation (e.g., work plane), and top part above the work plane. The quotient of the sky component of daylight C_s also accounts for objects obstructing the sky (Table 4.3). The quantities included in Longmore's formula are presented in Table 4.14.

Case Study Use Longmore's formula to calculate the DF_{av} for the room shown in Fig. 4.57.

$$A_{envelope} = 83.2\ m^2,\ A_{bottom} = 35.3\ m^2,\ A_{top} = 47.9\ m^2,$$
$$C_{s(\gamma_o=25°,\text{Table 3.3})} = 0.273$$

$$\rho_{av} = \frac{A_{ceiling} \cdot \rho_{ceiling} + A_{floor} \cdot \rho_{floor} + A_{wall} \cdot \rho_{wall} + A_w \cdot \rho_w}{A_{envelope}}$$
$$= \frac{20 \cdot 0.65 + 20 \cdot 0.28 + (43.2 - 4) \cdot 0.42 + 4 \cdot 0.1}{83.2} = 0.43$$

$$\rho_{bottom} = 0.34 \quad \rho_{top} = 0.48$$

$$DF_{av} = A_g \cdot \tau_v \cdot k_1 \cdot k_2 \left(\frac{C_s}{A_{bottom}} + \frac{C_s \cdot \rho_{bottom} + 0.05 \cdot \rho_{top}}{A_{envelope}(1 - \rho_{av})} \right) \cdot 100$$
$$= 4 \cdot 0.78 \cdot 0.85 \cdot 0.8 \left(\frac{0.273}{35.3} + \frac{0.273 \cdot 0.34 + 0.05 \cdot 0.48}{83.2(1 - 0.43)} \right) \cdot 100 = 2.2\%$$

Table 4.14 The quantities included in Longmore's formula for the average DF_{av}. In the case of multiple windows, the value of DF_{av} is calculated for each window and the individual daylight factors are summarized

symbol	unit	quanitiy	typical values
A_g	m^2	clear glazing area of the window	
$A_{envelope}$	m^2	total area of internal envelope of the space	
A_{bottom}	m^2	area of internal envelope of the space below work plane	
C_s	1	quotient of sky component of daylight	Tabel 3.3
τ_v	1	transmittance of the light of the glazing	0.55-0.70
k_1	1	window's frame factor	0.80-0.85
k_2	1	glazing clearance factor	0.85-0.90
ρ_{av}	1	average reflectance of the light of the structures (with area $A_{envelope}$)	0.45
ρ_{bottom}	1	average reflectance of the light of the structures below working plane (with area A_{bottom})	0.30-0.40
ρ_{top}	1	average reflectance of the light of the structures over work plane	0.50-0.60

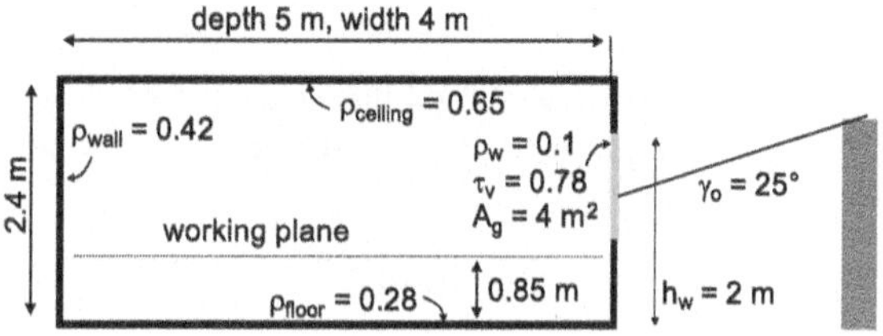

Fig. 4.57 Geometry of the room from the case study

The daylight levels in complex-shaped rooms can be determined with higher accuracy using software tools. Such tools enable the reflection of the light on the indoor surface to be modelled as specular, diffuse or spread when determining the internally reflected component. Different sky-luminance models can also be selected. If the CIE overcast model is used, the daylight factor can be determined as the ratio of the calculated illuminance of an arbitrary point inside the building and the illuminance of an external surface (Fig. 4.58).

Scaled architectural models can also be used for daylighting design (Fig. 4.59). A scale of ~1:200 is needed for a preliminary design, and scale ~1:50 for a highly detailed interior. Models must be built with an accurate geometric similarity and with the optical properties of the internal surfaces as close as possible to those of the planned building. The architectural model could be placed outdoors under the real sky conditions or under an artificial sky with lamps simulating the Sun's position, the distribution of the sky's luminance, and even the light colour temperature. The artificial sky can be designed in the form of a dome sky with a large number of lamps able to simulate the real sky conditions, or as a rectangular box with mirrors, where overcast sky conditions can be simulated.

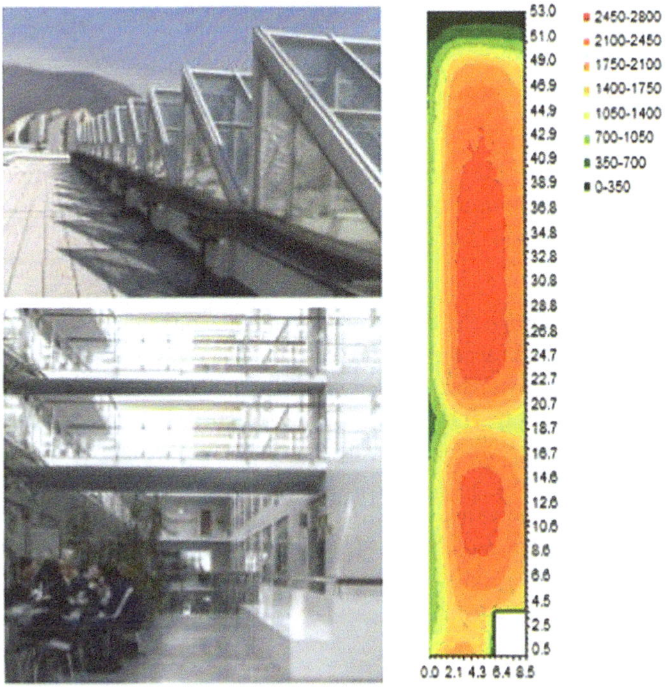

Fig. 4.58 Example of the result of a numerical simulation of the illumination level for the atrium in a public building. Lumen software was used [25, 38]

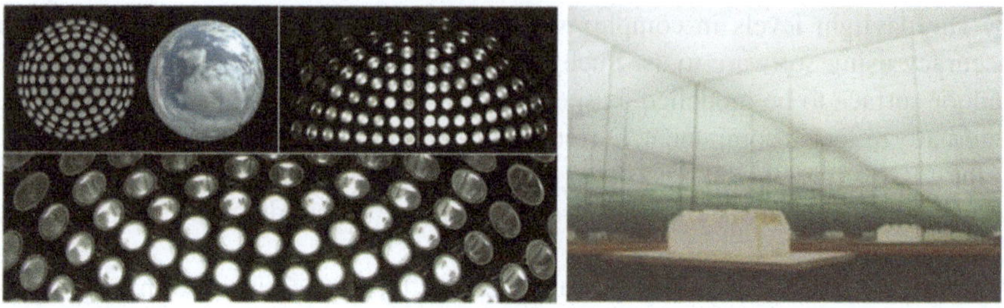

Fig. 4.59 Hemispherical (left) and box-type artificial sky simulator with mirrors (right). A rectangular box with mirrors can be used for daylighting analysis under the conditions of an overcast sky, the hemispherical artificial sky can simulate any sky luminance distribution [26, 27]

Fig. 4.60 Luminous flux sensor and lux meter

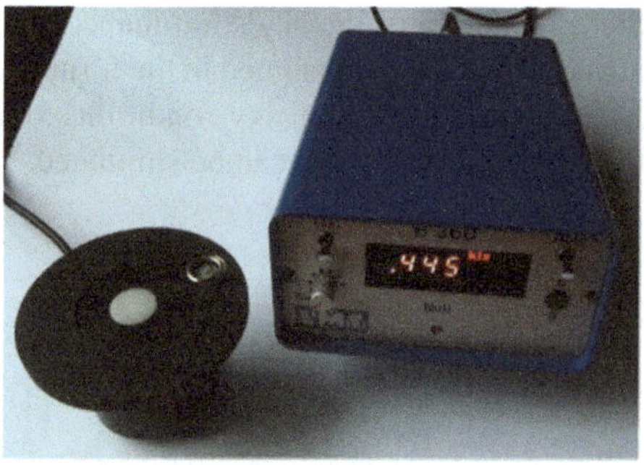

On existing buildings, the illumination level can be measured using lux meters and data loggers, as shown in Fig. 4.60. Such measurements are part of an indoor living-comfort assessment.

Case Study Determine the depth for a sidelight office that is 8 m long, 6 m wide and 2.5 m high according to the illuminance level of the work-plane 0.85 m above the floor if the required illuminance level E_H is 300 lx or 500 lx. The offices differ in the glazing-to-wall ratio and the glazing shape, and in the glazing light transmittance τ_v. The reflectance of the light of the envelope surfaces are: $\rho_{walls} = 0.75$, $\rho_{ceiling} = 0.85$ and $\rho_{floor} = 0.5$. Daylight illuminance should be determined using the CIE overcast-sky model and for an exterior horizontal illuminance $E_{H,e}$ 5000 lx [14]. The depth of the office regarding the required illuminance levels is presented in Fig. 4.61.

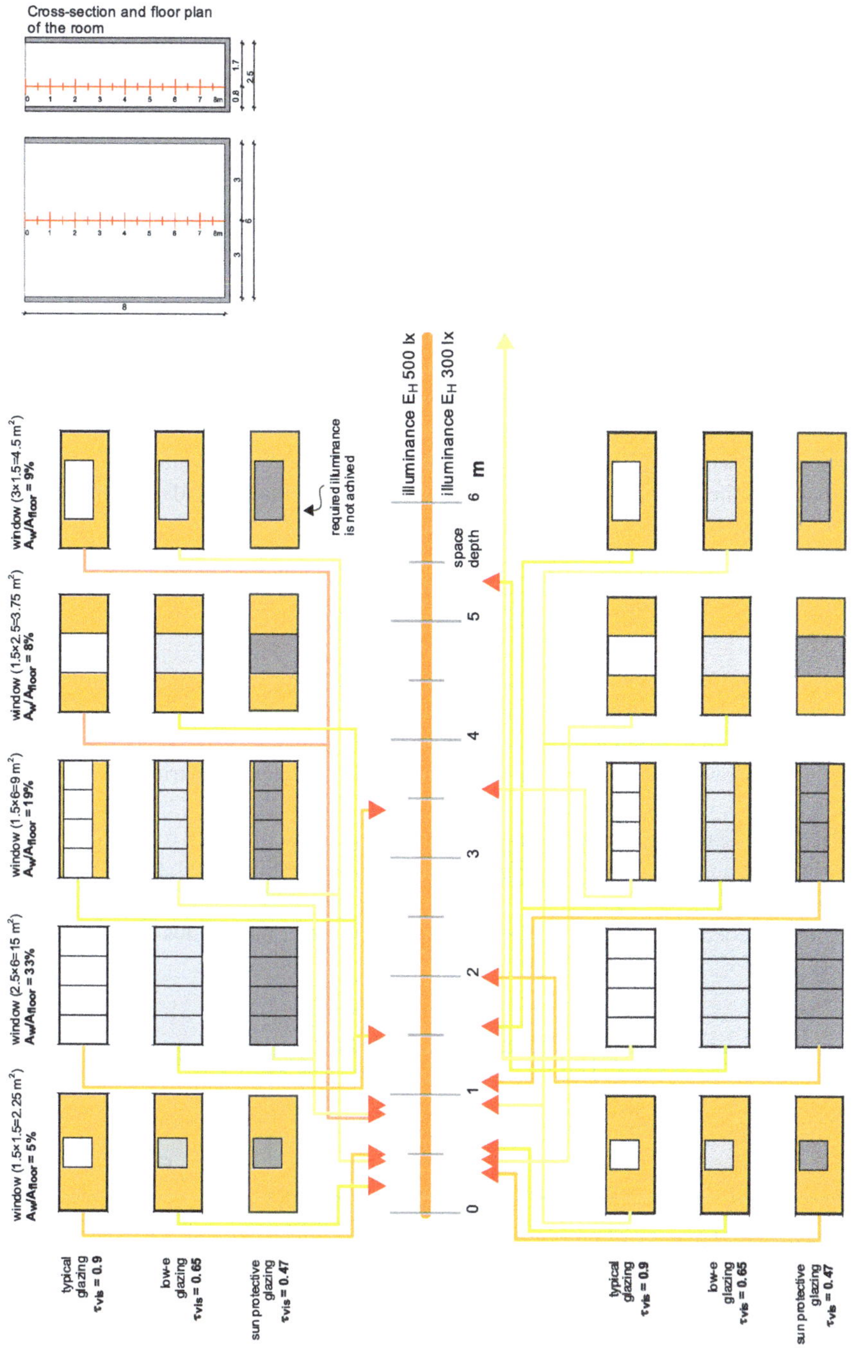

Fig. 4.61 Solution of the case study

4.7.4 Uniform Distribution of Daylight

To achieve highest level of visual comfort, workplaces in a room must be equally illuminated. Such conditions of uniform illumination are especially difficult to achieve in the case of single-side daylighted spaces shown in Fig. 4.62. The uniformity of the illumination is evaluated by the ratio of the illumination level $E_{\min}$ at the darkest workplace in the space to the average illumination level of all the workplaces E_{av}. The recommended $E_{\min}/E_{av}$ ratios are presented in Table 4.15.

The following measures will improve the uniformity of the illuminance:

- taller rooms make it possible to install taller windows and increase the reflection of light on internal surfaces, improving the uniformity of daylighting, especially in the critical second half of the room. The depth of the room should not exceed three times the height of the room or not exceed 2 times the height of the window-head. The following empirical relation can be used to test the uniformity of the illuminance in the preliminary design phase, and if the inequality holds, the uniformity of the space's illumination can be expected to be adequate:

$$\frac{\mathrm{a}}{\mathrm{b}} + \frac{\mathrm{a}}{\mathrm{h_w}} \leq \frac{2}{1 - \rho_{\mathrm{av},2}}$$

  where a is the depth of the room, b is the width of the room, h_w is the window-head height, and $\rho_{av,2}$ is the average reflectivity of the surfaces (walls, ceiling, floor) in the second half of the room;
- the windows placed on opposite facades, or by evenly distributed skylights (Fig. 4.62, middle, right);
- redirecting the light rays towards the ceiling and deeper into the room with reflective venetian blinds (Fig. 4.63), prismatic elements (Fig. 4.64) and holographic films;
- highly reflective light shelves installed on the inner or outer sides of the windows (Fig. 4.65).

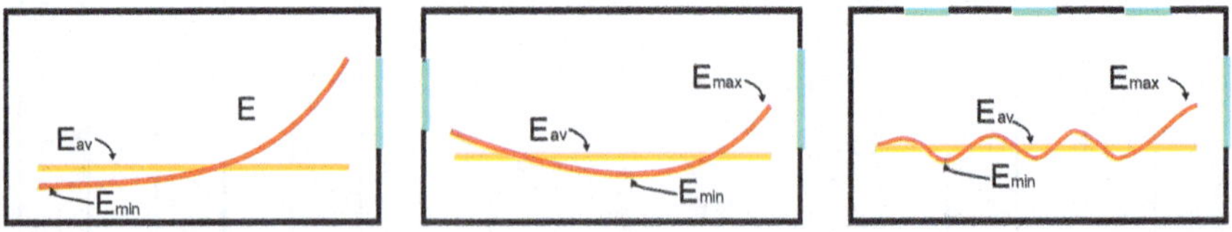

Fig. 4.62 The uniformity of illuminance in the space is determined by the minimum and the average illuminance of the work plane. Opposite windows and skylights considerably improve the uniformity of the daylight (middle, right)

Fig. 4.63 Reflective louvres in the upper third of the window redirect the light deeper into the room. The effect in the presented case is boosted by a highly reflective ceiling surface [28]

Fig. 4.64 Prismatic elements redirect the Sun's rays deep into the room, improving the uniformity of illuminance (left). Skylights in the form of light tubes can be up to 10 m long and conduct the light by multiple reflections several floors below (right) [28, 29]

Case Study Check the uniformity of the illuminance in a single-side daylighted space with height a 5 m, width b 4 m and the height of the window-head h_w 2 m. The average reflectivity of the surfaces in the second half of the room $\rho_{av,2}$ is 0.48.

$$\frac{a}{b} + \frac{a}{h_w} = \frac{5}{4} + \frac{5}{2} = 3.75 \leq \frac{2}{1 - \rho_{av,2}} = \frac{2}{1 - 0.48} = 3.85$$

The inequality holds, so we can assume the uniformity of the daylight to be adequate.

Fig. 4.65 Light shelves redirect the light deeper into the room. In the presented case the uniformity of the illuminance (E_{min}/E_{av}) improves from 1:6 to 1:3.5 [22]

Table 4.15 Requirements for the uniformity of illuminance of daylighted spaces

complexity of visual taks	E_{min} to E_{av} ratio
very complex visual taska	1 : 1.5
complex visual tasks (e.g. classrooms)	1 : 2.5
modearte complex visual tasks	1 : 3
simple visual tasks	1 : 6

4.7.5 *Discomfort Glare*

Glare is an irritating phenomenon in the visual sensation of the environment and one of the most common causes of visual discomfort. It occurs when a very bright patch occurs in the central field of vision. To avoid discomfort glare, the luminance of the observed surfaces should be below 500 cd/m^2 if such surfaces are seen in the central visual field or below 1000 cd/m^2 if the surfaces are seen in the peripheral field of view. Glare can be caused by daylight as well, but research has proven that almost twice-higher luminance will not cause discomfort glare in daylighted spaces. There are several types of glare: the direct, the reflected and the veiling glare:

- direct glare is caused by the sunlight entering the windows at lower solar altitudes, or the luminous flux emitted by electric lamps in the zone between 55° and 90° to the work plane's normal (Fig. 4.66), Direct glare can be prevented by using

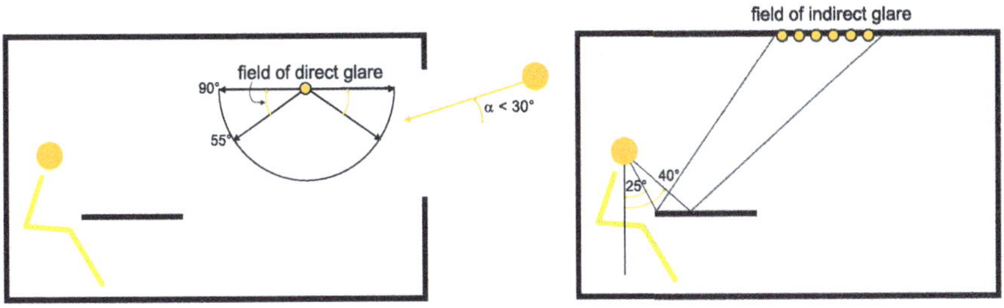

Fig. 4.66 Lamps and the daylight can cause direct and reflected glare

adjustable window shades and lighting fixtures that prevent the light from entering the offending zone;

- reflected or indirect glare is caused by the reflection of luminous flux from the illuminated surfaces into the central field of vision. The most uncomfortable is glare from surfaces such as computer monitors, glossy paper, floors with specular reflection or very brightly painted walls; reflected glare is reduced with diffuse surfaces that scatter the light, and by placing the lamps outside the offending zone where they would cause reflected glare;
- veiling or contrast glare. Even though our eyes can adapt to surfaces of different brightness, we find it difficult to look at surfaces with a large difference between the luminance of the task surface and its immediate surroundings (Fig. 4.67). The difference in luminance is defined by the luminance ratio, which should not exceed the recommended values shown in Table 4.16 [37]. The contrast glare can be reduced by selecting an adequate surface's light reflectivity. The reflectivity should be diffuse and at least 0.7 of the ceiling, at least 0.5 of the walls, meanwhile the floor surface should have a reflectivity of around 0.2.

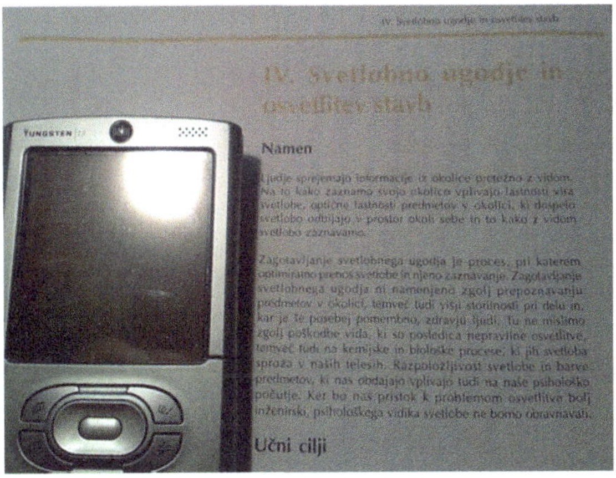

Fig. 4.67 High contrast caused by bright patch on the screen makes it difficult to read the text in the book

Table 4.16 Recommended luminance ratio according to the subject of the observation

luminance ratio	area in room	example
3:1	area where visual tasks are performed and immediate surroundings	book and desk
5:1	area of visual tasks and room	book and nearby areas
10:1	area of visual tasks and distants surroundings	book and distant wall
20:1	distant light source	window in distant wall

4.7.5.1 Evaluation of Discomfort Glare

How disturbing the perceived glare will be for an observer is assessed by various indicators. With these indicators the discomfort glare is estimated on the basis of the luminance of the glossy surface L_s (cd/m^2) and its relative size, which is defined by the spatial angle Ω (in steradians), the luminance of the immediate surroundings of the glossy surface L_b (cd/m^2) and the position index of the observer P (–).

The spatial angle Ω of the glossy surface is determined by a double integral, as shown in Fig. 4.16, or by a simplified approach, assuming that the glossy surface has the shape of a rectangle.

The position index of the observer P is defined by the height angle α and the direction angle γ between the centre of the glossy surface and the direction of view of the observer in the space. A higher value of the P index (Figs. 4.68) means that the sense of the glare caused by the glossy surface will be less likely.

Explanation A glossy surface is perceived as annoying because of glare if its luminance L_s is about 3000 cd/m^2 and the surface is located in the space in a way that the position index of the observer P is below 1, while the bright

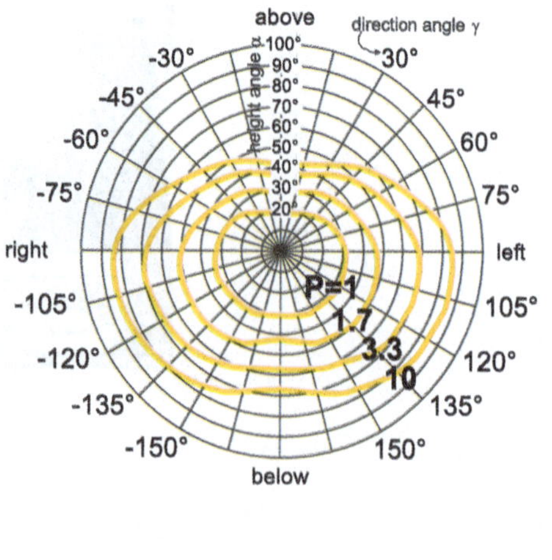

Fig. 4.68 The position index P of the observer is defined by the height angle α and the direction angle g between the direction of the view of the observer and centre of the observed object

surface is not disturbing if it is observed at position index P higher than 10, even if its luminance is ten-times higher (L_s 30,000 cd/m^2) [40].

For an assessment of the glare caused by luminaires, which are generally considered to have a small spatial angle Ω, the Unified Glare Rating (UGR) is proposed by the World Organization for Lighting CIE [42, 43]. The UGR is defined by the expression:

$$\mathrm{UGR} = 8 \cdot \log \frac{0.25}{\mathrm{L_b}} \sum_{\mathrm{i}=1}^{\mathrm{n}} \frac{\mathrm{L_s^2} \cdot \Omega}{\mathrm{P^2}} \quad [1]$$

The value of the UGR is determined by considering all n luminaires, located in the field of view of the observer ($P < 10$). The disturbing glare is than determined by considering the empirical values given in Table 4.17.

Case Study Determine the value of the UGR if the luminance of the observed object is L_s 7500 cd/m^2, the ambient brightness around the object L_b is 250 cd/m^2, the size of the object measured by a spatial angle Ω is 0.0037 sr and the observer position index P is 4.5.

$$\mathrm{UGR} = 8 \cdot \log \frac{0.25}{\mathrm{L_b}} \sum_{\mathrm{i}=1}^{\mathrm{n}} \frac{\mathrm{L_s^2} \cdot \Omega}{\mathrm{P^2}} \quad [1]$$

$$\mathrm{UGR} = 8 \cdot \log \frac{0.25}{250} \sum_{\mathrm{i}=1}^{\mathrm{n}} \frac{7500^2 \cdot 0.0037}{4.5^2} = 8.15$$

Given the scale in Table 4.17, we can conclude that the observed object does not cause discomfort glare.

Even in daylight rooms, the bright surfaces of windows or sun-exposed surfaces can cause discomfort glare. As these bright areas are generally larger than the surfaces of the luminaires and the luminous flux is detected in the upper and lower eye-vision areas, the discomfort glare caused by daylight is determined by Daylight

Table 4.17 Proposed values of the Unified Glare Rating (UGR) for an assessment of the discomfort glare caused by luminaire

UGR	glare
< 13	not perceived
33 - 22	perceived, not disturbing
23 - 28	disturbing
28	unbearable

Glare Probability (*DGP*) [41]. It is defined by the expression:

$$DGP = 5.87 \cdot 10^{-5} \cdot E_v + 9.18 \cdot 10^{-2} \log\left(1 + \sum_{i=1}^{n} \frac{L_{s,i}^2 \cdot \Omega_i}{E_v^{1.87} P_v^2}\right) + 0.16 \ [1]$$

where E_v (lx) is the illumination of the apparent vertical plane at the height of the eye, while $L_{s,i}$ (cd/m^2), Ω_i (sr) and P_i (–) are the luminance and spatial angle of the *i*-th glossy surface and the observer position index towards the *i*-th glossy surface. Whether the glare is annoying is determined by the limit *DGP* values defined in Table 4.18. As the position of the sun in the sky changes, so does the brightness of the interior surfaces. Therefore, the glare caused by daylighting should be checked throughout the year.

Case Study An observer has a view pointing to a window of size 1.2 × 1 m ❶, so that he/she also sees a sunlit floor surface of size 0.5 × 1.2 m ❷ (Fig. 4.70). The work place is 4 m away from the wall with the window and 2.5 m away from the sunlit floor. The spatial angles of the glossy surfaces are Ω_1 0.0736 sr, Ω_2 0.0929 sr, and the observer's position indexes are $P_1 = 1$ (towards window) and $P_2 = 1.6$ (towards floor). With the VELUX Daylight Visualizer 2 software (Fig. 4.69), the luminances of the glossy surfaces were determined as $L_{s,1} =$ 4900 cd/m^2 and $L_{s,2} = 2100$ cd/m^2, as well as the luminance of the immediate surroundings $L_{b,1} = 80$ cd/m^2 and $L_{b,2} = 42$ cd/m^2. The illumination of the apparent vertical plane in the eye area of the observer E_v is 550 lx.

$$\mathrm{DGP} = 5.87 \cdot 10^{-5} \cdot \mathrm{E_v} + 9.18 \cdot 10^{-2} \log\left(1 + \sum_{i=1}^{n} \frac{\mathrm{L_{s,i}^2} \cdot \Omega_i}{\mathrm{E_v^{1.87} P_v^2}}\right) + 0.16 \ [1]$$

$$\mathrm{DGP} = 5.87 \cdot 10^{-5} \cdot \mathrm{E_v} + 9.18 \cdot 10^{-2} \log\left(1 + \frac{\mathrm{L_{s,1}^2} \cdot \Omega_1}{\mathrm{E_v^{1.87} P_1^2}} + \frac{\mathrm{L_{s,2}^2} \cdot \Omega_2}{\mathrm{E_v^{1.87} P_2^2}}\right) + 0.16 \ [1]$$

$$\mathrm{DGP} = 5.87 \cdot 10^{-5} \cdot 550 + 9.18 \cdot 10^{-2}$$

$$\log\left(1 + \frac{4900^2 \cdot 0.0736}{550^{1.87} \cdot 1^2} + \frac{2100^2 \cdot 0.0929}{550^{1.87} \cdot 1.6^2}\right) + 0.16 \ [1]$$

$$\mathrm{DGP} = 0.301$$

Given the *DGP* values in Table 4.18, it can be concluded that at the time of the observation, daylight does not cause discomfort glare as perceived by the observer.

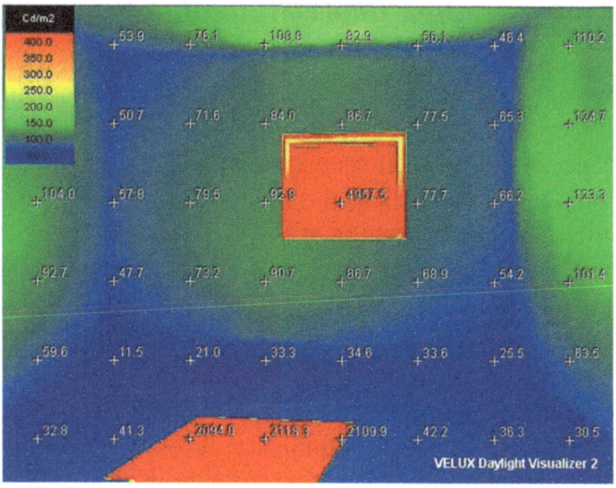

Fig. 4.69 Luminance of the surfaces in the space from the case study

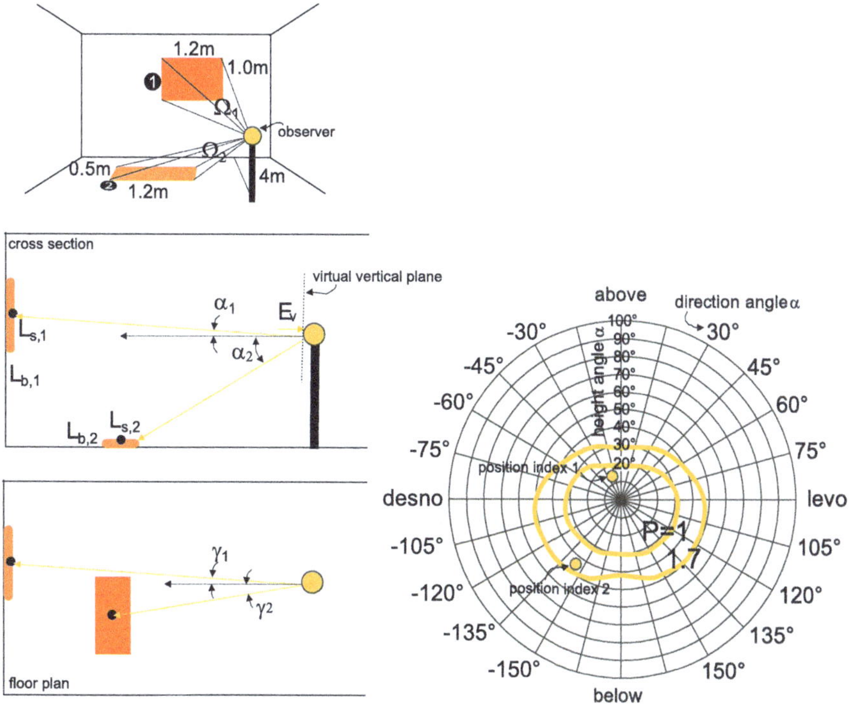

Fig. 4.70 Height angles and direction angles from the case study (left); position index of the observer P towards the window and the sunlit floor (right)

4.7.6 Relation Between Illumination and the Colour of Light

We have already discussed the light colour temperature (T_C, CCT) and the colour-rendering index (CRI) in the previous chapters. Nevertheless, visual comfort also depends on the relations between the light colour temperature and the illumination level of the surfaces in the room because the rods and the cones in the eye react differently when looking at brighter or darker surroundings. A light source with a high colour temperature (e.g., T_C 6000 K) appears white if its luminance is high, and bluish if the luminance is lower. In order for the surface of the light source to be perceived as white, the light colour temperature should be reduced. This also means that the light colour temperature of an electrical light source must be adopted to the required illumination level to maintain the same visual comfort. The Kruithof curve (Fig. 4.71) relates the illuminance level to the most suitable colour temperature of the light, showing that at a lower required illumination level, the colour temperature of the light should be warm and that at a higher required illumination level, the light source should emit cold light (Table 4.19). Contemporary research does not, however, confirm Kruithof's findings entirely.

Table 4.18 Proposed values of the Daylight Glare Probability (*DGP*) for an assessment of the discomfort glare in daylight spaces

DGP	glare
< 0.35	not perceived
0.35 - 0.40	perceived, not disturbing
0.40 - 0.45	disturbing
> 0.45	unbearable

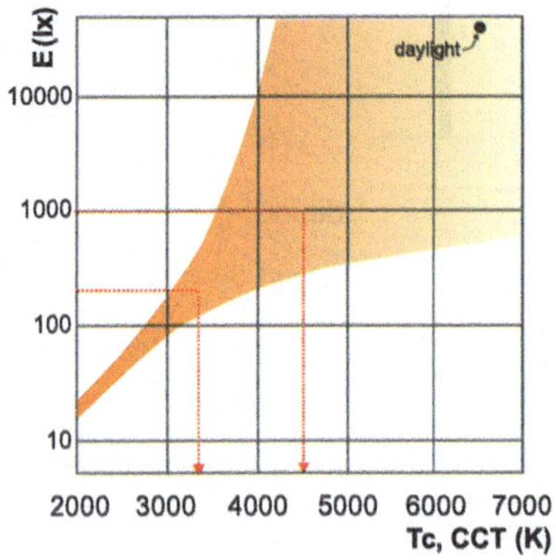

Fig. 4.71 The Kruithof curve relates the illuminance E to the light colour temperature T_C or T_{CT}. The shaded area represents the suitable values of light-colour temperature versus the illuminance E. The same colour perception is achieved when looking at a space illuminated to 200 lx by a 3300 K light source, or at a space illuminated to 1000 lx by a 4500 K light source [30]

Table 4.19 The perception of the space depends both on the illuminance level E and the colour temperature T_C or CCT of the light emitted by an electric light source

illuminance level (lx)	T_C, CCT < 3000 K warm colour	3000 - 5000 K neutral	> 5000 K cold colour
< 500	pleasant		
1000		pleasant	
> 3000			pleasant

4.7.7 *Assessment of the Visual Comfort with the VELUX Daylight Visualizer 2 Software Tool*

Advance software tools, such as VELUX Daylight Visualizer 2, enable a realistic visualisation of a building's interior as a virtual naturally or electrically illuminated environment. The software tool allows complex 3D models of the spaces to be imported from architectural software tools such as Auto CAD, ArchiCAD or Sketchup and facilitates the rapid modelling of 3D spaces and evaluating the interior lighting for scenarios with different layouts, vertical and roof-window openings, at different days and hours over the year, as well as for different sky conditions. The tool also makes it possible to select the surface colours, materials and textures, as these parameters influence the illumination through the inter-reflection of the light. The program has been verified by the CIE validation procedure.

VELUX Daylight Visualizer 2 displays the surface illuminance (lx), the surface luminance (cd/m^2), and the daylight factor (–).

Case Study An attic apartment with and without skylights was modelled. Figure 4.74 shows a rendering of a south-west-oriented room on March 21st at noon for overcast sky conditions. The building is located at a latitude of 46° N.

Isolux lines are shown in Fig. 4.72 for a horizontal plane 0.85 m above the floor. Comparing isoluxes to external illuminance, the average daylight factor DF_{av} of the observed plane can determine that in the side-lit room the DF_{av} is 1.8% and in the room with additional skylight windows the DF_{av} is 6.4%.

The risk of the glare can also be analysed. Figure 4.73 shows a rendered image of the apartment in daylight on July 21st at 10 a.m. for clear sky conditions and the luminance of the indoor surfaces at that time. The luminance of window jambs, a table and a patch of the floor exceed 2000 cd/m^2, so these surfaces could cause glare when seen from the observer's stand point. Besides a determination and visualization of the illuminance and the surface luminance of the indoor environment, the VELUX Daylight Visualizer 2 also produces an animation showing the penetration of direct sunlight into the room throughout the day (Fig. 4.75).

Fig. 4.72 Isolux lines shown at a height of 0.85 m above floor; in the apartment with wall windows (left) and for the apartment with additional skylight windows (right) for the CIE overcast sky and an outdoor illuminance of the horizontal plane equal to 5000 lx

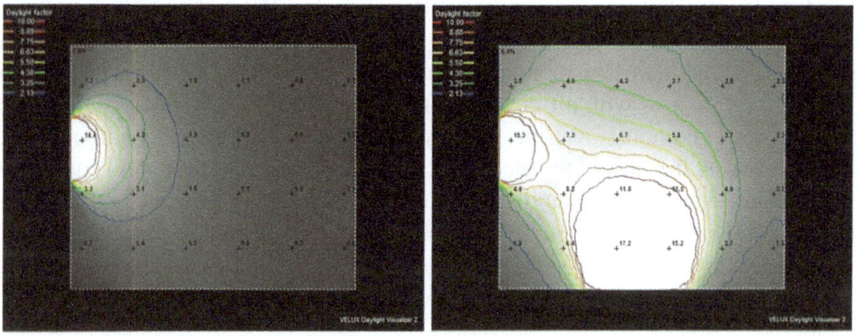

Fig. 4.73 Rendering of the attic apartment illuminated by direct daylight on July 21st at 10 a.m. in a city with a latitude L of 46° (left); luminance of the indoor surfaces at the same time (right)

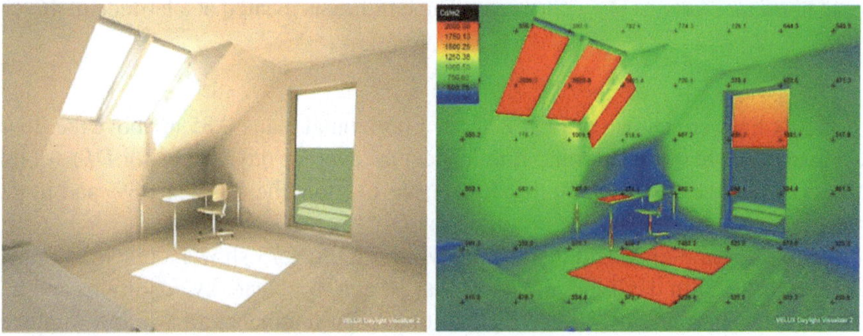

Fig. 4.74 Rendering of an attic apartment with and without skylight windows on March 21st at solar noon. The CIE overcast sky model was used

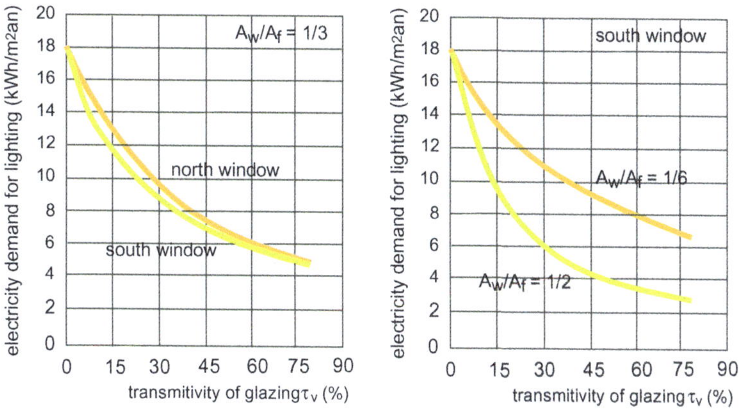

Fig. 4.75 Electricity consumption for artificial lighting in a office with a window in the south or the north façade, according to the transmittance of the light τ_v of the glazing; electricity demand was determine for the ratio of glazing surface area to floor surface area A_w/A_f is 1/3, the illumination level of 500 lx and installed specific luminaire power 8 W/m^2 (left); electricity consumption in a office with windows of different size on the south façade (A_w/A_f = 1/6 and 1/2); the same boundary conditions were assumed as in the previous case

4.8 Energy Demands for Lighting

Electrical lighting has a large impact on the overall energy demand of buildings. Lighting accounts for 5–10% of the total electricity consumption in households and up to 60% in non-residential buildings, which amounts to 10–20% of the total energy demand (Table 4.20). According to the Energy Performance of Buildings Directive [49] an indoor-electrical-lighting building service system must be included in the building's energy efficiency performance assessment (Table 4.21). The energy efficiency of the electrical lighting can be improved with the following measures:

- high-light-transmittance glazing should be installed to increase the amount of daylight. It is common for national legislation to specify the minimum light transmittance of glazing (e.g., in Slovenia τ_v must be at least 0.5);

Table 4.20 Electricity demand in buildings for lighting accounts for 5–60% of the total electricity consumption, depending on the type of building [29]

building type	share of consumption of electricity for lightening (%)
residential buildings	5 to 10
schools	10 to 15
industrial buildings	15
hospitals	20 to 30
office buildings	50 to 60

Table 4.21 Typical electric power of luminaries, calculated per m^2 of floor area, and the electricity consumption for lighting. The illumination level 500 lx is assumed and electric lighting is used during 20% of the office hours [32]

artifical light source	specific electric power (W/m^2)	electricity consumption (kWh/m^2)
incandescent light bulb	25	9.6
halogen lamp	20	7.6
fluorescert lamp	6	2.3

- the rooms should be designed with an adequate and uniform daylighting factor DF_{av}. Studies have shown that daylighting does not affect the energy consumption for lighting if DF_{av} is less than 2%, meanwhile the electricity use could be reduced by up to 50% if DF_{av} is 5% and by up to 80% if DF_{av} is12%. These savings can only be realised if the electric lighting is properly controlled. In the case of a DF_{av} above 10%, automatically controlled exterior shading must be installed to reduce the risk of overheating and glare;
- luminaires with high luminous efficacy should be used; the best available technologies emit up to 140 lm/W;
- up to 20% of electricity-demand decrease can be achieved by employing an optimal geometrical shape and high reflectivity of the light fixtures;
- lamps should be controlled by a smart control system and should enable dimming according to the level of daylight and the presence of occupants. Modern electronic devices can reduce the luminous flux of fluorescent lamps and LEDs down to 3% of the nominal luminous flux for a 90% decrease in electricity use. In this way, the consumption of energy for lighting in buildings over a year can be reduced by up to one-third.

4.8.1 Requirements Regarding the Energy Efficiency of Electric Lighting

Electricity use for lighting must be included in the assessment of the energy efficiency of a building. It is common that electricity demand is not limited to a certain amount of the annual electricity consumption; however, the installed power of the electric lamps per m^2 of useful building floor area is often constrained. Table 4.22 shows some typical values.

Knowing the installed power of luminaires, the energy use for lighting can be determined by taking into account the availability of daylight and the number of operating hours and the control of the electric lighting system. The standard EN 15193 [35] introduced the *LENI* (Lighting Energy Numerical Indicator) method for calculating the energy consumption for lighting per 1 m^2 of useful floor area of the

Table 4.22 Maximum allowed specific installed power of electric lights P_{el} in different buildings [33]

building type	maximum allowed specific installed power of electric lights (W/m^2)
parking garages	3
residential buildings	8
department stores, shopping centres	9
health centres, hotels, offices	11
gyms, museums, sport halls	12
conference centres, hospitals, schools	13
restaurants, libraries, industrial buildings	14
theatres	17

building A_u. *LENI* is determined as the sum of the values for n rooms using the following expression:

$$\mathrm{LENI} = \frac{\overbrace{\sum_{i=1}^{n} P_p \cdot t_p}^{\sim 6 \cdot A_u} + \sum_{i=1}^{n} P_n \cdot \left(t_d \cdot F_D \cdot F_O + t_n \cdot F_O\right)}{A_u \cdot 1000} \left[\frac{\mathrm{kWh}}{\mathrm{m^2 a}}\right]$$

where P_p is the total electrical power (W) of the emergency lighting and control devices installed in the building, P_n (W) is the total electrical power of the installed luminaires in the i-th room, t_p is the total number of operating hours in a year, t_d and t_n are the numbers of operating hours in the daytime (t_d) and night-time (t_h) and n is the number of zones in the building. Typical values for t_d and t_n according to the type of the building are given in Table 4.23. The daylight-dependency factor F_D depends on DF_{av}, and F_O is the coefficient that accounts for the average occupancy of rooms in the building. *LENI* can be calculated for the whole year or for each month. In this case, t_d and t_n are adapted to the average monthly length of the daytime and night-time periods of operation. The daylight-dependency factor F_D is calculated as:

$$F_D = 1 - F_{D,s} \cdot F_{D,c} \ [1]$$

where $F_{D,s}$ is the daylight-supply factor (–) shown in the Table 4.24, and $F_{D,c}$ is the electric lighting control factor (–) in Table 4.26. The occupancy-dependency factor F_O is given in Table 4.25.

Table 4.23 Typical number of building-occupancy hours per year during the daytime (t_d) and the night-time (t_n) according to the types of the buildings. The evolution of t_d and t_n have to be determined by considering the actual day length and working time [34, 35]

building type	t_d (h/a)	t_n (h/a)
residential buildings	0	1500
offices	2250	250
schools	1800	200
hospitals	3000	2000
restaurants	1250	1250
sport buildings	2000	2000
shopping centres	3000	2000
industrial buildings	2500	1500

Table 4.24 Daylight supply factor $F_{D,s}$ versus the average room daylight factor DF_{av}

daylight supply factor $F_{D,s}$ (1)		
average daylight factor	illuminance of room	
	300 lx	500 lx
DF_{av} < 2%	0.70	0.50
DF_{av} 2-3%	0.83	0.70
DF_{av} > 3%	0.89	0.82

Table 4.25 Occupancy factor F_O or different types of buildings

building type	F_o (1)
residential buildings, offices, schools, stores, industrial buildings, sport buildings (manual switching)	1.00
offices, schools (automatic switching)	0.90
hospitals	0.80
hotels	0.70

Table 4.26 Luminaire's control factor $F_{D,c}$ versus the average daylight factor DF_{av} and the luminaire control techniques

average daylight factor (%)	luminaire's control factor $F_{D,c}$ (1)		
	luminaire's control	illumination level	
		300 lx	500 lx
DF_{av} < 2%	manual switching	0.00	0.00
	automatic switching	0.09	0.00
	automatic switching with dimming	0.86	0.77
DF_{av} 2-3%	automatic switching	0.40	0.09
	automatic switching with dimming	0.91	0.86
DF_{av} > 3%	automatic switching	0.61	0.36
	automatic switching with dimming	0.95	0.91

Case Study 1 Determine the consumption of electrical energy for lighting an office with a 20 m^2 floor area. The specific electrical power of the luminaires is 11 W/m^2. The required illuminance level of the work plane is 300 lx and the average daylight factor DF_{av} is 3.5%. The lamps are not dimmable and controlled based on the occupants' presence. The factors $F_{D,s}$, $F_{D,c}$ and F_O are taken from Tables 4.24, 4.25, 4.26 and 4.27.

$$P_n = 11\left(\frac{W}{m^2}\right) \cdot A_u\left(m^2\right) = 220\ W; \quad F_{D,s} = 0.89; \quad F_{D,c} = 0.61; \quad F_O = 0.90$$

$$F_D = 1 - F_{D,s} \cdot F_{D,c} = 1 - 0.89 \cdot 0.61 = 0.45$$

$$\begin{aligned} LENI &= \frac{6 \cdot A_u + \sum_{i=1}^{n} P_n \cdot \left(t_d \cdot F_D \cdot F_O + t_n \cdot F_O\right)}{A_u \cdot 1000} \\ &= \frac{6 \cdot 20 + 220 \cdot (2250 \cdot 0.45 \cdot 0.90 + 250 \cdot 0.90)}{20 \cdot 1000} = 12.5\ \frac{kWh}{m^2 a} \end{aligned}$$

Case Study 2 The shopping mall in Fig. 4.76 is illuminated by skylights. The designed illumination level for a horizontal surface 0.85 m above the floor is 900 lx. Dynamic simulations using the TRNSYS code were performed to determine the energy demand for heating, cooling and lighting. This energy demand is expressed as a specific value per 1 m^2 of shopping-mall floor area. Table 4.27 shows the specific energy demand depending on the surface area

Fig. 4.76 The shopping mall with a useful floor area of 1500 m^2 is illuminated with skylights. The design illuminance level is 900 lx [36, 39]

Table 4.27 Annual specific energy demand for heating, cooling and lighting of shopping mall presented in Fig. 4.76 depending on size of skylights; the case without skylights are shown as the case with $DF_{a,v}$ 0%

size of skylights (% of floor)	DF_{av} (%)	heating (kWh/m^2a)	cooling (kWh/m^2a)	lighting (kWh/m^2a)	specific CO_2 emissions (kg/m^2a)
0.0	0.0	50.3	4.3	70.8	60.2
1.6	1.3	50.8	8.0	49.3	49.6
2.7	1.8	51.3	10.8	41.3	46.1
4.9	2.8	51.1	16.5	33.3	45.2

of the skylights. The impact of the daylight is evident. By introducing the emissions of the greenhouse gas CO_2 per kWh of delivered energy carrier, different delivered energy carriers (district heating, electrical energy for cooling and lighting) can be summarized. The skylights with a size of 3% of the useful area A_u of the shopping mall were proposed (and built) as the optimal solution according to the total yearly specific CO_2 emissions.

References

1. Behling, S., Behling, S.: SolPower, Die Evolution der solaren Architektur; Prestel, München, D (1996)
2. Eye Test; Bokan Technologies, iPhone Application
3. Schwartz, D.: Building Physics. Hochschule Liechtenstein, LI (2007)
4. Kladnik R.: Physics for university students—acoustic and optic, DZS, Ljubljana, 1989 (in Slovenian language)
5. Medved, S., Novak, P.: Environment Protection and Renewable Energy Sources. University of Ljubljana, Faculty of Mechanical Engineering, Ljubljana (2000) (in Slovenian language)
6. Commission internationale de l'Eclairage proceedings, 1931, Cambridge University Press, Cambridge, UK (1932)
7. Baker, N., Fanchiotti, A., Steemers, K.: Daylighting in Architecture: A European Reference Book. James & James, London, UK (1993)
8. Medved, S.: Solar Engineering. University of Ljubljana, Faculty of Mechanical Engineering, Ljubljana (1994).(in Slovenian language)
9. Keller, B.: Bautechnologie II, Bauphysik, ETH, Zürich, CH (2001)
10. Hegger, M., et al.: Energy Manual, Edition Detail, Basel, CH (2008)
11. www.satellite-light.com; Technical data (1.1.2014)
12. Equation of Time, iPhone Application
13. www.audemarspiguet.com; Commercial data (1.1.2014)
14. Arkar, C., Medved, S.: Daylighting, Computer Code. University of Ljubljana, Faculty of Mechanical Engineering, Ljubljana (2001)
15. Tsangrassoulis A., et al.: A method to investigate the potential of south-orientated vertical surfaces for reflecting daylight onto oppositely facing vertical surface under sunny conditions. Solar Energy **66**(6) (1999)
16. Wilkinson, A.M.: Illuminance at a Single Position. University of Bath, UK (2004)

17. Medved, S.: Building Physics. University of Ljubljana, Faculty of Architecture, Ljubljana, Lecture Notes (2009).(in Slovenian language)
18. Littlefair, P.: Measurements of the luminous efficacy of daylight. Light. Res. Technol. **13** (1990)
19. Medved, S., Arkar, C.: Applied lighting technologies for urban buildings. In: Environmental Design of Urban Buildings, Earthscan, UK (2006)
20. SunSeeker, iPhone Application
21. Rovan, D.: Sun Duration Analysis; Seminar Work (supervisor: prof. Sašo Medved). University of Ljubljana, Faculty of Architecture, Ljubljana (2008) (in Slovenian language)
22. Medved, S., et al.: Numerical and In-situ Assessment of Thermal Response and Daylight of Typical Office in Office Building HIT. University of Ljubljana, Faculty of Mechanical Engineering, Ljubljana (2001).(in Slovenian language)
23. DIN 5034-4:1997 Tageslicht in Innenräumen - Teil 4: Vereinfachte Bestimmung von Mindestfenstergrößen für Wohnräume
24. Longmore, J.: Daylighting: a current view. Light and Lighting, UK, vol. 68 (1975)
25. Černe, B., Medved, S., Vidrih, B.: Impact of Building Integrated PV on Daylighting and Thermal Comfort in Courtyard of Office Building HIT. University of Ljubljana, Faculty of Mechanical Engineering, Ljubljana (2005).(in Slovenian language)
26. University College Dublin Artificial Skies: buildings design for well-being (2014)
27. CLEAR (Comfort and Low Energy Architecture); The Low Energy Architecture Research Unit (LEARN) of London Metropolitan University, UK, 2009.
28. Bauer, M., et al.: Green Building. Callwey, München (2007)
29. www.solartube.com; Technical data (1 Jan 2014)
30. www.wikipedia.org (1 Jan 2014)
31. VELUX Daylight Visualizer 2; 2010; www.velux.si
32. Bodart, M.: Global energy savings in offices buildings by the use of daylight. Energy Build. **34** (2002)
33. Rules on thermal insulation and efficient energy use in buildings; Ministry of the Environment and Spatial Planning, R Slovenia (2002) (in Slovenian language)
34. Rules on efficient use of energy in buildings with a technical guideline; Ministry of the Environment and Spatial Planning, R Slovenia (2010) (in Slovenian language)
35. EN 15193-1:2005 Energy performance of buildings—Energy requirements for lighting—Part 1: Specifications, Module M9
36. Medved, S., Dolinar, J.: Energy demand in Mercator shopping malls Mercator. University of Ljubljana, Faculty of Mechanical Engineering, Ljubljana (2004).(in Slovenian language)
37. Lechner, R.: Heating, Cooling, Lighting, Design Methods for Architects. Wiley, New York, USA (2001)
38. www.reflex.si; Technical data (1 Jan 2014)
39. www.kristal.si; Technical data (1 Jan 2014)
40. Dubois, M.C., Sharples, S., Du, J.: Visual effects of light. IDES-EDU, IEE for sustainable future (2013)
41. Spring, K.R., Fellers, T.J., Davidson, M.W.: Human vision and color perception. Olympus Lifescience (2019)
42. Kim, W., Kim, J.T.: A Formula of the position index of a glare source in the visual field. In: SHB2010—3rd International Symposium on Sustainable Healthy Buildings, Korea (2010)
43. Wienold, J., Christoffersen, J.: Evaluation methods and development of a new glare prediction model for daylighting environments with the use of CCD cameras. Energy Build. **38**, 743–757 (2006)
44. CIE Discomfort Glare in the Interior Lighting, Commission International de l'Eclairage (CIE), Technical Committee TC-3.13 (1992)
45. Window 7.0, Berkeley Lab, University of California, USA
46. Darula, S., Christoffersen, J., Malikova, M.: Sunlight and insulation of building interiors. Energy Procedia **78**, 1245–1250 (2015)
47. Tregenza, P., Loe, D.: The Design of Lighting. Routledge (2013)

48. Hopkinson, R. G., Longmore, J., Petherbridge, P.: An empirical formula for the computation of the indirect component of daylight factor. Trans. Illum. Eng. Soc. **19**(7_IEStrans), 201–219 (1954)
49. Directive 2010/31/EU of the European Parliament and of the Council of 19 May 2010 on the energy performance of buildings

Chapter 5
Building Acoustics and Noise Control in Buildings

Abstract Emitting and perceiving sound are the most important ways that people use to exchange information. Sound is a phenomenon related to mechanical waves and oscillations that propagate through compressible/elastic matter, such as air or building materials, in the form of sound waves. However, human hearing also senses disturbing sounds, referred to as noise. Noise is constantly generated by the people and equipment in buildings and their surroundings, affecting well-being, productivity and health. As most of our time is spent in buildings, ensuring the proper level of acoustic comfort and protection from all types of noise is an important function of any building. There are two important tasks for the discipline of acoustics in the scope of building physics, associated with the acoustic-related comfort in buildings: room acoustics deals with the quality of sound perceived by the listeners in rooms containing sound sources, while building acoustics deals with the protection of listeners from the transmission of the noise in the surroundings, as well as between the rooms in a building, with adequate sound insulation. The first task focuses on the design of the architectural form of the rooms and the sound-absorbing properties of interior surfaces that define the reverberation time of the rooms, while the building acoustics focuses on the design and evaluation of sound insulation against airborne and impact sound.

Learning Objectives In this chapter you will learn about:

- the physical fundamentals of sound;
- the perception of sound and the physiological units for measuring sound;
- the characterization, measurement and exposure limits for environmental noise;
- the criteria and methods for the design and verification of room acoustics;
- the design and verification of the sound insulation of building structures against airborne sound;
  the impact of sound propagation in buildings and the impact sound insulation of building structures.

S. Medved, *Building Physics*, Springer Tracts in Civil Engineering,
https://doi.org/10.1007/978-3-030-74390-1_5

5.1 Symbols for Quantities and Units

A	Equivalent sound-absorption area (m^2)
A_o	Reference equivalent absorption area of the room (m^2)
A_f	Equivalent sound-absorption area for sound frequency f (m^2)
α_f	Absorption coefficient of sound frequency f (–)
α_{abs}	Absorption coefficient of sound absorber (–)
B	Room correction factor for impact sound (dB)
c	Speed of sound (m/s)
C_p	Room correction factor (dB)
C	Spectrum adaptation term for none white or pink noise (dB)
C_{tr}	Spectrum adaptation for traffic noise (dB)
χ	Bulk modulus (N/m^2)
D	Distance between walls in the room (m)
D_{nT},	Standardized sound-level difference (dB)
$D_{nT,w}$, $D_{nT,w,\min}$	Weighted standardized sound-level difference, minimum level (dB)
ΔL_{eq}	Reduction of sound (noise) pressure level (dB)
$\Delta L_{n,w}$	Weighted impact sound-reduction index (dB)
E	Modulus of elasticity (N/m^2)
ε	Proportion of area of holes in sound absorber (%)
f	Frequency of sound waves (s^{-1}, Hz)
f_n, f_1,f_r, f_c	Frequency of harmonics, frequency of the first harmonic (1), resonant frequency (r), critical frequency (c) (s^{-1}, Hz)
I, I_o	Sound intensity, reference sound intensity (W/m^2)
K_1, K_2,K_3	Adjustment factor for impulse noise (1), for noise with accented tones (2), for noise emitted by means of transportation (3) (dB)
l_L, l_o	Sound bridge length (L), reference length of sound bridges (o) (m)
L, H, h	Noise-barrier-related dimensions (m)
$L_{eq,A,F}$, $L_{eq,A,I}$, $L_{eq,A,S}$	Equivalent sound-pressure level, "A" weighted and fast (F) averaged equivalent sound level, impulse (I) averaged equivalent sound level, slow (S) averaged equivalent sound level (dB(A))
L_p	Sound-pressure level, loudness of sound (dB)

(continued)

(continued)

$L_{p,A}$, $L_{p,B}$, $L_{p,C}$, $L_{p,D}$	Sound-pressure level weighted with "A" curve (dB(A)), weighted with "B" curve (dB(B)), weighted with "C" curve (dB(C)), weighted with "D" curve (dB(D)),
L_w	Sound power level (dB)
L_I	Sound intensity level (dB)
$L_{eq,A,F,T}$, $L_{eq,A,F,day}$, $L_{eq,A,F,d}$, $L_{eq,A,F,n}$, $L_{eq,A,F,den}$	Equivalent sound-pressure level "A" weighted, fast averaged in time period T; between 0–24 h (day), between 6 am and 10 pm (d), between 10 pm and 6 am (n), over the year (den) (dB(A))
$L_{1,T}$;$L_{99,T}$	Peak sound equivalent pressure level that exceed 1% of time period T; 99% of time period (dB)
L_r, $L_{e,T}$	Assessment sound-pressure level, time period averaged (dB)
L'_n, $L'_{n,\max}$	Level of normalized impact sound, permissible level (dB)
$L'_{n,w}$, $L'_{n,w,\max}$	Weighted normalized impact sound-pressure level, permissible level (dB)
m'	Mass per unit of the building structure (kg/m^2)
λ, λ_s	Wavelength, wavelength of structure oscillation (s) (m)
p	Air pressure (Pa)
p_{ref}	Reference sound pressure (Pa)
p_o	Sound-pressure amplitude (Pa)
p_{RMS}	Effective, RMS (Root Mean Square) sound pressure (Pa)
r	Distance (m)
R, $R_{\min}$	Sound-reduction index of building structure for airborne sound, minimum sound-reduction index (dB)
R'	Apparent sound-reduction index (dB)
R_w R'_{w}, $R'_{w,\min}$	Weighted sound-reduction index, weighted apparent sound-reduction index, minimum weighted apparent sound-reduction index (dB)
$R'_{w,r}$, $R'_{w,res}$	Weighted apparent sound-reduction index, weighted apparent sound-reduction index of heterogeneous building structure (dB)
R_L	Correction term of flanking sound propagation (dB)
$R_{L,w}$	Weighted correction term of flanking sound propagation (dB)
ρ	Density (kg/m^3)
S, S_o	Surface area, reference surface area (m^2)
W, W_o	Sound power, reference sound power (W)

(continued)

(continued)

$W_{incident}$, $W_{reflected}$ $W_{transmited}$	Sound power incident to the structure, reflected from the surface of the structure, transmitted through the structure (W)
T	Period of sound waves (s), time period for which equivalent sound pressure is determine (h:min)
T_{60}, $T_{60,f}$	Sabine's reverberation time of the room at sound frequency 500 Hz, for octave frequency ranges (60, f) (s)
$T_{60,o}$	Reference Sabine's reverberation time of the room (s)
T_{30}, $T_{30,f}$	Reverberation time determined at decrease of sound-pressure level by 30 dB at sound frequency 500 Hz, for octave frequency ranges (s)
V	Volume of the room (m^3)
Z	Acoustic impedance (Ns/m^3)

5.2 Sound and Acoustics

Sound is a phenomenon related to mechanical waves that are transmitted through compressible/elastic matter as sound waves causing a small difference in pressure and density in the matter. The sound is a form of energy. The science of acoustics deals with the generation, propagation and detection of sound waves. The word acoustics comes from the Greek word *akoustikos*, meaning "ready to hear". Two subfields of acoustics are especially important for building design: physical acoustics and architectural acoustics. Physical acoustics describes the sound's transmission phenomena, such as reflection, diffraction and absorption, while architectural acoustics studies the insulation with respect to the transmission of sound, and its undesired form noise, between the surroundings and the building, as well as between the rooms in a building. One subfield of architectural acoustics is room acoustics, which studies how sound emitted by the source of sound in the room will be perceived by the listeners, while building acoustics deals with the sound insulation of building structures. Architectural acoustics aims to ensure a suitable level of acoustic comfort in buildings used for living and working.

5.2.1 Types of Mechanical Waves

Waves can be divided into transverse and longitudinal waves. In both cases we are dealing with mechanical waves that can only be transmitted through matter and where the position of the particles of matter changes periodically, causing a small

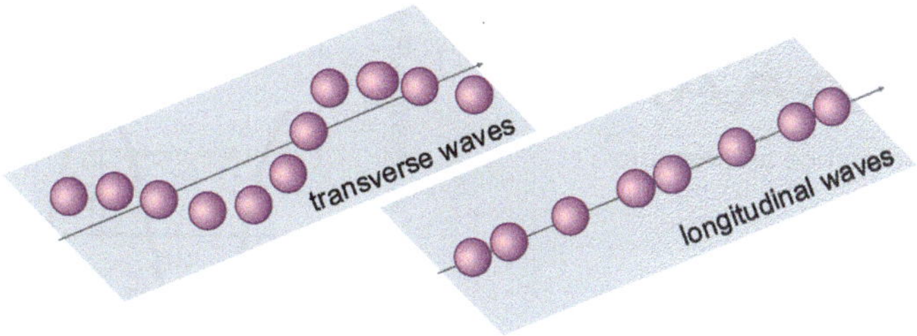

Fig. 5.1 Mechanical waves are transmitted through matter by the transverse (left) or longitudinal (right) oscillation of particles

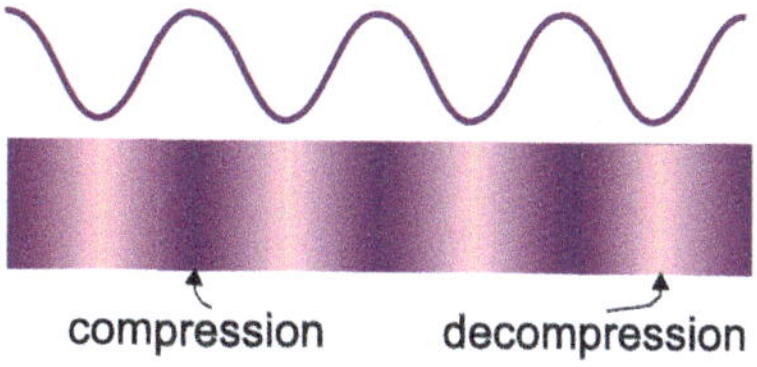

Fig. 5.2 Sound waves are longitudinal. In air or other compressible matter, sound waves cause the periodic spatial pressure differences known as sound pressure

difference in the pressure and density. If the particles oscillate perpendicularly to the direction of a wave, we are talking about transverse waves (Fig. 5.1, left). Transverse waves only propagate through solids, but not through fluids (i.e., gases and liquids). Transverse waves cannot be detected by our sense of hearing. Longitudinal waves, on the other hand, are able to propagate through any medium that exhibits at least some compressibility. As is the case with transverse waves, the particles of matter are constantly moving, but this time their motion is parallel to the direction of the wave's propagation (Fig. 5.1, right). Regions of compression and decompression are formed in the medium, which are detected as a pressure difference (Fig. 5.2). Sound is transmitted through the air and other media in the form of longitudinal waves.

The regions of compression and decompression formed in the air during the transmission of sound waves are sensed as a pressure increase or a pressure decrease against the reference air pressure of 10^5 Pa (Fig. 5.3, left). They occur periodically and the distance between two neighbouring regions of compression is known as the wavelength λ. In sound phenomena, the wavelength is measured in m. Pressure variations that take place in the matter as a consequence of sound transmission are known as the sound pressure and are measured in Pa. The smallest sound-pressure level that can still be detected is 2×10^{-5} Pa, while sound pressures in excess of 20 Pa will cause pain in the ears.

The maximum pressure difference from the reference pressure is known as the sound-pressure amplitude, designated p_o. As the sound pressure is constantly changing between positive and negative amplitudes, the time-averaged sound pressure would be 0. Therefore, it is not a useful measure. As a result, sound pressure is expressed as the effective or RMS (Root Mean Square) sound pressure p_{RMS} (Fig. 5.3, right), given by:

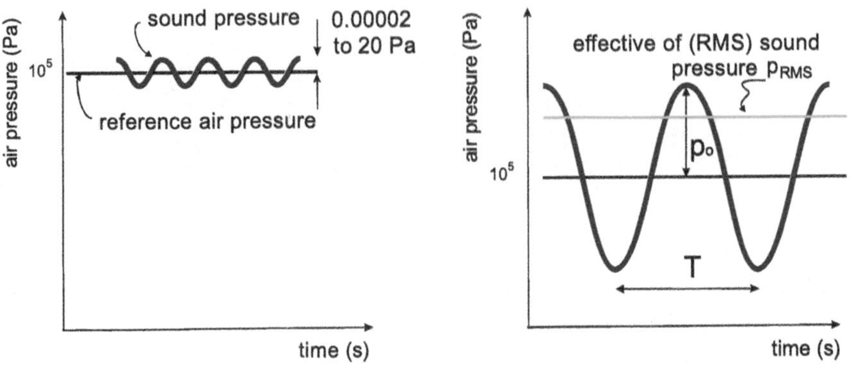

Fig. 5.3 The difference between the instantaneous and reference air pressure is defined as the sound pressure (left); sound pressure is expressed as the effective sound pressure or RMS (root mean square) sound pressure (right)

$$p_{RMS} \equiv p_{ef} = \frac{p_o}{\sqrt{2}} [\text{Pa}]$$

The threshold of human hearing is known as the reference sound pressure, designated p_{ref}.

Sound waves are a periodic phenomenon, defined by the sound-pressure amplitude p_o (Pa) and frequency f (s^{-1} or Hz). The frequency tells us how many compressions and decompressions there are in a unit of time. The frequency also determines the period T (s), or the time between two consecutive instances when the sound pressure assumes the same value:

$$T = \frac{1}{f} [s]$$

The wavelength λ (m) is a property of the sound wave describing the distance between two neighbouring points of equal sound pressure. It is given by:

$$\lambda = \frac{c}{f} \left[\frac{m \cdot s}{s} = m \right]$$

where c is the speed of a sound wave propagating through the medium (m/s). The speed of sound depends on the media and for solids it is a function of the material's elasticity, measured with the modulus of elasticity E (N/m^2) (Table 5.1). The speed of the propagation of sound in liquids depends on the compressibility, given by the bulk modulus χ (N/m^2), and in gases it is proportional to the square root of the absolute temperature.

Table 5.1 Speed of sound in different media [1]

	speed of sound(m/s)
air 0°C	332
air 15°C	341
water	1480
steel	4800 – 5000
glass	5100 – 5500
concrete	3500 – 4000
rubber	50

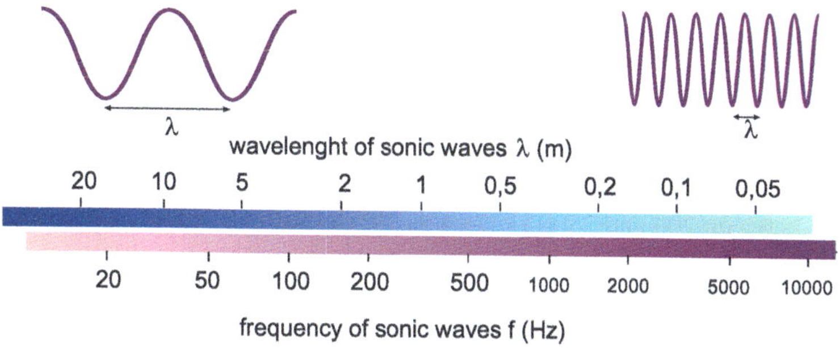

Fig. 5.4 Sound waves frequency f versus sound waves wavelength λ

Explanation The physics of propagation is different for light and sound: light is electromagnetic waves that can also travel through a vacuum (e.g., through outer space), while sound is transmitted by waves spreading through matter. The wavelength λ of light and sound waves have different orders of magnitude: light has wavelengths between 0.38×10^{-6} and 0.76×10^{-6} m, while the wavelength of sound waves assumes values between 0.05 and 20 m, corresponding to the order of magnitude of buildings and the objects inside them (Fig. 5.4).

This also explains the phenomena specific to the propagation of sound in buildings, such as diffraction and resonance.

5.2.2 *Types of Sound*

The types of sound can be defined in different ways. In acoustics, the sounds are evaluated and classified according to the frequency of sound and the sound spectrum. In the first case, the classification is based on the threshold sound frequencies that humans can hear, and in the second case the sounds are evaluated by the power of sounds emitted by the sound source at different frequencies. In this way the source's sound spectrum is described.

5.2.2.1 Types of Sound According to the Sound Frequency

According to the frequency, the sound is divided into:

- infrasound ($f < 16$ Hz). Humans are unable to hear it, but it drives our internal organs into oscillation and could affect the wellness and health of humans (and other species). Frequencies under 10 Hz could cause headaches and influence the functioning of the heart. Such frequencies are generated by machines and heavy transport vehicles, and they also occur in nature during storms and earthquakes (Fig. 5.5);
- audible sound (16 Hz $< f <$ 20,000 Hz). The sound frequencies that humans can perceive, although the range of audible sounds depends on the age and the physical and psychological conditions of the listener;
- spoken sounds. Humans are able to make sounds with frequencies between 50 and 8000 Hz (8 kHz), while the sounds emitted by speech are in the frequency range between 200 and 2500 Hz;
- ultrasound ($f >$ 20,000 Hz). Like infrasound, humans are unable to hear ultrasound. Some animals emit and sense ultrasound for the purposes of echolocation (bats, dolphins). Ultrasound propagates through water (and therefore also through our bodies) practically without losses (Fig. 5.6). Ultrasound can also be directed, and this property is used in sonars. Ultrasound kills airborne microorganisms and is used for sanitization. Ultrasound makes particles of solid matter oscillate rapidly and disintegrate, and this feature is used to remove surface oxide layers and fats. Ultrasound is used in industry for examining the homogeneity of materials, as well as for welding plastic materials and metals.

The field of acoustics in building physics is limited to studying the propagation of sound frequencies between 100 and (3150) 4000 Hz, while the acoustics of noise

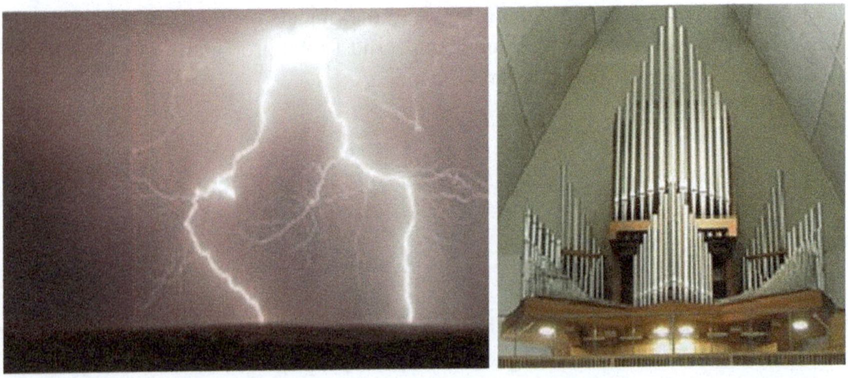

Fig. 5.5 Infrasound generated in storms travels hundreds of kilometres, because it is not attenuated in the atmosphere (left). The pipe organ is the only symphonic instrument that produces infrasound (right) [2]

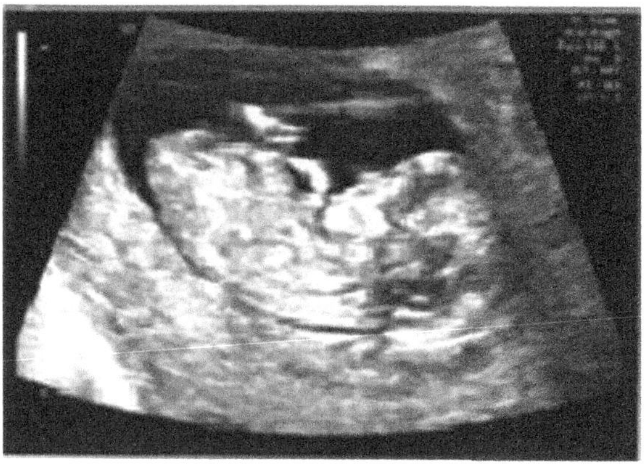

Fig. 5.6 Ultrasound waves have high sonic energies due to their high frequency, but it is not harmful because of the low effective sound pressure. Because ultrasound waves are transmitted through the water (and therefore also through our bodies) almost without losses, ultrasound is used in medicine [2]

control deals with the transmission of sounds in the frequency range from 50 to 10,000 Hz.

5.2.2.2 Types of Sound According to the Sound Spectrum

We have described how the propagation of sound causes variations in the pressure in the medium through which the sound waves are propagating. We also mentioned that human hearing can sense sound pressures greater than 2×10^{-5} Pa and that sound pressures in excess of 20 Pa will be perceived as painful. As the range is very wide, a logarithmic scale was adopted using the units of bel and decibel (dB, one tenth of a bel). Sound pressure measured in decibels is known as the sound-pressure level L_p (dB) and this is what is often referred to as the loudness of sound (Fig. 5.7). The decibel is a physical unit and its minimum or reference value 0 dB corresponds to the reference sound pressure $p_{ref} = 2 \times 10^{-5}$ Pa. The relation between the sound pressure and the sound-pressure level is given by:

$$\begin{aligned} L_p &= 10 \cdot \log \frac{p_{RMS}^2}{p_o^2} = 20 \cdot \log \frac{p_{RMS}}{p_o} = 20 \cdot \log \frac{p_{RMS}}{(2 \cdot 10^{-5})} \\ &= 20 \cdot \log p_{RMS} + 94[\text{dB}] \end{aligned}$$

where p_{RMS} is the effective sound pressure or RMS (root mean square) sound pressure.

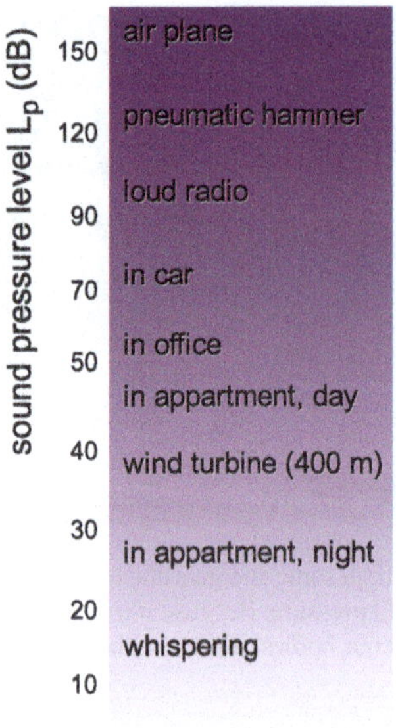

Fig. 5.7 Typical sound-pressure levels L_{pn} an indoor environment or emitted by different sound sources

Case Study Determine the sound-pressure level L_p for a sound pressure that is painful to human ears ($p_{RMS} = 20$ Pa).

$$\mathrm{L_p} = 20 \cdot \log\ \mathrm{p_{RMS}} + 94 = 20 \cdot \log\ 20 + 94 = 120\ \mathrm{dB}$$

The sound spectrum emitted by a sound source correlates with the sound-pressure level and the frequency. While the sound pressure is described in the time domain, the sound spectrum is presented as a function of the sound wave's frequency. According to the sound spectrum, we distinguish pure tones, sounds consisting of several harmonic waves and noise.

A tone consists of sound waves at a single frequency f_1. The variation of the sound-pressure with time can be described with a single sine or cosine curve. Such waves are known as harmonic waves. The sound spectrum of a pure tone consists of a single line at the fundamental frequency f_1, while the sound-pressure level equals the effective sound pressure squared $p_{RMS}{}^2$ (Fig. 5.8). From now on we will omit the subscript RMS when referring to the effective sound pressure.

If the sound is composed of several harmonic oscillations, the frequencies of the harmonic oscillations ($f_2, f_3, f_4 \ldots$) are a multiple of the fundamental frequency f_1, e.g., $f_2 = 2 \times f_1, f_3 = 3 \times f_1$ etc., and these frequencies are known as overtones. The sound spectrum consists of several sound-pressure levels L_{p1}, L_{p2}, L_{p3} … at harmonic frequencies f_1, f_2, f_3 (Fig. 5.9). Vocal cords generate harmonics when

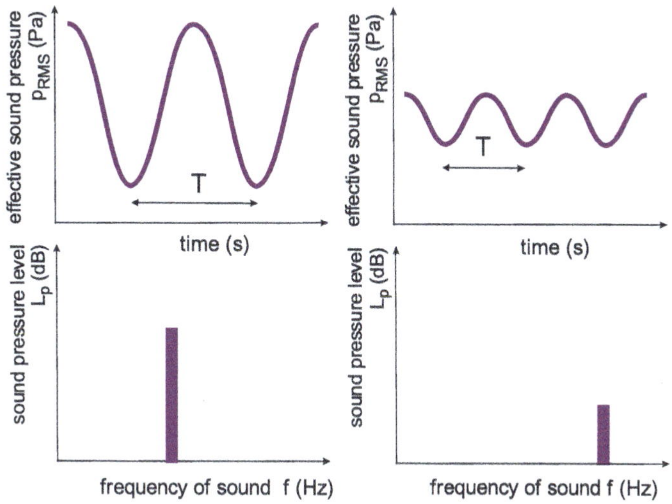

Fig. 5.8 Sound pressure and sound spectrum of two tones; the second tone has a lower amplitude of sound pressure p_{RMS} and shorter period T; because of that the sound spectrum of this ton is characterized by lower sound-pressure level L_p and higher frequency f

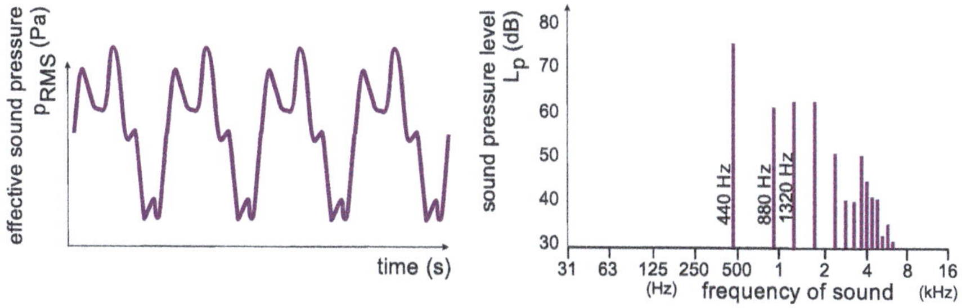

Fig. 5.9 Change in sound pressure p_{RMS} with time when tone "*a*" is pronounced (left), the sound spectrum of tone "*a*" (right). The sound spectrum of tone "*a*" consists of the fundamental frequency (440 Hz) and higher harmonics

articulating vowels. The timbre of a complex sound consisting of a fundamental tone and higher harmonics is determined by the sound-pressure-level ratio of the fundamental frequency to the overtones.

The frequency spectrum of consonants consists of very diverse, non-harmonic frequencies. Such sound is known as noise (Fig. 5.10). White noise is a special type of noise, distinguished by a frequency spectrum where the sound-pressure amplitude is the same at any frequency (Fig. 5.11, left). In the scope of building physics, the acoustic properties of rooms are measured using a sound source emitting white or pink noise. Pink noise is similar to white noise, the difference being that the sound-pressure level decreases linearly versus the frequency.

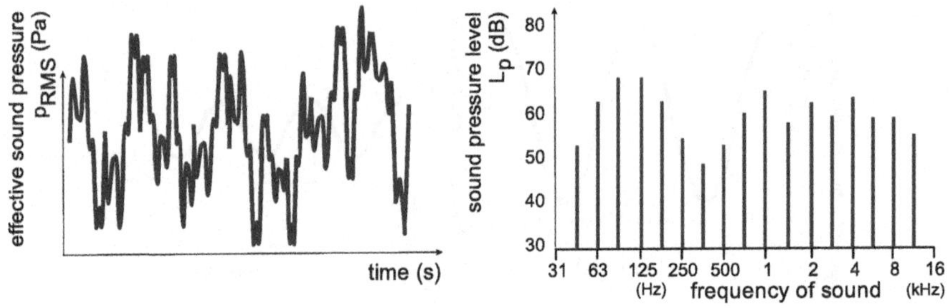

Fig. 5.10 Sound pressure p_{RMS} and sound spectrum for noise

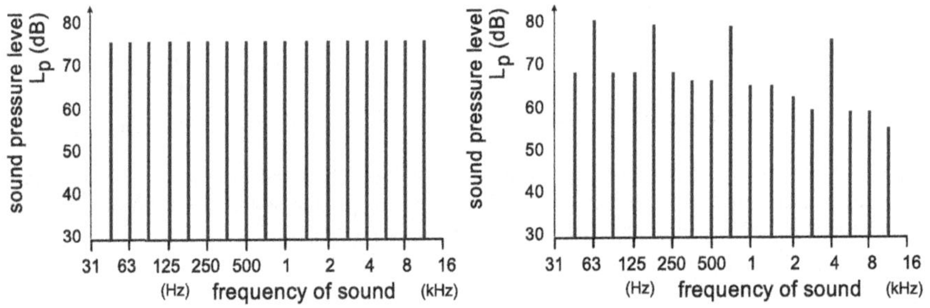

Fig. 5.11 Sound spectrum of white noise (left) and sound spectrum of noise with accented tones (right)

Physically, noise is distinguished by a characteristic sound spectrum. Physiologically, noise is useless, incomprehensible and unwanted sound waves. It can be detrimental to health and the impact of noise depends on the listener's mood, health, age and sex. The health-related issues can be a consequence of certain sound frequencies and/or the sound loudness. Especially disturbing is impulse noise and the noise that contains accented tones. Impulse noise is the noise with short periods of loudness that exceeds the sound-pressure level of permanent noise by more than 10 dB. The noise with accented tones occurs if the level of accented tones exceeds the noise level at two neighbouring frequencies by more than 5 dB (Fig. 5.11, right).

5.2.3 Sound-Spectrum Frequency Ranges

The sound pressure produced by sources of sound and noise in the environment and in buildings is not constant over time. The sound-pressure level (the loudness) L_p is therefore determined by time averaging over short time periods. The way the measured sound-pressure levels are time-averaged is a default property of sound-level meters. The F (fast, denoted $L_{p,F}$) and S (slow, $L_{p,S}$) time averaging are used by convention. As people are especially sensitive to noise containing accented tones, precise sound-level meters also enable the I (impulse, $L_{p,I}$) time averaging.

For a constant sound source, the measured sound-pressure levels $L_{p,F}$, $L_{p,S}$ in $L_{p,I}$ will be identical. Nevertheless, such cases will only be theoretical in engineering practice. In practice, different methods of time averaging influence the measured noise level of the same real source as shown in Fig. 5.12. This is why the method of time averaging must always be declared when presenting the results of acoustic measurements. Simple instruments (Fig. 5.13, left) measure the sound-pressure level over the whole range of audible sound frequencies, using only F or S time averaging.

But as humans do not perceive different sound frequencies equally, the sound-pressure levels in building physics must be determined at different sound frequencies. The whole audible sound frequency range is divided into predefined frequency bands.

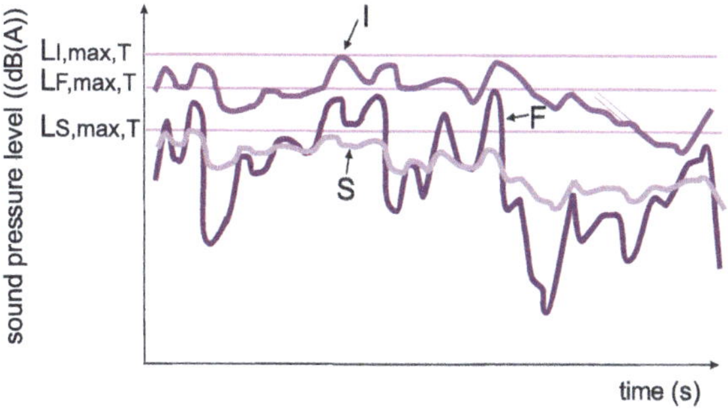

Fig. 5.12 Measured sound-pressure level depends on the method of the time averaging (F, S or I) [3]

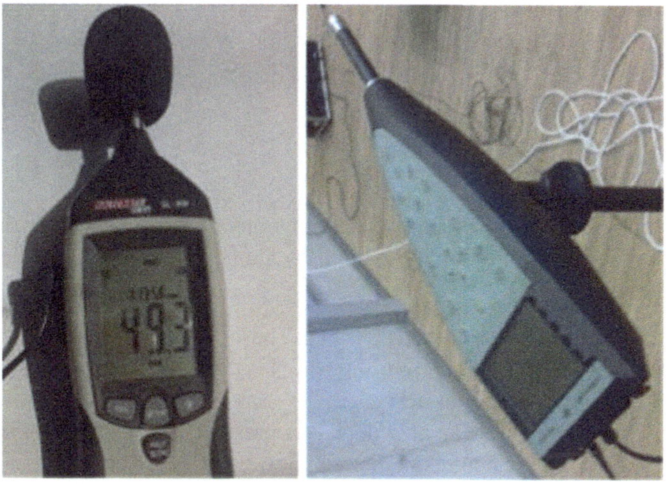

Fig. 5.13 Simple sound-pressure level meters evaluate the sound's loudness over the whole frequency range (between 20 and 16,000 Hz) and only offer "F" and "S" time averaging (left). Even state-of-the-art meters evaluate the sound-pressure levels only in discrete frequency bands [4] (right)

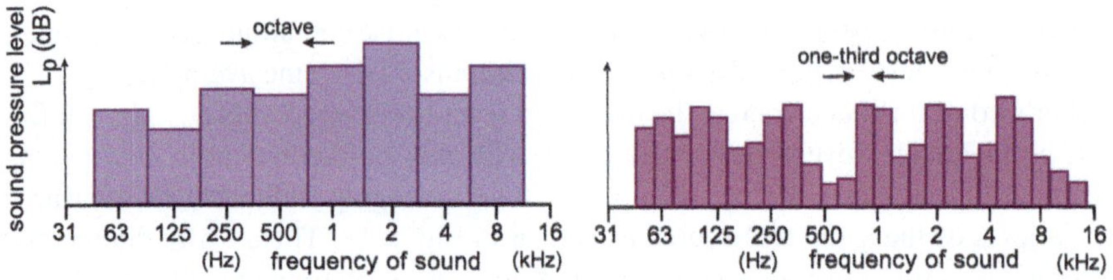

Fig. 5.14 In room and building acoustic problems, octave (left) and tertiary (one-third octave) frequency bands are used

When the central frequency of each band is twice the central frequency of the previous band, we are talking about octave bands:

$$f_{i+1} = 2 \cdot f_i [Hz]$$

In the scope of building physics, the audible sound levels are given for octave bands with the following central frequencies: 31.5, 63, 125, 250, 500, 1000, 2000, 4000, 8000 and 16,000 Hz (Fig. 5.14, left). Another option is to use tertiary frequency bands (one-third octave) (Fig. 5.14, right), where the subsequent central frequency is calculated as:

$$f_{i+1} = \sqrt[3]{2} \cdot f_i \cong 1.26 \cdot f_i [Hz]$$

The central frequencies of one-third octave bands are therefore: 31.5, 40, 50, 63, 80, 100, 125, 160, 200, 250, 315, 400, 500, 630 … 12,500, 16,000 Hz. Advanced instruments measure the sound-pressure level in tertiary bands (Fig. 5.13, right).

5.3 Perception of the Sound

Hearing organs convert the mechanical oscillations of the air into electric signals, which are in turn transmitted into the brain by the nerves (Fig. 5.15). The organ of hearing consists of:

- The outer ear with the auricle and ear canal. These two collect and direct the sound to the tympanic membrane, amplifying waves in the frequency range between 125 and 4000 Hz, which we emit in speech.
- Middle ear. The eardrum can only detect air-pressure differences, so the pressure on both sides of the tympanic membrane must be equalised over the nasal cavity and the Eustachian tube. The eardrum converts the air-pressure variations into mechanical oscillations, which are then transmitted to the inner ear by the hammer, anvil and stirrup bone.

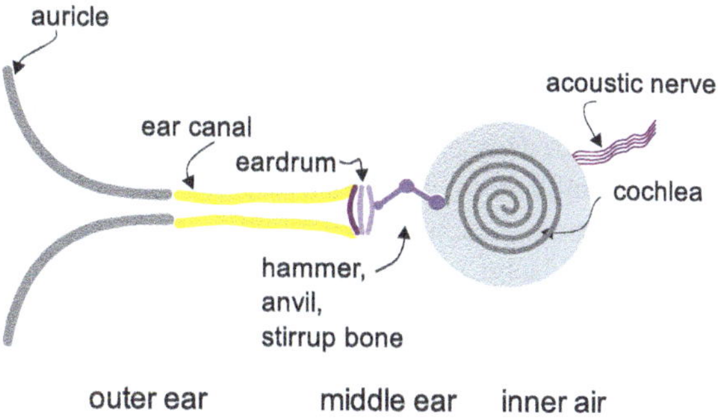

Fig. 5.15 A person's organ of hearing consists of the outer, middle and inner ears [5]

- The inner ear consists of the cochlea and the acoustic nerve; the receptor cells of the organ of Corti convert mechanical oscillations into electrical signals.

All the elements of the organ of hearing have evolved to be the most efficient at sensing sounds delivered by speech and at the same time protect us from the intensity of sounds that are less important for spoken communication.

5.3.1 *The Effects of Sound and Noise on Wellness and Health*

Sound is not only important for spoken communication. Music with a constant and easy rhythm can help to boost concentration when reading and studying. Music also improves the productivity and is often listened to in the background. Sound also provides therapeutic effects, as it influences the heart rhythm and breathing dynamics, and eases pain. Nevertheless, the evaluation of effects on human wellness and health mainly focuses on the harmful effects of noise. These include aural effects such as hearing impairment and hearing loss, and extra-aural effects that affect the whole organism:

- noise influences the productivity, impairing concentration and disrupting communication (Table 5.2, left);
- noise damages the organ of hearing, temporarily or permanently. Temporary hearing disorders are experienced after exposure to a noise louder than 120 dB

Table 5.2 Noise affects well-being and productivity at work

noise level (dB)	type of work
up to 45	demanding mental work
up to 50	routine mental work
up to 65	office work
up to 70	precise physical work
up to 80	routine physical work

noise level (dB)	consiquences of exposure
65	psychological impact
90	loss of hearing from long therm exposure
100	temporary loss of hearing by short therm exposure
120	pain threshold
150	immediately loss of hearing

Therefore, sound-pressure levels (noise levels) must be kept below a maximum value depending on the type of work (left). Possible consequences of exposure to different noise levels (right) [5]

for a couple of minutes or louder than 135 dB for a couple of seconds. Temporary disorders are perceived as tinnitus. Permanent hearing loss is a consequence of mechanical damage to the eardrum or parts of the middle or inner ear, caused by long-term daily exposure to noise levels over 85 dB (Table 5.2, right);

- noise influences the cardiovascular and autonomic nervous system, possible consequences include high blood pressure, irregular heartbeat, metabolism disorders, headache, people become irritable, tired, sleep disorders continue even after the exposure to noise ceases.

5.3.2 *Physiological Units for Evaluating Sound Loudness*

As mentioned before, the human ear exhibits a logarithmic sensitivity to sound intensity, but it also perceives differently different frequencies of equal intensity. We may be able to hear frequencies between 16 and 20,000 Hz, but our ears are much more sensitive to the characteristic frequencies of speech (125–4000 Hz), as other frequencies disrupt our ability to recognise speech. Furthermore, a sound-pressure level between 30 and 75 dB is necessary for speech recognition (Fig. 5.16). For comparison, the instruments in a symphonic orchestra emit sounds in the frequency range between 20 Hz and 10 kHz, and their loudness is between 30 and 100 dB [1].

To account for the non-linear perception of the loudness of the sound, a physiological unit was introduced to correlate the sound-pressure level L_p and the frequency of the sound. This unit is called the phon. Phon and sound-pressure level are equal at 1000 Hz and different at any other frequency. Phons can be used to define the absolute

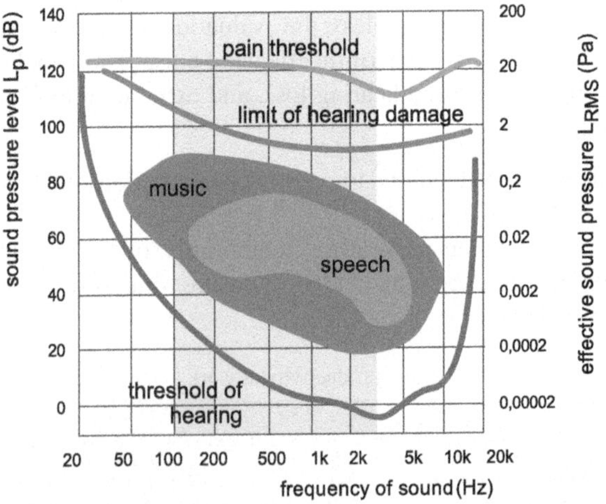

Fig. 5.16 The range of frequencies and sound-pressure levels L_p caused by speech and by symphonic orchestra. The bottom phon-curve represents the threshold of hearing and the top phon-curve represents the pain threshold

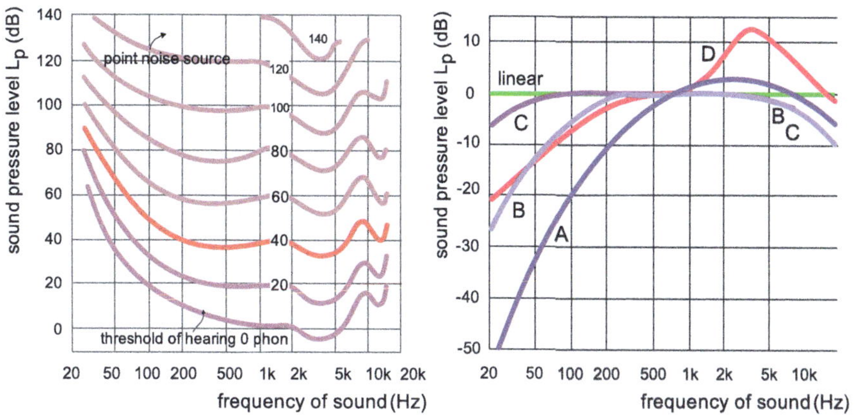

Fig. 5.17 The phon-curves describe the perception of the sounds with the same sound-pressure level L_p, but different frequencies; e.g., a sound with frequency 50 Hz is audible only when L_p is greater than 41 dB (left); weighting curves are used to adjust the sound-pressure level caused by difference sound sources to perceived loudness of the sound, known as weighted sound-pressure level; to determine the A-weighted sound-pressure level, 35 dB must be subtracted from the measured sound-pressure level (designated "Lin—linear") at 50 Hz and 8 dB at 500 Hz; the A-weighted sound-pressure level measured by fast time averaging is designated $L_{p,A,F}$ (in short L_A) and measured in dB(A) (right)

threshold of hearing and the pain threshold. The absolute threshold of hearing represents the frequency-dependent 0 (zero) phon-curve, which indicates the minimum sound-pressure level that can be heard at different frequencies of sound. At 1000 Hz it equals 0 dB, and at frequencies lower or higher than 1000 Hz the sound source must be louder for us to be able to hear it, for example, 10 dB at 200 Hz (Fig. 5.17, left). As the frequency rises over 1000 Hz, we are able to hear the sound source better and the threshold of hearing lies below 0 dB. As the frequency increases even further, the curve assumes a rising trend once again. The minimum threshold of pain, i.e., 110 dB, is located around 4000 Hz. Other iso-phon curves can be plotted between the absolute hearing threshold curve (0 phon) and the pain threshold curve (120 phon) (Fig. 5.17, left).

Various weighting curves have been developed to evaluate sound as it is actually perceived. They are derived from the sound-pressure levels measured in phons, yet some of them also account for the typical frequencies emitted by certain sources. Airplanes, for example, emit high-frequency noise. The A-weighting curve is commonly used in acoustics and corresponds to the 40-phon curve, the B-weighting curve corresponds to the 70-phon curve, and the C-weighting curve corresponds to the 120-phon curve used to evaluate a very loud sound. The D-weighting curve incorporates the weighting of higher frequencies. The weighting curves are shown in Fig. 5.17 (right). With regard to the weighting used, the sound-pressure levels are expressed as L_A, L_B, L_C and L_D and the corresponding units are dB(A), dB(B), dB(C) and dB(D), respectively.

5.4 Characterization of Sound Sources

Sound is a form of energy, and accordingly the main property of a sound source is its sound power, defined as the quantity of sonic energy emitted by the sound source in a unit of time. It is measured in W. The sonic energy is transmitted from the source to the surroundings as sonic flux. The sound power of different sources varies widely: speech releases a couple mW, musical instruments tens of W, while the sound power of a rocket engine is 100,000 W (Table 5.3). Sound sources can be isotropic, radiating the same sonic flux in all directions, or directed. The lowest sound power still detectable by humans is 10^{-12} W, a value known as the reference sound power W_o. The sound intensity I is defined analogously to specific heat and light flux density—it is calculated as the sound power transmitted per 1 m^2 of virtual surface area in space through which the sound waves are propagating, and measured in W/m^2. The minimum sound intensity that is still audible is 10^{-12} W/m^2, or the reference sound intensity I_o. The sound intensity is proportional to the effective sound pressure (RMS) squared and inversely proportional to the acoustic impedance Z of the medium transmitting the sound.

$$I = \frac{p_{RMS}^2}{\rho \cdot c} = \frac{p_{RMS}^2}{Z}\left[\frac{Pa^2 \cdot m^3 \cdot s}{kg \cdot m} = \frac{N^2 \cdot m^3 \cdot s}{m^4 \cdot kg \cdot m} = \frac{W}{m^2}\right]$$

The acoustic impedance Z (Ns/m^3) is a physical property of matter describing the acoustic resistance to the propagation of sonic energy.

Case Study Determine the sound intensity corresponding to the effective sound pressure $p_{RMS} = 21$ Pa if the sound is transmitted by air. Note that such a value of sound pressure is just under the pain threshold.

Table 5.3 Sonic power of different sound/noise sources

sound source	source sonic power (W)
jet plane	100.000
simphony orchestra	1
lighting	1
rock group(Deep Purple)	0,1
bike, accelerates	0,1
car on the highway	0,01
lawn mower	0,001
kitchen blender	0,001
alarm clock	0,0001
toilet flushing	0,00001
hear dryer	0,000001
breathing	0,00000000001

$$I = \frac{p_{RMS}^2}{\rho \cdot c} = \frac{21^2}{1.27 \cdot 341} \approx 1 \frac{W}{m^2}$$

For a spatially isotropic point sound source, the sound intensity can be determined from the source's sound power and the surface area through which the sound waves are propagating, in this case the surface area of a sphere with radius r:

$$I = \frac{W}{S} = \frac{W}{4 \cdot \pi \cdot r^2} \left[\frac{W}{m^2} \right]$$

The ear exhibits logarithmic sensitivity to the large range of audible sound intensities I—between 10^{-12} and 1 W/m^2. This means that the loudness of a sound source emitting threshold sound intensities will only be perceived as 12 times louder. As is the case with the sound pressure, the source's sound power and sound intensity are evaluated on the decibel (dB) scale as the sound power level L_w and the sound intensity level L_I, respectively, taking into account the reference sound power W_o 10^{-12} W and the reference sound intensity I_o 10^{-12} W/m^2. L_w and L_i amount to 0 dB for the reference values and for an arbitrary sound power and sound intensity the levels are given by:

$$L_W = 10 \cdot \log \frac{W}{W_o} = 10 \cdot \log \frac{W}{10^{-12}} = 10 \cdot \log W + 120 \ [dB]$$

$$L_I = 10 \cdot \log \frac{I}{I_o} = 10 \cdot \log \frac{I}{10^{-12}} = 10 \cdot \log I + 120 \ [dB]$$

If the sound is transmitted through the air having an acoustic impedance approximately 400 Ns/m^3, the sound-pressure levels and sound-intensity levels are practically identical, $L_p = L_I$. Note that the sound-pressure level is a scalar quantity, while the sound-intensity level is a vector quantity. Therefore, by measuring the sound-intensity level, we are able to determine the spatial distribution of the sonic flux coming from a non-isotropic sound source.

5.4.1 *Transmission of Sound Through Open Space*

5.4.1.1 Sound-Pressure Level as a Function of the Distance from the Sound Source

Let us assume the sound waves are propagating through an open space without obstructions. In the case that the sound source is a point source, meaning the sound propagates uniformly through all three dimensions of space (Fig. 5.18, left), the

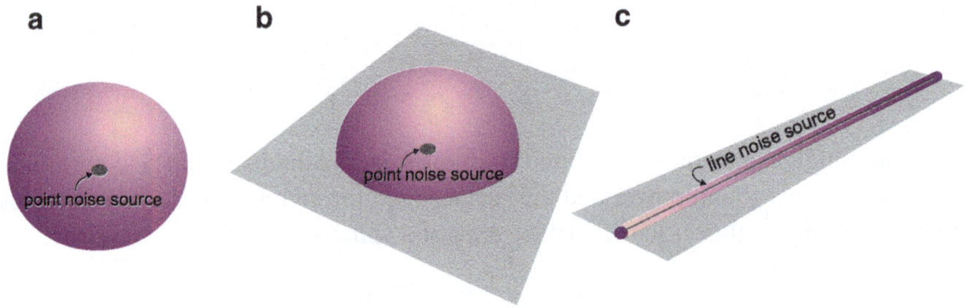

Fig. 5.18 The shape of 3D space into which sound waves are emitted from **a** a free-standing isotropic point source, **b** an isotropic point source near the ground and **c** a line source such as road and railroad traffic

sound-pressure level at a given distance *r* (m) from the source with sound power *W* (W) equals:

$$L_{p,(r)} = L_{I,(r)} = 10 \cdot \log \frac{I}{I_0} = 10 \cdot \log \frac{W}{S}\frac{S_o(= 1m^2)}{W_o} = 10 \cdot \log \frac{W}{W_0}\frac{1}{4 \cdot \pi \cdot r^2}$$
$$= L_w - 20 \cdot \log(r) - 10 \cdot \log(4 \cdot \pi) = L_w - 20 \cdot \log(r) - 11 \text{ [dB]}$$

Sound sources are often located near the ground and sound waves propagate only into a half-sphere sphere (Fig. 5.18, middle). The virtual surface through which the sound propagates is a half-sphere, and the sound-pressure level at an arbitrary distance *r* from the source is:

$$L_{p,(r)} = L_{I,(r)} = 10 \cdot \log \frac{I}{I_0} = 10 \cdot \log \frac{W}{S}\frac{S_o(= 1m^2)}{W_o} = 10 \cdot \log \frac{W}{W_0}\frac{1}{2 \cdot \pi \cdot r^2}$$
$$= L_w - 20 \cdot \log(r) - 10 \cdot \log(2 \cdot \pi) = L_w - 20 \cdot \log(r) - 8 \text{ [dB]}$$

A common source of noise in the outdoor environment is railway or road traffic, both being line sound sources (Fig. 5.18, right). In these cases, the sound propagates through the space in the form of a half-cylinder and the reduction in sound-pressure level with distance *r* from the traffic route is given by:

$$L_{p,(r)} = L_{I,(r)} = 10 \cdot \log \frac{I}{I_0} = 10 \cdot \log \frac{W}{S}\frac{S_o(= 1m^2)}{W_o} = 10 \cdot \log \frac{W}{W_0}\frac{1}{\pi \cdot r}$$
$$= L_w - 10 \cdot \log(r) - 10 \cdot \log(\pi) = L_w - 10 \cdot \log(r) - 5 \text{ [dB]}$$

Explanation It follows that as the distance *r* doubles from a point or line sound source emitting into a half-space, which is the case in most engineering problems, the sound-pressure level drops by $20 \times \log(2 \times r/r) = 20 \times \log(2)$

= 6 dB for the point source and by 10 × log (2 × r/r) = 10 × log(2) = 3 dB for the line source.

Case Study A rocket engine has the sound power of 100,000 W. Will the people observing the Space Shuttle launch from a distance of 5 km be able to withstand the noise without wearing earmuffs?

$$L_w = 10 \cdot \log\ 100,000 + 120 = 170\ dB$$

The rocket engine releases the noise into a half-space—in the atmosphere above the ground, so the sound pressure at the observers' position will be:

$$L_{p,(5000)} = L_w - 20 \cdot \log(r) - 8 = 170 - 20 \cdot \log(5000) - 8 = 88\ dB$$

Solution: The noise level is under the pain threshold and earmuffs will not be necessary.

The sound-power level L_w is a property of the sound source, while the sound-pressure level L_p must always be declared with respect to the distance from the source. If the sound-pressure level at a distance r_1 is known, the sound-pressure level at an arbitrary distance r_x can be determined with:

$$L_{p,2} = L_{p,1} - 20 \cdot \log\left(\frac{r_2}{r_1}\right) [dB]$$

Fig. 5.19 Fan in the condenser of a split-cooling unit is a source of environmental noise

Case Study The condenser of a cooling unit installed on the façade of a building (Fig. 5.19) produces noise with a sound-pressure level $L_p = 45$ dB measured at a distance of 1 m. Can the cooling unit operate continuously during the night if the neighbouring building is situated 8 m away? The sound waves are transmitted into a half-space.

$$L_{p,2} = L_{p,1} - 20 \cdot \log\left(\frac{r_2}{r_1}\right) = 45 - 20 \cdot \log\left(\frac{8}{1}\right) = 27 \text{ dB}$$

The sound-pressure level does not reach 40 dB (see Sect. 4.4.3), so the operation of the cooling unit will not be perceived by the neighbours as disturbing.

5.4.1.2 Multiplication of Sound Sources

The sound-pressure level caused by two identical sound sources is determined by doubling the sound intensity I of sources and the reference sound intensity I_o as:

$$L_{p,2} = 10 \cdot \log \frac{2 \cdot I}{I_0} = 10 \cdot \log \frac{I}{I_0} + 10 \cdot \log 2 = L_{p,1} + 3 \text{ dB}$$

where $L_{p,1}$ is the sound-pressure-level results from a single sound source. Two sound sources having the same sound intensity therefore cause the sound-pressure level to increase by 3 dB (Fig. 5.20, left). Such a change in the sound-pressure level would hardly be perceived by the listener (Table 5.4). In the case that there are n sound sources, the sound-pressure level increases by $10 \times \log(n)$. If there are several sound sources with different sound intensities, the sound-pressure level is given by the following expression:

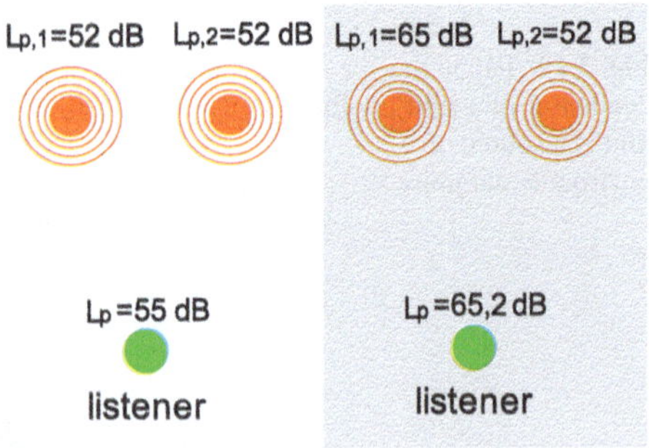

Fig. 5.20 Resulting effect of multiple sound sources is not determined by simple addition

Table 5.4 Audibility of differences in sound-pressure level [6]

audiable sound-pressure level difference L_p (dB)	how the difference is perceived
1 dB	is not perceive
3 dB	barely nocited
6 dB	as small difference
9 dB	as large difference

$$L_p = 10 \cdot \log\left(10^{\frac{L_{p,1}}{10}} + 10^{\frac{L_{p,2}}{10}} + \ldots + 10^{\frac{L_{p,n}}{10}}\right)[\mathrm{dB}]$$

Here, $L_{p,1}$, $L_{p,2}$ ... $L_{p,n}$ in dB are the sound-pressure levels caused by the each of n sound sources. According to the audibility of the difference in the sound-pressure level, the additional quieter sound source (Fig. 5.20, right) will not ne noticed (Table 5.4) by the listener.

5.4.2 Averaging the Sound Level Over Longer Periods

The measurements of the sound-pressure level L_p in the outdoor environment as well as in buildings are conducted over longer periods of time. The evaluation of the sound level also accounts for the dynamic properties of the sonic energy. The dynamics is evaluated using the "F" (fast) and "S" (slow) setting on the sound meter, while the "I" setting is used when measuring an impulse sound. The time-averaged sound-pressure level over longer time periods (e.g., seconds of hours) is known as the equivalent sound level L_{eq} (also: equivalent noise level) and designated using the letter of the sound meter's time-weighting function used (i.e. $L_{eq,F}$, $L_{eq,S}$, $L_{eq,I}$). The equivalent sound-pressure level L_{eq} express the constant sound level over a certain period T that carries the same sonic energy as real sound waves in the same period of time. Usually, the equivalent sound level is evaluated using the A-weighting curve to show what loudness of the sound will actually be perceived by the listeners and values are designated as $L_{eq,A,F}$ or $L_{eq,A,I.}$ Due to the fact that the sound level is expressed using a logarithmic scale, the time-averaging of all i instantaneous sound levels $L_{i,A,F}$ which we measure in the period T, must be done taking into account all n measured sound-pressure level values:

$$L_{eq,A,F} = 10 \log\left(\frac{1}{n}\sum_{i=1}^{n} 10^{0,1 \cdot L_{i,A,F}}\right) \quad [\mathrm{dB(A)}]$$

The measured sound-pressure levels $L_{i,A,F}$ can also be used to determine the sound pressures that have only been exceeded for a certain fraction of the time period T. For example, the peak sound level $L_{1,T}$ is a value that has been exceeded by 1% of the duration of the period T. Similarly, $L_{99,T}$ is the sound-pressure level exceeded for 99%

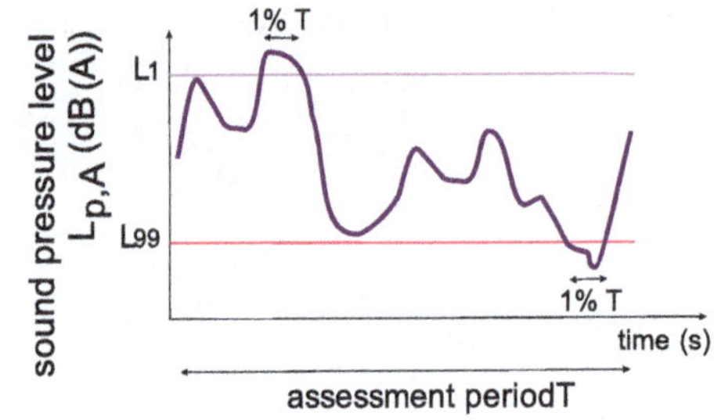

Fig. 5.21 How the peak-sound level $L_{1,T}$ and background-noise level $L_{99,T}$ are determined

of the period T, known as the background noise level (Fig. 5.21). When analysing the exposure to noise in an urban environment, the noise pollution is analysed as equivalent sound-pressure levels for the daytime between 6 am and 10 pm $L_{eq,A,F,d}$ (briefly $L_{eq,d}$) and for the night-time between 10 pm and 6 am $L_{eq,A,F,n}$ (briefly $L_{eq,n}$) or averaging over the whole day $L_{eq,A,F,day}$ ($L_{eq,day}$ The European Environment Noise Directive [7] even introduced a method for the evaluation of environmental noise over the whole year as the day-evening-night equivalent sound-pressure level L_{den}).

5.4.3 *Noise in the Urban Environment*

The environment in developed countries is exposed to serious noise pollution, especially in urban areas. Research has shown that several hundreds of millions of homes in Europe are over-exposed to environmental noise (Fig. 5.22, left). Because of that, the European Environment Noise Directive introduced a requirement for noise

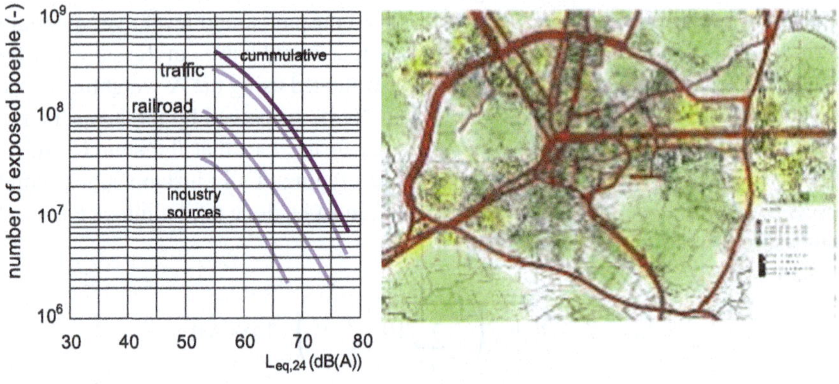

Fig. 5.22 Total number of citizens in the European Community exposed to different equivalent A-weighted sound-pressure levels $L_{eq,A,F,day}$ in dB(A). Values over 65 dB indicate excessive noise exposure [8] (left). More than 50,000 citizens of Ljubljana live in areas where the equivalent A-weighted sound-pressure level exceeds 65 dB. Noise pollution is the greatest in the city centre and in areas close to arterial roads carrying over 20,000 vehicles per day [9] (right)

mapping in all cities with more than 250,000 inhabitants. As an example, the Municipality of Ljubljana conducted noise measurements in the city centre and in residential communities. It was discovered that the daily equivalent noise level is greater than 65 dB(A) in the city centre and next to the arterial roads carrying over 20,000 vehicles per day. It is here that approximately 50,000 citizens of Ljubljana live (Fig. 5.22, right). As a general rule, the noise load in residential neighbourhoods is acceptable if the daily equivalent noise level $L_{eq,day}$ does not exceed 55 dB(A). The introduction of new noise sources is prohibited in environments where $L_{eq,day}$ reaches 65 dB(A). Noise levels $L_{eq,day}$ over 75 dB(A) call for active measures to reduce the sound-pressure level.

5.4.3.1 Sources and Propagation of Environmental Noise

There are various sources of environmental noise, but usually traffic and industrial facilities contribute the most to noise pollution. Environmental noise affects homes in the outdoor environment and can be transmitted inside buildings. The exposure to community noise depends on:

- the source's sound power;
- the type of noise source (whether it is a point-noise source or a line-noise source);
- the acoustic properties of the natural and built environments;
- the man-made obstacles placed between the noise source and the listeners that have an impact on the propagation of sound waves.

We have shown in Sect. 5.4.1.1 how the sound-pressure level reduces with the distance from the source. Nevertheless, we have not accounted for the sound attenuation in the air and close to surfaces. As it is transmitted through the air, a part of the sonic energy is converted into heat by the friction between the air particles with different velocities and interactions with the terrain. The drop in the sound-pressure level is frequency dependent and the attenuation is greater for higher sound-wave frequencies (above 2000 Hz), while the low-frequency sound waves ($f < 100$ Hz) are transmitted almost without any losses. The sound-pressure level at a frequency of 1000 Hz is reduced by only 5–8 dB per 1 km, while it is reduced by 50–60 dB per 1 km at frequencies above 4000 Hz (Fig. 5.23, right). The characteristics of road-traffic noise are such that the sound-pressure level drops in the open environment by 0.5 dB for every 100 m between the source and the listener. Air humidity affects the attenuation more than the air temperature. Sound attenuation due to interactions between the sound waves and the terrain varies with the height. The sound-pressure level drops 200 m from the noise source by 5 dB at a height of 1 m above the terrain, while only 1 dB at a height of 20 m.

Trees can help reduce the exposure to noise, but only if the plantation is wide enough. A tree plantation narrower than 20 m has negligible impact on noise pollution, while a 200 m-wide forest can reduce the sound-pressure level by more than 10 dB (Table 5.5).

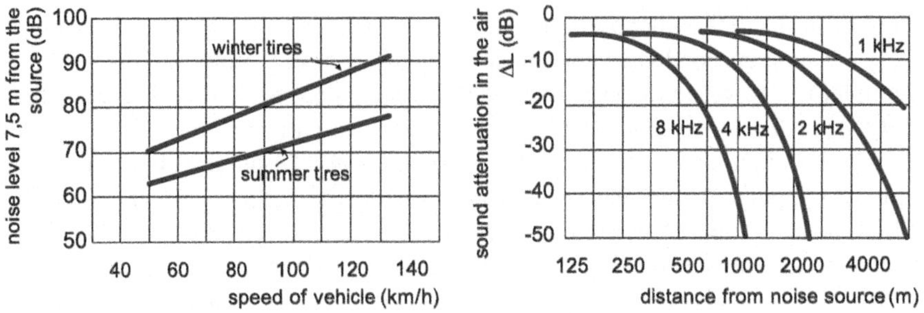

Fig. 5.23 Noise emissions from road vehicles depend on the tyre type and increase with vehicle speed. This is why the speed limit is lowered during the night time on highways near residential neighbourhoods (left). Sound attenuation in open space versus the distance from the source at different sound frequencies (right)

Table 5.5 Influence of tree-plantation width on the reduction of the sound-pressure level

tree plantation width (m)	reduction of noise level ΔL_p (dB)
<10 m	0
10 – 20 m	1
> 200 m	10 – 12

5.4.3.2 Decreasing Exposure to Environmental Noise

The most effective way to reduce exposure to environmental noise is noise reduction at the source. In the case of most problematic road traffic noise, the effective measures are laying low-noise-emission asphalt, the use of quieter tyres and, above all, speed limits (Fig. 5.23, left). All these measures, including the natural processes of sound attenuation, are often not enough to protect against excessive exposure to environmental noise. This is why noise barriers must be used in urban and industrial environments (Fig. 5.24). Noise barriers are based on the following physical principles:

- a noise absorption in the barrier;
- multiple reflections of sound waves between the sound barriers;
- redirecting the sound waves, thus creating an acoustic shadow.

The size of a noise barrier and its acoustic properties are not the only parameters that affect noise emissions, because noise protection also depends on the frequency of sound waves. As the sound waves hit a noise barrier, they are redirected in a way that depends on the ratio of the sound wave's wavelength λ to the size of the barrier L. This phenomenon is known as the deflection of sound waves. Sound waves with a long wavelength λ (= low-frequency sound waves) bend around the barrier, considerably reducing the area of the acoustic shadow (Fig. 5.25, left), while high-frequency sound waves with small wavelengths are deflected so that the acoustic shadow expands (Fig. 5.25, right). Computer simulations and in-situ experiments are used to evaluate the efficiency of noise barriers. Nevertheless, the nomograms shown in Figs. 5.26 and 5.27 can still be useful for initial planning.

Fig. 5.24 Noise barrier along a road consists of a noise-absorbing part (panels with a porous material in the foreground) and a noise-reflecting part (glass panels) [10]

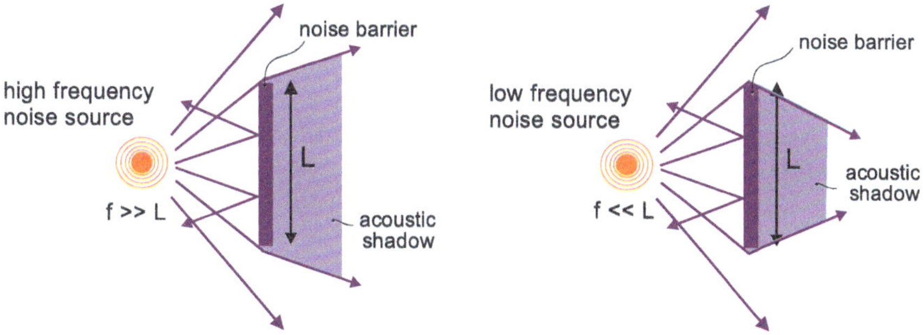

Fig. 5.25 Noise barriers are less efficient for low-frequency noise as the sound waves bend at the edge of the barrier and reduce the area of the acoustic shadow

Case Study Use the nomogram from Fig. 5.26 to determine the noise emission at a frequency of 750 Hz at the position of the listener. The equivalent weighted level of noise generated by the source is 75 dB(A). The distances are A 10 m, B 35 m and C 40 m.

The reduction of the equivalent weighted sound-pressure level $L_{eq,A}$ at a frequency of 750 Hz read from the nomogram is 18.5 dB(A). The noise emission at this frequency perceived by the listener will be:

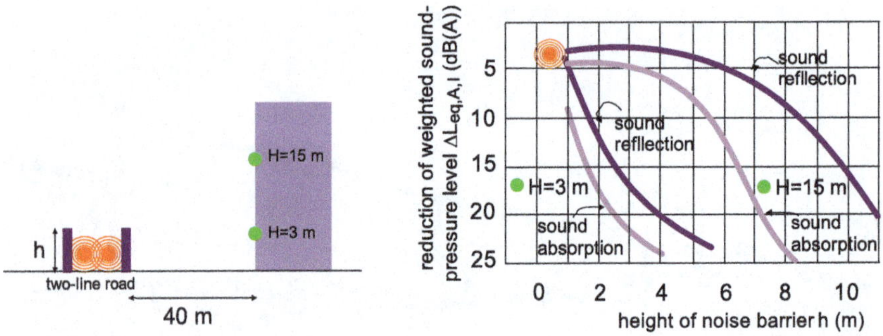

Fig. 5.26 Reduction of equivalent sound-pressure level next to the façade of a building standing 40 m away from a two-way traffic route with both side noise barriers: The reduction of the equivalent sound-pressure level depends on the noise barrier's height *h* and acoustic properties (either the noise barrier absorbs or reflects the sound waves). Case study: if the equivalent assessment sound-pressure level at the noise source L_r is 75 dB(A) ($L_{eq,A,I}$ + 5 dB for road traffic noise) and the equivalent sound-pressure level at the facade of the building should not excide 65 dB(A), the required noise reduction $L_{eq,A,I}$ must be 10 dB(A). For a third-floor apartment (H = 15 m) this can be ensured by a noise barrier with a height *h* 5.5 m if the noise barrier is designed to absorb the sound waves or 8.5 m if the sound waves are reflected from the surface of the sound barrier (e.g., if made by glass). At the same time, the sound-pressure level at the facade of a first-floor apartment (H 3 m) will be below 50 dB(A) [6]

$$L_{eq,A,F,listener} = L_{eq,A,F,source} - 18.5 = 56.5 \text{ dB}$$

From the nomogram it can be determined that the noise emission drops by 13 dB(A) at a sound frequency of 100 Hz and by 24 dB(A) at a sound frequency of 2500 Hz.

5.4.3.3 Measuring and Evaluating Environmental Noise

It is common that levels of protection against environmental noise are defined by national or municipal laws. As an example, four levels of noise protection are presented in the "Decree on the assessment and management of environmental noise" and in the "Decree on limit values for environmental noise indicators", published by Slovenian Ministry for spatial planning and environment:

- I. level of noise protection covers the areas demanding an increased protection against noise. These are natural areas used for tourism and recreation, the immediate surroundings of hospitals, health resorts and rehabilitation hospitals, as well as natural parks;

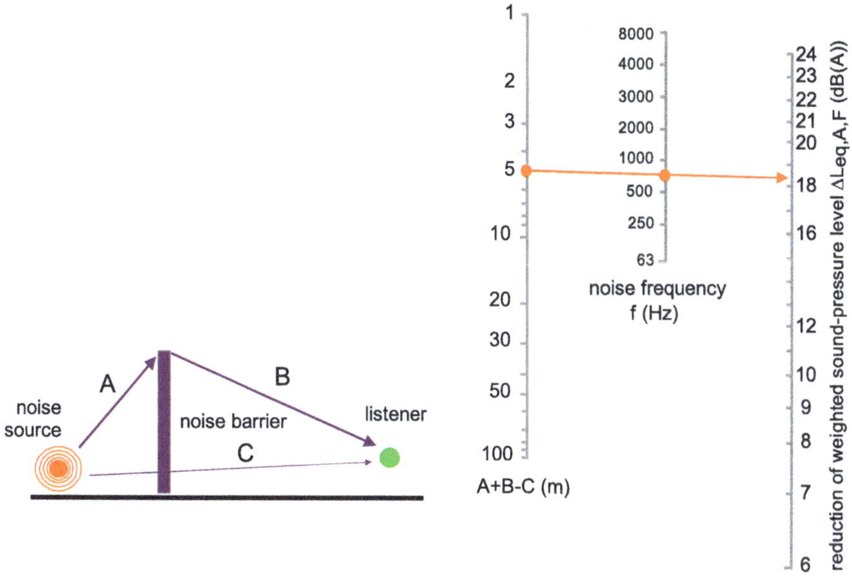

Fig. 5.27 Nomogram for determination of the frequency-dependent reduction of the equivalent sound-pressure level according to the geometry of the noise barrier [11]

- II. level of noise protection applies to areas where no new activities that would result in increased noise emissions are allowed. These areas include residential neighbourhoods, wider areas around schools and hospitals, playgrounds and public parks;
- III. level of noise protection applies to areas of permitted developments that disturb the environment due to noise emissions to a lesser extent. These are areas intended for agricultural, service or hospitality activities. Commercial and office buildings can be built in in these areas;
- IV. level of noise protection is in force in areas where heavier noise sources are allowed. These areas include no residential buildings and are only suitable for industrial activities and public utilities.

For each level of noise protection the maximum allowable equivalent sound-pressure levels are set for different periods of the day. The typical evaluation periods are presented in Sect. 5.4.2 and in Table 5.6. If an environment is characterized by impulse noise, impulse-weighting ($L_{eq,A,I,}$) must be used. Critical values of the equivalent sound-pressure level have been defined as well. If the equivalent level exceeds the critical one, noise emissions must be reduced either by reducing noise emissions at the noise source or by installing noise barriers.

On-site, the instantaneous sound-pressure level $L_{p,A,F}$ or $L_{p,A,I}$ is measured for the periods T of the day and the equivalent noise level $L_{eq,A,F,T}$ is determined and equivalent sound-pressure level is then used to determine the assessment sound-pressure level $L_{r,T}$ for each time period using the following expression:

Table 5.6 Permitted equivalent sound-pressure levels according to the type of noise protection areas; these values can be used in building design process or compared to the in-situ measured values to evaluated noise pollution in the environment [12, 13]

type of noise protection level	permited equivalent weighted sound-pressure levels (dB(A))	permited equivalent weighted peak sound-pressure levels (dB(A))	such noise exposure is convinint for
1. level	$L_{eq,A,F,d} = 50$ $L_{eq,A,F,n} = 40$	$L_{1,A,F,e} = 60$ $L_{1,A,F,n} = 60$ $L_{1,A,F,d} = 75$	natural environment, rehabilitation hospitals, recreation areas
2. level	$L_{eq,A,F,d} = 55$ $L_{eq,A,F,n} = 45$	$L_{1,A,F,e} = 65$ $L_{1,A,F,n} = 65$ $L_{1,A,F,d} = 75$	hospitals, spas, residential areas, tourist facilities
3. level	$L_{eq,A,F,d} = 60$ $L_{eq,A,F,n} = 50$	$L_{1,A,F,e} = 70$ $L_{1,A,F,no} = 70$ $L_{1,A,F,d} = 85$	mixed residental areas, schools, recreation faclities, sports building
4. level	$L_{eq,A,F,d} = 75$ $L_{eq,A,F,n} = 65$	$L_{1,A,F,e} = 90$ $L_{1,A,F,n} = 90$ $L_{1,A,F,d} = 90$	shopping centres, entertaiment parks, heavy industry

Note The designation "day" relates to 24 h period, "d" as daytime to the period between 6 am and 6 pm, "e" as evening to the period between 6 and 10 pm, and "n" as night-time to the period between 10 pm and 6 am of the following day; "I" averaging is used if impulse noise is characteristic to the environment ($L_{eq,A,I}$)

$$L_{r,T} = L_{eq,A,F,T} + \sum K_i$$

where K_i is the sum of the adjustment factors. For example, K_1 is the adjustment factor for the impulse noise, K_2 is the adjustment factor for the accented tones in the noise and K_3 is the adjustment factor for the means of transportation. K_1 assumes values 0 (impulse-less noise) or 5 dB in the case of impulse noise, the K_2 adjustment factor is 0 (noise without accented tones) or 5 dB and the K_3 adjustment factor is 0 dB for traffic and − 5 dB for railroad transportation because it was found that railroad noise is less disturbing. The adjustment factors can also be prescribed with other values, and other adjustment factors can also be defined.

The assessment sound-pressure level $L_{r,T}$ is then compared with the maximum-permissible value of the equivalent sound-pressure level defined according to the noise-protection level and assessed time periods. Typical values are stated in Table 5.6.

Example Figure 5.28 shows an office building situated near a traffic route with a sound-pressure measuring device next to the façade (in the circle). The instantaneous sound-pressure levels $L_{p,A,F}$ measured during the daytime and night-time are presented in Fig. 5.29. In the figure, the calculated equivalent sound-pressure levels $L_{eq,A,F,d}$ (68 dB(A)) and $L_{eq,A,F,n}$ (49 dB(A)) are shown, as well as the peak levels L_1 (74 dB(A)), which are exceeded for 1% of the duration of the measurement.

Fig. 5.28 Office building and position of sound meter. *Note* The sound meter was installed 0.5 m from the façade. As the sound is reflected from the façade, a proposed value of 3 dB(A) was subtracted from the measured value of $L_{p,A,F}$

5.5 Acoustic Comfort in Buildings

Acoustic comfort in buildings is achieved when the quality perception of speech and sound is ensured and noise is reduced to a level when it is not disturbing. The first requirement is met by the appropriate planning of room acoustics, based on analyses of the sound transmission from the source (speaker, instrument, orchestra) in the room to the listener. The second requirement is met by appropriate sound insulation against the noise coming from the environment and adjacent rooms. Building service systems such as mechanical ventilation can also be an important source of noise in buildings.

5.5.1 Room Acoustics

Room acoustics are defined by the way the sound waves propagate from the source to the listener. The main requirement for good room acoustics is that speech should be understood correctly. Based on the acoustic properties of rooms, we distinguish two opposite cases. An empty space in which sound waves on the peripheral surfaces are completely absorbed is called an "anechoic room" (Fig. 5.30, left). In such a space, we do not hear sound reflections when speaking, which can be embarrassing for

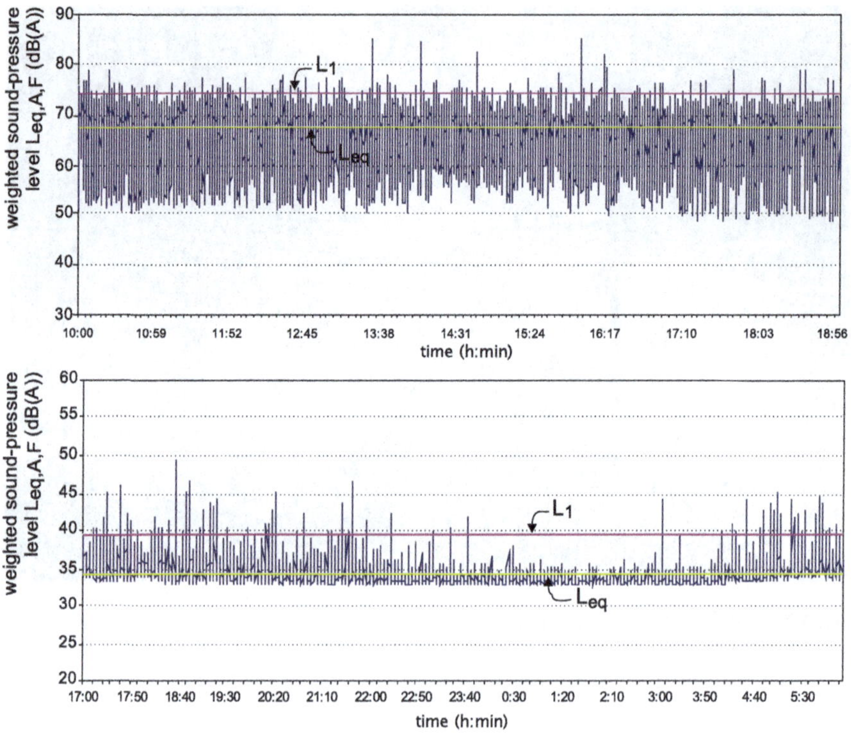

Fig. 5.29 Measured instantaneous A-weighted sound-pressure levels $L_{p,A,F}$ and equivalent A-weighted sound-pressure levels $L_{eq,A,F,d}$ and $L_{eq,A,F,n}$ on a weekday

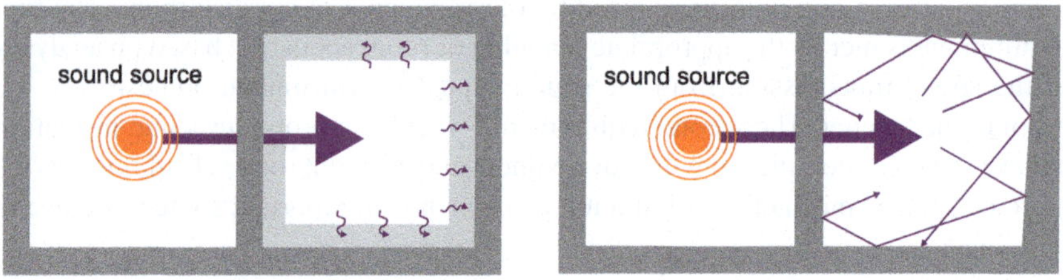

Fig. 5.30 Scheme of anechoic room (left) and reverberation room (right)

the speaker. That's why such spaces are called "deaf rooms". The absorptivity of the sound energy on all the walls of such a space must be greater than 99% for all frequencies emitted by the sound source. Anechoic rooms are used to determine the sound power field at different distances from the source, since the sound comes only from the sound source. The acoustic opposites of deaf rooms are "reverberation chambers" (Fig. 5.30, right). In them, all the surfaces are ideally sound reflecting. The walls are usually not parallel to avoid standing waves at certain frequencies. Reverberation chambers are used to measure the sound absorption of acoustic elements, which are placed on the walls in the reverberation chamber, and we determine the decrease in the sound pressure at the same source compared to the empty reverberation chamber. The spaces in the buildings are neither anechoic rooms nor reverberation chambers. Because of that, good room acoustics can only be achieved when different acoustic phenomena are known or predicted well. Due to this phenomenon, different acoustic areas are formed in the room, affecting the perception of sound in different areas of the room.

5.5.1.1 Behaviour of Room Acoustics

In open spaces with a sound source, the sound-pressure level decreases with the distance from the source. Indoors, however, different acoustic areas are formed due to the reflections of the sound waves on boundary and internal surfaces. Thus, in the vicinity of a sound source or a large reflective surface, the sound wave is not merely transverse and therefore sound-pressure fluctuations occur (Fig. 5.31). Even a small change in the position of the listener has a strong effect on the sound-pressure level. This area, called the direct sound field, extends from the source to a distance approximately equal to the minimum wavelength of the sound waves emitted by the source (for example, at a frequency of 20 Hz, that is ~ 5 m). Due to the reflections of the sound waves, a diffuse sound field is created in more remote areas of room, called reverberation areas, characterized by an equal sound-pressure level, regardless of the distance from the sound source. The area between the direct and reverberation fields is filled by an early reflection field, which is characterized by a decrease in the sound-pressure level of 6 dB with each doubling of the distance between the sound source and the listener. The acoustic properties of rooms are experimentally determined in this acoustic field.

Room acoustics mainly deal with problems in sound perception and recognition, which are influenced by two sound phenomena: echo and resonance. Echo is a phenomenon that occurs when the same sound source is perceived by the listener "several times" with a certain time lag. It is a consequence of the fact that we hear the sound emitted directly by the source sooner and as sound reflected from the

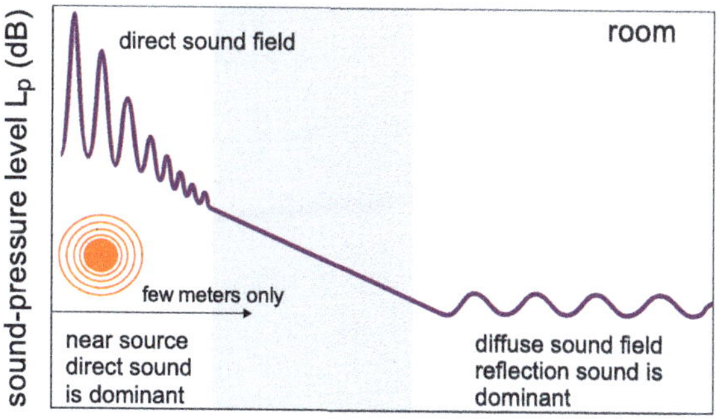

Fig. 5.31 Typical sound fields generated by the sound source placed indoors

surfaces in the room with a time lag. The echo is not perceived if the time between the detection of direct and reflected sound is, in general, shortened to 60 ms, or in case of speech recognition, about 100 ms.

If we consider that the speed of sound in the air is about 340 m/s, a delay of 100 ms means a difference in the path of the sound wave of 30 m (in Fig. 5.32 this is shown as the distance between r_o and r_1). Therefore, echo will be a problem, especially in large rooms. In such spaces the time delay could be reduced by sound-reflective surfaces installed near the sound source and in front of the listener, and by surfaces that absorb sound behind the listener.

Another disturbing phenomenon in the perception of sound in rooms is related to standing sound waves. This phenomenon occurs when two walls in a high space are parallel and sound-reflecting and the distance between them is equal to the wavelength of the sound wave λ. The incident and reflected sound waves are amplified by interference in certain places in space, and weakened in others (Fig. 5.33), causing resonance between the upward and downward sound waves. In the "sound nodes", which are spaced apart by λ/2, the sound pressure is the highest, and in the middle between the sound nodes the sound pressure is the lowest. Because the sound nodes are stationary in the space, the listener will perceive very different levels of sound pressure at different places in the room (Fig. 5.34). The frequency of sound waves at which a standing sound wave will occurs is called the first resonance frequency of a space and is determined by the expression:

$$\frac{\lambda}{2} = \frac{c}{2 \cdot f_1} \rightarrow f_1 = \frac{c}{2 \cdot D} \text{ [Hz]}$$

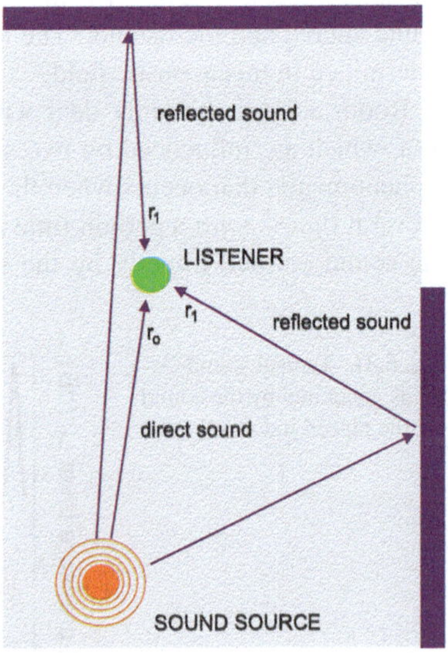

Fig. 5.32 Reflection occurs when the same sound source is detected several times with a certain time lag

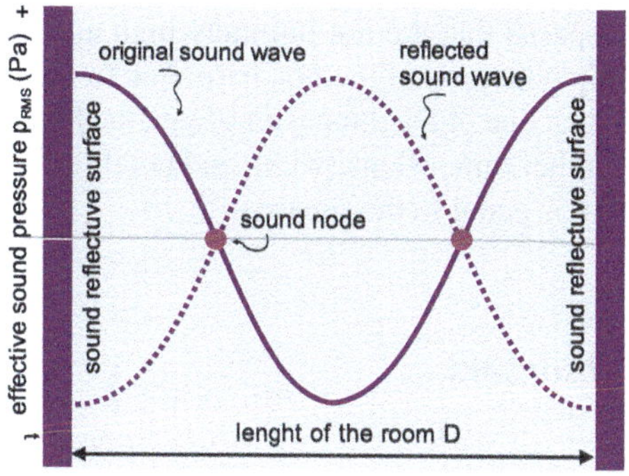

Fig. 5.33 In the case of standing sound field, stationary spatial high and low sound-pressure regions are formed in the room

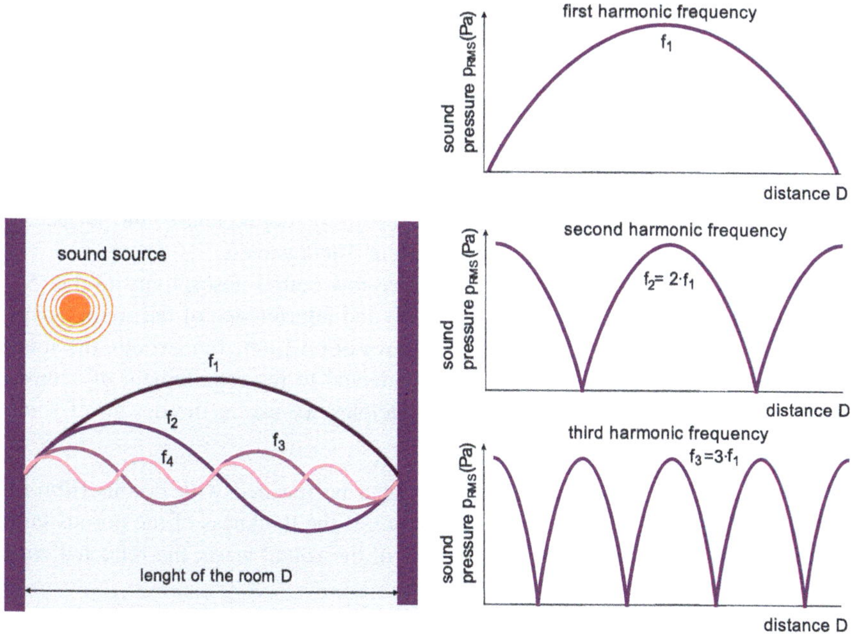

Fig. 5.34 A standing wave or resonance occurs in space at a wavelength of the sound wave equal to the first resonance frequency of space and its multiples, called higher harmonics

In the expression, c is the speed of sound in the air (m/s), f_1 is the first harmonic frequency of the space (Hz) and D is the distance between parallel walls in the space in m. Standing waves also occur in multiples of n (2, 3, 4, …) of the first harmonic frequency as higher harmonics.

$$\mathrm{f_n} = \frac{\mathrm{n} \cdot \mathrm{c}}{2 \cdot \mathrm{D}} [\mathrm{Hz}]$$

In the case of real spaces that are not infinitely high and have multiple perpendicular walls, the multiple factor n of the first harmonic frequency is affected by the room's aspect ratio. When one of the dimensions (e.g., the length of the room) is in a 2:1 ratio to the size of other dimensions (width and height), the multiple of the zero resonance frequencies n is equal to the square of 2.

5.5.2 Acoustic Absorbers

The echo and interference of sound waves in rooms decrease the acoustic quality. That is why the sound in the room can be perceived after the sound source has stopped emitting. Nevertheless, the reflections of the sound waves are attenuated, as the surfaces partially reflect and absorb the incoming sound waves. The ratio between the sound power of the reflected $W_{reflected,f}$ and the incoming sound waves $W_{incident,f}$ (Fig. 5.35) defines the sound-absorption coefficient α_f and is determined with the expression:

$$\alpha_f = \frac{W_{incident,f} - W_{reflected,f}}{W_{incident,f}} [-]$$

The index f emphasizes that the sound-absorption coefficient of the surfaces is different at different frequencies of the incident sound waves.

As a rule, "bare" building structures have low sound absorptivity (Table 5.7) for all sound frequencies that can cause echo and interference of the sound. Using sound absorbers installed on building structures or positioned in a room, the sound absorption of different frequencies can be adapted to the requirement of acoustic comfort (Fig. 5.36). The main physical principles we use in the design of sound absorbers are:

- the absorption of high-frequency sound can be reached with porous (fibrous) materials installed on the building structure. If the thickness of the porous layer is equal to a quarter of the wavelength λ of the sound wave, the reflected wave

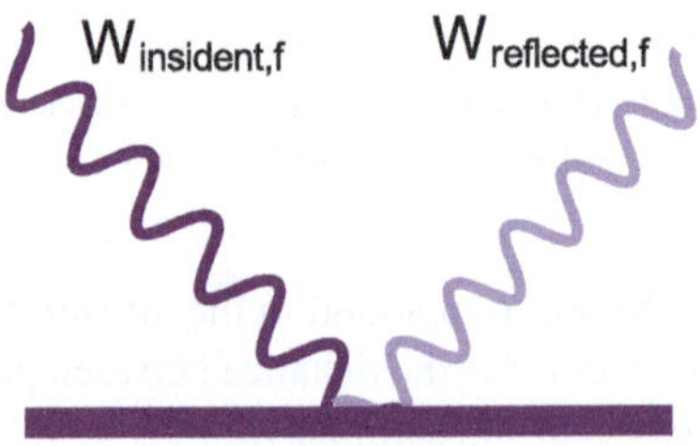

Fig. 5.35 Sound-wave absorptivity α_f is defined by the ratio of the sound intensity of the reflected and incident sound waves. The absorptivity of the surface α_f varies greatly at different frequencies of the sound wave. We use this feature in the design of the sound environment

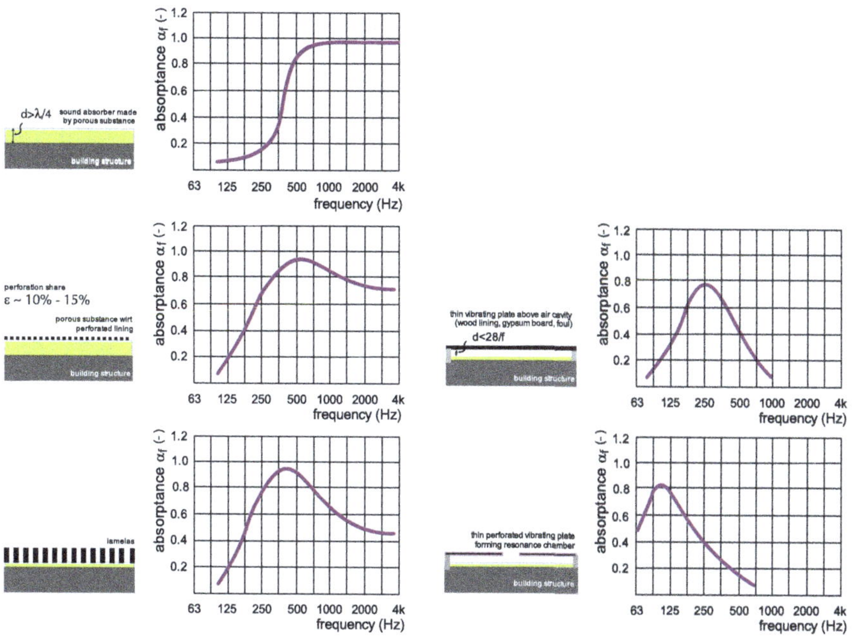

Fig. 5.36 Different physical principles and differently designed sound absorbers are used to adapt the sound-absorption coefficient α_f to the requirements of acoustic comfort [6]

is "subtracted" from the incident wave and therefore totally absorbed. Since, in practice, suitable cladding thicknesses are a few cm, only sound waves with frequencies above 1000 Hz can be absorbed in this way (e.g., sound waves with a frequency of 2500 Hz have a wavelength of about 12 cm, so the thickness of the porous material cladding required to absorb this frequency of sound is ~ 3 cm);

- cladding on the spacers acts as membranes and absorb low-frequency sound. The sound-absorption coefficient depends on the weight of the cladding, the distance between the cladding and the building structure and the type of filler that completely or partially fills the gap between the cladding and the building structure;
- perforated claddings on spacers or on a layer of porous material absorb the mid-range of sound frequencies. Such a sound absorber is also called a Helmholtz resonator. The sound-absorption coefficient can be adapted using the proportion of the surface area of the holes (ε), the distance of the perforated cladding from the building structure and the substance in the gap (this can be air or a porous substance).

In practice, sound absorbers have complex shapes, so their sound-absorption coefficients are determined experimentally. A sample of a sound absorber with an area of ~ 10 m^2 is placed in the reverberation chamber and the reverberation time (see Sect. 5.5.3) is measured in octave frequency ranges. The frequency-dependent

Table 5.7 Sound absorptivity of some building materials, claddings/lining and blocks [14]

materials, linings building block	sound absorption coefficient α_f (-) octave band frequency range (Hz) 125	250	500	1000	2000	4000
concrete	0.02	0.02	0.03	0.04	0.05	0.05
brick	0.02	0.03	0.03	0.04	0.05	0.07
wood on wall or floor	0.03	0.04	0.04	0.05	0.05	0.05
gypsum board	0.02	0.03	0.04	0.05	0.06	0.08
marble, ceramic tiles	0.01	0.01	0.02	0.02	0.02	0.03
PVC, rubber	0.02	0.03	0.03	0.04	0.04	0.05
perforated fiberboard	0.26	0.88	0.99	0.91	1	1
needle carpet	0.03	0.04	0.06	0.20	0.30	0.40
knotted carpet	0.03	0.06	0.10	0.30	0.50	0.60
window, glazing	0.12	0.10	0.05	0.04	0.02	0.02
window, open	1	1	1	1	1	1
door, wooden	0.12	0.10	0.05	0.05	0.05	0.05
gypsum board in suspended ceiling	0.29	0.10	0.05	0.04	0.07	0.09
sprayed cellulose fibers	0.08	0.29	0.75	0.98	0.93	0.76
gypsum board, perforated, 10 cm in front of a concrete wall, with 3 cm mineral wool	0.30	0.69	1	0.81	0.66	0.62
chair, wooden	0.05	0.05	0.05	0.05	0.08	0.05
chair, upholstered	0.15	0.25	0.25	0.15	0.18	0.30
curtins	0.10	0.20	0.30	0.40	0.50	0.50
water surface pool	0.00	0.00	0.13	0.15	0.20	0.25

Details of the various sound absorbers are provided by the manufacturers

sound-absorption coefficients of the element α_f are determined by comparing the measured reverberation time in the reverberation chamber without a sample and with the absorber sample using the Sabine method (see Sect. 5.5.3). According to the sound absorptance, sound absorbers are classified into five classes of sound absorptivity (A–E, Fig. 5.37).

Explanation Due to the simple calculation of the reverberation time using Sabine's method and the relatively small size of the sample of the sound absorber that can be installed in the reverberation chamber, the sound-absorption coefficient can be greater than 1 at a certain frequency, which is contrary to law of the conservation of energy. Therefore, in this case we assume that the value of α_{sab} equals to 1. Instead, sound-absorption coefficients, frequency-dependent equivalent sound-absorption areas A_f of the sample can be defined as well.

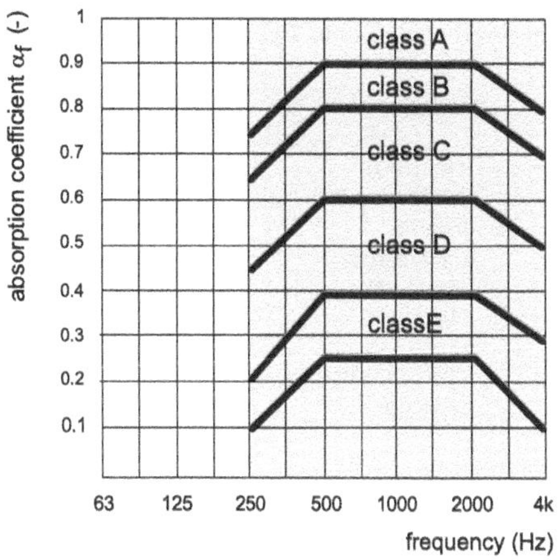

Fig. 5.37 According to the SIST EN ISO 11654 standard, sound absorbers are classified according to their frequency-dependent sound absorptivity α_f into five classes of sound absorptivity

5.5.3 Reverberation Time

After the sound source is interrupted, the sound-pressure level L_p decreases, but not at the same rate in different spaces. These phenomena can be expressed by the reverberation time. Reverberation time is the acoustic property of a particular space and is defined by the time that elapses from the moment that the sound source is interrupted to the moment when the sound level $L_{p,f}$ at frequency f decreases by 60 dB (Fig. 5.38). It is labelled with $T_{60,f}$ and is measured in seconds. It is common that the reverberation time (T_{60}) is stated for a sound frequency of 500 Hz. A too short reverberation time for the room gives the feeling that the room does not respond to sounds, and with a too long reverberation time, disturbing echoes are created. Reverberation times between 0.7 and 1.2 s, 1.2–1.7 s in concert halls and between 1.3 and 2 s in churches are the best for speech perception (Table 5.8) When

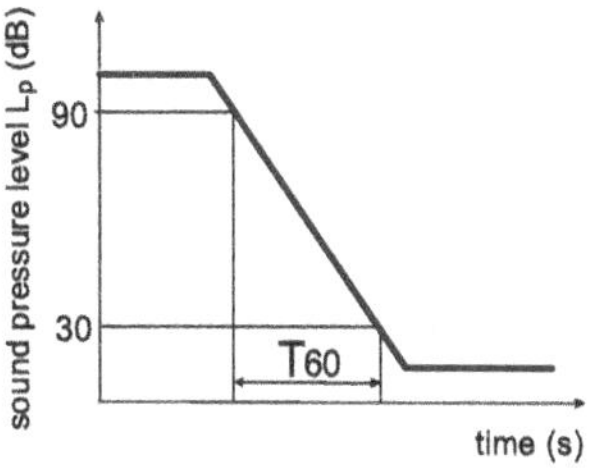

Fig. 5.38 Reverberation time T_{60} is the time that elapses between the moment when a strong sound source is switched off in the room and the moment when the sound-pressure level decreases by 60 dB. As a sound source we use a speaker and a sound generator that emits "white" or "pink" noise

Table 5.8 Recommended reverberation time T_{60} in rooms with different purposes or different sound sources

type of the space	recomended T_{60} (s)
classroom	1.0
conference hall	1.0
thearter	1.0
multi-purpose auditorium	1.3 to 1.5
church	1.4 to 1.6
opera	1.4 to 1.6
philharmonic hall	1.8 to 2.0
chatedral	3.0 or more

Values are for sound with a frequency of 500 Hz

designing room acoustics, we determine the reverberation time $T_{60,f}$ with a simplified calculation based on the Sabine method. The basis for the calculation is the equivalent absorption area A_f (m^2) of all the surfaces in the room, including occupancies. It is determined by the known surfaces of structures S (m^2) and frequency-dependent sound-absorption coefficients α_f.

$$A_f = \alpha_{f,1} \cdot S_1 + \alpha_{f,2} \cdot S_2 + .. + \alpha_{f,n} \cdot S_n \left[m^2\right]$$

Therefore, the equivalent absorption area of a room will not be equal in the case of different sound frequencies. According to the Sabine method, the reverberation time depends on the volume of the room V (m^3) and the equivalent absorption area A_f (m^2) in the following way:

$$T_{60,f} = 0.163 \frac{V}{A_f} [s]; \; T_{60} = 0.163 \frac{V}{A_{f(500\ Hz)}}$$

As rooms are occupied, an additional equivalent absorption surface S_k (m^2) of residents should be added to the room's equivalent absorption surface, taking into account the values from Table 5.9.

The sound-absorption coefficient depends on the frequency of the sound waves. Therefore, the reverberation time will differ at different sound frequencies. The recommended $T_{60,f}$ values for music listening rooms are 10% longer for each octave

Table 5.9 Equivalent absorption surfaces of people on the premises

equivalent sound absorption area S_k (m^2)	sound absorption coefficient α_f (-) octave band frequency range (Hz)					
	125	250	500	1000	2000	4000
person in a choir 0.5 m^2	0.15	0.25	0.40	0.50	0.50	0.60
visitor, walking person 0.65 m^2	0.30	0.45	0.60	0.65	0.75	0.75
musician 2 m^2	0.45	0.65	0.85	1	1	1

lower than 250 Hz and 10% shorter for each octave above 2000 Hz. In rooms intended for speech perception, the optimum reverberation times $T_{60,f}$ are shorter by 10% for each octave below the frequencies 250 Hz and for each octave above 2000 Hz (Fig. 5.39). An adequate reverberation time must be provided for at least 2/3 of the occupancy positions in the room and for each octave range with mid-frequencies between 63 and 4000 Hz.

It is important to note that Sabine's term applies to rooms with a volume of up to approximately 500 m^3, without emphasized dimensions and with evenly distributed sound absorbers. Since these criteria are often not met, the reverberation time is determined by in-situ measurements.

Explanation It is common that the reference reverberation times T_{60} are defined for a frequency of 500 Hz. The proposed values are shown in Table 5.8. It is also common that for residential rooms with a volume up to 200 m^3, the reverberation time should be 0.7 s, and for rooms with a volume between 200 and 500 m^3, 0.8 s with deviation of $\pm$20% is allowed. For larger spaces, we need to use more precise calculation methods and a validation with experiments.

Case Study 1 Determine the reverberation time of a small conference hall 10 m long, 6 m wide and 3.2 m high shown in Fig. 5.40 with concrete walls with a sound absorption coefficient of 0.03. The window in the room has an area of 20 m^2. The sound absorption coefficient of the glazing is 0.05. In the middle of the room there will be a group of 20 seated people. The equivalent

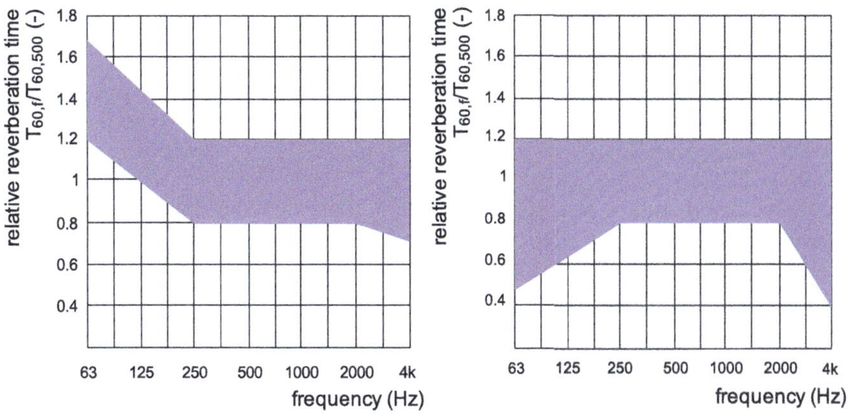

Fig. 5.39 Optimum reverberation time in rooms with different purposes or different sound sources. For mid-size residential rooms (left) and larger spaces and music halls (right) [6]

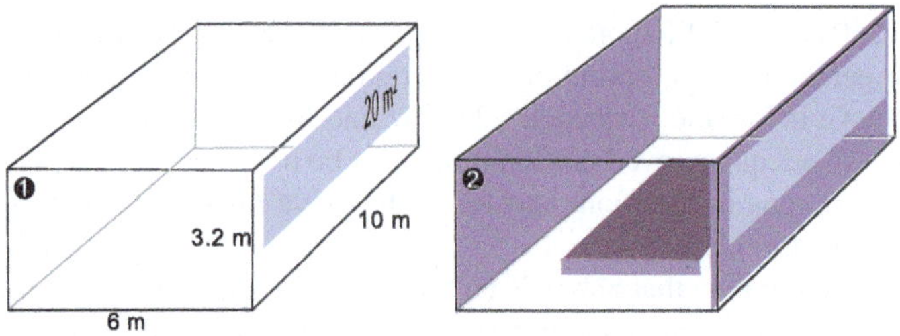

Fig. 5.40 Conference hall analysed in the Case Study 1

area of the listener is 0.65 m^2 and its sound absorptivity is 0.60. Found out if the reverberation time suitable for a conference room, after the back wall is cladded with wood fibre cladding with a sound absorption coefficient of 0.75, and a curtain with a sound absorption coefficient of 0.30 is placed in front of the entire wall with windows. All the absorption coefficients are given for a frequency of 500 Hz.

In the first case, the equivalent absorption surface A_1 and the reverberation time T_{60} is equal to:

$$\begin{aligned} A_1 &= 6 \cdot 3.2 \cdot 2 \cdot 0.03 + 10 \cdot 6 \cdot 2 \cdot 0.03 + 10 \cdot 3.2 \cdot 0.03 \\ &\quad + (10 \cdot 3.2 - 20) \cdot 0.03 + 20 \cdot 0.05 = 7.07\ \text{m}^2 \end{aligned}$$

$$T_{60,500} \equiv T_{60} = 0.163 \frac{V}{A_1} = 0.163 \frac{10 \cdot 6 \cdot 3.2}{7.07} = 4.4\ \text{s}$$

and in a furnished room T_{60} will be:

$$\begin{aligned} A_2 &= 6 \cdot 3.2 \cdot 2 \cdot 0.03 + 10 \cdot 6 \cdot 2 \cdot 0.03 + 10 \cdot 3.2 \cdot 0.75 \\ &\quad + 10 \cdot 3.2 \cdot 0.3 + 20 \cdot 0.65 \cdot 0.6 = 46.15\ \text{m}^2 \end{aligned}$$

$$T_{60} = 0.163 \frac{V}{A_2} = 0.163 \frac{10 \cdot 6 \cdot 3.2}{46.15} = 0.68\ \text{s}$$

The reverberation time of the meeting room will be appropriate. The required reverberation time that ensures good speech perception in the conference hall is $0.7 \pm 20\%$ s. So, between 0.56 and 0.84 s.

Table 5.10 Sound-absorption coefficients of built blocks in the lecture room from case study 2

material, block	sound absorption coefficient α_f (-) octave band frequency range (Hz)					
	125	250	500	1000	2000	4000
concrete	0.02	0.02	0.03	0.04	0.05	0.05
chair, wooden	0.05	0.05	0.05	0.05	0.08	0.05
gypsum board	0.02	0.03	0.04	0.05	0.06	0.08
perforated fiberboard	0.26	0.88	0.99	0.91	1.00	1.00

Table 5.11 Results for Case Study 2: required sound-absorption area of the ceiling of the lecture room $A_{f,ceiling}$ and required sound-absorption coefficients α_f of the ceiling lining

material, block	Sabin's equivalent sound absorption area A (m²) octave band frequency range (Hz)					
	125	250	500	1000	2000	4000
floor (20 m²)	0.4	0.4	0.6	0.8	1.0	1.0
chairs (100 m²)	5.0	5.0	5.0	5.0	8.0	5.0
walls, gypsum boards (146 m²)	2.9	4.4	5.8	7.3	8.4	11.7
walls, perforated board (30 m²)	7.8	26.4	29.7	27.3	30	30
total	16.1	36.2	41.1	40.4	47.8	47.7
required sound absorption area of the ceiling (m²)	81.7	61.6	56.7	57.4	50.0	50.1
required sound absorptance coefficient α_f of the ceiling (-)	0.68	0.51	0.47	0.48	0.42	0.42

Case Study 2 Determine the required class of sound absorber installed on the ceiling of the lecture room with a volume of 480 m³ and a floor area of 120 m². The lecture hall is 4 m high, 12 m long and 10 m deep. The floors are concrete; the chairs are placed on an area of 100 m² of the floor. The walls are partly built with bricks lined with gypsum boards with area of 146 m², and partly, with area of 30 m², with bricks having a cladding of fibrous materials. The sound-absorption coefficients of the building blocks in the lecture room are listed in Table 5.10. The adequate reverberation time of the lecture room at 500 Hz is 0.8 s.

The required equivalent absorption area of the room must be:

$$T_{60} = 0.163\frac{V}{A} \rightarrow A = 0.163\frac{480}{T_{60}} = 0.163\frac{480}{0.8} = 97.8\ \text{m}^2$$

The results are presented in Table 5.11 as the required equivalent absorption area of the ceiling $A_{f,ceiling}$. To ensure such an equivalent absorption area of the ceiling, the ceiling lining must have sound-absorption coefficients α_f listed in Table 5.11.

From Fig. 5.37 it can be read that the ceiling surface will need to have Class-E sound absorptivity.

In complex and larger spaces, the reverberation time $T_{60,f}$ is evaluated using in-situ experiments. The sound source that emits "white" or "pink" noise is placed in the room (Fig. 5.41). The sound-pressure level is measured with a microphone and analysed in tertiary frequency. Because the background sound (for example, noise coming into the space from the outdoor environment) is often so large that we cannot ensure a reduction in the sound level by 60 dB after sound source is switched off, the reverberation time $T_{60,f}$ is determine by doubling the time it took to reduce the sound-pressure level by 30 dB ($T_{30,f}$) [15, 16]:

Fig. 5.41 Plečnik's lecture room at the Faculty of Architecture, University of Ljubljana. An isotropic source of pink noise that emits sound evenly in all directions was used as the sound source (top, right). Sound-pressure level was measured at two places, in the corner of the first and the last row (right)

$$T_{60,f} = 2 \cdot T_{30,f}\ [s]$$

Case Study 3 [4] The reverberation time for the lectures of the Faculty of Architecture, University of Ljubljana, was determined using an in-situ experiment in an empty room. A noise source that emits "pink" noise evenly in all directions was installed (Fig. 5.41, top right) in the speaker's position. The sound-pressure level was measured in two places, in the first and in the back row of seats (Fig. 5.41, bottom). The reverberation time was determined for a frequency range 50 and 10,000 Hz.

Solution: The decrease of the sound-pressure level $L_{p,f}$ after the source was switched off is presented in Fig. 5.42. The reverberation times $T_{60,f}$ determined by doubling $T_{30,f}$ are shown in Table 5.12. T_{60} at a frequency of 500 Hz at the position of the first row is 0.88 and in the last row 0.98 s. This is ideal, given the size and purpose of the room. The reverberation time $T_{60,f}$ at other frequencies is adequate as well, since it increases at sound frequencies below 250 Hz and decreases at sound frequencies above 5000 Hz in relation to the reference value T_{60} (Fig. 5.39, left).

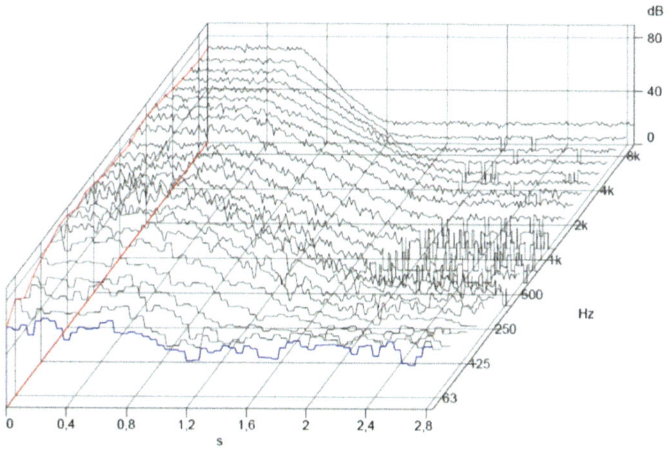

Fig. 5.42 Measured reduction of sound-pressure level at medium frequencies of the tertiary frequency ranges after switching off the sound source

Table 5.12 Measured reverberation times in tertiary frequency bands

frequency (Hz)	reverberation time (s) place 1	place 2	frequency (Hz)	reverberation time (s) place 1	place 2
50	1.46	1.43	800	0.89	0.88
63	1.44	1.47	1000	0.94	0.97
80	1.27	1.17	1250	1.02	1.00
100	1.36	1.19	1600	1.11	1.15
125	1.15	1.08	2000	1.26	1.22
160	1.36	1.28	2500	1.25	1.23
200	0.97	1.17	3150	1.16	1.20
250	0.99	0.92	4000	1.07	1.08
315	1.07	1.04	5000	1.01	0.96
400	1.01	1.01	6300	0.81	0.76
500	0.88	0.98	8000	0.7	0.67
630	0.99	0.98	10000	0.53	0.52

Values are the average values of five consecutive measurements [4]

5.5.4 *Acoustic Room Design*

An adequate reverberation time is just one of the conditions for a good acoustic environment in buildings. At the same time, we must also ensure that all the listeners, regardless of their place in the room, receive a sound of the same intensity. We also need to ensure an appropriate ratio between the direct sound coming from the source and the indirect sound reflected on the surfaces. The starting point for acoustic planning is the purpose of the room. Above all, it is important to know whether the room will be for listening to speech or music, and in this case how many and which sound sources will be in the room. Another factor in terms of acoustic comfort is the volume of the room that "belongs" to the individual listener (Fig. 5.43). Thus, the volume of the room in which we listen to speech should be around 4 m^3 per person, and not less than 8 m^3 per listener in a concert hall, as larger areas are needed to reflect the sound waves to create a diffuse sound field. The size of the room is also limited according to the volume of sound emitted by sound sources without

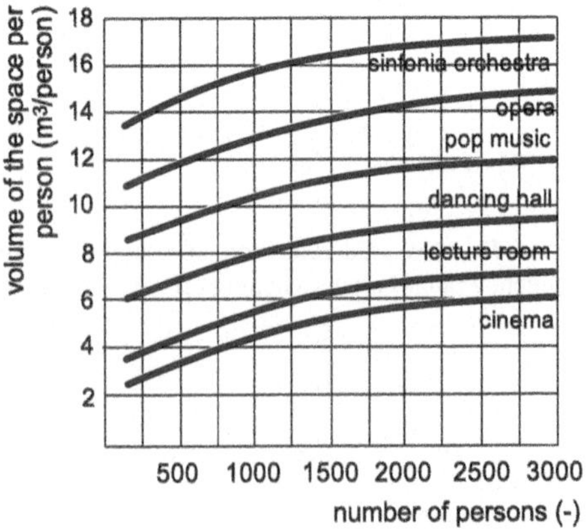

Fig. 5.43 Recommended volume of space per listener according to the purpose of the space

Fig. 5.44 In rooms with non-parallel walls, it is easier to avoid resonance (left). Sound absorbers ensure that the time difference between the moments when direct and indirect sound waves are detected is so short that sound monitoring is not disturbed. They should be installed on the ceiling and at the back of the room (right)

electronic devices. Thus, the lecture halls should not be larger than 3000 m^3, concert halls not larger than 10,000 m^3 and symphony halls no larger than 25,000 m^3 [17].

In addition to size, the shape of the room is also important. To avoid the occurrence of resonance, the room should have a ratio of sides in unequal multiples. Even better, the surfaces should not be completely parallel (a 5° offset is sufficient). The time difference between the moment when direct sound is detected (emitted by the sound source) and the moment when indirect sound is detected (reflected sound waves on the surfaces in room) should be less than 80–100 ms to avoid the echo effect. Listeners should receive sound in the range of $\pm$ 45° from the sound source. So, a trapezoidal shape of a larger space is more appropriate than a rectangular one. The sound absorbers should be installed on the ceiling and next to the back wall (Fig. 5.44), while sound reflectors should be placed above the sound source and above the remote listeners (Fig. 5.45). As a rule, concave elements that condense the sound field in a narrow area in the space should be avoided. If we want to extend the reverberation time, electronic devices must be installed.

5.5.5 *Other Indicators of Sound Comfort in Rooms*

To assess the sound comfort in larger and open rooms, such as ocean-type offices, we also use some other indicators to assess the acoustic comfort. The Articulation Index AI is defined by the proportion of spoken words of the speaker that the listeners recognize (Fig. 5.46). Because listeners' abilities to recognize sound differ, this is a subjective method. It is being replaced by methods in which a mechanical sound generator imitates human speech, which is detected by microphones placed in different parts of the room. The most widely used method is the rapid-speech transmission index (RASTI). Both the AI index and the RASTI index are represented by numbers between 0 and 1. The values of both indices should be at least 0.7.

Fig. 5.45 With sound reflectors, we direct the sound wave emitted by the source to the listeners. Sound reflectors are usually flat or convex, so as not to create a condensed field of sound pressure in the room. Reflectors are placed above the sound source (left) and above more distant listeners (right)

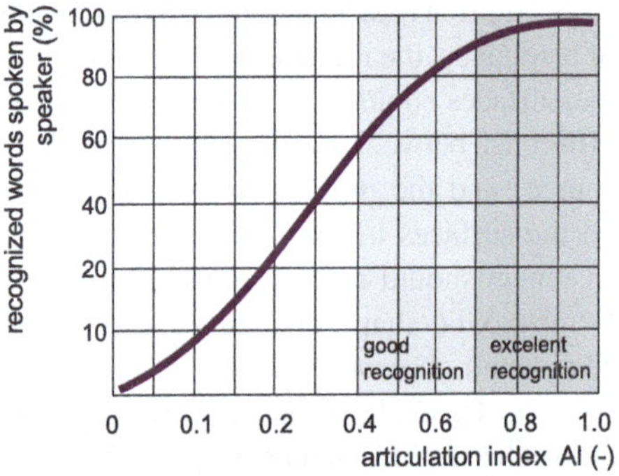

Fig. 5.46 The AI index is determined based on the recognized words spoken by the speaker. In speech rooms, an AI index value of at least 0.6 must be ensured

A very effective method for increasing speech recognition is sound masking. In this case, with an additional sound source, such as a number of speakers, we cause the disturbing sound, such as the speech of a roommate, not to be perceived as disturbing, as it "intertwines" with the planned background sound. An example of sound masking is listening to music while learning. Sound masking is mainly used in ocean-type offices, where the sound-pressure level in the place of the listener is either too low and the sound even from a very distant workplace can be detected, or it is too high and disturbs concentration at work. We use sound masking to ensure acoustic comfort in public spaces as well to ensure that a conversation between two people is not distinguished by other visitors.

5.6 Sound Propagation in a Building

Sound propagates in buildings as airborne sound through sound waves in the air and as structure-borne or impact sound by mechanical vibrations of building structures (Fig. 5.47), potentially causing a sound intensity that disturbs occupants at work and rest, or even causes temporary or permanent hearing impairment. To prevent such conditions in an indoor environment, building structures must have adequate sound insulation against airborne and impact sound that propagate from surrounding spaces and the outdoor environment.

The intensity of the disturbing sound known as noise, depends on the occupants' activities. Because activities differ during the day, the permissible A-weighted equivalent sound-pressure levels $L_{eq,A,F}$ differ according to the period of the day as well. Permissible values in living rooms and working spaces between 6 a.m. and 10 p.m. are expressed as day-time permissible values $L_{eq,A,F,d,\max}$, while values $L_{eq,A,F,n,\max}$ are permissible for the night-time period. The latter is defined by largest hourly value during the time period between 10 p.m. and 6 a.m. the next morning. In-situ equivalent sound-pressure values are averaged spatially, based on the measurements at 3 or 5 points in furnished rooms. The most common limit values of A-weighted sound-pressure levels that do not yet disturb the occupants are shown in Table 5.13.

The airborne sound that is generated by acoustic devices or other noise sources causes "background noise" that disturbs our speech perception. Besides the intensity of the background noise, the speech recognition depends on the distance between the speaker and the listener and is given in Table 5.14.

The structure-borne or impact sound originates from the mechanical force acting on the building structure and it propagates through structures by mechanical vibrations. This can be due to walking, bumps, or oscillations of devices installed in the room or in surrounding spaces. Part of the mechanical energy that is transmitted

Fig. 5.47 Sound propagates in buildings as airborne and structure-borne (impact) sound [18]

Table 5.13 Permissible A-weighted equivalent sound-pressure levels during day-time ($L_{eq,A,F,d,\max}$) and during the night-time ($L_{eq,A,F,n,\max}$) in rooms with different purposes

type of the room, building	permissible equivalent sound pressure level $L_{eq,A}$ (dB(A)) daytime	night-time (maximum during one hour interval)
residental building - living rooms	35	30
room in hotel or inn	35	30
hospital room	30	30
lecture room, reading room, cabinets, library	35	35

Note Permissible values are stated for the spaces with furniture; in unfurnished premises, the values can be higher by 5 dB(A) due to the sound absorptance [19]

Table 5.14 The comprehension of speech is influenced by the sound-pressure level of the background noise and the distance between the speaker and listener

sound pressure level L_p dB(A)	comprehension of speech according to the background sound and didtance between speaker and listener 2 m	4 m	6 m	10 m
30	good	good	good	good
40	good	good	good	good
50	good	possile	possible	poor
60	possible	poor	poor	difficult
70	poor	difficult	difficult	difficult
80	difficult	difficult	difficult	difficult
90	difficult	difficult	not possible	not possible
100	difficult	not possible	not possible	not possible

through building structures is converted at the building structures' surfaces into sound waves, causing an increased intensity of the airborne sound. The intensity of the impact sound is presented as the level of normalized impact sound and is denoted by L'_n. Just like the sound-pressure levels of airborne sound, impact sound can also be perceived as noise at sound-pressure levels above those shown in Table 5.15. The influence of impact sound on the acoustic comfort is evaluated by the permissible, normalized, impact sound-pressure level $L'_{n,\max}$. In-situ levels of impact sound are determined with a standardized sound source that mimics walking (Fig. 5.63). The procedure by which the normalized and weighted values of the impact sound are found from the measurements of the sound-pressure level is shown in the Sect. 4.6.6.

Table 5.18 Case study of the required sound-insulation properties of selected building structures

separating building structure	sound insulation requirement parameter	
wall between the apartments	R'_w	52 dB
wall without door between common staircase or corridor	R'_w	52 dB
wall between the apartment and elevator shaft	R'_w	52 dB
wall between the industrial zone and apartment	R'_w	62 dB
wall between the apartment and less noisy engine room	R'_w	57 dB
wall between room in hotel	R'_w	52 dB
entrance door from common staircase	R'_w	37 dB
wall between hotel room and common staircase	$D_{nT,w}$	45 dB
mezzanine structure between the apartments	R'_w	52 dB
	$L'_{n,w}$	55 dB
mezzanine structure between the apartment and restauran bellow	R'_w	62 dB
	$L'_{n,w}$	58 dB
mezzanine structure between the apartment and restauran above	R'_w	62 dB
	$L'_{n,w}$	43 dB

the transmission of structure-borne impact sound. Examples of common permissible values are shown in Table 5.18. Nevertheless, different values could be required in the national regulative documents.

Explanation The methods for determining the weighted sound-reduction index R_w, weighted standardized sound-level difference $D_{nT,w}$ and weighted normalized impact sound-pressure level $L_{n,w}$ are presented in the Sects. 4.6.3, 4.6.4 and 4.6.6.

Explanation: For building structures in which airborne and structure-borne sound are normally transmitted, both requirements are defined.

5.6.3 *Determination of the Sound-Reduction Index of Building Structures (R, R′ and R′$_w$)*

The sound insulation of building structure can be determined by numerical models or, more often, by experiments. In the case of a laboratory experiment the building structure is installed between two spaces in a way that the sound energy between the spaces is transmitted only through the tested structure and not along the other walls (Fig. 5.52).

In one of the spaces an isotropic sound source that emits "white" or "pink" noise is installed, creating the sound-pressure level L_1, and in the other one the sound-pressure level L_2 caused by sound's transmission through the tested structure is measured. This space must have the property of a reverberation chamber (see Sect. 4.5.1.1), so the

Fig. 5.52 Schematic representation of laboratory test conditions for determining the sound-reduction index *R* of a building structure

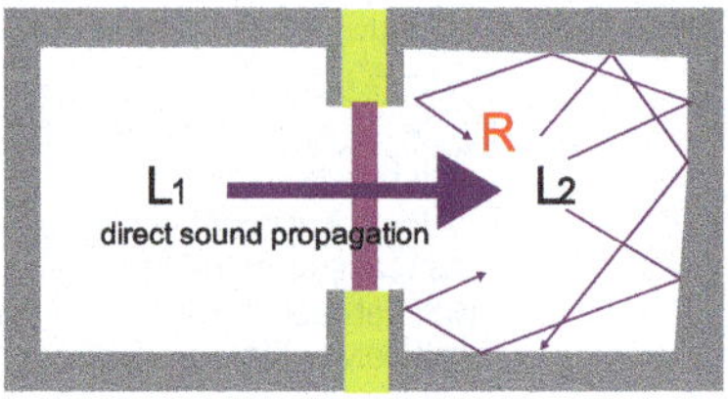

sound field is diffuse (uniform) and we can measure the sound-pressure level L_2 at any point in the space.

The sound-reduction index *R* (dB) of the building structure is equal to the difference between the sound-pressure level in the space with the sound source L_1 and the sound-pressure level L_2 in the sound-receiving space:

$$R = L_1 - L_2 [dB]$$

The sound-pressure levels L_1 and L_2 are measured at discrete frequencies of the sound waves emitted by the sound source. These can be selected as the midrange of the octave or tertiary bands, normally between 100 and (3150) 4000 Hz. An example of the sound-reduction index *R* for single and double glazing is shown in Fig. 5.53.

The transmission of airborne sound, however, is slightly different under in-situ conditions, compared to the laboratory experiments. In buildings, the sound is also transmitted through the joints as flanking sound. A more realistic in-situ sound-reduction index of the building structure is determined similarly as in case of laboratory test, nevertheless flanking sound transmission and the acoustics of the furnished

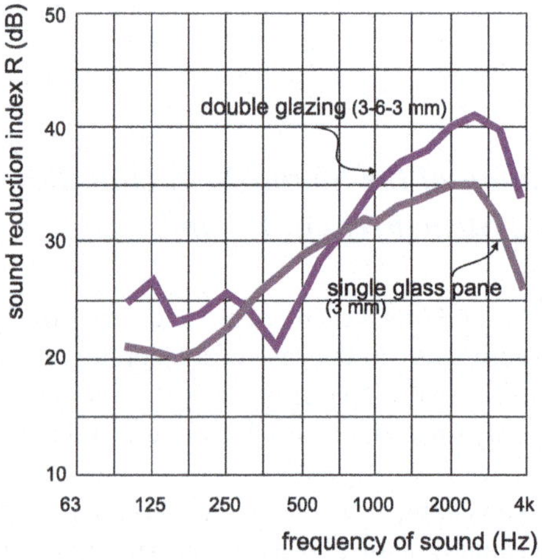

Fig. 5.53 Sound-reduction index *R* of single 3 mm-thick glass pane and double glazing consisting of double 3 mm glass pane and 6 mm-thick air gap. Values were determined by laboratory experiment in tertiary frequency ranges [18]

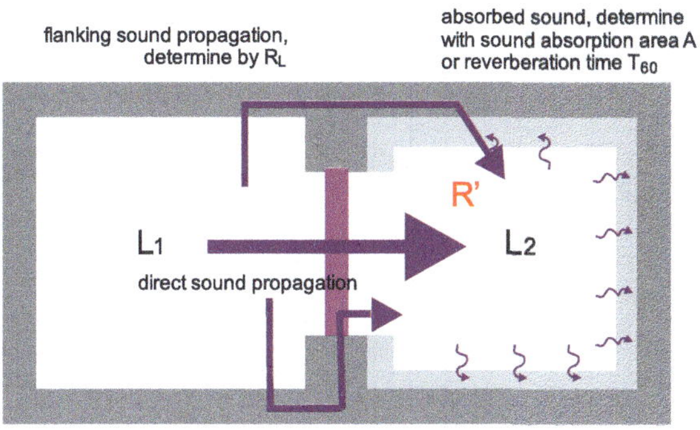

Fig. 5.54 Schematic representation of the conditions under the apparent sound-reduction index R' of a building structure is determined

receiving room are taken into account. The sound-reduction index measured "in-situ" is designated as R'(dB) and presented as the apparent sound-reduction index.

The apparent sound-reduction index R' of a building structure is evaluated using the pressure level L_2 in the sound receiving space, which is lower compared to the reverberation chamber in laboratory test conditions due to the absorption of sound in the walls (Fig. 5.54). The amount of absorbed sound energy depends on the surface area of the structure S_o (m^2), for which the sound insulation is determined and the equivalent absorption surface area A (m^2) of the remaining walls in the sound receiving space, which equals to $\Sigma\alpha_{i.}S_i$ according to Sabine's approach (see Sect. 4.5.1). The apparent sound-reduction index R' is determined by correcting the measured sound-pressure level L_2 with respect to the absorption of sound energy in the elements of the sound receiving room:

$$\mathrm{R}' = \mathrm{L}_1 - \mathrm{L}_2 + 10 \cdot \log\left(\frac{\mathrm{S}_\mathrm{o}}{\mathrm{A}}\right) \quad \mathrm{R} = \mathrm{L}_1 - \mathrm{L}_2 - \mathrm{R}_\mathrm{L} + 10 \cdot \log\left(\frac{\mathrm{S}_\mathrm{o}}{\mathrm{A}}\right)[\mathrm{dB}]$$

while the sound-reduction index R is determined by considering the correction term of flanking sound propagation R_L (dB) with an approximated value of 2 dB for walls, ceiling and windows and 5 dB for doors. Both the sound-reduction index R and the apparent sound-reduction index R' of the building structure are measured at discrete midrange frequencies of the octave or tertiary bands in the range between 100 and (3150) 4000 Hz.

For a simpler assessment of the sound insulation of building structures, the sound-reduction index R and apparent sound-reduction index R' are shown with one figure as the weighted sound-reduction index R_w (dB) and the weighted apparent sound-reduction index R'_w(dB). The procedure for the determination is the same for both indexes and is shown for the apparent sound-reduction index below [20]:

- the values of the apparent sound-reduction index R' of the building structure, measured in tertiary frequency ranges, are compared with the reference sound-insulation curve N for airborne sound (Fig. 5.55). The reference curve follows the

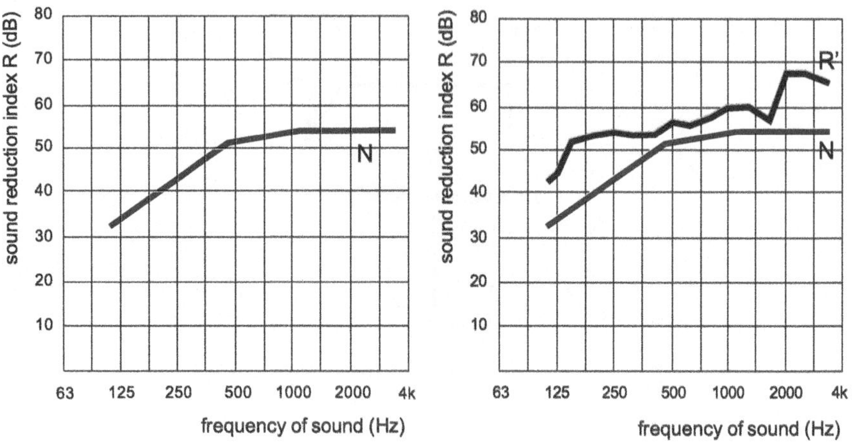

Fig. 5.55 Reference sound insulation curve N for airborne sound is compared with values of previously determined apparent sound-reduction index R' of the building structure

weighted curve A of the sound-pressure level; therefore, a lower sound-reduction index is required at lower than at higher sound frequencies;

- the reference sound insulation curve N is moved by 1 dB against the curve of the measured apparent sound-reduction index R' and for each tertiary frequency range the difference in dB between both curves is calculated ($\Delta = R' - N$) and summarized over the whole frequency range. The only difference when R' is lower than N is taken into account, so the difference between the curves in dB is always negative;
- the last shift of the reference curve N, before the sum of the differences ($\Sigma\Delta$) becomes less than $-$ 32 dB, determines the final position of the reference curve of sound insulation N' (Fig. 5.56, left). An additional condition is that the absolute sum ($\Sigma\Delta$) divided by the number of frequency ranges (e.g., 16) does not exceed 2 dB;
- at a sound frequency 500 Hz the weighted apparent sound-reduction index R'_w of the building structure for airborne sound is determined according to the final position of the reference sound-insulation curve N' (Fig. 5.56, right).

Sound-reduction indexes R and weighted sound-reduction index R_w are determined in the same way, but in this case based on the data from laboratory experiments. For selected building structures they are shown in Table 5.19.

Glazing is one of the weakest elements regarding sound insulation against outdoor (municipality) noise. In the case of double glazing, which is still the most commonly used today, the sound-reduction index can decrease due to the resonance of the glass panes. The resonant frequency can be controlled by filling the gap between the glass panes with a heavy gas instead of the air. In the past, sulphur hexafluoride (SF_6) was used and the sound insulation in the mid-range and high noise frequencies was increased by up to 20 dB. The use of SF_6 gas is banned today, as it is a very impactful

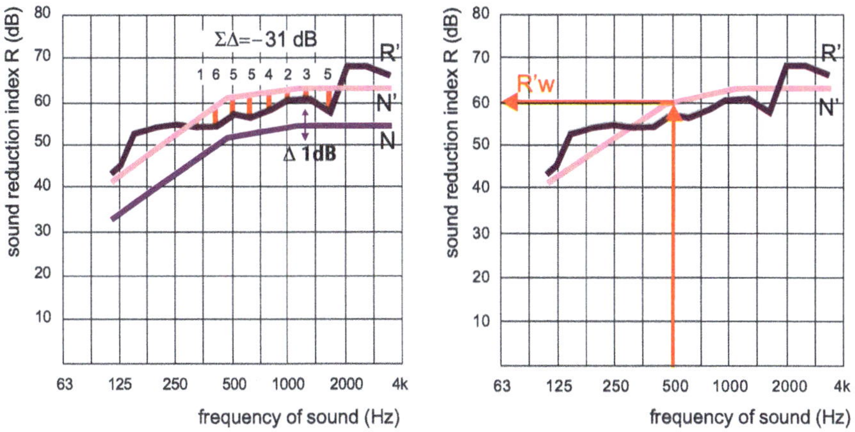

Fig. 5.56 Reference sound-insulation curve N is moved by 1 dB relative to the curve of the measured apparent sound-reduction index R' until the sum of the differences ($\Sigma\Delta$) becomes less than—32 dB (left). The weighted apparent sound-reduction index R'_w is read from the final position of the reference curve (N') at a sound frequency of 500 Hz (right)

Table 5.19 Sound-reduction index R and weighted sound-reduction index R_w of selected building structures [21, 22]

building structure	R' (dB)						R'w (dB)
	125 Hz	250 Hz	500 Hz	1000 Hz	2000 Hz	4000 Hz	
drywall structure 2×13 mm, gypsum plasterboard on wooden beams	17	31	33	40	38	36	33
drywall structure 2×13 mm, gypsum plasterboard on wooden beams with 50 mm fibrous thermal insulation	15	30	34	44	46	41	37
drywall structure 2×26 mm, gypsum plasterboard on wooden beams	25	34	41	51	48	50	41
drywall structure 2×13 mm, gypsum plasterboard on flexible supports without sound bridges	23	28	39	46	54	44	39
drywall structure 2×13 mm, gypsum plasterboard on flexible supports without sound bridges with 50 mm fibrous thermal insulation	29	38	45	52	58	50	48
drywall structure 2×13 mm, gypsum plasterboard on metal beams	22	27	43	47	37	46	39
drywall structure 2×13 mm, metal beams, 50 mm fibrous therma insulation	26	41	52	54	45	51	45
200 mm concrete wall	36	44	50	54	58	56	53
100 mm brick wall	32	34	40	47	55	61	45
44mm thick wooden door	14	19	23	18	17	21	19
44mm thick massive wooden door	29	31	31	31	39	43	34

greenhouse gas. Instead, glazing today is made of laminated multilayer glass pane or by glass panes with different thicknesses. The sound-reduction index of triple-glazing is higher, but only if the sum of the glass thicknesses and the sum of the gaps between the panes are greater than that of double-glazed glazing (Fig. 5.57). Examples of the sound insulation properties of contemporary glazing are given in Table 5.20. The sound insulation of the windows is also influenced by the frame

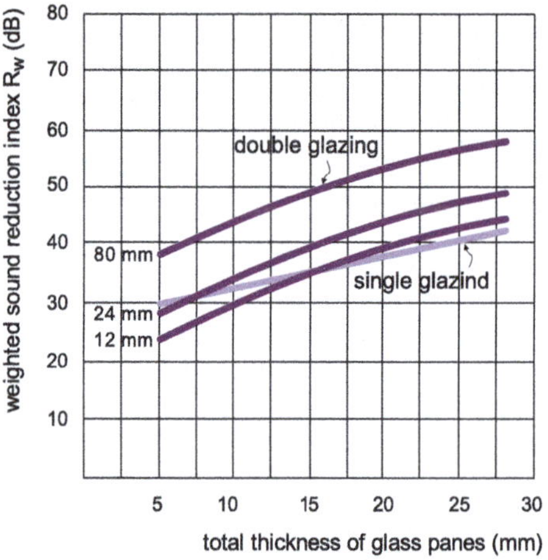

Fig. 5.57 Weighted sound-reduction index R_w of single and double glazing with different thickness of glass panes and width of the gap between glass panes [23]

Table 5.20 Sound insulation properties of selected glazing characterized by the sound-reduction index R and the weighted sound-reduction index R_w [21, 24, 25]

type of glazing	sound reduction index R (dB) octave band frequency range (Hz)						R_w (C,C $_{tr}$)
	125	250	500	1000	2000	4000	(dB)
single glass pane4 mm	17	20	26	32	33	26	29 (-2, -3)
single glass pane6 mm	18	23	30	35	27	32	31 (-2, -3)
single glass pane10 mm	23	26	32	31	32	39	33 (-2, -3)
double glazing 4 mm / (6-16$^{\text{air gap thickness}}$) mm / 4 mm	21	17	25	35	37	31	29 (-1, -4)
double glazing 8 mm / (6-16$^{\text{air gap thickness}}$) mm / 6 mm	20	21	33	40	36	48	35 (-2, -6)
double glazing 10 mm / (6-16$^{\text{air gap thickness}}$) mm / 6 mm	24	24	32	37	42	43	35 (-1, -3)
triple glazing 4 mm/ 16 mm / 4 mm / 16 mm / 4 mm	21	22	33	39	37	47	35 (-2, -4)
double glazing 10 mm / 12 mm / 8,8 mm $^{\text{2×4 laminated}}$							42 (-2, -5)
double glazing 10 mm / 16 mm / 16,8 mm $^{\text{2×8 laminated}}$							45 (-1, -5)
double glazing 8,8 mm$^{\text{2×4 laminated}}$ / 20 mm / 10,8mm $^{\text{2×5 laminated}}$							47 (-2, -7)

Spectrum adaption terms C and C_{tr} are shown [23, 26]

and the tightness of the windows, but only if the weighted sound-reduction index of glazing R_w is above 40 dB [23].

Case Study The in-situ measured sound-reduction index R' of the ceiling structure between two apartments shown in Fig. 5.58. are listed in Table 5.21. Determine the weighted apparent-sound-reduction index R'_w for the building structure and verify that the structure meets the minimum requirements for the sound insulation of airborne noise. The minimum required weighted apparent-sound-reduction index $R'_{w,\min}$ is presented in Table 5.21.

Solution: The weighted apparent-sound-reduction index of airborne sound R'_w is equal to 60 dB (Fig. 5.59). Because the required minimum weighted apparent-sound-reduction index for such a type of building structure $R'_{w,\min}$ is 52 dB, the sound insulation of the building structure is adequate.

To achieve the appropriate acoustic comfort regarding the exposure to the airborne sound, building structures must have an adequate weighted apparent sound-reduction index R'_w. Table 5.22 presents the common required values for different types of building structures [27].

For some building elements, such as doors or windows, it is typical for manufacturers to state the weighted sound-reduction index for airborne sound R_w based on laboratory tests. The in-situ sound-reduction index of products will be lower because of the flanking sound propagation. The difference between 2 dB (for windows) and 5 dB (for entrance doors) can be assumed (Fig. 5.22, right).

The sound-reduction index R and the apparent sound-reduction index R' are determined in labs or in-situ by using a sound source that emits "white" or "pink" sound. The noise in real environment can have the property of an impulse noise and can contain accented tones (see Sect. 4.2.2). Such noise is more disturbing and because of that, the sound-reduction index of building structures must be increased to maintain the acoustic comfort. To compensate for such noise conditions when the weighted apparent-sound-reduction index is evaluated, the spectrum adaptation

Fig. 5.58 Case study building structure consist of four layers

	layers	d (cm)
1	Hard wood	1.5
2	Rock wool insulation	7
3	Concrete	25
4	Plaster	1

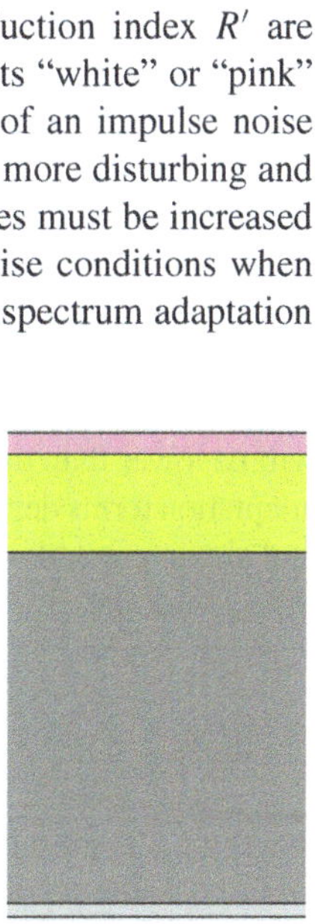

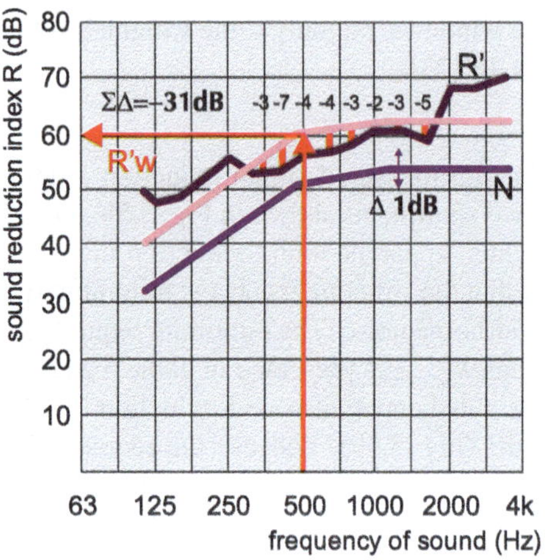

Fig. 5.59 Determination of the weighted apparent sound reduction index R'_w of case study building structure according to the final position of reference curve N'

Table 5.21 Apparent sound-reduction index R', values of reference curve N and procedure showing determination of weighted apparent sound-reduction index R'_w of the case-study building structure in Fig. 5.58

frequency (Hz)	100	125	160	200	250	315	400	500	630	800	1000	1250	1600	2000	2500	3150
R' (dB)	50	48	49	53	55	53	52	56	57	59	61	61	59	68	68	70
N (dB)	33	36	39	42	45	48	51	52	53	54	55	56	56	56	56	56
R' -(N+4dB)							-3						-1			
R'-(N+5dB)							-4	-1	-1				-2			
R'-(N+6dB)						-1	-5	-2	-2	-1		-1	-3			
R'-(N+7dB)						-2	-6	-3	-3	-2	-1	-2	-4			
R'-(N+8dB)						-3	-7	-4	-4	-3	-2	-3	-5			
R'-(N+9dB)						-4	-8	-5	-5	-4	-3	-4	-6			

terms C and C_{Tr} were introduced. The adaptation term C takes into account the extended frequency range (especially below 100 Hz) when the sound originating from an orchestra, radio or speech, while C_{Tr} is used to evaluate the apparent sound-reduction index of the structures exposed to traffic noise. Both terms are negative, which means that the weighted apparent sound-reduction index of building structures will be lower than the one determined using a "white" or "pink" sound source. Both adaptation terms depend on the building structure's weight and the windows' glazing, and are, in general, in the range between 0 and − 10 dB. The sound-reduction index of the building structure $R'_{w,\min}$ must be at least equal to the calculated weighted apparent-sound-reduction index $R'_{w,r}$ determined by the expression.

$$R'_{w,\min} \le R'_{w,r} = R'_w + C + C_{tr} = R_w - R'_d + C + C_{tr} = L_{eq,1} - L_{eq,2}[\mathrm{dB}]$$

where $L_{eq,1}$ is the actual equivalent sound-pressure level in the outdoor environment and $L_{eq,2}$ is the permissible equivalent sound-pressure level in the room according to the type of the space (Table 5.13). Because $L_{eq,2}$ is the default state for a typically

Table 5.22 Example of required weighted apparent sound reduction indexes $R'_{w,\min}$ of selected building structures (left); required weighted sound-reduction indexes $R_{w,\min}$ for doors and windows (right)

building structure	R'w (dB)
wall between apartment	52
wall between apartment and corridor	52
wall between apartment and elevator shaft	52
wall between apartment and mechanical room	57
wall between apartment and noisy restaurant	62
mezzanine structure between apartments	52
mezzanine structure between apartment and office	57
mezzanine structure between apartment and noisy restaurant	62

element	R_w (dB)	R'w (dB)
door to the apartment with entrance hall	37	32
door to the apartment toward living space	47	42
windows and balcony door (outdoor equivalent sound pressure level $L_{eq,A,F,day}$ 65 dB(A))	34	32
windows and balcony door (outdoor equivalent sound pressure level $L_{eq,A,F,day}$ 75 dB(A))	44	42

When installed in buildings the weighted apparent-sound-reduction indexes R'_W will be lower due to the flanking sound propagation [19]

furnished room, the absorption of the sound in the sound receiving room is not taken into account. If the weighted sound-reduction index $R_{w,\min}$ is required instead $R'_{w,\min}$, *a* correction term of the flanking sound $R'_{L,w}$ (dB) must be included. If no actual data about $L_{eq,1}$ are available, the permissible values defined for a different noise-protection area (Table 5.6) can be used.

Case Study A concrete wall with thickness of 16 cm has a weighted sound-reduction index R_w 61 dB, C – 1 dB and C_{Tr} –6 dB. The correction term of the flanking sound propagation $R'_{L,w}$ (dB) is 2 dB.

In the case that such construction is a partition wall between two apartments, the calculated weighted apparent sound-reduction index $R'_{w,r}$ is 58 dB (61 dB – (2 dB) + (– 1 dB)), while in the case of envelope structure exposed to the traffic noise $R'_{w,r}$ is 53 dB (61 dB – (2 dB) + (– 6 dB)).

5.6.4 *Weighted Standardized Sound-Level Difference $D_{nT,w}$ of Building Structures*

With the measured of the sound-pressure level in the sound source room L_1 and in the sound receiving room L_2, we can also determine the standardized sound level difference D_{nT}, provided by the building structure instead of apparent sound-reduction index R'. It is defined by the expression:

$$D_{nT} = L_1 - L_2 + 10 \cdot \log\left(\frac{T_{60}}{T_{60,o}}\right) [dB]$$

where L_1 (dB) is the sound-pressure level in the sound source room, L_2 (dB) is the sound-pressure level in the sound receiving room, while the last term in the equation represents the correction factor of sound absorption in the sound receiving room. In this case, the correction is based on the actual reverberation time of the sound receiving room T_{60} and the reference reverberation time $T_{60,0}$. $T_{60,0}$ is presented in Table 5.22, and it is equal to 0.5 s in residential buildings. Because L_1 and L_2 are measured in-situ, D_{nT} also includes flanking sound propagation ($R'_{L,w} = 0$ dB) [28]. The relation between sound-reduction index R' and standardized sound level difference D_{nT} can be found as follows:

$$R' = L_1 - L_2 + 10 \cdot \log\left(\frac{S_o}{A}\right) \rightarrow L_2 = L_1 - R' + 10 \cdot \log\left(\frac{S_o}{A}\right)$$

$$D_{nT} = L_1 - L_1 + R' - 10 \cdot \log\left(\frac{S_o}{A}\right) + 10 \cdot \log\left(\frac{T_{60}}{T_{60,o}}\right) [dB]$$

$$D_{nT} = R' - 10 \cdot \log\left(\frac{S_o}{A}\right) + 10 \cdot \log\left(\frac{T_{60}}{T_{60,o}}\right) [dB]$$

$$R' = D_{nT} + 10 \cdot \log\left(\frac{S_o}{A}\right) - 10 \cdot \log\left(\frac{T_{60}}{T_{60,o}}\right) = D_{nT} + C_p [dB]$$

Taking into account Sabine's equation for calculating the reverberation time, the correction factor C_p is equal to.

$$C_p = 10 \cdot \log\left(\frac{S_o}{A}\right) - 10 \cdot \log\left(\frac{T_{60}}{T_{60,o}}\right) = 10 \cdot \log\left(\frac{S_o \cdot T_{60,o}}{A \cdot T_{60}}\right)$$
$$= 10 \cdot \log\left(\frac{S_o \cdot T_{60,o}}{A \cdot 0.163 \frac{V}{A}}\right)$$

$$C_p = 10 \cdot \log\left(\frac{S_o \cdot T_{60,o}}{0.163 \cdot V}\right) = 10 \cdot \log(S_o) + 10 \cdot \log\left(\frac{T_{60,o}}{0.163}\right) - 10 \cdot \log(V) (dB)$$

Table 5.23 Reference reverberation times according to CH SIA-Norm 18; standard EN ISO 717-1 states only the reference reverberation time of 0.5 s for residential buildings [6]

volume of the V (m^3)	$T_{60,o}$ (s)
up to 100	0.5
between 100 in 2500	$\frac{\sqrt{V}}{20}$
over 2500	2.5

where S_o (m^2) is the area of the building structure for which we determine the apparent sound-reduction index R', and V (m^3) is the volume of the sound receiving room. The reference reverberation time $T_{60,0}$ (s) for different buildings are given in Table 5.23.

For some building constructions, national regulative defines permissible weighted standardized sound level difference $D_{nT,w,\min}$ instead of weighted apparent sound-reduction index $R'_{w,\min}$. For example, the weighted standardized sound level difference between the living area of the hotel room and the hallway must be higher than 45 dB. In such cases, minimum calculated weighted apparent sound-reduction index $R'_{w,\min}$ could be determined with correction factor C_p:

$$R'_{w,r} = D_{nT,w} + C_p[dB]$$

The method of determination of weighted standardized sound level difference $D_{nT,w}$ is the same as presented for weighted apparent sound-reduction index R'_w (see Sect. 5.6.3).

Case Study Determine the required, calculated, weighted apparent-sound-reduction index $R'_{w,r,\min}$ of an external wall with an area of 12 m^2. The equivalent sound-pressure level $L_{eq,1}$ outdoors is 55 dB and does contain accented tones ($C = 5$ dB). Flanking sound propagation contributes to the sound-pressure level in the sound-receiving room with ($R'_{L,w}$) 2 dB. The permissible equivalent sound-pressure level $L_{eq,2}$ in the room is 30 dB. The volume of the room is 200 m^3. The sound absorption in the sound-receiving room should be determined on the basis of the reverberation time T_{60} 0.63 s.

The reference reverberation time $T_{60,0}$ (Table 5.22.) and the correction factor C_p are:

$$T_{60,o} = \frac{\sqrt{V}}{20} = \frac{\sqrt{200}}{20} = 0.71 \text{ s}$$

$$C_p = 10 \cdot \log(12) + 10 \cdot \log\left(\frac{0.71}{0.63}\right) - 10 \cdot \log(200) = -5.8 \text{ dB}$$

The required, calculated, weighted apparent-sound-reduction index $R'_{w,r,min}$ is:

$$D_{nT,w} = (L_{eq,1} + 5) - L_{eq,2} = 55 + 5 - 30 = 30 \text{ dB}$$

$$R'_{w,r,min} = D_{nT,w} + C_p = 30 - 5.8 = 24.2\text{dB} \rightarrow 25\text{dB}$$

Solution: The weighted apparent-sound-reduction index of the building structure $R'_{w,r}$ must be at least 25 dB, while the weighted sound-reduction index of the building structure $R_{w,r}$ must be at least 27 dB ($R'_{L,w}$ is 2 dB). In the case that the building structure is exposed to outdoor noise caused by traffic, the weighted apparent-sound-reduction index of the building structure $R'_{w,r}$ must be increased by the adaptation term C_{Tr}.

5.6.5 Sound Insulation of Facades Consisting of Several Elements

The façades of buildings usually consist of several different building structures, such as walls and windows, each of them with its individual weighted sound-reduction index $R_{w,i}$ and surface area S_i (m^2). Additional elements, such as ventilation openings or grilles, are often installed in facades. It is common practice that the manufacturers of these elements state the weighted-standardized sound-level difference $D_{nT,w}$, that is calculated based on a normalized area of 10 m^2 [28]. In this case, the resultant, weighted sound-reduction index of the facade is determined using the expression:

$$R_{w,res} = -10 \cdot \log\left(\sum_{i=1}^{n} \frac{S_i}{S_1 + S_2 + S_3 + \ldots} 10^{\frac{-R_{w,i}}{10}} + \sum_{j=1}^{n} \frac{10}{S_1 + S_2 + S_3 + \ldots} 10^{\frac{-D_{nT,w,j}}{10}}\right) [\text{dB}]$$

where $S_1, S_2, S_3, \ldots S_n$ (m^2) are the areas of the ith facade structure, $R_{w,1}, R_{w,2}, R_{w,3} \ldots R_{w,n}$ (dB) are the weighted sound-reduction indexes of the i-th facade structures, the index i is the counter of structures in the façade and index j is the counter of additional elements in the façade with a standardized sound-level difference $D_{nT,w,j}$ (dB) of the jth element.

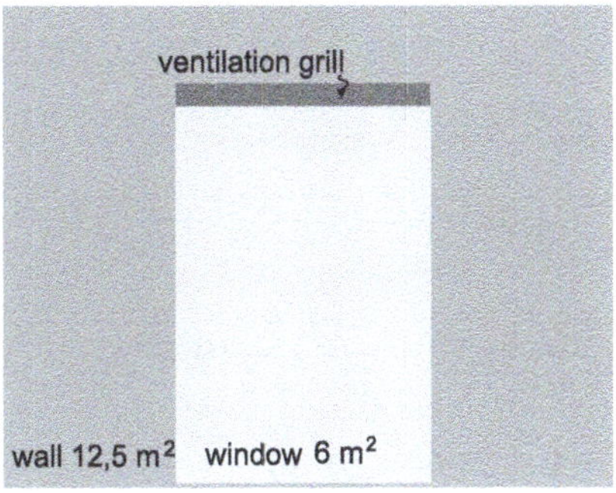

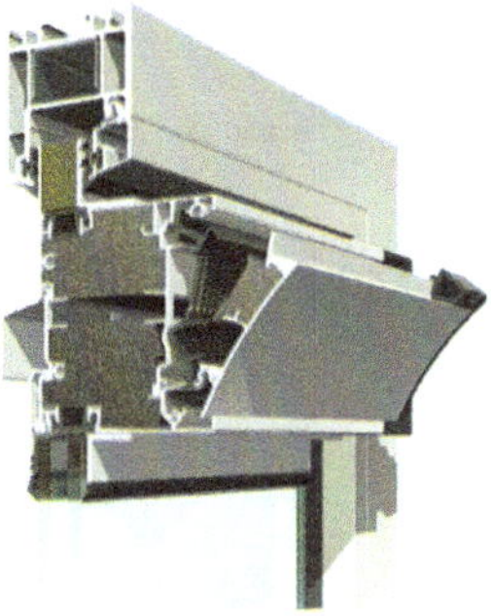

Fig. 5.60 Structure of the facade (left); ventilation grille with sound silencer; with the flap closed, $D_{nT,w}$ is 48, (0; − 2) dB, and with the flap open $D_{nT,,w}$ is 39, (0; − 2) (right) [29]

Example The façade of a room in a building along the road consists of a solid wall (with 1.5 cm of layered plaster, 19 cm solid brick, 12 cm thermal insulation and 1.5 cm finishing plaster) with weighted sound-reduction index $R_{w,1}$ 43, (0, − 3) dB with area of 12.5 m^2 and windows with area of 6 m^2 and weighted sound-reduction index $R_{w,2}$ 32, (− 1, − 3) dB. A ventilation grille for controlled natural ventilation is installed next to the windows (Fig. 5.60) The manufacturer states that the weighted standardized sound-level difference of the grille is $D_{nT,w}$ 39, (0, − 2) dB. The equivalent sound level of the municipality noise that contains accented tones $L_{eq,A,F}$ is 65 dB (A), the permissible equivalent sound-pressure level in the room $L_{eq,A,F,night}$ is 35 dB (A). The spectrum adaptation term of the room C_p is − 5.8 dB. Flanking sound propagation can be neglected ($R_{L,w} = 0$). Is the resultant weighted sound-reduction index $R_{w,r}$ of the façade adequate?

The weighted sound-reduction indexes of a wall, window and ventilation grille, considering the spectrum adaptation terms, equal:

$$R_{w,r,1} = R_{w,1} + C + C_{tr} = 43 + 0 + (-3) = 40 \text{ dB}$$

$$R_{w,r,2} = R_{w,2} + C + C_{tr} = 32 + (-1) + (-3) = 28 \text{ dB}$$

$$D_{nT,w,r} = D_{nT,w} + C + C_{tr} = 39 + (0) + (-2) = 37 \text{ dB}$$

The calculated, weighted sound-reduction index of the entire façade is:

$$R_{w,res} = -10 \cdot \log\left(\sum_{i=1}^{2} \frac{S_i}{S_1 + S_2} 10^{\frac{-R_{w,r,i}}{10}} + \sum_{j=1}^{1} \frac{10}{S_1 + S_2} 10^{\frac{-D_{nT,w,j}}{10}}\right) [\text{dB}]$$

$$R_{w,res} = -10 \cdot \log\left(\frac{S_1}{S_1+S_2}10^{\frac{-R_{w,r,1}}{10}} + \frac{S_2}{S_1+S_2}10^{\frac{-R_{w,r,2}}{10}} + \frac{10}{S_1+S_2}10^{\frac{-D_{nT,w,1}}{10}}\right)[\mathrm{dB}]$$

$$R_{w,res} = -10 \cdot \log\left(\frac{12.5}{12.5+6}10^{\frac{-40}{10}} + \frac{6}{12.5+6}10^{\frac{-28}{10}} + \frac{10}{12.5+6}10^{\frac{-37}{10}}\right) = 31.6 \rightarrow 31\ \mathrm{dB}$$

The weighted, standardized level difference of the façade $D_{nT,w,\mathrm{min}}$ taking into account that the noise has accented tones ($L_{eq,1}$ is in this case increased by 5 dB) must be at least:

$$D_{nT,w,min} = (L_{eq,1} + 5) - L_{eq,2} = 65 + 5 - 35 = 35\ \mathrm{dB}$$

and the required minimum calculated weighted, apparent-sound-reduction index of the façade:

$$R'_{w,r,min} = D_{nT,w} + C_p = 35 - 5.8 = 29.2 \rightarrow 30\ \mathrm{dB}$$

Since the resultant, weighted apparent-sound-reduction index of the façade $R'_{w,res}$ (31 dB) is greater than the required calculation, the weighted apparent-sound-reduction index $R'_{w,r,\mathrm{min}}$ (30 dB), the design of the façade with respect to sound protection against airborne municipality noise is adequate!

5.6.5.1 Detailed Evaluation of Flanking Sound Propagation

The apparent sound-reduction index values R' and R'_w, which are determined by "in-situ" measurements on the building, include all the mechanisms of sound transmission through building structures, including the influence of the so-called flanking sound propagation. The flanking sound appears, for example, at the joints and junctions of building structures and supporting elements (for example, I or C profiles) in lightweight building structures, and which can be interpreted as sound bridges. When we calculate the weighted apparent-sound-reduction index R'_w of building structures, the influence of sound bridges must be taken into account. Sound bridges are characterized with a sound insulation index R_L (or $R_{L,w}$) (dB) (Fig. 5.61). The resultant weighted apparent sound-reduction index $R'_{w,res}$ of the building structure that consists of several homogeneous parts and sound bridges is determined using the expression:

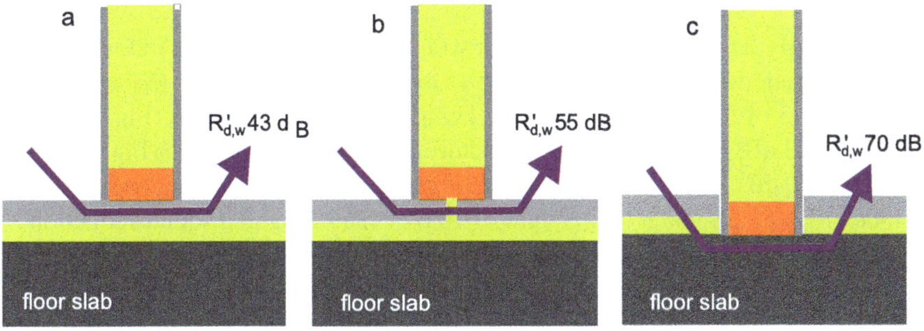

Fig. 5.61 Examples of sound bridges at different contacts of partition and mezzanine building structure and sound-insulation index $R_{L,w}$ (a—installation of partition on screed, b—screed discontinued by a layer of soft material (thermal insulation), c—installation of partition on mezzanine construction with interrupted screed) [6]

$$R'_{w,res} = -10 \cdot \log\left(\sum_{i=1}^{n} \frac{S_i}{S_1 + S_2 + S_3 + \ldots} 10^{\frac{-R_{w,i}}{10}} + \frac{I_o}{S_1 + S_2 + S_3 + \ldots} \sum_{k=1}^{q} I_{L,k} \cdot 10^{\frac{-R_{L,w,k}}{10}}\right) [\mathrm{dB}]$$

where S_1, S_2, S_3 (m^2) are the areas of the homogeneous parts of building structures, $R_{w,1}, R_{w,2}, R_{w,3}$ (dB) are the weighted sound-reduction indexes of homogeneous structures, index i is the counter of homogeneous building structures, k is the counter of sound bridges, $R_{L,w,k}$ is the corresponding weighted sound-insulation index of the k-th sound bridge, $I_{L,o}$ the reference length of the sound bridge (10 m) and $I_{L,k}$ the actual length of the kth sound bridge (m).

Fig. 5.62 Case study of composite building structure

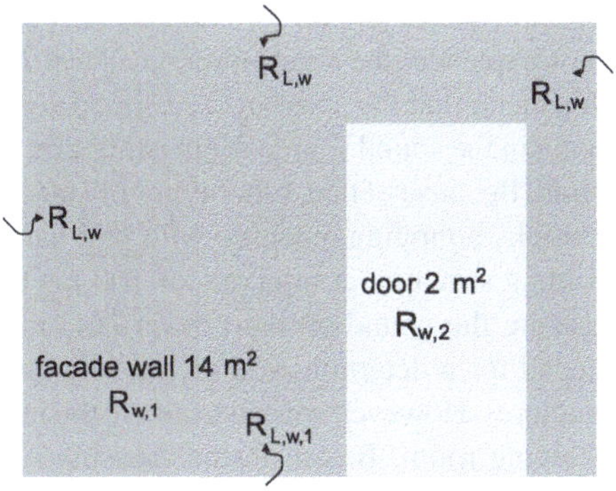

Case Study Determine the resultant, weighted sound-reduction index $R'_{w,res}$ of the façade wall shown in Fig. 5.65 ($S_1 = 14\ m^2$) with a door ($S_2 = 2\ m^2$) in the case of that building structure being installed on a screed (Fig. 5.62a; $R_{L,w,1}$ is 43 dB, $l_{L,1}$ is 4 m) or on a structure with interrupted screed (Fig. 5.64c; $R_{L,w,1}$ is 70 dB). The remaining sound bridges have a weighted sound-insulation index $R_{L,w}$ 52 dB and a total length of 10.6 m. The weighted sound-reduction index $R_{w,1}$ of the façade wall is 51 dB and of the door $R_{w,2}$ 25 dB.

$$R'_{w,res} = -10 \cdot \log\left(\sum_{i=1}^{2} \frac{S_i}{S_1 + S_2} 10^{\frac{-R_{w,i}}{10}} + \frac{l_o}{S_1 + S_2} \sum_{k=1}^{2} l_{L,k} \cdot 10^{\frac{-R'_{d,w,k}}{10}}\right) [dB]$$

$$R'_{w,res}(a) = -10 \cdot \log\left(\frac{(14-2)}{14} 10^{\frac{-51}{10}} + \frac{2}{14} 10^{\frac{-25}{10}} + \frac{10}{14}\left(4 \cdot 10^{\frac{-43}{10}} + 10.6 \cdot 10^{\frac{-52}{10}}\right)\right) [dB]$$

$$R'_{w,res}(a) = 31.9\ dB \quad R'_{w,res}(c) = 33.0\ dB$$

Solution: In the second case, the value of the resultant weighted sound-reduction index of the partition wall is decreased by ~ 1 dB.

5.6.6 Normalized Impact Sound-Pressure Level L'_n and $L'_{n,w}$

Structure-borne or impact sound propagates into the sound-receiving room as a consequence of the direct transfer of sound energy into the building structure. As a consequence, the sound-pressure level L_p in the sound-receiving room increases. This mean that the standardized sound source that emits "white" or "pink" airborne sound in the sound source room cannot be used to evaluate the propagation of impact sound. Because of that, other types of standardized impact sound source are used. For example, a tapping machine with five hammers weighing 0.5 kg, which strikes the building structure at intervals of 100 ms (Fig. 5.63). Therefore, we cannot directly measure the sound-pressure level (L_1) in the sound source room with the source needed for a determination of the sound-reduction indexes R or R' of the building structures. However, we can measure the impact sound-pressure level L_2 in the sound-receiving room. Based on the measured values, we define the normalized impact sound-pressure level L'_n (shorter sound level) and the weighted normalized impact sound level $L'_{n,w}$. In both cases, we consider that part of the sound energy that propagates from the impact sound source room, is absorbed in the sound-receiving room. If the influence of absorption is considered with the equivalent absorption surface A, the

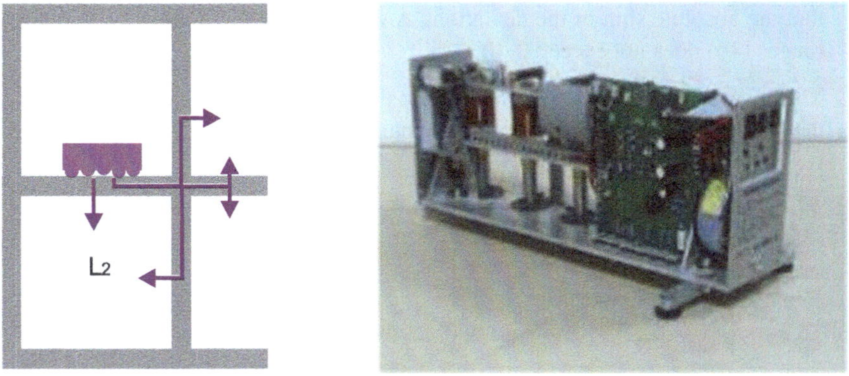

Fig. 5.63 Impact sound is transmitted to adjacent rooms through partition and mezzanine building structures (left). A standardised device for impact sound generation is placed directly on a building structure [27] (right)

result is given as L'_n. However, if the absorption of sound in the room with the listener is evaluated with the reverberation time of the room T_{60}, the result is reported as $L'_{n,T}$. The sound-pressure level in the sound-receiving room L_2 is measured in octave or tertiary frequency ranges between 100 and (3150) 4000 Hz. The normalized impact sound-pressure level L'_n is determined using the expression:

$$L'_n = L_2 + 10 \cdot \log\left(\frac{A}{A_o}\right)[\text{dB}]$$

where A ($\Sigma\, \alpha_i\, S_i$) in m^2 is the equivalent sound-absorption area of the sound receiving room determined according to Sabine's method, and A_o is the reference equivalent absorption area equal to 10 m^2. For a simpler assessment of the propagation of impact sound in building structures, the weighted normalized impact sound-pressure level $L'_{n,w}$ is used. It is defined as follows [30]:

- the measured values of the normalized impact sound-pressure levels L'_n in tertiary frequency ranges are compared with the reference curve of the impact sound level N (Fig. 5.64). The shape of the reference curve indicates the frequency-dependent values of the permissible impact sound level;
- the reference curve of the impact sound level N is moved by 1 dB towards the curve of the measured normalized impact sound-pressure levels L'_n and for each tertiary frequency range the difference in dB between them is added up. However, we add only those differences when L'_n is higher than N (Fig. 5.65, left);
- when the sum of the differences becomes less than − 32 dB we obtain the final shifted reference curve of the impact sound level N';
- at a frequency of 500 Hz, we read the level of the impact sound on the curve N'. This is the weighted normalized impact sound-pressure level of the building structure, which is denoted by $L'_{n,w}$ (Fig. 5.65, right). An additional condition is

that the absolute sum of the deviations ΔN divided by the number of frequency ranges (16) does not exceed 2 dB.

The lower the value of the weighted, normalized impact sound-pressure level $L'_{n,w}$ is, the better is the sound protection of the building structure against impact sound.

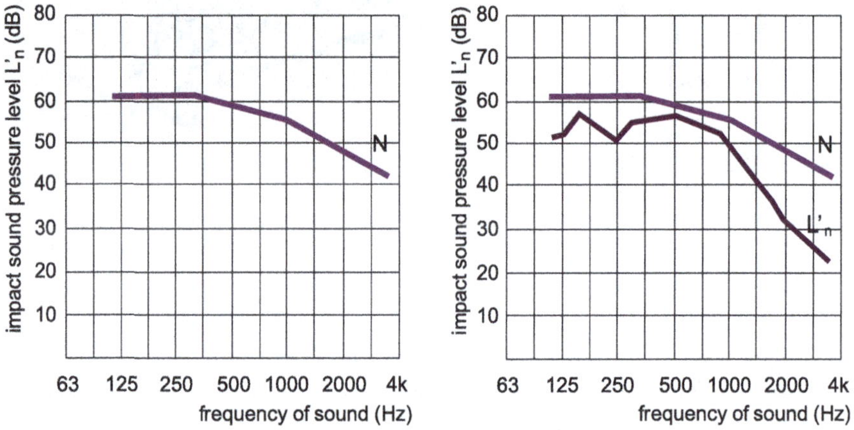

Fig. 5.64 The reference curve of the impact sound level N is compared with values of previously determined the curve of the measured normalized impact sound-pressure levels L'_n

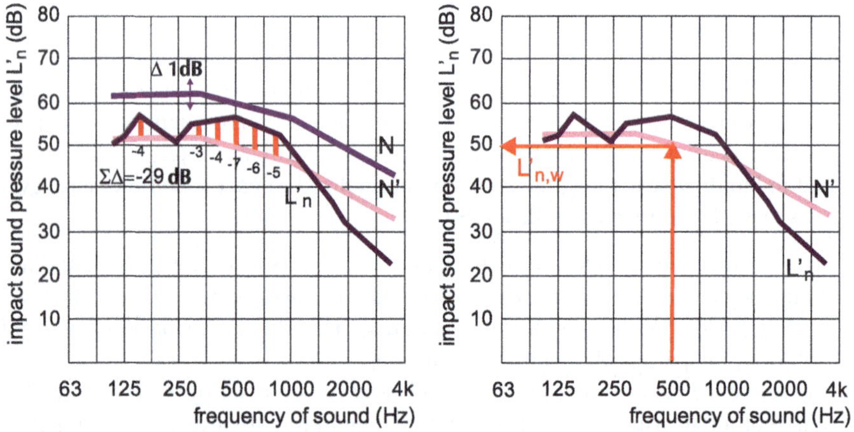

Fig. 5.65 Reference curve of the impact sound level N is moved by 1 dB against the curve of the measured normalized impact sound-pressure levels L'_n until the sum of the differences ($\Sigma\Delta$) becomes less than − 32 dB (left). The weighted normalized impact sound-pressure level $L'_{n,w}$ is read from the final position of the reference curve (N') at a sound frequency of 500 Hz (right)

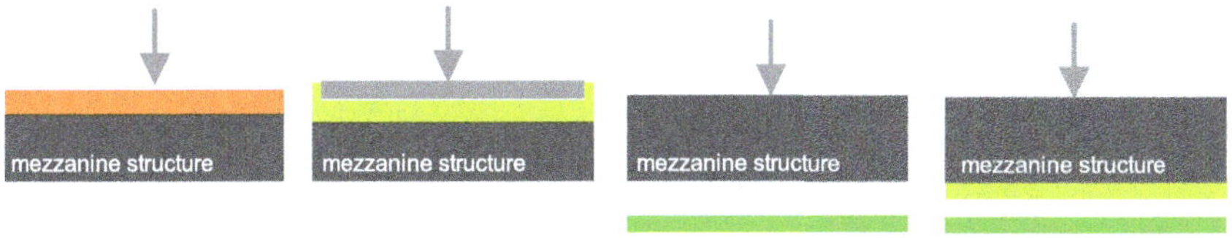

Fig. 5.66 Basic principles for increasing the sound insulation of inter-storey structures for impact sound: lining with elastic material (rubber, carpet), installation of soft material between the load-bearing plate and the floor covering (thermal insulation), suspended ceiling in the adjacent room, suspended ceiling in the adjoining room with sound absorber (soft, fibrous matter)

5.6.6.1 Improving the Impact Sound Insulation of Building Structures

Because mechanical waves are transmitted in rigid, massive building structures with very low energy losses, the most effective way to decrease impact sound propagation is to separate the load-bearing building structures from the impact sound sources with an elastic substance or air layer. The sound-protection options (Fig. 5.66) are as follows:

- installation of elastic walking cladding on the building structure;
- installation of soft material between the load-bearing structure and the floor covering; such an implementation is called a "floating floor";
- suspended ceiling with closed air space on the underside of the building structure in the adjacent room;
- suspended ceiling in the adjacent room with an additional layer of soft material that is a sound absorber.

The effect of the techniques for decreasing the normalized impact sound-pressure level L'_n (or $L'_{n,w}$) is evaluated by the impact sound-reduction index ΔL_n (or $\Delta L_{n,w}$) of the floor covering or absorption layers. Typical values for weighted impact sound-reduction index $\Delta L_{n,w}$ are given in Table 5.24.

In engineering practice, the weighted normalized impact sound-pressure level $L'_{n,w}$ of the basic structure is determined first and in the next step the value of $L'_{n,w}$ is reduced by the weighted impact sound-reduction index $\Delta L_{n,w}$ of the selected measure [31]. For concrete load-bearing structures of different thicknesses, the values of $L'_{n,w}$ are given in Table 5.25 (left). Since part of the sound energy is also absorbed in the sound-receiving room, this is taken into account by the room-correction factor B. It is determined on the basis of Sabine's reverberation time and the reference reverberation time $T_{60,0}$:

$$B = 10 \cdot \log(V) - 10 \cdot \log\left(\frac{T_{60,o}}{0.0163}\right) (\text{dB})$$

The reference reverberation time depends on the volume of the room V (m^3). The values are given in Table 5.25 (right). The calculated weighted normalized impact sound-pressure level $L'_{n,w,r}$ must be lower than the required values $L'_{n,w,\max}$:

Table 5.24 Weighted impact sound reduction index $\Delta L_{n,w}$ of selected floor covering or sound-absorption layers [6]

type of mezzanine structure lining, covering	ΔL_n (dB)
PVC covering (2 mm)	5
PVC covering with felt (3mm)	15
rubber (2,5 mm)	10
rubber (5 mm)	24
carpet, natural fibres	20
needle carpet	25
wood on concrete slab	16
massive wood on foil of fibrous matter - mineral or coconut fiber (5mm)	24
laminate flooring on cork I(20 mm)	6
laminate flooring on wood fibre board	16
laminate flooring on wood fiber board (20 mm) and on foil of fibrous material (5 mm)	28
floating floor - screed on wood fiber boards (12 mm)	15
floating floor - screed on polystiren (10 mm)	26
floating floor - screed on coconut boards (8 mm)	16
(14 mm)	22
floating floor - screed on mineral wool boards (15 mm)	31

Table 5.25 Weighted normalized impact sound-pressure levels of concrete mezzanine structures without claddings (left) [14]

thickness of mezzanine concrete slab (mm)	$L'_{n,w}$ (dB)
100	78
150	75
200	70
250	67
300	65

volume of the room V(m³)	$T_{60,0}$ (s)
up to 100	0,5
between 100 in 2500	$\frac{\sqrt{V}}{20}$
over 2500	2,5

Reference reverberation times according to CH SIA-Norm 181, standard EN ISO 3382-2:2008 states only the reference reverberation time of 0.5 s for residential buildings [6] (right)

$$L'_{n,w,max} \leq L'_{n,w} - \Delta L_{n,w} + B \text{ [dB]}$$

An overview of the maximum-permitted, weighted, normalized impact sound-pressure level $L_{n,w,\max}$ for selected building structures is given in Table 5.26.

Table 5.26 Maximum-permissible, weighted, normalized, impact sound-pressure levels $L'_{n,w,\max}$ for selected load bearing building structures

building structure	$L'_{n,w,max}$ (dB)
mezzanine structure between apartments	55
mezzanine structure between apartment and office	58
mezzanine structure between apartment and restauran	43

Case Study Determine whether a 200-mm-thick concrete floor slab with a cement screed on a mineral fibre slab ($\Delta L_{n,w}$ is 31 dB, Table 5.24) between apartments with a volume of 90 m^3 meets the requirements of the maximum-permitted, weighted, normalized impact sound-pressure level. This is 52 dB.

The reference reverberation time $T_{60,0}$ is equal to 0.5 s (Table 5.25, right) and the correction factor of the space B is:

$$\begin{aligned} \mathrm{B} &= 10 \cdot \log(\mathrm{V}) - 10 \cdot \log\left(\frac{\mathrm{T}_{60,o}}{0.0163}\right) \\ &= 10 \cdot \log(90) - 10 \cdot \log\left(\frac{0.5}{0.0163}\right) = 4.7\ \mathrm{dB} \end{aligned}$$

$$\mathrm{L}'_{n,w,max} = 55 \leq \mathrm{L}'_{n,w} - \Delta \mathrm{L}_{n,w} + B = 70 - 31 + 4.7 = 43.7\ \mathrm{dB}$$

Solution: According to the permissible, weighted, normalized impact sound-pressure level the design of the building structure is adequate.

References

1. Kladnik, R.: Physics for University Students—Acoustic and Optic, DZS, Ljubljana (1989). (in Slovenian language)
2. www.norsonic.com; Technical data (28.4.2014)
3. Bruil & Kjaer; Measuring Sound; Manual (2009)
4. Milavec, M.: Report on Experimental Evaluation Reverberation Time. Hidria Inštitut Klima, Godovič (2009). (in Slovenian language)
5. Čudina, M.: Engineering Acoustic. University of Ljubljana, Faculty of Mechanical Engineering, Ljubljana (2001). (in Slovenian language)
6. Keller, B.: Bautechnologie II – Bauphysik. ETH Zürich, CH (2001)
7. Directive 2002/49/EC of the European Parliament and of the Council of 25 June 2002 relating to the assessment and management of environmental noise
8. Stanners, D., Bourdeau, P.: Europe's Environment (1995)
9. Špes, M., Cigale, C., Lampič, B.: Environmental Problems in City of Ljubljana. University of Ljubljana, Faculty of Arts, Ljubljana (2002)
10. www.trimo.si; Technical data (1.1.2014)
11. Harris, D.: Noise Control Manual for Residential Buildings. McGraw-Hill, NY, USA (1997)
12. Decree on the Assessment and Management of Environmental Noise: Ministry of the Environment and Spatial Planning, R Slovenia. (in Slovenian language)
13. Decree on Limit Values for Environmental Noise Indicators: Ministry of the Environment and Spatial Planning, R Slovenia. (in Slovenian language)
14. Hohmann, R., Setzer, M.J., Wehling, M.: Bauphysikalische Formeln und Tabellen: Wärmeschutz - Feuchteschutz – Schallschutz; Werner Verlag, D (2004)
15. Bruil & Kjaer: Measurements in Building Acoustics (1.1.2014)
16. Bruil & Kjaer: Environmental Noise Propagation (1.1.2014)
17. Egan, M.D.: Architectural Acoustics. J. Ross Publishing, NY, USA (2007)

18. Merkel, J.: Sound Propagation in Windows; Seminar Work (supervisor prof. Sašo Medved). University of Ljubljana, Faculty of Architecture, Ljubljana (2007). (in Slovenian language)
19. Rules on Protection Against Noise in Buildings: Ministry of the Environment and Spatial Planning, R Slovenia. (in Slovenian language)
20. ISO 717-1:2013: Acoustics—Rating of Sound Insulation in Buildings and of Building Elements—Part 1: Airborne Sound Insulation
21. Hausleden, G., Saldanha, M., Liedl, P.: ClimaSkin, Callwey, München, D (2006)
22. Beckett, H., Godfrey, J.: Windows—Performance, Design and Installation. RIBA, London, UK (1974)
23. www.saint-goben-glass.com; Technical data (1.1.2014)
24. www.knauf.si; Technical data (1.1.2014)
25. www.reflex.si; Technical data (1.1.2014)
26. EN 12354-3:2017: Building Acoustics—Estimation of Acoustic Performance of Buildings from the Performance of Elements—Part 3: Airborne Sound Insulation Against Outdoor Sound
27. www.brd-nonoise.com/BRD; Technical data (1.1.2014)
28. EN 12354-1: 2017: Building Acoustics—Estimation of Acoustic Performance of Buildings from the Performance of Elements—Part 1: Airborne Sound Insulation Between Rooms
29. www.ronson.com; Technical data (1.1.2014)
30. ISO 717-2:2013: Acoustics—Rating of Sound Insulation in Buildings and of Building Elements—Part 2: Impact Sound Insulation
31. EN 12354-2:2017: Building Acoustics—Estimation of Acoustic Performance of Buildings from the Performance of Elements—Part 2: Impact Sound Insulation Between Rooms

Chapter 6
Buildings Fires and Fire Safety

Abstract Besides providing indoor living comfort, a building must also ensure the safety of its occupants. Natural disasters excluded, the safety of a building's occupants is most often threatened by fire. More than one-third of all fires in the world are related to fires in buildings, and currently more than 10 people per million of the population die in building fires. Due to the constant updating of fire-safety regulations, the number of deaths due to fires in buildings has declined by 65% over last 30 years [1]. Building physics in the field of fire safety focuses on a description of the physical fundamentals of the fires in buildings and the way that building materials and structures respond to a fire. A fire in a building occurs in several stages. In the first stage, which usually takes for up to several tens of minutes and is known as the fire growth stage, it is important to protect people as they are evacuating the building. In the second stage, with a fully developed fire, it is necessary for the building structures to retain their load-bearing capacity and prevent the spread of the fire to other building zones or the surrounding buildings. Thus, extinguishing a fire is less dangerous for the firefighters and leads to less damage to property. The fire safety in buildings is achieved with both passive and active measures. The passive measures are related to building materials and the building construction's response to the fire, while the active measures include the control of smoke and heat propagation in buildings by either the natural or forced ventilation of fire compartments. Other active measures include technical systems for detection, alarm and fire-extinguisher systems.

Learning objectives In this chapter you will learn about:

- the process of combustion, the pollutants emitted during a fire in a building, and the phases of the fire's development in buildings;
- the properties of the building materials and the building structures according to the response to the fire in different phases of the fire's growth;
- the ISO 834 and EUROCODE temperature-time curves at different load densities;
- the modelling, requirements and technical measures of heat and smoke extraction from buildings during the fire's growth phase.

S. Medved, *Building Physics*, Springer Tracts in Civil Engineering,
https://doi.org/10.1007/978-3-030-74390-1_6

6.1 Symbols for Quantities and Units

A, A_{en}, A_f, A_{str}	Area, area of opaque building structures in contact with flue gasses in fire compartment (en), of fire compartment (f), of building structure (str) (m^2)
α	Fire-growth factor (kW/s)
b	Thermal effusivity of building material ($J/m^2Ks^{0,5}$)
C_i, C_o	Hydraulic loss coefficient of inlet opening (C_i), at outlet opening (o) (i)
$C_{O2,air}$, $C_{O2,g}$	Mass fraction of oxygen in air, mass fraction of oxygen in flue gasses (%)
c_p, $c_{p,g}$	Specific heat capacity at constant pressure, of flue gases (p,g) (kJ/kgK)
χ	Combustion efficiency (1)
D	Pool-fire diameter (m)
Δp_i, Δp_o	Pressure difference at inlet opening (i), at outlet opening (o) (Pa)
g	Acceleration due to gravity (m/s^2)
H_i	Heat of combustion (MJ/kg)
H	Height of fire compartment (m)
h_o	Height of ventilation opening (m)
k_β	Fuel-burning-area constant (m^{-1})
l_o	Width of ventilation opening (m)
λ	Thermal conductivity (W/mK)
m	Mass (kg)
$\dot{m}_i$, $\dot{m}_o$, $\dot{m}_g$, $\dot{m}_{g(z)}$	Mass flow rate of air at inlet opening (i), of flue gases at outlet opening (o), of flue gases in fire compartment (g), of flue gases at height z in fire compartment g(z) (kg/s)
M_a, M_{O2}	Mole mass of air, of oxygen (g/mol)
$\dot{m}''_\infty$	Mass of burning liquid that burning per second from 1 m^2 of surface of the fire pool (kg/m^2s)
n	Air-exchange rate in fire compartment (h^{-1})
p_o, p_i	Air pressure outdoor (o), indoor (i) (Pa)
q, Q'	Specific heat released, specific heat (MJ/m^2)
$\dot{Q}$, $\dot{Q}_g$, $\dot{Q}_a$	Heat-release rate, heat flux transferred by flue gases (g), heat flux stored in building structures (a) (kW)
Q, Q'	Heat (MJ), specific heat (MJ/m^2)
r	Heat of vaporization for water (kJ/kg)
ρ, ρ_{air}, $\rho_i \rho_g$,	Density, of air, of indoor air (i), of flue gases (g) (kg/m^3)
R	Gas constant (J/Kmol)
t	Time (s, h)
T_i, T_o, T_g	Absolute temperature of indoor air (i), of outdoor air (o), of flue gases (g) (K)
θ_i, θ_g, $\theta_{g,max}$	Indoor air temperature in fire compartment (i), temperature of flue gases (g), maximum temperature of flue gases (g, max) (°C)

(continued)

(continued)

t^*	Normalized time (h)
t^*_{max}	Normalized time in the moment when $\theta_{g,max}$ is established (h)
t_{max}	Time at which $\theta_{g,max}$ is established (h)
VF	Ventilation factor ($m^{0,5}$)
$\dot{V}$	Volume flow rate (m^3/s, m^3/h)
v_w, v_i, v_o	Velocity of wind (m/s), of air at inlet opening (i), of flue gases at outlet opening (o) (m/s)
V	Volume (m^3)
z	Height from floor level to the bottom level of flue-gases layer in the fire compartment (m)

6.2 Combustion

Combustion is exothermic oxidation involving combustible materials and oxygen. It is a chain reaction, which can start if three elements of the so-called fire triangle are provided in the appropriate mixture: the fuel, the oxygen and the heat. In a building fire, the oxygen is provided by the indoor air and the heat by the ignition of the fuel in an inflammable state.

Combustible building materials are made primarily of carbon and hydrogen in long chains. Such materials are difficult to ignite, because the molecules must first break down into shorter, gaseous hydrocarbon molecules that ignite easily. The chemical reaction that takes place in building fires is known as pyrolysis. The theoretical process of combustion is described by the combustion stoichiometry, i.e., the mass balance of the reactants, that enters the combustion and the products of combustion in the form of heat and flue gases. In the stoichiometry, the quantities are measured in the number of moles.

Complete combustion is a theoretical process in which all the carbon from the fuel is oxidized to CO_2 and all the hydrogen to H_2O. For example, the following stoichiometric equation describes the complete combustion of methane (CH_4):

$$1\text{mol } CH_4 + \underbrace{2\text{mol}\left(O_2 + \frac{0.79}{0.21}N_2\right)}_{\text{air}} \rightarrow 1\text{mol } CO_2 + 2\text{mol } H_2O + 7.52\text{mol } N_2$$

The quantity of oxygen necessary for complete combustion is known as the stoichiometric amount. Complete combustion also generates a maximum amount of heat, with the heat of combustion determined using a calorimeter. Calorimeters measure the heat released during combustion with pure oxygen, known as the heat of combustion H_i, which is usually expressed in a normalized form per unit of material mass in MJ/kg (Table 6.1).

Table 6.1 Heat of combustion for some combustible building materials [2, 3]

material	H_i (MJ/kg)
polyethylene	43.6
polypropylene	43.4
polystyrene	39.2
polyurethane	25.0
PVC	16.4

The combustion that occurs in building fires and technical appliances (e.g., heating boilers) is always incomplete, releasing less heat than complete combustion. From the viewpoint of fire safety, it is important to note that in the case of incomplete combustion, not all the carbon is oxidized to CO_2, but rather into extremely dangerous carbon monoxide (CO), while a part of it stays in the form of small, airborne, respirable solid particles called soot. The combustion efficiency χ is expressed as the ratio of the heat released in incomplete combustion to the heat released in (theoretical) complete combustion. The typical values of χ in building fires are 0.7 for burning furnishings and 0.95 for spilled flammable liquids. Building materials are composed of different chemical elements, including heavy metals and halogens, and a simplified chemical reaction of complete combustion can be written as follows:

$$\text{building materials } [\mathrm{C, H, O, S, N, Cl, F, Br}] + \text{air} \xrightarrow{\text{products}} [\mathrm{CO_2, H_2O, SO_2, N_2, Cl_2, F_2, Br}]$$

and for incomplete combustion as:

$$\text{building materials } [\mathrm{C, H, O, S, N, Cl, F, Br}] + \text{air} \xrightarrow{\text{products}} \left[\mathrm{CO_2, CO, \underbrace{C}_{\text{soot}}, HC, SO_2, N_2, NO_x, HCN, HCl, HF, HBr}\right]$$

Many toxic or irritant gases are created in building fires, and it is the inhalation of these toxic and hot gases that does the most harm to people caught up in a building fire.

6.3 Dangers Associated with a Building Fire

In the first phases of a fire, people who are still present in a burning building are in danger because of the high temperatures and the harmful properties of the flue gases. The heat is transferred to people by the radiation emitted from hot objects, flames and gases, by convection from hot gases and smoke particles, and by conduction when hot objects are touched. The effects of fire on human health are as follows:

- the hazard of burns due to radiative heat transfer depends on the heat flux at the surface of the skin and the time of exposure; burns are expected when the heat flux exceeds 2.5 kW per m^2 of skin surface and the skin is exposed to such thermal radiation for more than three minutes; the victim suffers second degree (medium) burns if the skin is heated to more than 70 °C for a short while or to more than 50 °C for several tens of seconds; skin also suffers burns when in contact with objects heated to over 60 °C; inhalation of air at 200 °C causes permanent damage to the respiratory organs;
- combustion also consumes oxygen in the air; movement becomes difficult as the volumetric ratio of oxygen drops below 17%, onset of dizziness takes place at 15%; if the oxygen concentration drops even further, this is followed by loss of consciousness at 8% and death from asphyxiation at 6%;
- solid particles in smoke reduce visibility and make it difficult to escape from the building; walking speed is reduced by 90% when visibility drops to 2 m.

People in a building on fire are in great danger due to the different gases, aerosols and small particles (smoke) that originate as products of the combustion and move around the building's interior. The most dangerous of these pollutants are:

- carbon dioxide (CO_2) might not be considered a hazardous gas, but its presence in the air is important as it accelerates breathing in concentrations over 3% (by volume), which in turn causes other toxic gases to be inhaled; in rare cases the concentration will become high enough to cause a loss of consciousness (~10%);
- much more dangerous is carbon monoxide (CO), which combines with the haemoglobin in the blood to form carboxyhaemoglobin (COHb) and reduces the oxygen-carrying capacity of the blood; carbon monoxide is responsible for most of the deaths from asphyxiation in building fires;
- hydrogen cyanide (HCN) has a similar effect to carbon monoxide, halting cellular respiration; it is transferred by the blood, is water soluble and can also enter the body through the skin;
- hydrogen chloride (HCl), hydrogen fluoride (HF) and hydrogen bromide (HF) can be released when halogen-containing substances are burned; in low concentrations they irritate the mucous membranes of the respiratory system and the eyes, in high concentrations they become lethal; phosgene ($COCl_2$) is released from burning chlorine-containing substances; it causes the formation of toxic CO and HCl in the body and is lethal even in small air concentrations;
- sulphur dioxide (SO_2) is a product of the combustion of sulphur-containing substances, irritating mucous membranes in low concentrations and causing respiratory-system damage and death in higher concentrations;
- nitric oxide (NO) released from burning nitrates in synthetic fibres or varnishes causes similar effects to sulphur dioxide;
- burning hydrocarbons release formaldehyde (HCHO), a carcinogenic substance that irritates the mucous membranes at low concentrations and causes breathing problems and allergies at high concentrations, and acrolein (unsaturated aldehyde CH_2CHO), which also irritates the respiratory organs and can be lethal in high concentrations;

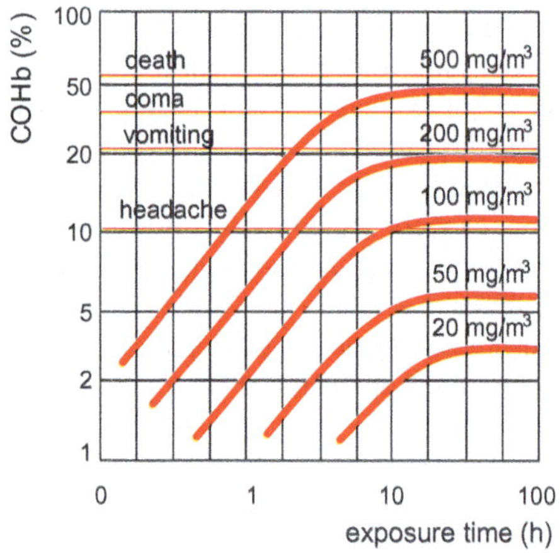

Fig. 6.1 The impact of flue gases on the health depends on their concentration in the environment and the time of exposure; the diagram shows the impact of carbon monoxide (CO), which causes the formation of carboxyhemoglobin (COHb) in the blood, which reduces the transfer of oxygen in the body [4]

- flue gases also contain aerosols, airborne solid or liquid particles of 2–10 μm in size; when inhaled, such solid particles (soot) accumulate in the lungs and are often the cause of death in fire victims.

Regardless of the pollutant, the impact on an individual person depends on the pollutant concentration and exposure time (Fig. 6.1).

The basic principle of fire safety in the early phases of a fire, before the flashover, is therefore to vent the flue gasses and smoke out of the building and away from the escape routes.

Explanation During burning several gases are released and the mixture of these gases is called flue gas, while the smoke, which is also released during burning, contains aerosols and solid particles (soot). Because of common practice, the term smoke will also be used for flue gases in the following chapters.

6.4 Fire Ignition and Dynamics

A fire is uncontrolled combustion in enclosures or outdoors. A fire begins with the ignition, which can be caused by an open flame from a lit match or a spark, or it can happen spontaneously due to the heating of inflammable substances. Ignition is followed by the growth stage. The rate with which the fire develops depends on the heat released and the way different substances burn. Fire develops rapidly in substances that burn with a flame (e.g., alcohols), because the flame spreads rapidly across the surface of the flammable substance. Smouldering combustion substances tend to develop slow fires, so the heat released in a unit of time is lower. However, such fires are very dangerous because the smoke contains more toxic substances. The incipient stage is usually shorter in structure fires, up to 30 min from the moment of ignition. The heat-release rate Q in the fire's growth phase is determined numerically using the "t-squared fire" approach:

$$\dot{Q} = \alpha \cdot t^2 \text{ [kW]}$$

where α is the fire-growth factor in kW/s^2 and t is the time from ignition in seconds. The values of the fire-growth factor were determined experimentally, and four fire-growth-rate classes were defined according to the acceleration of the fire (Table 6.2).

The fire-growth rate can also be assessed from the time needed for the heat-release rate to reach 1000 kW. If the fire develops slowly, this time is about 600 s, while in a rapid fire it can be as short as 75 s (Fig. 6.2). This heat-release rate also represents the limit value for the occurrence of flashover. Flashover is the phase of a fire where all the flammable surfaces in a compartment on fire burn simultaneously (Fig. 6.3). It happens not later than 30 min after the ignition, or when the temperatures of the

Table 6.2 The fire growth factor α; fire growth class is determined according to fire growth factor [1]

fire growth rate class	α (kW/s)	type of buildings, where to expect
slow	0.003	fireproof materials, massive furniture
medium	0.012	apartments
fast	0.047	hotels, hospitals, schools, offices
ultra fast	0.190	shopping malls, entertainment facilities

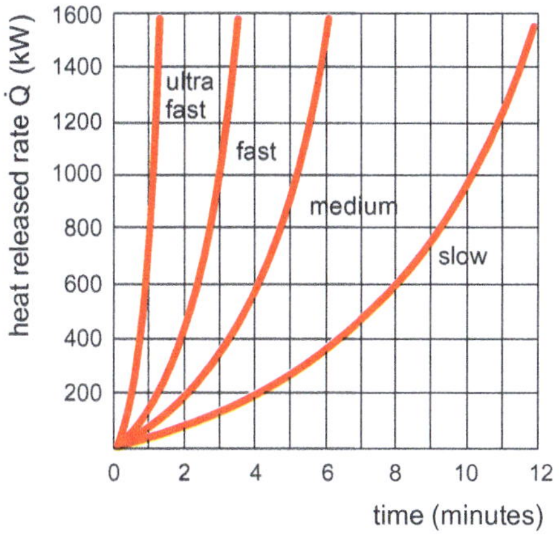

Fig. 6.2 The heat-release rate in a fire according to the fire-growth factors α [2, 3]

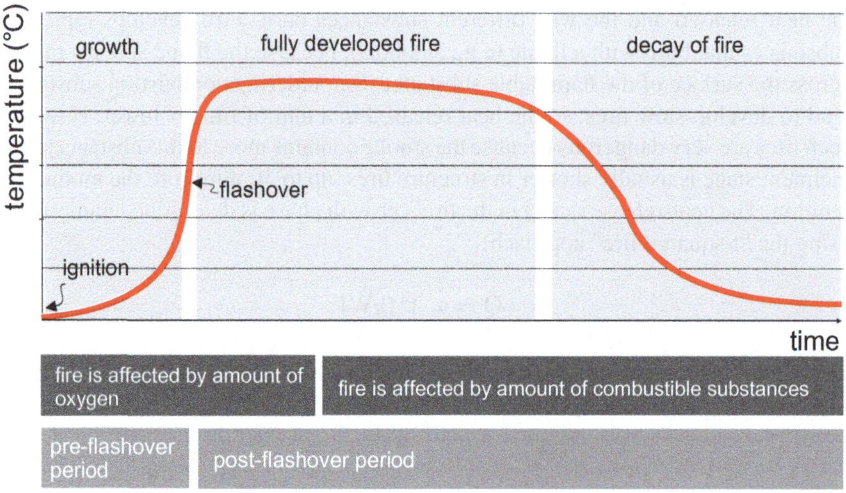

Fig. 6.3 Smoke-temperature variation in the stages of a compartment fire. In building physics a compartment fire is usually divided into the pre-flashover and post-flashover periods

room climb to 500–600 °C. The heat-release rate can also be taken as a measure of flashover: experience shows that flashover occurs when the heat-release rate exceeds 1000 kW, or 15–20 kW/m^2 of compartment area. These values must be interpreted as informative, while exact values are determined with laboratory fire tests (Fig. 6.4).

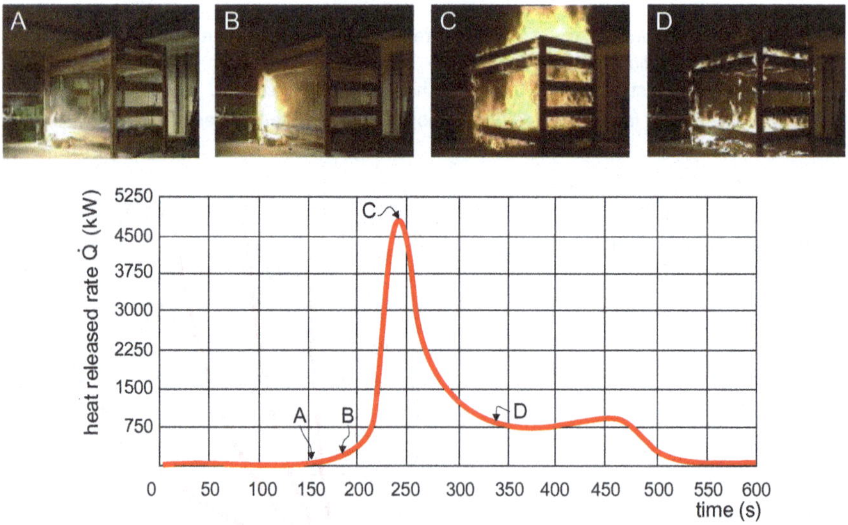

Fig. 6.4 A bunk bed at four stages of a fire test: fire growth stage (A, B), flashover (C) and decay of fire (D) [5]

It is important for fire-safety that everyone must evacuate the building on fire not later than the moment of flashover. Flashover is followed by the fully developed fire phase, when the heat-release rate reaches a maximum, the room temperatures rise to 700–1200 °C, and then remains relatively constant during this period. The last phase is the decay stage, when the heat-release rate and the temperatures in the room drop. A fire can continue to burn for several hours after the flashover.

Explanation If the fire develops slowly, oxygen deficiency and a decreasing flame temperature can make the fire transit directly to the decay stage, without the flashover and fully developed phases.

The other way that the phases of a fire in a building could be characterized is based on observing the elements of the fire triangle (Fig. 6.3):

- the initial period, when there is plenty of combustible materials in the building and the fire growth depends on the quantity of oxygen available for oxidation (ventilation-controlled fire);
- the period when the fire's development depends on the quantity of the remaining combustible materials (fuel-controlled fire).

Considering the building physics and the requirements for fire safety in the building, we distinguish the pre-flashover and post-flashover periods (Fig. 6.3). During the pre-flashover period, it is important that the building's occupants are able to evacuate safely, so we must consider the building materials' response-to-fire performance and find ways to prevent gases and smoke from filling the escape routes. After the flashover, the building structures must retain their load-bearing capacity and prevent the fire from spreading to adjacent compartments or buildings. Building-structure fire resistance is, therefore, crucial for fire safety in the post-flashover period.

6.5 The Response of Building Materials to Fire

In the following section the responses of the most commonly used building materials, i.e., brick, steel, concrete and wood, are presented.

Brick is an incombustible material and therefore does not increase the building's fire load. The first regulations on the compulsory use of brick for fire safety date back to the seventeenth Century, after the Great Fire of London (1666). Brick walls are rarely damaged in a fire, heated brick emits no gases or soot, and it also exhibits a high thermal inertia. A part of the heat released in a fire is therefore absorbed and stored in the brickwork. The fire resistance of a brick wall depends on the wall thickness, the height and the length of the building structure. Brick guarantees the best fire resistance for building structures (Fig. 6.5).

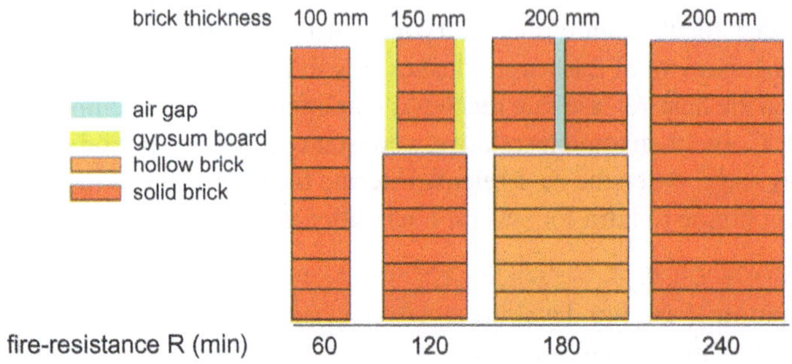

Fig. 6.5 Fire resistance R of brick building structures, the fire resistance of building structures is defined in Sect. 6.8.2 [3]

Fig. 6.6 Relative decrease of tensile and compressive strength for steel beams at elevated temperature [6]

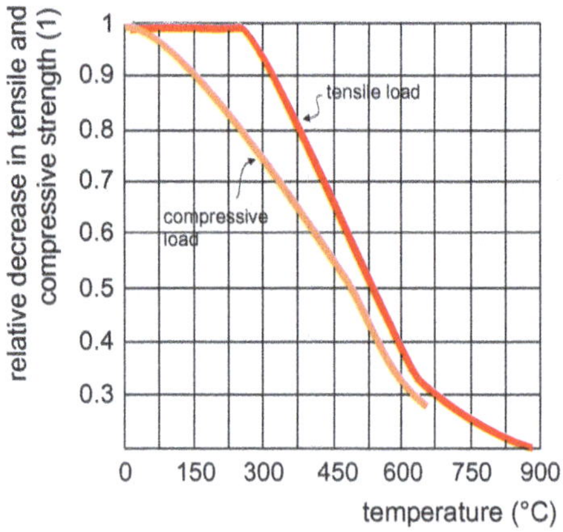

When steel structures are exposed to high temperatures in a fire, their compressive and tensile strengths are reduced, and they also undergo thermal expansion (Fig. 6.6). The compressive and tensile strengths of structural steel are almost halved when the steel is heated from room temperature to 550 °C. As a result of the high thermal conductivity and the ratio of the heat-transfer area to the structure volume, the whole steel beam heats up and expands rapidly (Fig. 6.7, left). A 10-m-long steel beam expands by 70 mm when it is heated to 550 °C. The fire resistance of steel beams can be improved with fire-protection materials, such as coatings, high-temperature-resistant thermal insulation, and gypsum or calcium silicate boards. These coatings are applied by painting or spraying (Fig. 6.7, right). Sprayed coatings usually contain gypsum with added perlite and cellulose or glass fibre. They are the cheapest solutions for fireproofing steel building structures. Uneven surfaces (such as bolt heads) can be covered too, but the coatings must be applied to a pre-prepared steel surface, and their rough surface can appear as visually unpleasant indoors. Steel structures can also be protected with intumescent coatings. In tests using standard temperature curves (Sect. 6.8.2), these coatings expand and their thickness increases 15–30 fold.

Fig. 6.7 Typical fire damage to an unprotected steel structure [7] (left); FireDam 200 coating is sprayed to improve the fire resistance of a steel building structure [8] (right)

They contain compounds that break apart at high temperatures, emitting gases and transforming the coating into a thick layer of thermal insulation with a fire-resistance period from 30 to 120 min. Stone wool is used as the thermal insulation for steel structures to satisfy the requirements for high-temperature resistance. If the thermal insulation layer is wrapped in a metal foil, the radiative heat transfer from the flames to the thermal insulation surface is reduced too. Gypsum boards are the easiest to install onto steel beams (Fig. 6.8). In order to ensure 60 min of fire resistance, the thermal insulation or gypsum boards have to be 15–20 mm thick.

The properties of concrete structures remain unchanged until heated up to 100 °C. At temperatures above 300 °C, so-called explosive spalling may occur in the concrete structure due to the heat- and mass-transfer taking place in the concrete. Heating makes the concrete structure expand, giving rise to mechanical stresses. Because the concrete is a porous material, it contains water molecules in the pores. When heated, the water in the pores boils and the water vapour molecules move around the structure. Since the thermal conductivity of concrete is relatively low (with λ between 0.5 and 0.8 W/mK), the inside of the concrete structure stays cooler for a longer time and the condensation of water vapour inside the pores occurs. This can lead to

Fig. 6.8 Steel beams are protected to improve the fire resistance R—spray gypsum plaster fireproofing, high-temperature resistant stone wool thermal insulation, gypsum boards

filling of the pores with water, which prevents the further transfer of water-vapour molecules. Additional water molecules increase the pressure inside the concrete structure. Furthermore, the concrete aggregates undergo chemical transformations above 500 °C. As a result of all these processes, the pressure inside the structure rises, which can lead to explosive spalling. Above 600 °C, the colour of the concrete changes and such structures must be removed after the fire. Water-vapour transfer inside the concrete can be blocked by adding polypropylene fibre, while the thermal stresses can be reduced with thermal insulation, as discussed in the section about fireproofing steel structures.

Timber is a combustible building material that increases the building's fire load. Nevertheless, it does exhibit some good properties that benefit the fire resistance of timber structures. It contains a substantial amount of water, which boils by absorbing the heat released in combustion, slowing down the initial stage of the fire. The thermal conductivity of wood is low, so the inside of a timber structure heats up slowly and the surface of the structure is charred. The charred layer has an even lower thermal conductivity and hinders the oxygen supply as well and the initial charring rate of 0.5–0.8 mm/min is halved. Because the material properties of the core of timber structures remain unchanged, the load-bearing capacity of timber is only affected by the reduced cross-section and the moment of inertia. In spite of that, the duration of the fire resistance for unprotected timber building structures is no longer than 30 min.

The fire resistance of timber building structures can be improved by modifying the process of pyrolysis by decreasing the temperature and preventing the oxygen supply to the burning wooden surface. In theory, pyrolysis (combustion) can be modified so that the cellulose (wood, $C_6H_{10}O_5)_n$ is decomposed to char (C) and water (H_2O), making the wood incombustible, but in practice the pyrolysis process is slowed down and the amount of flue gases and consequentially the heat release from a burning timber structure decreases (Fig. 6.9). Therefore, by slowing down the combustion, the fire-growth factor α is lowered. The fire resistant of timber structures can be enhanced by treating the wood with fire retardants based on boric or phosphoric acid and their salts. The timber is impregnated with the fire retardant over the entire thickness in a vacuum or overpressure chamber. Boric acid is also used to decelerate the combustion and the soothing of the cellulose thermal insulation. The

Fig. 6.9 Dricon® FRT fire retardant reacts with the combustible gaseous products of pyrolysis and transforms them to carbon (soot), CO_2 and water [9]

fire resistance of timber building structures can be improved by thermal insulation, like the one used to protect steel structures.

Gypsum boards were mentioned as a material suitable for improving the fire resistance of building structures made of any material. Gypsum or calcium sulphate ($CaSO_4 \cdot 2H_2O$) is a hydrate, i.e., a substance containing chemically bound water. The mass fraction of water in gypsum is approximately 21%. As the water boils off in a fire, an endothermic process takes place and the gypsum boards effectively slow down the fire's growth. The reaction takes place at temperatures between 125 and 225 °C. Once all the water is eliminated from the hydrate, the boards shrink and turn to dust. Glass fibre is added to fire-rated gypsum boards (type X) to ensure the mechanical load-bearing capacity up to 400 °C. Volcanic rock is added to type C gypsum boards, which expands when heated and preserves the board's shape, while maintaining the fire resistance.

Case Study A room has a floor area A_f of 16 m^2, the area of the walls and ceiling A_{str} is 60 m^2. All the structures are covered on the inside with gypsum boards with thickness d 18 mm; the density of gypsum ρ is1500 kg/m^3. The specific heat released in a fire is Q' = 250 MJ per 1 m^2 of room floor area. Determine the fraction of heat released in the fire that is used to evaporate the chemically bound water in the gypsum boards; the heat of vaporization for water is 2500 kJ/kg = 2.5 MJ/kg. The total heat released in the fire is:

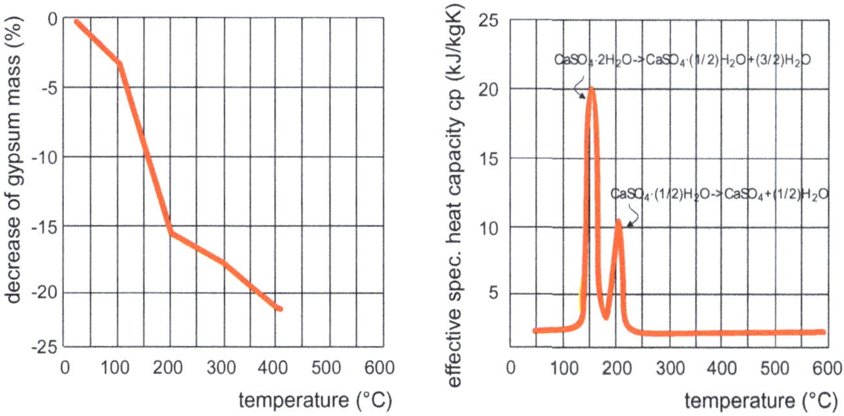

Fig. 6.10 Chemically bound water vaporises from heated gypsum and the gypsum board mass decreases in proportion to the mass fraction of water. The dehydration process ends at 400 °C [10] (left); Water vaporisation from gypsum can be seen through the temperature-dependent effective specific heat capacity c_p. The dehydration processes take place at 125 and 225 °C. The specific heat capacity at these temperatures is significant higher, which means that a lot more heat must be supplied (e.g., by fire) to the gypsum for a further increase of the temperature; as a consequence, the gypsum temperature barely changes until all the chemically bound water has vaporised [11] (right)

$$Q = Q' \cdot A_f = 250 \cdot 16 = 4000\left[\frac{MJ}{m^2}m^2 = MJ\right]$$

The mass fraction of water in gypsum is 21% (Fig. 6.10 shows the difference between the mass of the gypsum sample when heated from room temperature to 400 °C when it is completely dry), so the total chemically bound water in gypsum is:

$$m_{H_2O} = 0.21 \cdot A_{str} \cdot d \cdot \rho = 0.21 \cdot 60 \cdot 0.018 \cdot 1500$$
$$= 340\left[\frac{kg \cdot m^2 \cdot m \cdot kg}{kg \cdot m^3} = kg\right]$$

The fraction of heat released in the fire used to evaporate the water in gypsum is therefore:

$$x = \frac{340 \cdot 2.5}{4000}100 = 21.2\left[\frac{kg \cdot MJ}{kg \cdot MJ}\% = \%\right]$$

6.6 Fire Load of a Building

A building's fire load is defined by the heat release or the heat-release rate from a burning surface in a given moment. The heat release is measured in J, and the heat-release rate in J per second (W).

In the case of fire in buildings, the combustion process is out of control and the quantity of material burned in a unit of time depends on several parameters, so it must be determined with tests (Fig. 6.11). Materials, objects and devices in buildings are made of different substances, so the heat release in a fire is measured with special

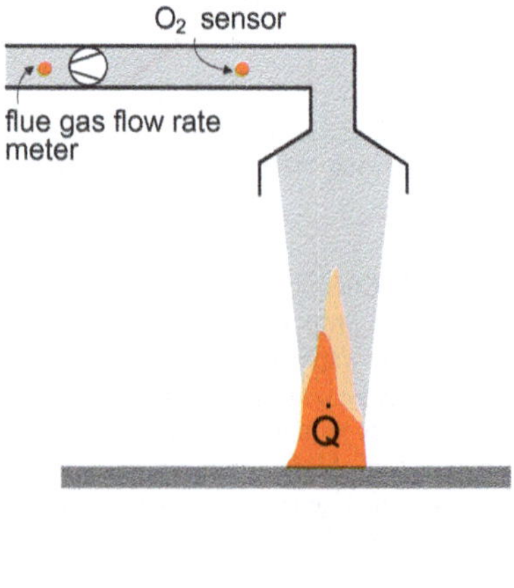

Fig. 6.11 Heat release rate $\dot{Q}$ from burning building materials and interior design elements (e.g., Fig. 6.4) is measured using oxygen consumption calorimetry [2, 12, 16]

tests, oxygen consumption calorimetry being the one most commonly used. For a variety of fuels, the heat release is proportional to the quantity of oxygen reacting in the oxidation. A total of 13,100 J of heat is released for every kg of oxygen involved in combustion. The heat-release rate in a fire is determined by measuring the quantity of flue gases collected in a canopy hood above the burning object, and the mass fraction of oxygen in the flue gases. The heat release is calculated by integrating the heat-release rate (heat flux) over time.

Case Study A burning timber structure releases 0.1 kg of flue gases every second. The mass fraction of oxygen CO_2 in flue gasses is 14.5%, and the mass fraction of oxygen in air is known to be $CO_{2,air}$ 23%. Find the instantaneous heat-release rate.

$$\dot{Q} = 13100 \cdot \left(CO_{2,air} - CO_{2,g}\right) \cdot \dot{m}_g = 13100 \cdot (0.23 - 0.145) \cdot 0.1$$
$$= 111.4 \left[\frac{kJ \cdot kg \cdot kg}{kg \cdot kg \cdot s} = \frac{kJ}{s} = kW\right]$$

Explanation The fractions of gas in the flue gases are usually declared by volume fraction. The volume fraction of oxygen in air is 20.9%. The mass fraction is determined by multiplying the volume fraction by the ratio of the air's molar mass ($M_{air} = 29$ g/mol) to the oxygen's molar mass ($M_{O2} = 32$ g/mol). The mass fraction of oxygen in the air is therefore $M_{O2}/M_{air} \times 20.9 = 23\%$.

The time-dependent heat-release rates $\dot{Q}$, determined with tests, can be used to build parametric models for certain combustible materials (Fig. 6.12). The ignition of liquid flues is a very common cause of fires in buildings. Liquids characteristically spill over a surface to form a pool, described by a diameter D. In the case of a pool fire, the heat-release rate $\dot{Q}$ is determined by the mass of burning liquid $\dot{m}''_{\infty}$ that every second burns off 1 m^2 of surface of a pool:

$$\dot{Q} = A \cdot \dot{m}''_{\infty} \cdot \left(1 - e^{-k_{\beta}D}\right) \cdot \chi \cdot H_i \left[\frac{m^2 \cdot kg \cdot MJ}{m^2 \cdot s \cdot kg} = \frac{MJ}{s} = MW\right] \tag{6.1}$$

where the constants k_{β} (m^{-1}) and χ (1) account for the size of the pool and the combustion efficiency, A is the area of the pool in m^2, and H_i is the heat of combustion of the spilled liquid fuel in MJ/kg (Table 6.3). In general, the $\dot{m}''_{\infty}$ (kg/m^{2}s) is more or less the same for any pool with a diameter D over 1.5 m.

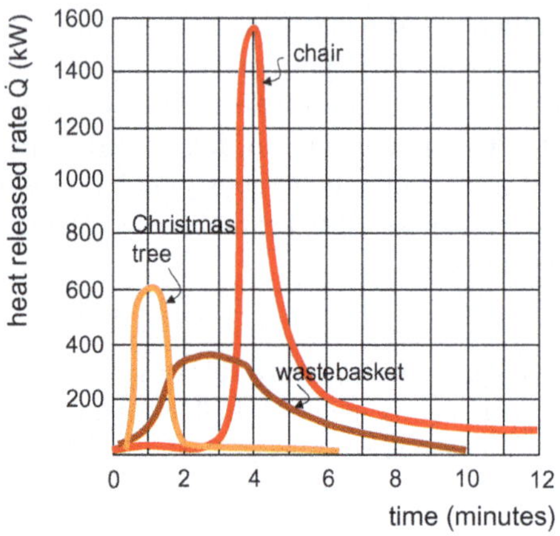

Fig. 6.12 Heat-release rate $\dot{Q}$ from burning interior design elements vs. time after the ignition. The total heat released from combustion equals the area under the heat release rate curve [1]

Table 6.3 Physical properties of some combustible liquids [2]

material	$\dot{m}''_\infty$ (kg/m²s)	k_β (m⁻¹)	H_i (MJ/kg)	ρ (kg/m³)
liquid hydrogen	0.017	6.1	120	70
methanol	0.017	∞	20.0	800
gasoline	0.055	2.1	43.7	740
kerosene	0.039	3.5	43.2	829
diesel fuel	0.035	1.7	42.5	940
heating oil	0.035	1.7	42.5	940
transformer oil	0.039	0.7	46.4	760

Case Study Find the heat-release rate from burning diesel fuel that was spilled over a garage floor. A total of 15 L of fuel were drained from the fuel tank and form a pool with a diameter D of 2.5 m. The efficiency of combustion $\chi = 0.95$. How long will the fuel burn? See Table 6.3 for the material properties of diesel fuel. The mass of spilled fuel and the pool area are:

$$m = V \cdot \rho = \frac{15}{1000} \cdot 940 = 14.1 \left[\frac{1 \cdot m^3 \cdot kg}{1 \cdot m^3} = kg \right]$$

$$A = \frac{\pi \cdot D^2}{4} = \frac{\pi \cdot 2.5^2}{4} = 4.9\ m^2$$

Solution: The heat-release rate $\dot{Q}$ from burning fuel is:

$$\dot{Q} = A \cdot \dot{m}''_\infty \cdot \left(1 - e^{-k_\beta D}\right) \cdot \chi \cdot H_i = 4.9 \cdot 0.035 \cdot \left(1 - e^{-1.7 \cdot 2.5}\right) \cdot 0.95 \cdot 42.5$$

$$\dot{Q} = 6.8 \text{ MW}$$

The duration of the fire (provided the fire does not spread to other combustible materials) is:

$$t = \frac{m}{\dot{m}''_{\infty} \cdot \left(1 - e^{-k_{\beta} D}\right)} = \frac{14.1}{0.035 \cdot \left(1 - e^{-1.7 \cdot 2.5}\right)} = 409 \text{ s}$$

There are different ways to define the heat-release rate for fires involving solid materials. For example, the charring rate describes the depth of material charred per minute (Table 6.4, left) and for elements with a known geometric shape, the heat-release rate is described using the specific heat q release per m^2 or kg of burning element (Table 6.5, left).

In engineering practice, the buildings are divided into fire enclosures (zones). A fire zone is a part of a building containing one or several rooms with a certain surface area (common over 200 m^2), designed so that the fire cannot spread to other compartments for a given time period. For every fire zone, the fire load is determined according to the expected heat release q in a fire using a five-levels scale. It is given as a specific quantity per m^2 of fire zone area. The fire-load-levels scale is presented

Table 6.4 Charring rates in millimetres per minute for various wood species [13] (left); Fire load scale and heat realised per 1 m^2 of fire-compartment floor area [14, 15] (right)

material	Δd (mm/min)
softwood	0.8
softwood, after burning for 10 minutes	0.6
hardwood	0.5
tropical wood	0.4

fire load level	q (MJ/m²)
very small	< 250
small	< 500
medium	< 1000
high	< 2000
very high	> 2000

Table 6.5 Heat releases during the combustion of 1 m^2 of building material or interior design element [3, 16] (left); Typical fire load expressed as the heat released during the fire per 1m^2 of building floor area in different building types [2] (right)

material	q (MJ/m²)
rubber	40
plywood	20
cork	25
PVC 17	
parquetry	20
papre	20
wool carpet	20
wooden door	410
window	510
wardrobe	810

building type	q (MJ/m²)
galleries	300-750
closed parking garages	400-800
museums	550
schools	550
hospitals	550
apartments	600
hotels	600-800
shops	800
offices	up to 1000
industrial facilities	up to 5500
warehouses	up to 50000

in Table 6.4 (right), and Table 6.5 (right) lists the specific heat q releases according to the type of building.

6.7 Characteristics of a Pre-flashover Fire

6.7.1 Building Materials Response-to-Fire Classification System

The combustibility of a material has to be determined with tests due to the complexity of the combustion processes and the unknown ingredients of nonhomogeneous building materials. The fire's growth phase from the ignition to flashover usually lasts less than 30 min and the temperature in the fire zone reaches between 500 and 600 °C. Building-material properties that are crucial during this phase are evaluated using the so-called response-to-fire performance. The test method used in Europe is called the Single Burning Item (SBI). Details can be found in [17]. A sample of tested material is inserted into the apparatus shown in Fig. 6.13 and heated with a gas burner. The response-to-fire performance is measured indirectly by monitoring the parameters of the flue gases. The 20-min test determines the heat-release rate (heat released per unit of time), the total heat released during the first 600 s of the test, the quantity of smoke (solid particles in the flue gases), and the size of the flaming debris falling from the material surface during the test. The measured values (Fig. 6.14) are presented using relative quantities: FIGRA (Fire Growth Rate) for the heat-release rate, SMOGRA (Smoke Growth Rate) for the smoke quantity and LFS for the falling flaming debris, among others. According to the Euroclass classification, building materials are divided according to the FIGRA (SBI) into seven response-to-fire performance classes: A1, A2, B, C, D, E and F. The classes are described in Table 6.6.

With respect to the smoke generated, building materials are divided into the following classes: s1 (minimum quantity of soot in the flue gases), s2 (mid quantity of soot) and s3 (not s1 or s2), and with respect to flaming debris falling from the surface they are classified as follows: d0 (no falling flaming debris), d1 (only short-lived (< 10 s) flaming debris) and d2 (no limitations) [17].

> **Explanation** A building material's response-to-fire performance is designated as follows: A2, s1, d0 for a gypsum board with a paper coating, and D, s2, d0 for most wood products with no fireproofing.

Fig. 6.13 In the test according to SBI (Single Burning Item) method, the corner of the apparatus is lined with specimens of tested material, which are then heated with a gas burner. The building material's reaction-to-fire class is determined by analysing the heat release and smoke soot, as well as by observing the flaming debris falling from its surface [17–19]

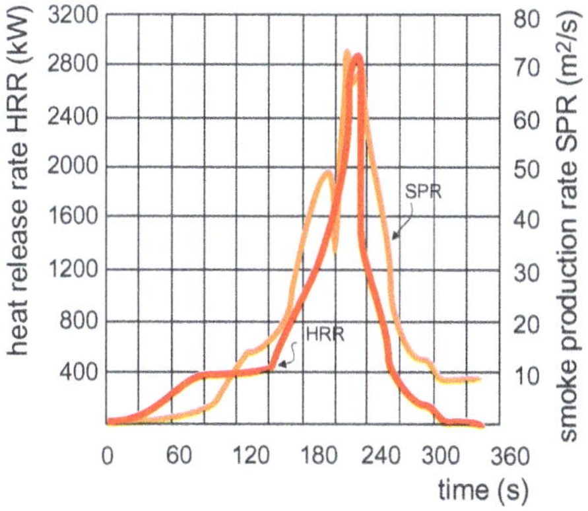

Fig. 6.14 SBI test results of heat-release rate (HRR) and smoke production rate (SPR). The HRR is determined by measuring the quantity of remaining oxygen in the flue gases, and the SPR is determined from the flue gases' light transmittance. The relative numbers FIGRA and SMOGRA are calculated according to the test results as basic indicators for the classification of a building material's response-to-fire. [12, 21]

Table 6.6 Building material response-to-fire performance classes

class	response to fire
A1	non-combustible materials; materials non-combustible by their material properties (stone, concrete, brick, metals)
A2	non-combustible materials; combustion takes place on a limited surface area; no significant heat release
B	materials difficult to ignite, no flashover
C	flashover after 10-20 minutes, mostly wood products treated with a fire retardant
D	flashover after 2-10 minutes, mostly wood products treated with a fire retardant
E	combustible materials that do not undergo SBI testing, flashover would take place sooner than in 2 minutes
F	materials not classified for response-to-fire performance

Fig. 6.15 Thermal insulation of residential buildings must have at least response-to-fire class B-d1. In the case of buildings taller than 22 m, the section of thermal insulation with response-to-fire class A1 or A2 must be applied with a width of at least 20 cm above the windows and doors and must be extended at least 30 cm beyond the edge of the window or door. Facades with an open air gap must be built with thermal insulation having response-to-fire class A1 or A2, s1, d0 to prevent the spread of the fire in the air cavity

Explanation In the regulations for fire safety in buildings the selection of building materials according to their response-to-fire performance is commonly requested. For example, external building walls, load-bearing structures and thermal insulation installed in residential, commercial and educational buildings with more than 1 ground + 3 floors must be made from A1 or A2 response-to-fire class materials (Fig. 6.15).

6.7.2 *Height of the Smoke Layer in a Compartment on Fire*

The fire zone can be divided into zones of permanent flame, sporadic flame and the smoke layer. Permanent flame is observed on the surface of burning materials. Occasionally, intense mixing of combustible gases and air also gives rise to sporadic flames above the permanent flame, usually with a frequency of 1–3 Hz. A plume of smoke is formed above the flames, rising and spreading across the room height. Smoke is hotter than indoor air and its density is lower than the density of the air surrounding the smoke plume, so smoke rises due to the natural buoyancy. The plume of smoke above the flame widens as the ambient air is entrained in the transversal direction (Fig. 6.16). Once the smoke reaches the ceiling, it starts to spread across it completely. From this moment on, the thickness of the layer of smoke increases.

According to the law of the conservation of energy, the heat added to gases increases their internal energy (temperature) and kinetic energy (pressure). Because of that, the flue gases expand. If the fire burns in a tight compartment, the smoke and heated air are unable to expand and the pressure in the compartment rises.

Case Study Find the pressure in an air-tight room on fire as the temperature of the smoke and the air inside reaches 40 °C (313 K). The air temperature in the room before the fire was 20 °C (293 K) and the air pressure was 1 bar (101 kPa).

Thermodynamic transformation of the gas mixture during which the volume remains constant is an isochoric process. The relation between the pressure and the temperature for an isochoric process is as follows:

$$\frac{p_o}{T_o} = \frac{p_1}{T_1} \rightarrow p_1 = \frac{T_1 \cdot p_o}{T_o} = \frac{313 \cdot 101}{293} = 107.9\ \text{kPa}$$

The pressure difference between the inside and outside is therefore 6.9 kPa. Comparing this result with the wind stagnation pressure (Sect. 2.2.6), it turns out the wind velocity v_w to yield the same pressure difference (assuming air density $\rho_{air} = 1.2\ \text{kg/m}^3$) would have to be:

$$(p_i - p_o) = \frac{1}{2}\rho_{air} \cdot v_w^2 \rightarrow v_w = \sqrt{\frac{2(p_i - p_o)}{\rho_{air}}} \approx 400\ \frac{km}{h}!$$

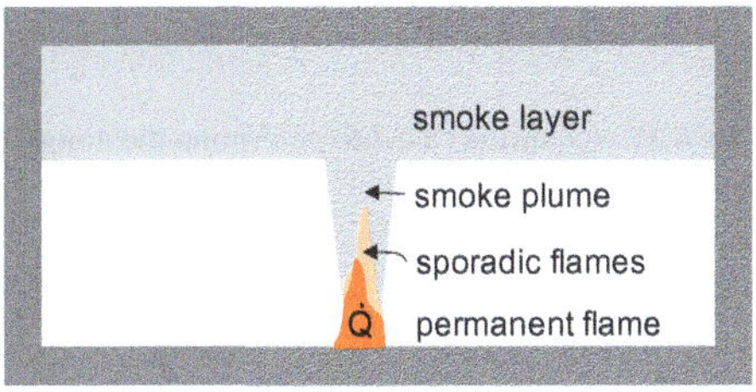

Fig. 6.16 Elements of fire in a fire compartment

Explanation The pressure differences in building fires are much smaller, because only a fraction of the heat released from the combustion is transferred to the smoke. A fraction of the heat (approximately one-third) is radiated from the flames and absorbed by the walls, while a part of the heat from the smoke is stored in the building structures. Nevertheless, the main reason why the real pressure differences are much lower is because the real buildings are not air-tight and some envelopment structures (such as glazing) will break when exposed to pressure differences of this magnitude between the inside and outside.

Considering fire safety, it is important that the smoke layer, which thickens from the fire compartment ceiling downwards during the fire-growth stage, remains above the space where people are evacuating long enough so that they can exit the building safely. The time from ignition to the moment when smoke layer descends to a given height z is determined by the quantity of smoke produced in the fire.

6.7.2.1 Amount of Smoke Produced in a Fire

Uncontrolled combustion is a complex phenomenon and experimental research is a very important tool for predicting the combustion process. Multi-parametric empirical models are built based on the results of experiments and this approach is also used to predict the quantity of smoke produced in a fire. Several models are available, e.g., according to Zukoski, Heskestad, McCaffrey and others [2]. The Zukoski model is the most commonly used in engineering practice. Here, the quantity of smoke is calculated from the heat-release rate $\dot{Q}$ (kW). As this is the initial stage of the fire, the heat-release rate is determined using the fire-growth factor α (Table 6.2). The quantity of smoke depends on the distance z from the base of the fire to the bottom level of the smoke layer (Fig. 6.17, left) because the flue gases are expanding and mixing with the air from the room, and Zukoski's expression gives the quantity of smoke realised according to the given height z:

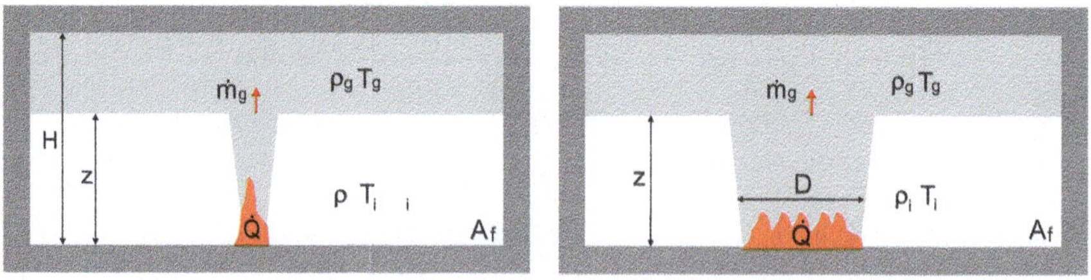

Fig. 6.17 Quantities used to determine the amount of the smoke in a fire using the Zukoski model (left) and according to the Thomas model (right) [1]

$$\dot{m}_{g(z)} = 0.21\left(\frac{\rho_i^2 \cdot g}{c_{p,i} \cdot T_i}\right)^{1/3} \cdot \dot{Q}^{1/3} \cdot z^{5/3}\ [kg/s] \tag{6.2}$$

where T_i (K) and ρ_i (kg/m^3) are the air temperature and density in the room, g is the acceleration due to gravity (m/s^2), $c_{p,i}$ is the specific heat capacity of air at constant pressure (kJ/kgK), $\dot{Q}$ is the heat-release rate (kW) and z is the height (m) above the base of the fire. Assuming the air temperature is 20 °C, the expression simplifies to:

$$\dot{m}_{g(z)} = 0.071 \cdot \dot{Q}^{1/3} \cdot z^{5/3}\ [kg/s] \tag{6.3}$$

Explanation The amount of flue gases generated by combustion is considerably smaller than the amount of smoke predicted by the Zukoski model, mainly due to the intense mixing of smoke and air in the fire compartment. Smoke from the combustion, amounting to 0.5–2% of the total amount of flue gases.

The heat-release rate $\dot{Q}$ can also be determined from the quantity of material $\dot{m}''_\infty$ burned off one m^2 of surface every second (Table 6.3) using Eq. 6.1. In the case that the fire load of a fire compartment is defined by the specific heat q release per m^2 of burning substance (Table 6.5, left) and the duration of the fire t is known, the heat-release rate $\dot{Q}$ can be determined with:

$$\dot{Q} = \frac{1000 \cdot q \cdot A_t}{t}\left[\frac{kW \cdot MJ \cdot m^2}{MW \cdot m^2 \cdot s} = kW\right]$$

For burning spilled liquids, the quantity of smoke can be calculated based on the burning surface diameter D (Fig. 6.17, right) using the Thomas model [2]:

$$\dot{m}_{g(z)} = 0.592 \cdot D \cdot z^{3/2}\left[\frac{kg}{s}\right] \tag{6.4}$$

6.7.2.2 Variation of Smoke-Layer Height with Time

Assuming the law of conservation of mass, the Zukovski model for the smoke quantity $\dot{m}_{g(z)}$, the "t-squared" rule for the heat-release rate $\dot{Q}$ and the air properties at T_i = 20 °C, the height up to the bottom of the smoke layer's z (Fig. 6.17, left) at time t from ignition can be determined with:

$$z = \left(\frac{3.06 \cdot 10^{-2} \cdot \alpha^{1/3} \cdot t^{5/3}}{A_t} + \frac{1}{H^{2/3}}\right)^{-3/2}\ [m]$$

The expression can be modified to find the time t when the bottom of the smoke layer reaches a given height z (Fig. 6.18):

$$t = \left(\frac{32.7 \cdot A_t}{\alpha^{1/3}}\left(\frac{1}{z^{2/3}} - \frac{1}{H^{2/3}}\right)\right)^{3/5} \quad [s]$$

where A_f (m^2) and H (m) are the floor area and height of the compartment, while α (kW/s^2) is the fire-growth factor.

Case Study An "ultra-fast" class fire (Table 6.2, fire-growth factor α = 0.19 kW/s^2) broke out in a manufacturing plant with a floor area A_f 1250 m^2 and height h 6.5 m. Find the smoke layer height z two minutes after ignition.

$$z = \left(\frac{3.06 \cdot 10^{-2} \cdot \alpha^{1/3} \cdot t^{5/3}}{A_t} + \frac{1}{H^{2/3}}\right)^{-3/2}$$

$$z = \left(\frac{3.06 \cdot 10^{-2} \cdot 0.19^{1/3} \cdot 120^{5/3}}{1250} + \frac{1}{6.5^{2/3}}\right)^{-3/2} = 5.3\ \text{m}$$

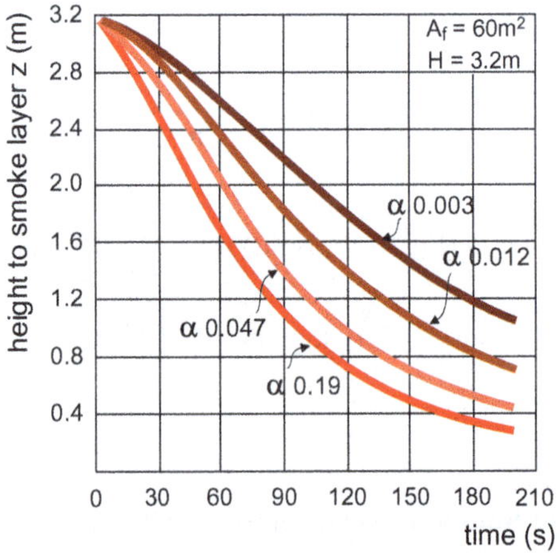

Fig. 6.18 The variation of height of the bottom level of the smoke layer z versus time for different fire-growth factors α in a fire-compartment with the floor area $A_f = 60$ m^2 and height $H = 3.2$ m.
Fig. 6.18 The variation of height of the bottom level of the smoke layer z versus time for different fire-growth factors α in a fire-compartment with the floor area $A_f = 60$ m^2 and height $H = 3.2$ m

Case Study A middle-growth-rate class fire broke out in a room with a floor area A_f 60 m^2 and height H 3.2 m (Table 6.2, $\alpha = 0.012$ kW/s). Determine the time after ignition when the smoke layer will descend to a height $z = 2$ m above the ground.

$$t = \left(\frac{32.7 \cdot A_f}{\alpha^{1/3}}\left(\frac{1}{z^{2/3}} - \frac{1}{H^{2/3}}\right)\right)^{3/5} = \left(\frac{32.7 \cdot 60}{0.012^{1/3}}\left(\frac{1}{2^{2/3}} - \frac{1}{3.2^{2/3}}\right)\right)^{3/5} = 78 \text{ s}$$

Explanation The expressions above are valid for rooms with small openings due to which the indoor air pressure is balanced with the outdoor air pressure for the duration of the fire ($dp/dt = 0$), while no smoke is exiting the room.

6.8 Characteristics of a Post-flashover Fire

6.8.1 Smoke Temperature in the Fire Compartment During the Post-flashover Phase of the Fire

The temperature in fire compartment increases during pre-flashover as well as at the beginning of the post-flash fire period, until there is a shortage of combustible materials. To determine the load-bearing capacity of the building structures as well as their heat-insulation capacity, it is important to know the temperature of the smoke at the surface of exposed building structures. Among the other approximation models, the prediction of the temperate in fire compartments could be made with a nominal temperature–time curve, also called "ISO 834" after the standard published by International Standard Organization (ISO) [22]. The empirical expression used to determine the smoke temperature θ_g based on the fire's duration t quoted in minutes is as follows:

$$\theta_g = 20 + 345 \cdot \log(8 \cdot t + 1) \left[{}^\circ C\right]$$

Explanation The "ISO 834" curve was determined empirically for typical building fires, but there is also a temperature curve for tunnel fires (the hydrocarbon curve) and an outdoor fire curve.

The "ISO 834" temperature–time curve does not predict that the temperature in the fire compartment will start to drop once the combustibles are depleted (Fig. 6.19). More complex computational models must be used in order to provide a more plausible prediction of the room or fire-compartment temperature θ_g as a function of time. The single-zone model is often used, which means that we assume that the smoke at uniform temperature θ_g in the fully developed stage fills up the entire fire compartment. The single-zone model is based on the energy balance in the unit of time (Fig. 6.20), including:

- the heat-release rate $\dot{Q}$ determined by the building's fire load q (Table 6.4, right) and the area of the fire compartment A_{com};
- the rate of heat accumulated in the building structures $\dot{Q}_a$, which depends on the smoke temperature θ_g, the building structure's thermal effusivity b (Table 6.7), calculated as $(\lambda \times \rho \times c)^{1/2}$, and the building structure's area A_s; (see also Sect. 1.7 on transient heat transfer);
- the radiative heat-transfer rate $\dot{Q}_s$ through the ventilation openings towards the outdoor environment, determined using the Stefan–Boltzmann law, assuming the smoke emissivity $\varepsilon_g \sim 1$;
- the heat-transfer rate $\dot{Q}_c$ removed from the building by smoke through the ventilation openings; it depends on the smoke flow rate estimated indirectly using the ventilation factor VF determined by the expression:

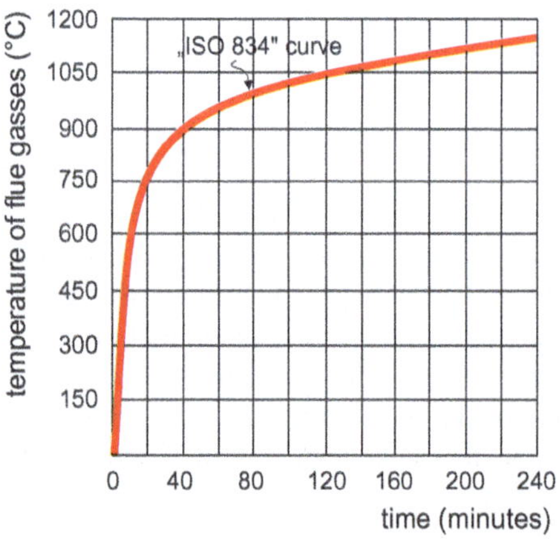

Fig. 6.19 Standard "ISO 834" temperature–time curve applied for an experimental evaluation of a building structure's response-to fire [22]

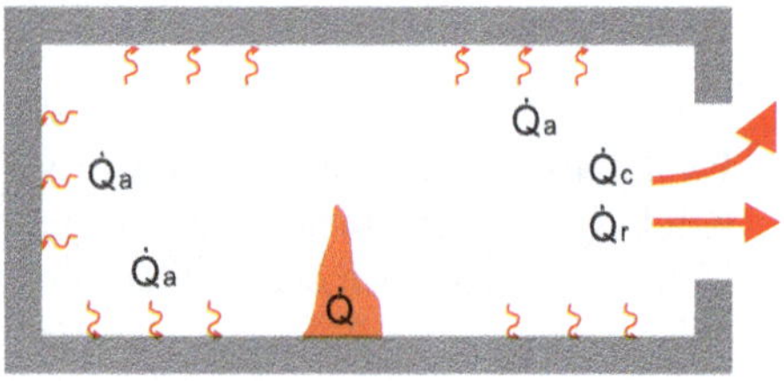

Fig. 6.20 Heat flows included in the energy balance for a single-zone fire-compartment in the fully developed stage of a fire

Table 6.7 Thermal effusivity b for a variety of building materials

material	thermal effusivity b (J/m^2 Ks$^{0.5}$)
concrete	2180
brick	1160
aerated concrete	250
wood	380
light sandwich elements	100

$$VF = \frac{\sum_{i=1}^{n} A_{o,i}\sqrt{h_{o,i}}}{A_{env}} = \frac{\sum_{i=1}^{n} l_{o,i} \cdot h_{o,i}^{\frac{3}{2}}}{A_{env}} \left[m^{\frac{1}{2}}\right]$$

where $A_{o,i}$ (m^2) is the i-th free opening area in the building envelope, $l_{o,i}$ (m) is the i-th opening free width and $h_{o,i}$ (m) is the i-th opening free height (Fig. 6.21, left) through which smoke passes to the outside, n is the numerator of the ventilation openings and A_{env} is the fire compartment's opaque envelope area. It is typical that the free-ventilation opening area amounts to 65–80 % of the surface of the opening in the fire-compartment envelope (e.g., a wall).

Explanation If a room or fire compartment is made of several different structures, the mean thermal effusivity is calculated with the following expression:

$$b = \frac{\sum_{j=1}^{n} b_j \cdot A_j}{\sum_{j=1}^{n} A_j} \left[\frac{J}{m^2Ks^{1/2}} = \frac{Ws^{1/2}}{m^2K}\right]$$

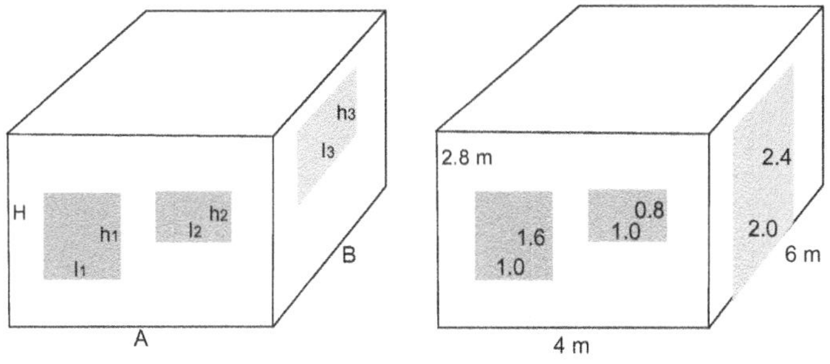

Fig. 6.21 Openings and fire-compartment dimensions used in the calculation of the ventilation factor VF (left), free openings and fire-compartment dimensions for the case study building (right)

The ventilation factor *VF* also influences the quantity of oxygen (air) necessary for combustion and supplied to the fire compartment from outside. It follows that *VF* decides whether the course of the fire will be determined by the quantity of oxygen (if $VF < 0.02\ m^{1/2}$) or the combustible materials (if $VF > 0.08\ m^{1/2}$). In the former case, the temperatures θ_g are lower and the fire burns longer, while in the latter case the fire is more intense (the maximum temperature $\theta_{g,max}$ will be higher and achieved sooner), but the duration of the fire is shorter. The most dangerous are the fires where the ventilation factor lies somewhere between these boundary values, as the combustion efficiency χ and the temperature θ_g are then the greatest.

Case Study Estimate whether the limiting factor for fire in the compartment shown in Fig. 6.21 (right) will be the quantity of oxygen or the combustible materials in the room.

$$VF = \frac{\sum_{i=1}^{n} l_{o,i} \cdot h_{o,i}^{\frac{3}{2}}}{A_{env}}$$

$$= \frac{l_1 \cdot h_1^{\frac{3}{2}} + l_2 \cdot h_2^{\frac{3}{2}} + l_3 \cdot h_3^{\frac{3}{2}}}{(2 \cdot A \cdot B + 2 \cdot A \cdot H + 2 \cdot B \cdot H) - (l_1 \cdot h_1 + l_2 \cdot h_2 + l_3 \cdot h_3)} \left[m^{\frac{1}{2}}\right]$$

$$VF = \frac{1 \cdot 1.6^{\frac{3}{2}} + 1 \cdot 0.8^{\frac{3}{2}} + 2 \cdot 2.4^{\frac{3}{2}}}{(2 \cdot 4 \cdot 6 + 2 \cdot 4 \cdot 2.8 + 2 \cdot 6 \cdot 2.8) - 7.2} = 0.11\ m^{\frac{1}{2}}$$

Solution: Ventilation factor *VF* is greater than $0.08\ m^{1/2}$, which means the fire will be limited by the quantity of combustibles.

Based on the single-zone model and considering the typical fire-related properties of buildings and the conditions governing the growth of a fire in a building, the European Committee for Standardization has proposed the EUROCODE method. This method determines the temporal variation of the temperature in a fire compartment for the periods of fire growth and decay. The length of the growth period t_{max} (h) is determine by the moment when the flue gases reach the highest temperature:

$$t_{max} = \frac{0.13 \cdot 10^{-3} \cdot q}{VF}\ [h]$$

where *VF* ($m^{1/2}$) is the ventilation factor of the openings and *q* (MJ/m^2) (Tables 6.4, right and 6.5, right) is the specific heat in MJ realised from 1 m^2 of fire-compartment area A_{com}. In the case that the effect of heat storage in the building structures differ from the reference value of the thermal effusivity *b* (1160 $J/m^2Ks^{1/2}$) and the ventilation factor *VF* differs from the reference one (0.04 $m^{1/2}$), the length of the growth period of the fire is determined using the normalised time $t^*_{max}(h)$, given by:

$$t^*_{max} = t_{max} \cdot \left(\frac{VF}{b}\right)^2 \left(\frac{1160}{0.04}\right)^2 \; [h]$$

Knowing the normalized time t^*_{max}, the maximum temperature $\theta_{g,max}$ in the fire compartment at the end of the fire-growth stage is given by:

$$\theta_{g,max} = \theta_i + 1325\left(1 - 0.324 \cdot e^{-0.2 \cdot t^*_{max}} - 0.204 \cdot e^{-1.7 \cdot t^*_{max}} - 0.472 \cdot e^{-19 \cdot t^*_{max}}\right) \left[^\circ C\right]$$

where θ_i is the initial temperature in the room. The temperature θ_g rise from the initial temperature θ_i to the maximum temperature at the end of the growth period $\theta_{g,max}$ is determined by replacing the t^*_{max} in the expression for $\theta_{g,max}$ with the observation time t^* (h) ($t^* \leq t^*_{max}$).

After reaching $\theta_{g,max}$, the intensity of the fire begins to decay. The temperature in the fire compartment during the fire-decay period ($t^* > t^*_{max}$) is determined by one of the following equations:

$$\theta_g = \theta_{g,max} - 625\left(t^* - t^*_{max}\right) \left[^\circ C\right] \text{ at } t^*_{max} \leq 0.5 \text{ h} \quad (6.5)$$

$$\theta_g = \theta_{g,max} - 250\left(3 - t^*_{max}\right)\left(t^* - t^*_{max}\right) \left[^\circ C\right] \text{ at } 0.5 < t^*_{max} < 2 \text{ h} \quad (6.6)$$

$$\theta_g = \theta_{g,max} - 250\left(t^* - t^*_{max}\right) \left[^\circ C\right] \text{ at } t^*_{max} \geq 2 \text{ h} \quad (6.7)$$

Figure 6.22 shows the temperature in the fire compartment with different fire loads q for different ventilation factors VF (0.04 and 0.12 $m^{1/2}$). In both cases the building structure's thermal effusivity b of 1160 $J/m^2Ks^{1/2}$ was taken into account. It can be seen that in the first case, the limited amount of oxygen (as VF is lower) causes a lower maximum temperature of the smoke $\theta_{g,max}$ while the fire lasts longer.

Case Study Use the EUROCODE method to find the maximum temperature θ_g in a room with the fire load $q = 400$ MJ/m^2 and the ventilation factor $VF = 0.04$ m$^{1/2}$, if the building structures are made of concrete ($b = 2180$ J/m^{2}K s$^{1/2}$), brick ($b = 1160$ J/m^{2}K s$^{1/2}$) or timber ($b = 380$ J/m^{2}K s$^{1/2}$). The temperature in the room θ_i before the fire starts was 0 °C.

The time t_{max} at which the growth period of fire ends is equal for all cases:

$$t_{max} = \frac{0.13 \cdot 10^{-3} \cdot q}{VF} = \frac{0.13 \cdot 10^{-3} \cdot 400}{0.04} = 1.30 \text{ h} = 1:20 \text{ h}:\text{min}$$

For the room with concrete structures, the normalized time t^* is equal to:

$$t^*_{max} = \frac{0.13 \cdot 10^{-3} \cdot q}{VF}\left(\frac{VF}{b}\right)^2 \left(\frac{1160}{0.04}\right)^2 = \frac{0.13 \cdot 10^{-3} \cdot 400}{0.04}\left(\frac{0.04}{2180}\right)^2 \left(\frac{1160}{0.04}\right)^2$$
$$= 0.37 \text{ h}$$

and the maximum temperature $\theta_{g,max}$ at the end of the growth phase of the fire will be:

$$\begin{aligned}\theta_{g,max} &= 0 + 1325\left(1 - 0.324 \cdot e^{-0.2 \cdot t^*_{max}} - 0.204 \cdot e^{-1.7 \cdot t^*_{max}} - 0.472 \cdot e^{-19 \cdot t^*_{max}}\right) \\ &= 0 + 1325\left(1 - 0.324 \cdot e^{-0.2 \cdot 0.37} - 0.204 \cdot e^{-1.7 \cdot 0.37} - 0.472 \cdot e^{-19 \cdot 0.37}\right) \\ &= 781^{\circ}C\end{aligned}$$

The temperature in the room in the period of fire decay is determine by Eq. (6.5) in the case of a room with concrete structures, with Eq. (6.6) in the case of a room with structures made of bricks and with Eq. (6.7) in the case of timber structures.

Solutions for all three cases are shown in Fig. 6.23. The maximum temperatures $\theta_{g,max}$ will be 1280 °C in the building with timber structures, 965 °C in the building with brick structures, and 780 °C in the building with concrete structures due to the higher heat accumulation in the building structures. For the same reason, the temperature in fire compartments built with concrete walls is higher for a longer period (Fig. 6.24).

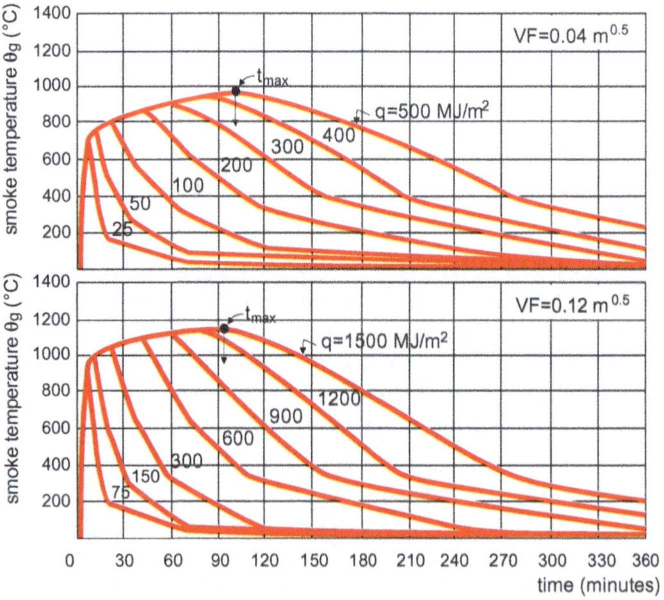

Fig. 6.22 Fire-compartment EUROCODE temperature θ_g-time curves for different specific heat releases q (MJ/m^2), above for ventilation factor $VF = 0.04$ m$^{1/2}$, below for ventilation factor $VF = 0.12$ m$^{1/2}$. In both cases it was assumed that the thermal effusivity of the building structures is 1160 J/m^{2}Ks$^{1/2}$

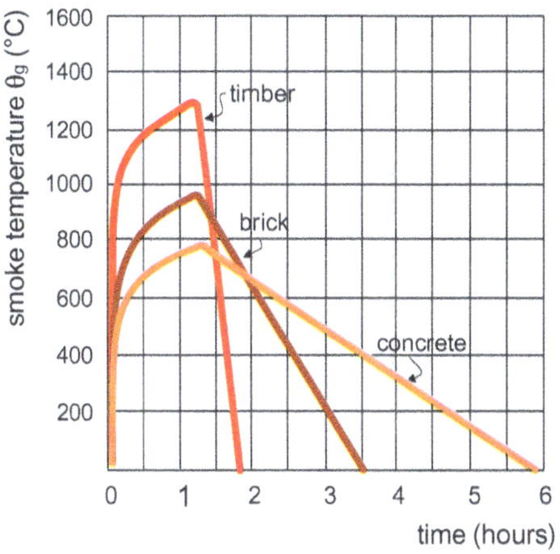

Fig. 6.23 EUROCODE temperature–time curves θ_g in the fire compartment determined for the case study

Fig. 6.24 Fire test furnace with gas burners at the side. The tested building structure is inserted air-tight in the front opening. The temperature in the furnace is increased by burners according to the "ISO 834" temperature–time curve [22]

6.8.2 Building Structures' Response to Fire and the Structures' Fire-Resistance Classification

In the case of fire, building structures must retain their mechanical load-bearing capacity, prevent the spread of fire to neighbouring buildings and protect people from the heat and smoke. These tasks are the basis for setting requirements relating to the fire resistance of the building structures. Building structures are made of different materials and elements, so it is common in engineering practice that their response to the fire has to be determined using tests. The tests are conducted in a fire test furnace with one 3 × 3 m wall built from the tested building structure. Gas burners are installed around the circumference of the furnace to heat the air inside the furnace and they are controlled in a way that the flue-gas temperature θ_g in the furnace increases in compliance with the standard "ISO 834" temperature–time curve.

The building structure's response to the fire is described with several specific ratings that tell us for what period of time after the start of the test the building structure retains its:

- load-bearing capacity; designated R as resistance;
- integrity, determined as the building structure's ability to prevent the passage of smoke through cracks in the structure and the ignition of the unexposed (opposite) side of the structure; designated E as integrity;
- thermal insulation, measured by the temperature on the side of the building structure; the time period until the maximum temperature on the opposite surface of the structure reaches 140 °C is measured; designated I as insulation.

The building structures' fire-resistance classification uses a system of time classes expressed in minutes:

- (REI/RE/R/EI/E) (15, 30, 45, 60, 90, 120, 180, 240, 360)

For example, a building structure with a response to the fire REI 120 will maintain load-bearing, tightness for flue gases and keep the opposite surface (towards a neighbouring room) below 140 °C for 120 min.

The fire-resistance properties of building structures can also be extended with other ratings, such as W, which evaluates the building structure's resistance to radiative heat transfer, the ability to prevent gas leakage S, and C for self-closing doors. These ratings are commonly stated for particular structures, for example, doors between fire compartments (Fig. 6.25).

Case Study National legislation on the fire safety in buildings in Slovenia regulates the minimum fire resistance R for load-bearing structures. For structures that separate fire compartments and prevent the spread of the fire, an adequate integrity rating E and the insulation rating I for building structures

Fig. 6.25 Sprinkler heads; red fluid in the glass bulb indicate that the sprinkler system will activate at approximately 77 °C [23] (left); sprinkler system for automatic fire extinguishing in an underground garage; each sprinkler head covers approximately 20 m^2, demanding a complex network of pipes for water supply (right)

Table 6.8 Example of building structure fire resistance requirements [14] according to the type and height of the building

building type	building height				
	P+1	P+2	P+3	P+4	P+6
residential	$R60^a$ $EI60^b$	$R60^b$ $EI60^b$	$R60^b$ $EI90^d$	R60 EI90	R60 EI90
public	$R30^a$ $EI30^a$	$R60^b$ $EI60^b$	$R90^d$ $EI90^d$	R90 $EI90^d$	$R90^c$ $EI90^c$
schools	$R30^a$ $EI30^a$	$R60^b$ $EI60^b$	$R90^d$ $EI90^d$	R90 $EI90^d$	$R90^c$ $EI90^c$
hotels	$R30^a$ $EI30^a$	$R60^b$ $EI60^b$	$R90^d$ $EI90^d$	R90 $EI90^d$	$R90^c$ $EI90^c$
shops	$R30^a$ $EI30^a$	$R60^b$ $EI60^b$	$R90^d$ $EI90^d$	R90 $EI90^d$	$R90^c$ $EI90^c$
industrial q =< 1 MJ/m^2	R30 $EI30^a$	R60 $EI30^a$	R90 $EI60^b$	R90 EI60	R90 EI60
garages	$R30^a$ $EI30^a$	$R30^b$ $EI30^a$	$R60^b$ $EI60^b$	R60 EI60	R60 EI60

a—timber structures are permitted; b—timber structures with non-combustible coatings are permitted; c—an automatic sprinkler fire-suppression system must be installed d—fire resistance R can be lowered by one class if an automatic sprinkler fire-suppression system is installed

are required additionally. Table 6.8 shows an example of the requirements in the Slovenian fire-safety regulations. The regulations require that doors with a minimum fire resistance EI 30—CS [meaning doors having a self-closing mechanism (C) and increased smoke-tightness (S)], are installed on the boundaries of the fire compartment. The rules on fire safety in buildings allow the minimum R-rating for building-structure fire resistance to be degraded by one class (e.g., from 240 to 180) if there is a sprinkler system for fire extinguishing installed in the building. Sprinkler systems are among the active building fire-protection measures, which include systems and devices for automatic fire detection and suppression, as well as transporting smoke and heat away from the building. Automatic fire-suppression systems operate by preventing the development of the fire by blocking the fuel supply, lowering the flame temperature, or preventing the oxygen supply. Automatic fire-suppression systems are distinguished by the extinguishing agent used: water or water mist, foam, gases or powder.

Sprinkler systems are used for automatic fire extinguishing by spraying the water in residential and commercial buildings, hotels and hospitals (Fig. 6.23). They are designed as a horizontal pipeline network connected to the public water supply, a pressure booster pump or a water storage tank with capacity that can provide enough water during the growth phase of the fire. The supply pipeline is installed under the ceiling or in plenum space. Each pipeline branch ends with a sprinkler head, having a fluid-filled glass bulb. This fluid expands when heated by the hot flue gases in the case of fire, breaking the glass and allowing the water to exit the nozzle. The sprinkler-activation temperatures can be different and are coded by the colour of the bulb fluid. The most common red glass bulbs break when exposed to 77 °C for several tens of seconds. Each sprinkler head covers 20 m^2 of floor area, and the water consumption for fire suppression is between 3 and 5 L per minute and m^2 of fire-protected area. In wet pipe systems, the water is always kept under pressure inside the pipeline. Such systems are suitable for rooms where there is no risk of freezing, otherwise the pipes must be heated using electric cables. Only sprinkler heads situated in areas of fire compartment on the fire are activated. A high system water pressure and special spraying nozzles enable fire suppression with a water mist (HI-FOG systems). The water droplet diameter is 20-times smaller than for ordinary sprinkler systems (approx. 50 μm instead of 1 mm), while their surface area is 400-times larger. This is the reason why these micro-droplets are quicker to boil and the amount of heat removed is significantly larger. In dry-pipe sprinkler systems utilized in unheated buildings, the pipes are filled with pressurized air. Once a sprinkler head opens, the air pressure in the pipes drops and the water enters the system. Such systems supply a spray of water with some delay because the system must first be vented through activated sprinkler heads and the pipelines are subject to corrosion. Using copper or stainless-steel pipelines or using nitrogen gas to pressurize the system could be solutions to avoid the risk of corrosion.

For special applications, automatic foam fire-suppression systems can be used, most often to extinguish fires on inflammable liquids. The foam covers the liquid surface, preventing the oxygen supply. Their design is like water-based sprinkler systems and a foam concentrate is added to the water, which reacts with the air to form the foam. The same physical principle of fire suppression is also used in gaseous fire-suppression systems. These systems work by flooding the burning surfaces with an inert gas such as CO_2, halon or nitrogen. These gases leave no trace and are therefore used in mission-critical applications or when the interior must be protected. Dry chemical powder fire-suppression systems cover burning surfaces with a dry agent, carried by an inert gas (CO_2 or N_2). The agent must be non-hygroscopic, so it stays dry and does not conduct electricity. Such systems are suitable for extinguishing fires on electrical installations.

6.8.3 Smoke Control in a Fire Compartment Using Natural and Forced Ventilation

Note The extraction of smoke from the fire compartment is not only related to the extraction of toxic flue gases, but also with the equally important heat dissipation released in a fire and therefore the two problems are addresses at the same time.

In the fully developed fire that follows the flashover (Fig. 6.3) the heat-release rate is constant and the amount of the smoke formed by combustion can be treated as constant as well.

One of the tasks for building fire-safety designers is to ensure the smoke is extracted from the fire compartment, so the smoke layer stays high enough to ensures that residents will be able to escape from the burning building in time. By extracting the smoke, the heat load of the building decreases because the heat is transferred by hot gases to the outside.

There are two main techniques for extracting smoke from the compartment on fire (Fig. 6.26): ventilation based on natural buoyancy effects, and forced ventilation with mechanical systems. Both techniques are effective only if the supply of outdoor air below the smoke layer is ensured at the same time. According to the law of conservation of mass, the mass flow rate $\dot{m}_i$ of the air entering the zone under the smoke layer through the ventilation inlet openings is equal to the mass flow rate of smoke formed by combustion $\dot{m}_g$ and the mass flow rate of smoke exiting the compartment $\dot{m}_o$:

$$\dot{\mathrm{m}}_\mathrm{i} = \dot{\mathrm{m}}_\mathrm{g} = \dot{\mathrm{m}}_\mathrm{o} \left[\frac{\mathrm{kg}}{\mathrm{s}}\right]$$

Considering the law of the conservation of energy, the heat rate $\dot{Q}$ realized by burning is partly transferred by heat rate $\dot{Q}_g$ of flue gases that exit the fire compartment

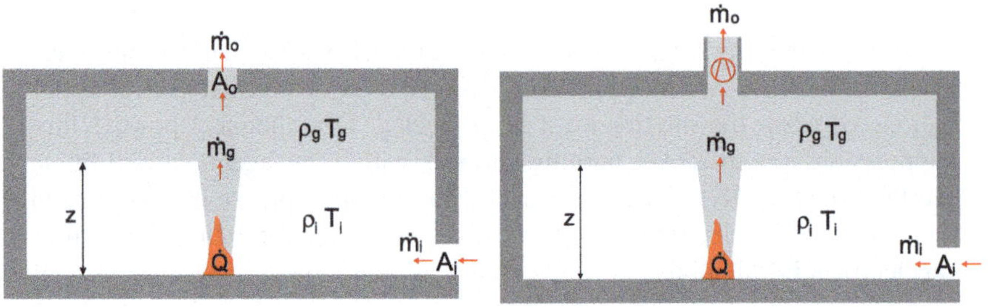

Fig. 6.26 Mass flows in the case of natural (above) and forced (bottom) smoke ventilation. The heat release rate $\dot{Q}$ during the fully developed stage is constant and so are all the mass flow rates: of smoke $\dot{m}_g$, of inflow air $\dot{m}_i$, and of outgoing flue gas and air mixture $\dot{m}_o$; as consequence, the depth of the smoke layer is constant as well as

towards the surroundings and partly stored as heat rate $\dot{Q}_a$ in the fire-compartment building structures.

The heat flow rate $\dot{Q}_g$ is proportional to the mass flow of smoke exiting the building $\dot{m}_o$, while the heat flow rate $\dot{Q}_a$ is proportional to the building structure's thermal effusivity b. The law of the conservation of energy can therefore be written as:

$$\dot{\mathrm{Q}} = \dot{\mathrm{Q}}_{\mathrm{g}} + \dot{\mathrm{Q}}_{\mathrm{a}} = \dot{\mathrm{m}}_{\mathrm{o}} \cdot \mathrm{c}_{\mathrm{p,g}} \cdot \left(\theta_{\mathrm{g}} - \theta_{\mathrm{i}}\right) + \frac{\mathrm{b} \cdot \mathrm{A}_{\mathrm{env}}}{\sqrt{\pi \cdot \mathrm{t}}} \cdot \left(\theta_{\mathrm{g}} - \theta_{\mathrm{i}}\right) \ [\mathrm{W}]$$

where $\dot{m}_o$ (kg/s) is the mass flow of smoke exiting the room, $c_{p,g}$ (kJ/kgK) is the specific heat capacity of the smoke, θ_g is the temperature of the smoke and θ_i (°C) is the temperature of the indoor air in the layer below the smoke, which is equal to the temperature of outdoor air entering the fire compartment, t (s) is the time after ignition, and A_{env} (m^2) is the surface of the opaque building structures in contact with the smoke. The smoke temperature is derived from the above equation of energy conservation:

$$\theta_{\mathrm{g}} = \theta_{\mathrm{i}} + \frac{\dot{\mathrm{Q}}}{\dot{\mathrm{m}}_{\mathrm{o}} \cdot \mathrm{c}_{\mathrm{p}} + \frac{\mathrm{b} \cdot \mathrm{A}_{\mathrm{env}}}{\sqrt{\pi \cdot \mathrm{t}}}} \ [\mathrm{K}]$$

The smoke density ρ_g can be determined by the ideal gas law:

$$\rho_{\mathrm{g}} = \frac{\mathrm{p} \cdot \mathrm{M}}{\mathrm{R} \cdot \mathrm{T}_{\mathrm{g}}} = \frac{353}{\theta_{\mathrm{g}} + 273} \left[\frac{\mathrm{N} \cdot \mathrm{kg} \cdot \mathrm{K} \cdot \mathrm{mol}}{\mathrm{m}^2 \cdot \mathrm{mol} \cdot \mathrm{J} \cdot \mathrm{K}} = \frac{\mathrm{N} \cdot \mathrm{kg} \cdot \mathrm{K} \cdot \mathrm{mol}}{\mathrm{m}^2 \cdot \mathrm{mol} \cdot \mathrm{N} \cdot \mathrm{m} \cdot \mathrm{K}} = \frac{\mathrm{kg}}{\mathrm{m}^3} \right]$$

where, p is the pressure in Pa (N/m^2), M (kg/mol) is the molar mass, R (J/K mol) is the gas constant and θ_g (°C) is the smoke temperature. We have ascertained in Sect. 2.2.5 that the fluid flow in a natural ventilated enclosure depends on the pressure differences between the indoor and outdoor air pressure at the cross-sections of the inlet Δp_i and outlet openings Δp_o. In the case of natural ventilation of the enclosure, the difference in air pressure occurs due to the different temperatures of the outdoor air θ_o, the air in the compartment below the smoke layer θ_I and the temperature of the smoke θ_g. A pressure gradient in the natural ventilated fire-compartment is shown in Fig. 6.27. The pressure difference at the inlet (lower) opening Δp_i is determined with Bernoulli's equation (Sect. 2.2.5), assuming the outdoor-air temperature is equal to the indoor-air temperature. Using the continuity equation, the velocity of inlet air v_i can be substituted with the inlet air mass flow rate $\dot{m}_i$.The pressure energy losses are accounted for with the hydraulic loss coefficient C_i. Typical values of C_i are between 0.6 and 0.7.

$$\Delta p_i = \frac{1}{2} \cdot \rho_p \cdot v_i^2; \ \dot{m}_i = C_i \cdot \rho_p \cdot A_i \cdot v_i \rightarrow \Delta p_i = \frac{\dot{m}_i^2}{2 \cdot \rho_p \cdot C_i^2 \cdot A_i^2} \ [\text{Pa}]$$

The pressure difference at the outlet opening equals the difference between the hydrostatic pressure difference Δp_b and the pressure difference at the inlet opening Δp_i:

$$\Delta p_o = g \cdot (\rho_p - \rho_g)(H - z) - \Delta p_i \ [\text{Pa}]$$

Bernoulli's equation is also used to determine the mass flow rate of smoke $\dot{m}_o$ leaving the room:

$$\dot{m}_o = C_o \cdot \rho_g \cdot A_o \cdot v_o = C_o \cdot \rho_g \cdot A_o \cdot \sqrt{\frac{2 \cdot \Delta p_o}{\rho_g}} = C_o \cdot A_o \cdot \sqrt{2 \cdot \rho_g \cdot \Delta p_o} \left[\frac{\text{kg}}{\text{s}}\right]$$

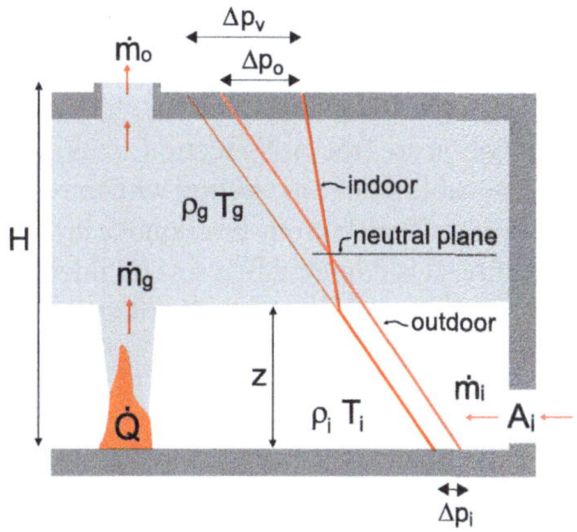

Fig. 6.27 Mass flow rates in the case of natural smoke ventilation depend on the gradient of pressure inside the fire compartment and on the outdoor side of openings; smoke flows to environment if the pressure difference at the outlet opening is greater than that at the inlet opening—$\Delta p_o > \Delta p_i$

In this expression, C_o is the hydraulic loss coefficient at the outlet opening, the value of which is like C_i.

Explanation Bernoulli's equation for kinetic energy assumes the outdoor air velocity is 0 m/s.

By solving the above equations, the size of cross-sectional areas of the inlet A_i and outlet A_o openings can be determined to ensure the required height z of the bottom level of the smoke layer under stationary conditions during the period of the fire. Because all the variables are all mutually related, the problem must be solved by iteration. The solution procedure is as follows:

- select the desired height z of the bottom level of the smoke layer;
- select A_o and A_i, both areas are usually the same, as a rule of thumb amounting to at least 1% of the fire-compartment floor area A_f;
- select the hydraulic-loss coefficients C_i and C_o, indicative values are between 0.6 and 0.7;
- determine the smoke mass flow rate of smoke $\dot{m}_{g(z)}$ using Eqs. (6.2, 6.3) or (6.4)
- determine Δp_i;
- determine the temperature of the smoke θ_g;
- determine the density of the smoke ρ_g; at temperature θ_g;
- calculate Δp_i; a **negative value means natural ventilation is impossible** and a larger intake opening area A_i must be selected;
- determine the smoke mass flow rate through the outlet opening $\dot{m}_o$;
- the calculation is over when $\dot{m}_o \sim \dot{m}_g$;
- otherwise enlarge the areas A_i and A_o and repeat the calculation.

The results of the calculation are free cross-section areas of the inlet and outlet openings A_i and A_o that enable smoke extraction with mass flow rate $\dot{m}_g$ that is high enough to maintain a constant bottom-level smoke layer height z during the stage of the developed fire. It is common that several inlet and outlet openings with total area A_i and A_o are installed, evenly distributed around the perimeter (inlet openings) or on the ceiling (outlet openings) of the fire compartment.

Case Study 1 Determine the necessary inlet and outlet opening areas A_i and A_o for the natural ventilation of smoke from a shopping mall with floor size 25 × 60 m and height 6 m, so the smoke layer does not descend under a height z 3 m. The building structures are made of dense aerated concrete (b = 600 kJ/m^{2}s$^{0.5}$). The floor area A_f of the mall is 1560 m^2. Assume the fire has

Table 6.9 Results of Case study 1

A_i m^2	A_o m^2	Δp_i Pa	θ_g °C	ρ_g kg/m^3	Δp_o Pa	$\dot{m}_o$ kg/s
10	10				neg	
10	11	2.9	67.9	1.04	4.0	20.2
11	11	3.2	67.9	1.04	3.7	21.4
12	12	3.2	67.9	1.04	3.7	23.3
12	14				neg	
14	14	3.2	67.9	1.04	3.7	27.2
15	15	3.2	67.9	1.04	3.7	29.1

engulfed 20% of the mall's floor area (0.2 A_f) and the duration of the fire is 15 min. The opening hydraulic-loss coefficients C_i and C_o are 0.7. The area of the shopping centre is smaller than 2000 m^2, which means the mall can be treated as a single-zone fire compartment.

The building structure's total area in contact with the smoke is $(H - z) \times 2 \times (25 + 60) = 510$ m^2.

Table 6.5 (right) provides the specific heat release q per unit surface area for shops (800 MJ/m^2). Taking the duration of the fire, we can determine the heat-release rate $\dot{Q}$ and the amount of smoke $\dot{m}_g$ to be released during steady state of the developed fire:

$$\dot{Q} = \frac{1000 \cdot q \cdot A_t}{t} = \frac{1000 \cdot 800 \cdot 0.2 \cdot 1560}{15 \cdot 60} = 2.773 \cdot 10^5 \text{ [kW]}$$

$$\dot{m}_{g(z)} = 0.071 \cdot \dot{Q}^{1/3} \cdot z^{5/3} = 0.071 \cdot \left(2.773 \cdot 10^5\right)^{1/3} \cdot 3^{5/3} = 28.8 \left[\text{kg/s}\right]$$

The results of the iteration solutions for various opening sizes A_i and A_o are presented in Table 6.9.

Final solution: The required inlet and outlet opening areas A_i and A_o are 15 m^2. Figure 6.28 (left) shows areas of A_i and A_o if a higher value of z is required.

Case Study 2 Determine the influence of the fire-compartment height H and the designed smoke layer height z on the size of opening for natural ventilation of the smoke. Except for the height H of the fire compartment, the data are taken from the previous case. The solution is shown in Fig. 6.28 (right).

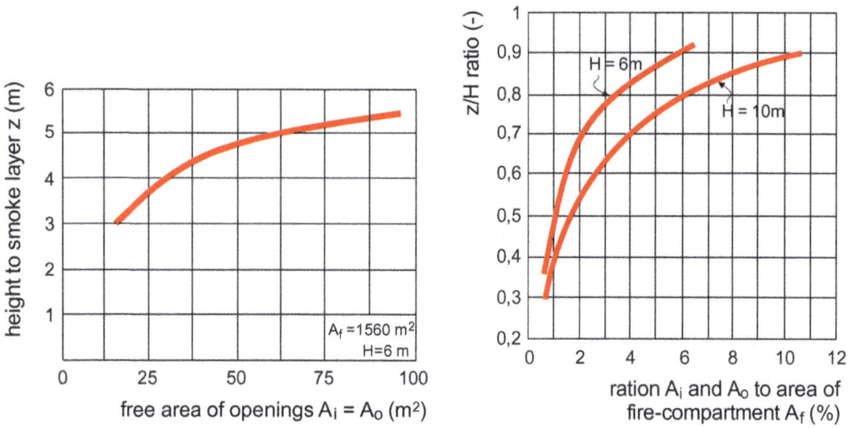

Fig. 6.28 Inlet and outlet opening (total) areas A_i and A_o required to maintain the bottom level of the smoke layer at a constant height z for the case-study shopping mall (left); dimensionless smoke-layer height z/H vs. the ratio of inlet (and outlet) ventilation opening to floor area of fire compartments ($A_i/A_f = A_o/A_f$) expressed in percentages, for case-study shopping mall. The ratio z/H is determined for two height H of the shopping mall. Other parameters are listed in the case-study building description (right)

Simplified method for determining the size of the openings for the natural ventilation of smoke In the pre-design phase of the building planning, rule-of-thumb guidelines can be used for the design of the openings for the natural ventilation of smoke [15]:

- fire compartments of up to 1600 m^2 can utilize natural smoke ventilation if the free compartment height is at least 4 m;
- the aerodynamic area ($C_o \times A_o$) of smoke-ventilation outlet openings must be at least 3% of the compartment's gross area A_f;
- inlet openings, for which the size will not be calculated in detail, must have the same aerodynamic area ($C_o \times A_o$) as the outlet openings ($C_i \times A_i$).

Outlet ventilation openings are equipped with smoke flaps that must open automatically by mechanical, pneumatic or hydraulic means (Fig. 6.29). The inlet openings must be controlled synchronously with the state of the outlet openings. In normal conditions, controlled openings can be used for natural ventilation of the building and for night-time cooling by ventilation. In fire compartments with areas over 1600 m^2 and a free height of less than 4 m, natural buoyancy will be insufficient to remove the smoke. In this case a mechanical smoke-extraction system with high-temperature-resistant fans must be installed, despite being more expensive and occupying part of the space in the building (Figs. 6.30 and 6.31). The fans can be placed on the fire-compartment envelope, e.g., on the roof, or the smoke can be extracted through a network of ducts. In this case, the duct elements' fire resistance must equal that of the

Fig. 6.29 Smoke flaps with automated opening mechanisms [24]

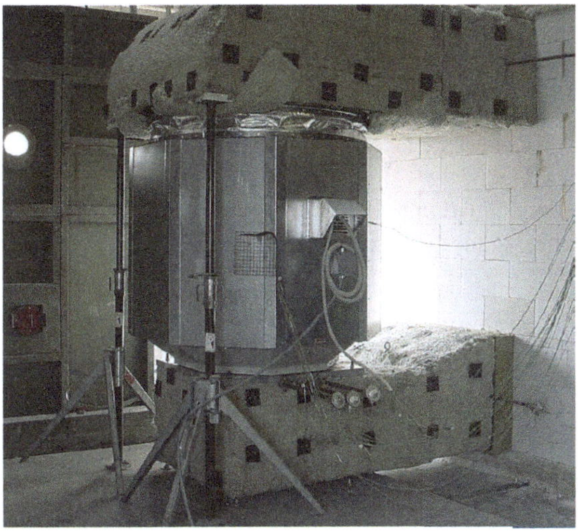

Fig. 6.30 As is the case with building structures, the fire resistance of the fans used for smoke extraction has to be tested too. Fire-resistant fans are divided into classes: a F200 class fan operates for at least 120 min with a smoke temperature up to 200 °C, while a F842 high-temperature resistant fan must operate for at least 30 min with the smoke temperature up to 842 °C [25]

building structures in the compartment (e.g., EI 60). Furthermore, smoke dampers must be installed to prevent any passage of smoke between the compartments.

Mechanical smoke-ventilation systems require a control based on smoke and/or heat detectors. The detectors operate by measuring the temperature, the optical transmittance of the air, or the CO_2, CO or soot content. The fans are sized with the following procedure:

Fig. 6.31 Five fans are installed on top of a commercial building atrium (left) to extract smoke in the case of a fire (mid, left). Inlet openings with automatically controlled flaps are installed in the ground floor of the atrium

- select the design height of the bottom level of the smoke layer z;
- determine the smoke quantity $\dot{m}_g$ [Eqs. (6.2 or 6.3) or (6.4)];
- determine the smoke temperature θ_g;
- determine the smoke density ρ_g;
- determine the smoke volume flow rate $\dot{V}_o$ that must be extracted from fire compartment and pick a suitable fan or several fans (Fig. 6.31) from the manufacturers' catalogues:

$$\dot{V}_o = \frac{\dot{m}_g}{\rho_g}\left[\frac{kg \cdot m^3}{s \cdot kg} = \frac{m^3}{s}\right]$$

- use the continuity equation to determine the inlet opening size A_o, so the smoke velocity at outlet v_o does not exceed 3 m/s.

$$\dot{V}_o = v_o \cdot A_o \rightarrow A_o = \frac{\dot{V}_o}{v_o}\left[\frac{m^3 \cdot s}{s \cdot m} = m^2\right]$$

Simplified method for mechanical smoke-ventilation and heat-extraction system design

In the literature [15] is it required that a mechanical ventilation system must provide a minimum air-exchange rate n of 10 h^{-1} in industrial buildings, at least n 12 h^{-1} in garages and at least n 20 h^{-1} in the spaces without windows. In the spaces with a high density of occupancies, the extraction mass flow rate of the smoke must be at least 36 m^3/h per 1 m^2 of floor area of the fire compartment A_f.

Case Study For the shopping mall discussed in the previous case study, this means that in the case of the mechanical ventilation of smoke, fans must provide the volume flow rate of extracted smoke equal to:

$$\dot{V}_o = \frac{A_f \cdot 36}{3600} = \frac{26 \cdot 60 \cdot 36}{3600} = 15.6 \quad \left[\frac{m^2 \cdot m^3 \cdot h}{h \cdot m^2 \cdot s} = \frac{m^3}{s}\right]$$

The outlet and inlet openings must be sized so that the air velocity v_o and v_i in the free cross-section of the opening does not exceed 3 m/s:

$$A_o = A_i = \frac{\dot{V}_o}{C_o \cdot v_o} = \frac{15.6}{0.7 \cdot 3} = 7.5 \quad \left[\frac{m^3 \cdot s}{s \cdot m} = m^2\right]$$

Advanced methods for mechanical smoke-ventilation system design

Fire-growth predictions for complex structures are made using software tools, based on the numerical solutions to differential equations and accounting for the laws of the conservation of mass, energy and momentum. The software makes it possible to calculate the 3D-temperature, velocity, pressure, gas and solid-particle concentration fields in a building at any given moment during the fire-growth period (Fig. 6.32). This solving method is known as CFD or computational fluid dynamics. The development of computer hardware and software have made it possible to use these tools even with basic engineering knowledge.

6.9 General Requirements of Building Fire Safety

The regulations for building fire safety are part of the laws on fire safety covering the protection of people, animals, property and the environment against fire and explosion, in accordance with the principles of international law.

In general, fire safety in buildings provides the basic requirements, including:

- the prevention of the spreading of fire to adjacent structures;
- the maintenance of the load-bearing capacity of the building structures during the period of the fire and controlling the spread of smoke and heat through the buildings;
- the plan for escape routes, fire-detection and fire-alarm systems;
- the firefighting equipment and providing access for firefighters.

Despite the common requirements of fire safety in buildings, detailed planning requires a knowledge of the national standards and requirements defined in the national regulations.

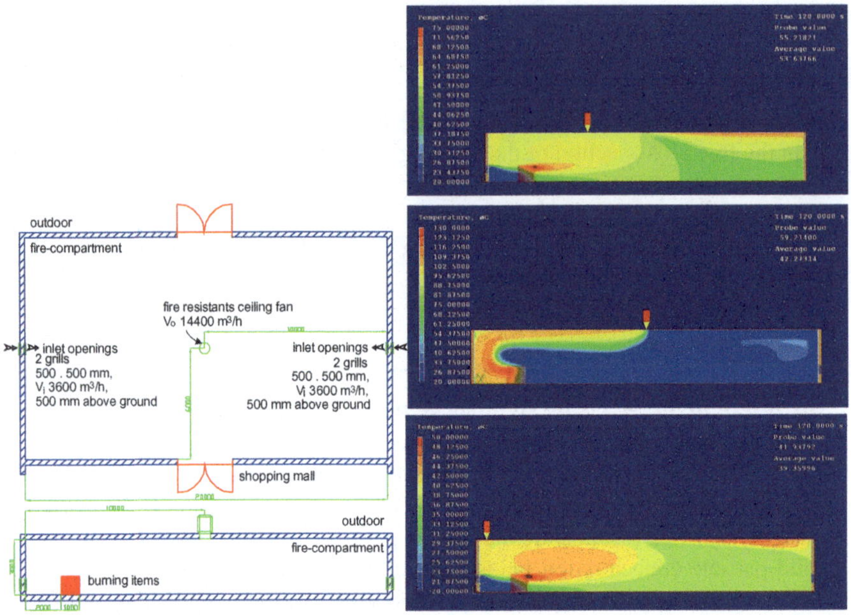

Fig. 6.32 CFD temperature analysis for a warehouse fire 120 s after the ignition. Fire growth factor $\alpha = 0.0127$ kW/s^2 was assumed in the "t-squared" fire-growth model. The locations of inlet the opening and the position of the extraction fan are shown on the warehouse floor plan (left). Transient CFD simulations were performed for the case of single side inlet openings (left ones, near the object on fire) (right, top) and right ones away from object on fire (right, centre) and for double side inlet openings (right, bottom). It is clear that temperatures in warehouse are much lower at the observed time in the case of double side inlet openings [24]. *Note* temperature scales on the left figures vary

References

1. Modern Building Alliance—Safe and Sustainable Construction with Plastics, Fire Safety Statistics (2015)
2. Karlsson, B., Quintiere, J.G.: Enclosures Fire Dynamics. CRC Press, Boca Raton, Florida, USA (2000)
3. Keller, B.: Bautechnologie II, Bauphysik, Teil 1. ETH Zürich, CH (2001)
4. Medved, S.: Introduction in Environmental Engineering. Faculty of Mechanical Engineering, University of Ljubljana (2002)
5. Building and Fire Research Laboratory, Gaithersburg, MD, ZDA; www.bfrl.nist.gov (Access 1 Jan 2014)
6. Žarnič, R.: Materials properties. Faculty for Civil Engineering and Geodesy, University of Ljubljana (2003) (in Slovenian language)
7. www.bfrl.nist.gov (Access 1 Jan 2014)
8. Firedam (2000). www.barrierst.com (Access 1 Jan 2014)
9. www.archchemicals.com (Access 1 Jan 2014)
10. Manzello, S.L., Gann, R.G., Kukuck, S.R., Lenhert, D.B.: Influence of gypsum board type (X or C) on real fire performance of partition assemblies. Fire Mater. **31** (2007)
11. Crauer, S.M., Friday, O.M., White, R.H., Sriprutkiat, G.S.: Mechanical properties of gypsum board at elevated temperatures. Fire Mater. 2003 Conference, USA (2003)

12. Myllymaki, J., Baroudu, D.: Prediction of smoke production and heat release by convolution model. Technical Research Centre of Finland, ESPOO F (1999)
13. Njankouo, J.M., Dotreppe, J.C., Franssen, J.M.: Fire resistance of timbers from tropical countries and comparison of experimental charring rates with various models. Constr. Build. Mater. (June, 2005)
14. Rules on fire safety in buildings. MOP (2005) (in Slovenian language)
15. Technical guidelines TSG-1-001:2007 Fire-safety in buildings, Ministry for environment protection (2007) (in Slovenian language)
16. Steckler K.D.: Heat Release Rate for Materials and Products Using an Oxygen Consumption Calorimeter, NFPA, Quimcy, MA, USA (1998)
17. SBI; European Standard—Reaction to fire test for building products—Building products excluding flooring exposed to the thermal attack by single burning item. EN 13823:2002, CEN, Brussels, B (2002)
18. Mielo, R.V., Satte, B.: The Single Burning Item (SBI) test method—a decade of development and plans for the near future
19. Lehner, S.: European fire classification of construction product, New test method »SBI«, and introduction of the European classification system into German Building regulations; MPA-Otto-Graff-Institute, D (2005)
20. www.koxka.com/English/Documents/Euroclasses%20and%20reaction%20to%20fire.pdf (Access 1 Jan 2014)
21. Ohlemillar, T.J., et al.: Measurement needs for fire safety. NIST, USA (2000)
22. ISO 834 Fire resistant Tests, Element of Building Construction, International Organization for Certification (1975)
23. SP Technical Research Institute of Sweden (Access 1 Jan 2014)
24. Pilko, A.: Environment friendly mechanical fire extinguishing systems for buildings; Diploma work (supervisor assoc. prof. Jurij Modic); Faculty of Mechanical Engineering, University of Ljubljana (2000) (in Slovenian language)
25. Venko, S.: Roof top fire resistant fans; Diploma work (supervisor assoc. prof. Sašo Medved); Faculty of Mechanical Engineering, University of Ljubljana (2007) (in Slovenian language)
26. Egan, D.: Građevinske konstrukcije i požar, Građevinska knjiga, RS (1990)
27. web.jjay.cuny.edu/~tflan/documents/101docs/FIS101HumanBehavior-Fire.pdf (Access 1 Jan 2014)
28. Purser, D.A.: Toxic product yields and hazard assessment for fully enclosed design fires. Polym. Int. **49** (2000)
29. Kavčič, T.: Building fires and impact on occupant's health, Seminar work (supervisor: assoc. prof. Sašo Medved). Faculty of Architecture, University of Ljubljana (2007) (in Slovenian language)

Chapter 7
Urban Environment and Local Climate

Abstract More than 80% of the population in developed countries live in urban areas. Urban-area developments in the past century have led to major changes in land use and, what is even more problematic, it also affects the physical, chemical and biological processes occurring in the nature. One of the most obvious consequences of urbanization is the appearance of microclimatic conditions, described by the intensity of an urban heat island, and the specific air flow patterns in urban neighbourhoods that influence the thermal response of the urban environment as well as the spread of air pollutants. These conditions are a consequence of construction technologies, built materials and modern lifestyle, including the increasing levels of energy use and anthropogenic pollution. Sustainable architectural and urban planning can contribute significantly to the mitigation of urban heat islands and increased urban biodiversity, while improving the indoor environment's quality and the energy efficiency of buildings. In this chapter the phenomena of the urban climate are described and techniques for the mitigation of an urban heat island are presented. The empirical models and the dedicated computational fluid dynamics (CFD) tools for modelling the thermal response of the urban environment are shown. The mitigation of urban heat islands and street canyons with the use of vegetated building structures, greened building blocks and city parks is discussed.

Learning Objectives In this chapter you will learn about:

- how the climate in the urban environment differs from the neighbouring rural areas;
- the thermal response and air-flow patterns in urban canyons;
- mitigating the phenomena of urban heat islands with greened building blocks, vegetated building structures, water surfaces and city parks;
- the impact of the predicted global climate change and urban heat islands on thermal comfort in the urban environment.

S. Medved, *Building Physics*, Springer Tracts in Civil Engineering,
https://doi.org/10.1007/978-3-030-74390-1_7

7.1 Symbols for Quantities and Units

$CHII$	Street canyon heat-island intensity (°C)
$UHII$	Urban heat-island intensity (°C)
$\Delta Q'_{C,nd}$	Increased specific energy needs for cooling (kWh/m^{2}a)
F_{sky}	Sky view factor (1)
f_{LAI}	Leaf area index (1)
φ_e	Relative humidity of outdoor air (%)
GWP	Global warming potential (-)
h	Height of buildings in the street canyon (m)
λ	Wavelength of electromagnetic radiation (m)
OHH	Overheating hours (h/year)
P	City population (–)
θ_{sky}	Sky temperature (°C)
θ_e	Temperature of outdoor air (°C)
$\Delta\theta_e$	Average drop in air temperature (°C)
θ_i	Indoor air temperature (°C)
θ_{PET}	Physiological equivalent temperature (°C)
$v_{(10)}$	Reference wind velocity 10 m above the ground (m/s)
w	Width of street canyon (m)

7.2 Environmental Spheres and the Urban Sphere

The Earth's environment is divided into elementary parts called environmental spheres. The most important for our study is the biosphere—the environmental sphere where life has developed. The biosphere is the zone placed between the atmosphere (air), hydrosphere (water) and lithosphere (soil). Through the evolution of the life on the Earth, a natural balance was established between the living and non-living nature. Biosphere organisms exchange matter and energy with the other environmental spheres, and all these spheres together are referred to as the eco sphere. Populations of different species known as ecosystems grow and live in communities adapted to their living environment. Biomes are created by grouping ecosystems. Examples of such biomes are tundra, tropical forests and deserts. Biomes are transformed primarily by global climate changes. The atmospheric greenhouse effect has caused the average temperature in the twentieth century to rise by 0.6 °C [1], the largest increase in the last 1000 years. For example, in Slovenia the air temperature has been rising steady for the last 50 years. The temperature increase has been the

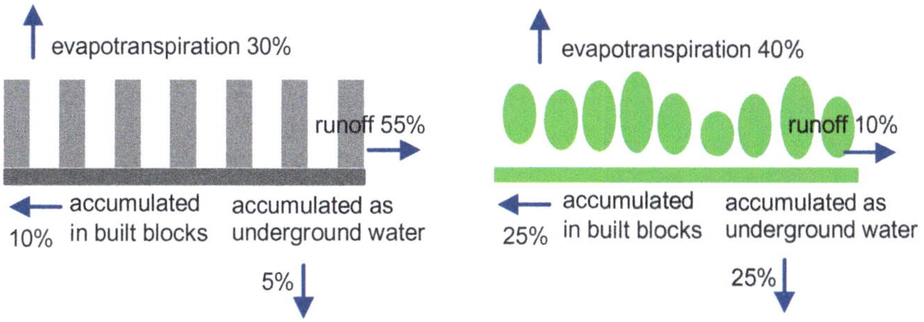

Fig. 7.1 Water balance in the urban (built) and rural (natural) environments. Because of lower evapotranspiration and precipitation accumulation in the soil and in the underground aquifers, water outflow is much higher in the urban environment. This can cause floods and drinking-water shortages and increase the urban heat-island intensity

greatest in the larger cities like Maribor (+ 1.7 °C) and Ljubljana (+ 1.4 °C), and somewhat smaller in rural towns such as Kočevje and Postojna (+ 0.7 °C) [2].

Demographic studies show that today 55% of the world's population lives in urban areas, and it is expected that this number will increase to almost 70% by 2050. Even now in North America, Latin America and Europe more than three-quarters of the population lives in the cities. The urban environment has been changing ever since the industrial revolution for a multitude of reasons, including the rapid expansion of cities that absorb their suburbs. The size and the number of megacities with more than 10 million people has grown rapidly by size and population in the last century. It can be seen that use of local materials and traditional construction techniques involving timber, clay or straw typical for vernacular architecture were replaced by "modern" building materials (concrete, steel, and glass) and unified world-wide architectural trends The consequences of continuous urban expansion and the specific way of planning modern cities have led to the emergence of a new environmental sphere—the urban sphere. The urban sphere is characterized by increased temperatures and pollutant concentrations compared too the rural environment. Built areas affect water-flow patterns. Much less precipitation is stored in natural systems such as greened building blocks, soil and underground aquifers and less water is transported by the evapotranspiration of plants. As a consequence the runoff of precipitation in the urban environment increases from 10 to 55% compared to the rural environment (Fig. 7.1). This leads to the congestion of water-collection systems and floods and a lack of high-quality drinking water.

7.3 Climatic Characteristics of the Environment

The European continent is special in that it exhibits a great variety of landscapes shaped by climatic characteristics. More than 30 climate zones have been identified in Europe, formed under the influence of precipitation, surface-water evaporation, water

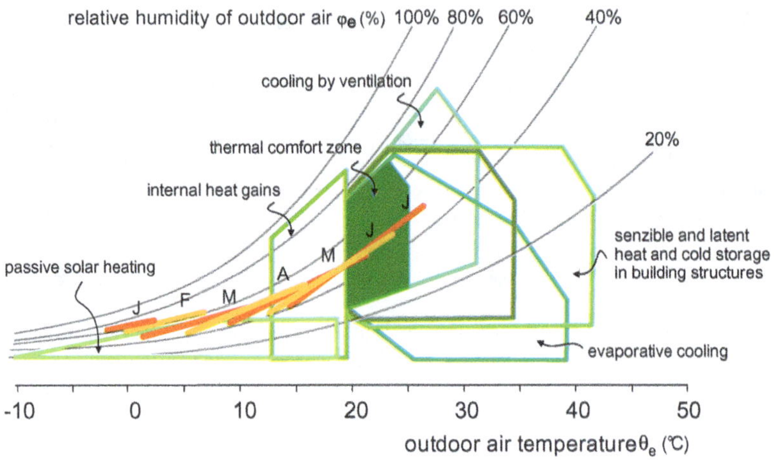

Fig. 7.2 Bioclimatic chart for Ljubljana (Koppen-Geiger climate classification Cfb) with the lowest and highest average monthly temperatures and the relative humidity of the outdoor air in the first seven months (Jan–Jul), techniques of natural heating and cooling with which adequate indoor thermal comfort can be ensured without the use of technical building systems are shown

surface flows, wind and solar irradiation. The local climate characteristics therefore depend on several factors that can be presented in different ways. The Koppen-Geiger climate classification is a widely used classification system. Classification includes five main groups (A tropical, B dry, C temperate, D continental and E polar) and sub classification based on the average annual temperature and precipitation [3]. Bioclimatic charts are also often used during the planning of a building. Such charts (Fig. 7.2) show average or extreme air temperatures θ_e and air humidity φ_e or other environmental properties for certain periods, e.g., months in a year. Also plotted in a bioclimatic chart are the indoor thermal comfort zone and the zones of local climate conditions, where the indoor thermal comfort can be realised with passive heating and free-cooling techniques. These techniques include passive solar heating, the use of environmental heat and cold, cooling by enhanced ventilation during the night-time, evaporative and radiative cooling and heat and cold accumulation in building structures [4].

7.4 Local Climatic Characteristics of the Urban Environment

Urban areas (Fig. 7.3) exhibit specific micro-climate conditions that are evident through the following phenomena [5–7]:

- heat islands are formed. The accumulated solar energy in the urban environment is noticeably larger due to the denser construction and modern building methods. As

Fig. 7.3 Submerging suburban settlements, high density of buildings, replacing the vegetation with built building stocks, high energy demand and large emission of pollutants are the main reasons for the emergence of urban heat islands and street heat canyons

a consequence the temperatures of the air and surfaces in the cities are higher than in the rural environment. The maximum diurnal difference between air temperatures in the urban and rural environments determines the urban heat-island intensity (*UHII*) (°C). Besides the increased accumulation of the heat in building blocks and decreased evaporative cooling by vegetation and lost water bodies the *UHII* is exacerbated as radiative cooling is reduced because the streets are being made narrower and the buildings taller;

- the local wind speed decreases and the air flow patterns change in the narrower streets and dense settlements. This has an impact on the air and the surface temperatures as well as on the dilution of pollutants emitted into the atmosphere. In this way urban street canyons with micro-climate conditions are formed;
- solar irradiation decreases because the sun's rays scatter on pollutant gas molecules and solid particles. As a consequence of more polluted air in the urban environment, the potential of daylight and the number of hours of sunshine is lower compared to the rural environment;
- the amount of precipitation and the water-flow regime change because solid particles in the atmosphere act as condensation nuclei. The precipitation is more intense, but the water runs off quickly because the vegetated areas are smaller and less water is captured and stored.

Although all phonemes that affect the urban micro-climate are important and must be carefully addressed in the process of urban planning and building design, the heat-island formation and mitigation will be discussed in detail later on because *UHII* is probably the most exposed problem related to outdoor and indoor environment thermal comfort and the energy efficiency of the buildings.

7.4.1 Urban Heat Islands

It is common that the temperatures in modern cities are higher than in the countryside areas surrounding them. This phenomenon is known as the urban heat island and meteorologists have been warning about it for more than a century. Urban heat islands appear at any time of the year, but extreme summer conditions affect the quality of life and energy use in cities the most. Figure 7.4 presents typical summer-time maximum daily air temperatures over different parts of urban and rural areas. The greatest difference between urban and rural air temperatures during the day defines the urban heat-island intensity. *UHII* depends on the climatic and the topographic characteristics of the site, the city shape and the urban building blocks, (for example, the size and shape of water courses and parks), the current weather conditions, such as clouds, air humidity and wind speed. The heat-island intensity also depends on the ratio of building height to street width and the albedo of built materials.

Several factors that affect the heat-island intensity must be considered in urban planning:

- the radiative properties of urban built blocks. Surfaces of built blocks absorb more solar radiation due to a lower albedo (a measure of the reflectivity with respect to solar radiation, Table 4.4) and multiple reflections of sunlight off the surfaces in narrow urban streets. It was discovered that the temperature of bright surfaces with an albedo over 0.7 is up to 45 °C lower compared to dark surfaces with an albedo under 0.2 (Fig. 7.5). Accounting for the multiple reflections of sun rays, the average albedo of the urban environment turns out to be in the range 0.15–0.30 for large European and US cities, and in the range 0.45–0.60 for rural settlements [8];

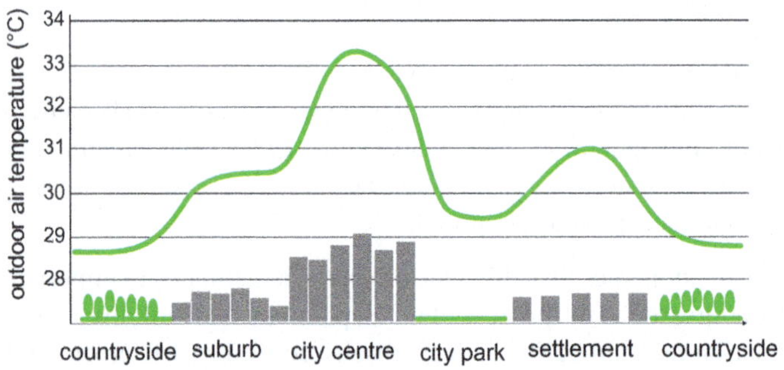

Fig. 7.4 Heat islands occur in cities regardless of their size. They defined by the maximum daily difference in air temperatures in the urban, sub-urban and rural environments and represented by the urban heat-island intensity (*UHII*) [3]. *UHII* can reach 10–12 °C in megacities and 3–4 °C in cities with a population below 500,000 inhabitants. Heat-island intensity could be stated according to the lowest diurnal temperatures as well. These phenomena only reach 2–4 °C, but could significantly affect the efficiency of the night-time cooling of buildings [adapted from 7 and 8]

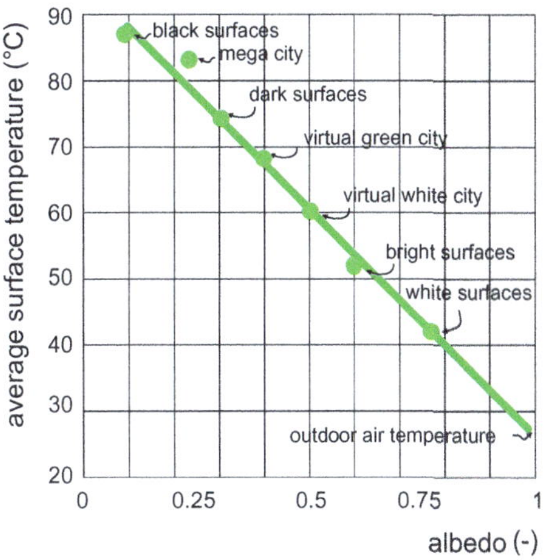

Fig. 7.5 Temperatures of some built materials and average temperatures in the cities under clear sky conditions according to their albedo. Higher temperature in the cities are a consequence of the multiple solar-irradiation reflections in narrow streets [adapted from 8]

- less longwave thermal radiation is emitted into the sky from surfaces of the urban-environment, leading to a reduced radiative cooling effect as a consequence of the tall buildings and narrow streets with a lower sky view factor F_{sky} (Table 1.23). The sky in cities is more often cloudy and in this case the sky temperature θ_{sky} (Sect. 1.7.2) is much higher than in clear sky conditions. The photochemical smog in the atmosphere increases the backwards longwave thermal radiation emitted by the atmosphere to the urban-environment, because the greenhouse effect of the atmosphere (see explanation) is more pronounced in urban environments [7];
- less vegetated built blocks and more "water-tight" surfaces where precipitation run-off without time-lag, reduce the possibility of latent heat accumulation, while at the same time more sensible heat is stored in the built blocks as they have a higher specific heat capacity compared to natural surfaces. The former causes higher daytime, and the latter higher night-time, air and surface temperatures of urban environment compared to the countryside;
- heat transfer by convection on surfaces of urban built blocks is lower due to the lower wind speeds. The surface of the dark sun-lighted sidewalk cools down by 30 °C when the wind speeds above the sidewalk increase from 0 to 15 m/s [7];
- larger anthropogenic heat sources, more heat generators and air-conditioning equipment that extract heat to the environment are present in urban areas. The density of the anthropogenically induced heat flux in large cities is 50–200 W per m^2 of urban area and the air temperature in the streets rises by up to 6 °C when the air-conditioning in apartment buildings is turned on [7].

Table 7.1 Most important anthropogenic greenhouse gases. The main culprit for the emissions are the burning of fossil fuels and natural gas, farming, landfills, refrigerants, SF_6 and construction

greenhouse gas	decomposition of gas (years)	global warming potencial (GWP)		
		20 years	100 years	500 years
methane (CH_4)	12	62	23	7
nitrous oxide (N_2O)	114	275	296	156
fluorinated hydrocarbons	14	3300	1300	400
perfluorinated compound	260	9400	12000	10000
sulfur hexafluoride (SF_6)	3200	15100	22200	32400
carbon dioxide (CO_2)	-	1	1	1

Explanation The Sun is the most important energy source for our planet. The energy flux emitted from the Sun's photosphere is determined by the Stefan–Boltzmann law (Sect. 1.5). Solar irradiation travels through space and reaches the Earth with no transmission losses as short-wavelength electromagnetic energy flux (with predominant wavelengths $\lambda = 3$ μm The part of the solar irradiation, defined by albedo of Earth's surface, is reflected back into the Space. Earth's albedo is approximately 0.3, which means that 30% of the solar irradiation is reflected without affecting the planet's temperature. The remaining part of the solar irradiation is absorbed and heats up the Earth, which itself is a thermal emitter and emits long-wavelength irradiation (predominantly with wavelengths $\lambda > 3$ μm towards space, the temperature of which is close to absolute zero (0 K). The Earth's equivalent temperature determined from the energy balance of the received, reflected and emitted energy fluxes will be about 255 K or − 18 °C, which is considerably lower than the planet's real mean temperature (288 K or + 15 °C). The reason lies in the Earth's atmosphere's ability to trap a part of the long-wavelength irradiation emitted by the Earth's surface towards space. This phenomenon is known as the greenhouse effect, because the property of the high transmittance of the short-wavelength solar irradiation and absorptance of the long-wavelength thermal irradiation is also a characteristic of glass. The atmospheric greenhouse effect has made it possible for life to develop on Earth. Also responsible for the greenhouse effect, in addition to water vapour, are CO_2, methane and some other gases of natural origin. However, the anthropogenic greenhouse emissions during the past century have grown to the point where they started to amplify the atmospheric greenhouse effect. The most important anthropogenic greenhouse gases are listed in Table 7.1.

Not all greenhouse gases contribute to the atmospheric greenhouse effect to the same degree, and their lifespan in the atmosphere varies too. The impact of each greenhouse gas is measured using the global warming potential (GWP) in a given time period. The values of the GWP of most severe anthropogenic greenhouse gases are listed in Table 7.1 [9].

The total greenhouse effect of the greenhouse gases emitted into the atmosphere is defined by the sum of the products of the mass of each greenhouse-gas emission and its GWP defined according to the time scale (Table 7.1). A time scale of 100 years is commonly used and six greenhouse gases are commonly included in world-wide environmental protection studies (Table 7.1). The total greenhouse effect is expressed as the carbon dioxide equivalent ($CO_{2,eq}$). The increased greenhouse effect of the atmosphere caused by mankind is one of the main concerns of the sustainable development and the reasons for climate changes, and this is bound to affect the urban microclimate too.

7.4.1.1 Modelling the Urban-Heat-Island Intensity

Air-temperature analyses for the past 50 years show warming for every city: 2.9 °C for New York, 3 °C for Tokyo, 3.5 °C for Moscow and as much as 6.5 °C for Shanghai. A number of empirical models for predicting *UHII* have been developed. Oke [8] is considered as a pioneer of urban micro-climate research. He devised a model for calculating *UHII* depending on the logarithm of populations for the cities with populations over 1 million, assuming no-wind conditions. The buildings in US cities are taller and the construction is denser, which means that the *UHII* is higher for American than European cities (Fig. 7.6). Oke [4] upgraded his model to account for the wind speed measured in the suburbs in front of the city:

$$UHII = \frac{P^{0.25}}{(4 \cdot v_o)^{0.5}} [°C]$$

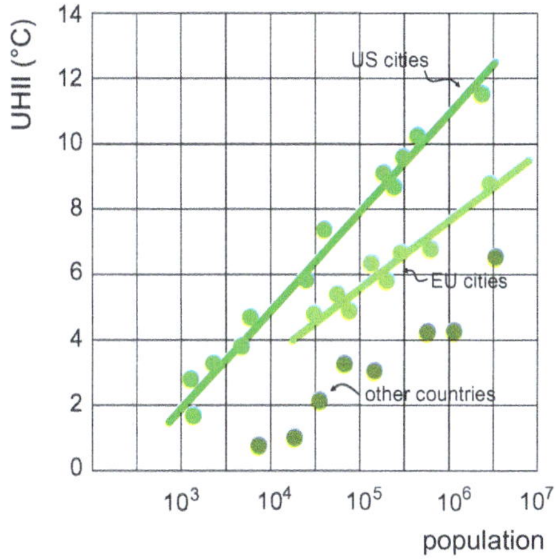

Fig. 7.6 *UHII* of US and European cities; *UHII* depends on the urban architecture and energy demand in buildings [8]

where P is the population of the city and v_{10} is the wind speed in the suburbs at a height of 10 m (m/s). Santamouris [7] analysed some smaller cities and discovered that the *UHII* is lower for South American and Asian cities of comparable size due to more modest living standards and the needs of the population. Oke [8] also introduced the street shape into his *UHII* model, describing the street shape by the ratio of the height h to the width w:

$$UHII = 7.54 + 3.97 \cdot \ln \frac{h}{w} [°C]$$

7.4.2 Air-Flow Patterns in the Urban Environment

The characteristics of buoyancy-driven air flow as well as wind velocity are different in the urban, suburban and countryside canopy layers. As a result of a specific urban-environment topography and heat-island effect, the air velocities are lower in urban areas compared to the countryside. Urban topography divides the air mass into two distinct layers (Fig. 7.7): the bottom layer spanning from the ground to the top of the tallest building is known as the canopy layer, and the layer above extending from the top of the highest building to the undisturbed air mass flows over the city that is called the boundary layer.

Due to the iteration between temperature-driven buoyancy and the wind-driven air flow, the spatial air velocity in the urban environment has to be determined using numerical methods and computational fluid dynamics (CFD). Figure 7.8 shows an example of such a study for different forms of settlements. With known velocities the natural ventilation of street canyons can be determine and expressed by air exchange rates or by the age of the air. With known locations and amount of pollutant sources,

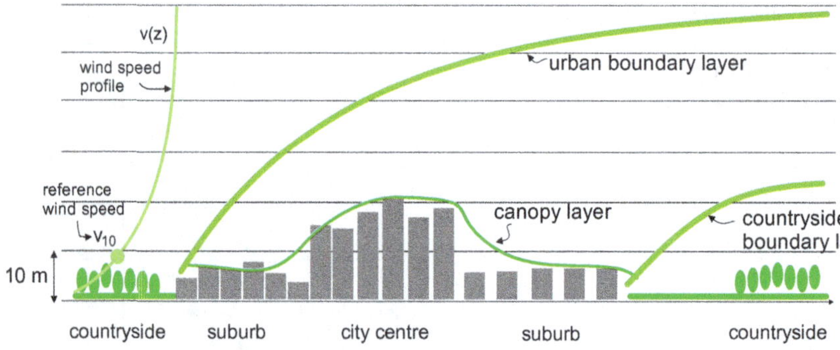

Fig. 7.7 Canopy and boundary layers of air mass define the air flow patterns in the urban environment [adapted from 8]

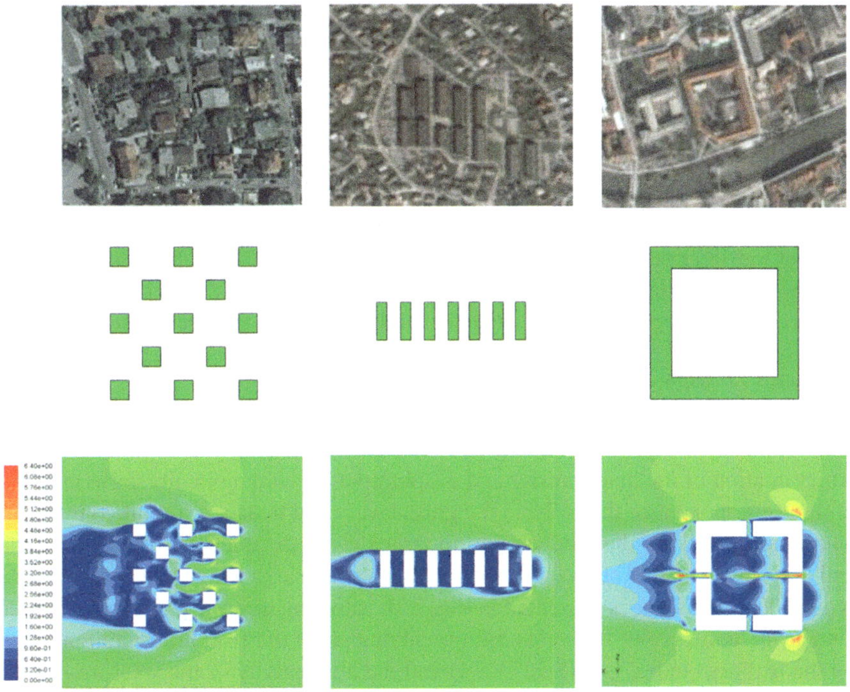

Fig. 7.8 An example of CFD analyse of air velocities modelled using CFD: in the chessboard, strip and atrium type of settlements; air velocities are shown for the horizontal plane 1.5 m above the ground for the reference air velocity of 4 m/s in the open flat environment before the settlement in the direction of the x axis [10]

the concentration of pollutants and the impact of pollutants on the health of the inhabitancies can be predicted.

7.4.3 Mitigation of Urban Heat Islands

The urban heat island affects the outdoor as well as indoor living comfort and the energy use, especially for cooling of the buildings. The intensity of the urban heat islands can be mitigated by implementing different approaches and measures that should be considered in the urban planning process. The following measures have proven to be the most effective:

- Reducing the absorbed solar radiation by shading the building envelope. In this way the radiative temperature decrease that improves the outdoor thermal comfort and at the same time, the equivalent air temperature θ_{eq} (Sect. 1.7.2) is reduced and as consequence, the heat flow passing through the building envelopment is reduced.

- Reducing the amount of the sensible heat stored in built blocks by selecting materials or paintings with a higher albedo [11–16].
- Increasing the amount of the heat stored in the form of latent heat in the hydrosphere and moist air and take advantage of the evaporative cooling of the air on vegetated or water areas or on the leaves of plants and trees in the city parks [17–19]

Research has confirmed that the latter measures are the most effective. A model for heat transfer over vegetated surfaces is presented in Sect. 1.5.4. A similar model can also be used to analyse water bodies, dividing them into several layers by depth and determining the amount of stored heat and water-body surface temperature. The effect of evaporative cooling of the air flowing over the surfaces of the water body or grass area of different lengths versus the reference air velocity v_{10} (m/s) downwind of the park is shown in Fig. 7.9. The average air-temperature drop $\Delta\theta_e$ is defined for a 20 m-thick layer above the ground, i.e., for the canopy layer in urban areas. An evaluation of the heat transfer and evaporative-air cooling in city parks consists of three components—the soil and the grass layers and trees are simplified by using empirical models for the prediction of the tree height, the diameter of the tree trunk and the crown size and shape according to the tree species and age. An example for a typical deciduous tree is presented in Fig. 7.10, showing how the properties of a tree change up to the age of 100 years. Empirical models of the evapotranspiration based on the leaf area index f_{LAI} (Sect. 1.5.4) and the absorptivity of the solar irradiation on the leaves in the tree crown, considering the growth/vegetation period, are also needed. The average air-temperature drop $\Delta\theta_e$ along the city park is presented in Fig. 7.11 according to the tree-planting density (left) and the age of the trees (right). The park should be at least 80 m long and the tree density should be at least 50

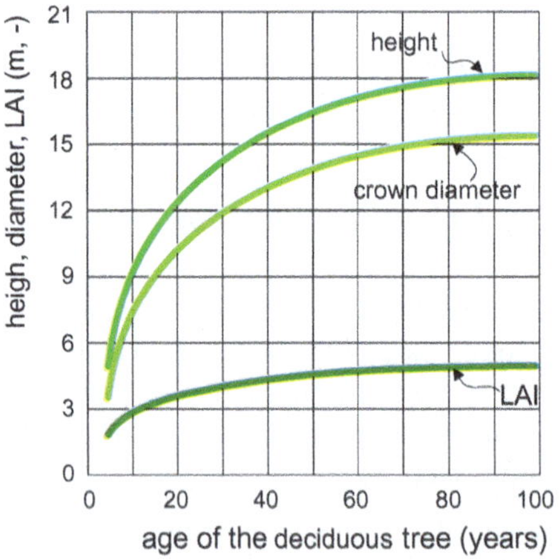

Fig. 7.9 Decrease of the air temperature $\Delta\theta_e$ in the wind direction in the canopy layer above the water body and the grass surface due to evaporative cooling versus air velocity v_o at a height of 10 m on a sunny day [10]

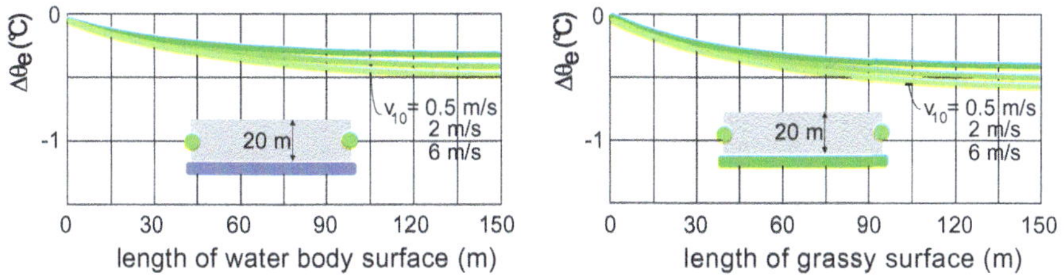

Fig. 7.10 Typical height of the deciduous tree, the diameter of the crown and leaf area index in the first hundred years of age [10]

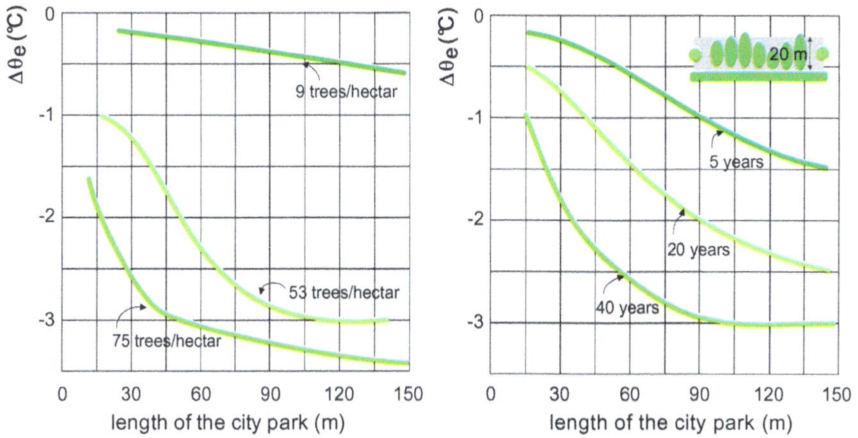

Fig. 7.11 Decrease of the air temperature $\Delta\theta_e$ in the wind direction in the canopy layer of city park at wind speed 2 m/s versus length of the city park according to the tree-planting density (lest) and the age of the trees (right) [10]

trees per hectare to ensure the air-temperature decrease in the canopy layer is $\Delta\theta_e \sim$ 2.5 °C.

Case Study Vegetated urban areas in the urban environment not only affect the microclimate in cities, but also significantly increase the retention of precipitation (Fig. 7.12). Therefore, the loads on the municipality's drainage system are reduced, especially during extreme weather conditions such conditions are becoming more common due to the global climate changes (Fig. 7.13).

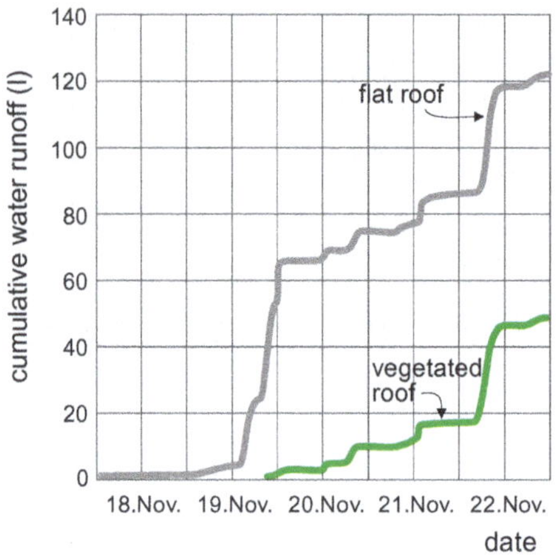

Fig. 7.12 Difference in the outflow of precipitation from tile-covered and vegetated rooves with the Urbanscape green-roof system (Fig. 7.13) [20]

Fig. 7.13 In the Urbanscape green-roof system, a layer of plants (sedums) is laid on top of a layer of fibrous mineral thermal insulation in which water is accumulated and plants are supplied with nutrients. 20–40 L of water are retained in a layer 20–40 mm thick

7.4.4 Temperature Conditions and Thermal Comfort in Urban Canyons

Urban canyons are elements of the urban environment where means of transportation, extracted heat from buildings, the high absorptance of solar irradiation that is caught in a maze of narrow and high streets and weak airing caused the phenomena of the urban-canyon heat island. Micro-climate conditions in urban canyons define

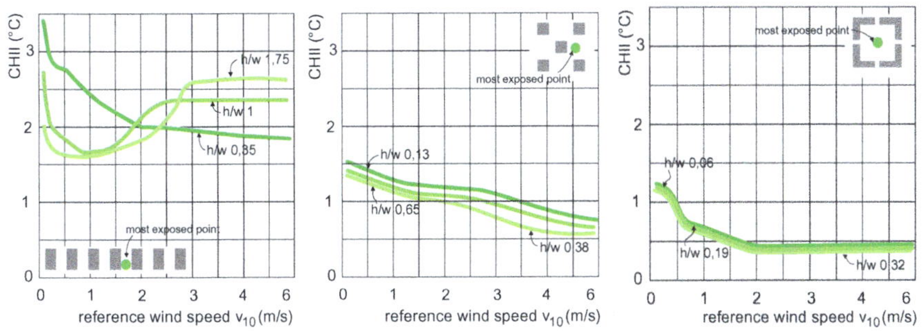

Fig. 7.14 Canyon heat-island intensity (*CHII*) at the most exposed point a in street canyon 1.5 m above the ground versus wind speed in the open environment before the settlements and the ratio between the height (*h*) and width (*w*) of the street canyon; in the strip type (left), chessboard type (middle) and in atrium type (right) settlement [6, 21]

the thermal comfort in the outdoor environment, the energy demand of buildings and the dilution of pollutants originating in the canyon or transferred from the neighbourhood. The local temperatures and velocities in a canyon depend to a large extent on its geometric form and celestial orientation. Therefore, CFD software tools are required for a comprehensive analysis of the canyon's micro-climate. As a rule of thumb, the canyon heat-island intensity (*CHII*) exceeds the urban heat-island intensity (*UHII*), while the airing of the canyon follows from interactions between the canyon's ratio of height *h* to width *w*, the air speed upwind of the city and the built blocks' surface temperatures affecting the buoyancy flow of warm air. Figure 7.14 presents the *CHII* in settlements of different forms (strip type, chessboard type, atrium type), assuming the albedo of canyon's surfaces 0.3. The *CHII* is shown for the most exposed point in each settlement.

After that the local air temperatures and the air humidity, the surface temperatures of the built blocks and the air velocity in the urban canyon are determined and the Sun's current position and cloudiness are taken into account, the physiological equivalent temperature θ_{PET} (Sect. 1.8.2) can be calculated and used as an integral indicator of the thermal comfort in an urban street canyon.

The same measures as pointed out in the case of *UHII* mitigation will decrease the *CHII* as well. As in the case of the *UHII* mitigation, the trees planted in street canyons will have the greatest impact on the *CHII*. Besides that, the thermal comfort is better on narrower streets, where the buildings provide the shade. Figure 7.15 shows the results of CFD-modelled air temperatures in a strip-type street canyon at different street h/w ratios. The temperatures were determined for a reference wind velocity of 1 m/s and a wind direction along the street canyon. Higher wind speeds only affect θ_{PET} for a *h*/*w* ratio of less than 0.5.

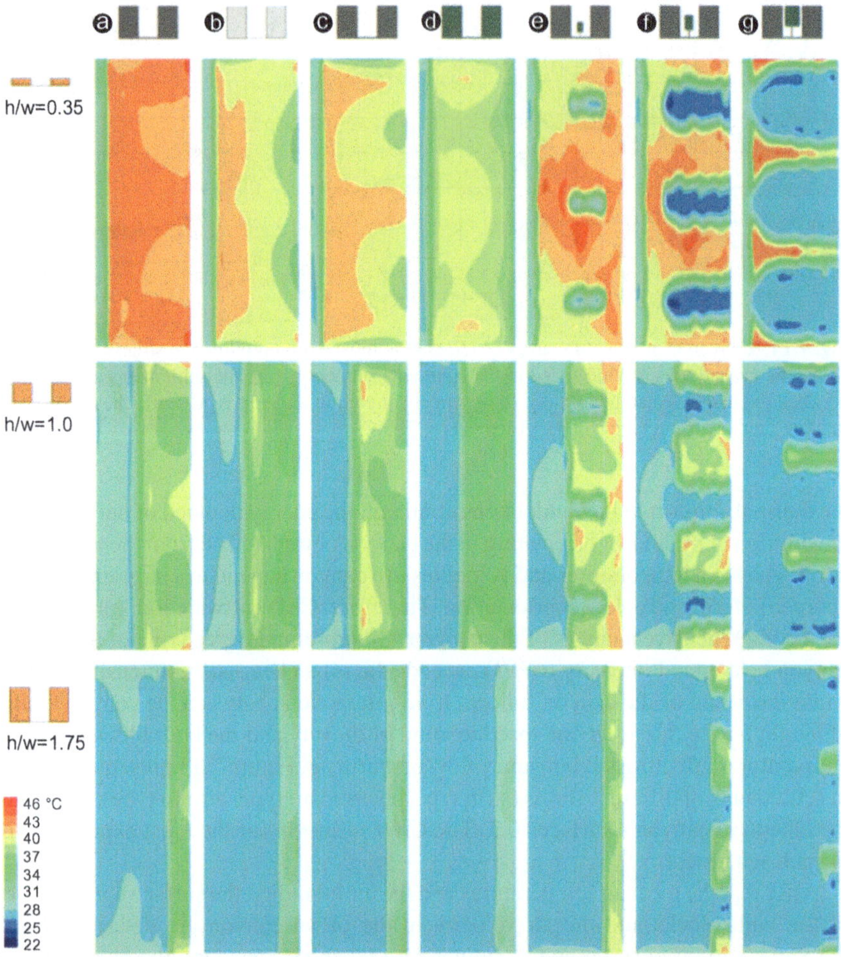

Fig. 7.15 Physiological equivalent temperature θ_{PET} in the strip-type street canyons on a clear summer day 1.5 m above the ground versus canyon height-to-width ratio (*h*/*w*); θ_{PET} are shown (from left to right) for **a** the case of dark canyon surfaces (building facades and ground; albedo 0.3), **b** bright canyon surfaces (albedo 0.7), **c** for dark canyon surfaces and grassy ground, **d** for street canyon with total vegetated surfaces (albedo 0.4), **e–g** for dark canyon with grassy ground and trees of different size planted in the middle of the canyon; results can be used for planning of the pavement position and for forecasting the impact on pedestrian health due to possible thermal shock [21]

Trees planted in the street canyons decrease the perceived temperatures mainly because they block the solar irradiation. Although the evapotranspiration process causes leaves to have only a few K higher temperature than the air around the tree, despite the high absorption of solar irradiation, water vapour released by the tree increases the humidity of the air. Besides that, the airing of the canyon is weaker if trees are planted and the concentration of air pollutants can increase. With CFD and purpose-made computer-simulation tools such as ENVI-met [21], the thermal comfort and concentrations of the most severe pollutants, e.g., CO_2, NOx and PM can be modelled. Figure 7.16 shows such a case study. The difference between the street canyon's and the rural surrounding's air temperature ($\Delta\theta_e$), the relative air humidity ($\Delta\varphi_e$) and the concentration of CO_2 (ΔCO_2) emitted by cars in street canyons with different h/w ratios are shown. From the case study, it can be seen how important an advance assessment of the micro-climate conditions and the air quality in street canyons is, as some measures that reduce *CHII* can lead to a deterioration in the quality of the outdoor environment.

7.4.5 Impact of Global Climate Changes and Urban Heat Islands on the Indoor Comfort and Energy Use in Urban Buildings

The Intergovernmental Panel on Climate Change (IPCC) [1] developed several future anthropogenic greenhouse-gas emission scenarios on the basis of predicted socioeconomic development, accounting for population growth and fossil-fuel use. According to these scenarios, the global temperatures on the planet will rise by 1.5–6 °C by the year 2100, causing severe global climate changes [22, 23]. The predicted global climate change and urban heat islands will have a synergic effect on the quality of life and energy use in urban buildings. Using available Typical Reference Year databases (TRY, a database of hourly values for meteorological quantities), we can combine the effect of predicted global changes with the urban micro-climate models. By using corrected TRYs as a boundary condition in buildings' thermal response modelling, several indoor-comfort and energy-efficiency indicators of buildings could be developed. For example, the number of hours in which the indoor-air temperatures in free-running (not cooled) buildings will exceed the limit of indoor thermal comfort, or increased energy use for cooling per m^2 of living space in mechanically cooled buildings [24]. Figure 7.17 shows the results of such analyses for buildings designed to meet the requirements of passive houses.

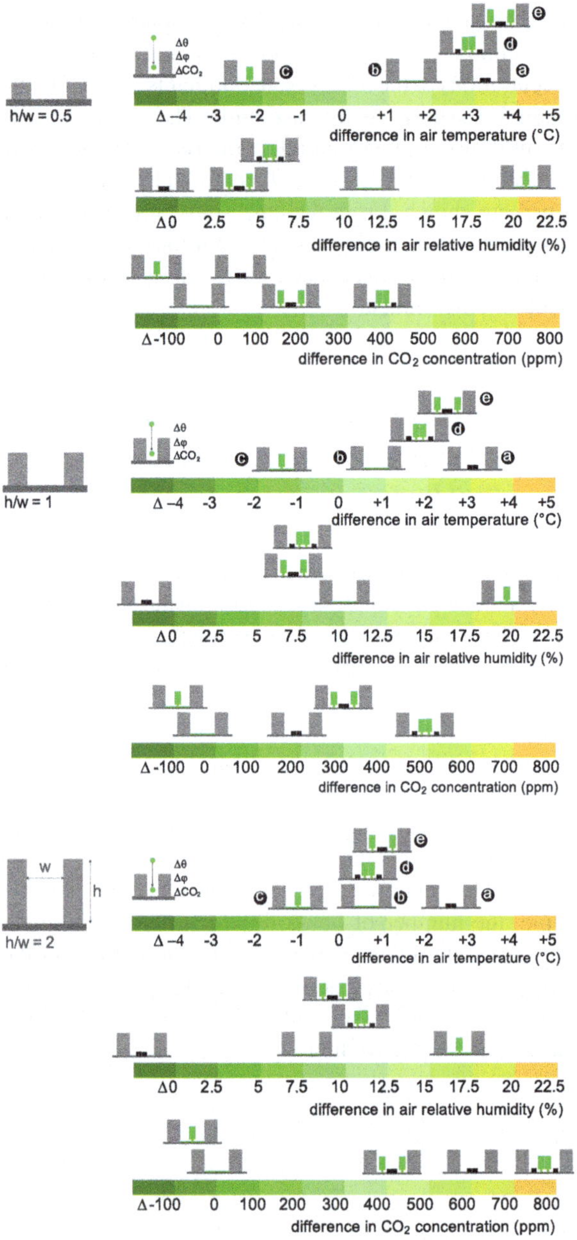

Fig. 7.16 Difference between the street canyon's and rural surrounding's air temperature ($\Delta\theta_e$), relative air humidity ($\Delta\varphi_e$) and concentration of CO_2 (ΔCO_2) at the level of pedestrian. Cases are shown for: **a** the street canyon with road, **b** the street canyon with grassy ground, **c** street canyon with a tree-lined avenue, **d** street canyon with two-way traffic and tree-lined avenue along the sides of the street and **e** the street canyon with two-way traffic and tree-lined avenue in the middle of the canyon. Three examples are presented according to the street canyon height-to-width (*h/w*) ratio [25]

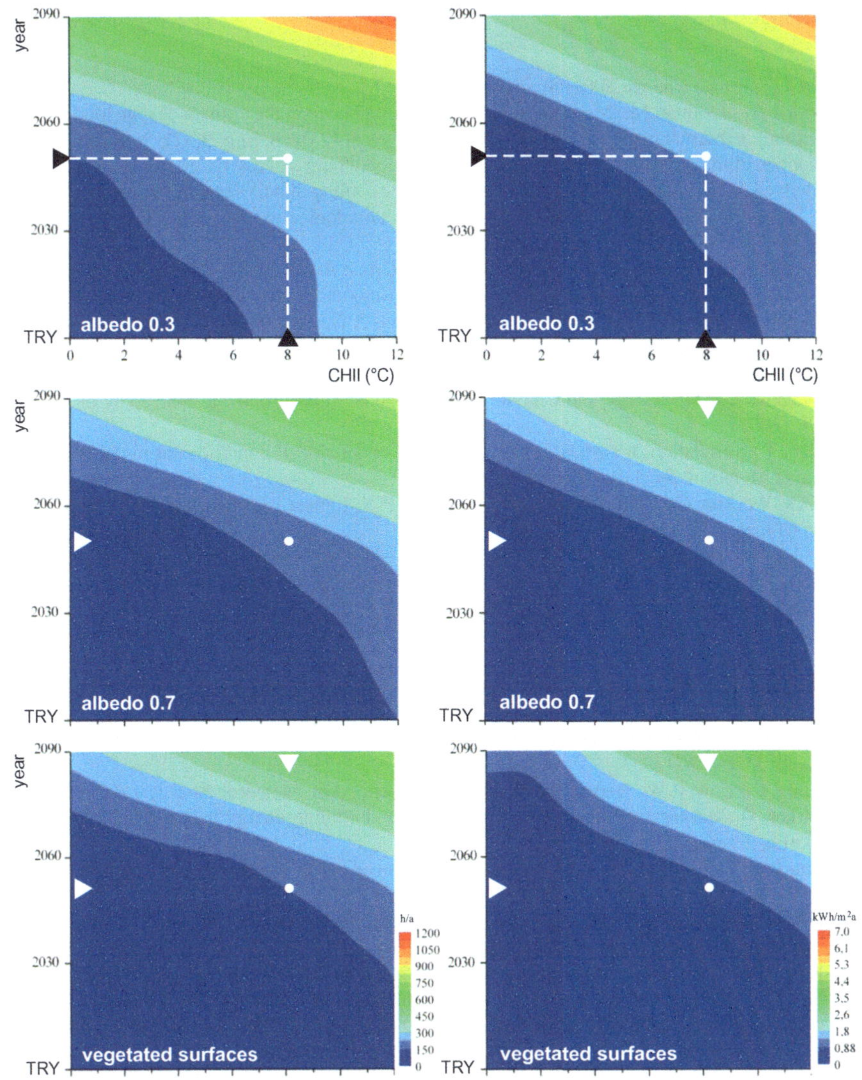

Fig. 7.17 Number of overheating hours *OHH* (h/year) when indoor-air temperature exceeds the thermal comfort limit (θ_i = > 26 °C) (left) and predicted increase in specific energy needs for cooling ($\Delta Q_{C,nd}$ in kWh/m^{2}a) (right) versus street-canyon heat-island intensity *CHII* from present to 2090; values are shown for the dark building blocks (albedo 0.3), light coloured building blocks (albedo 0.7) and street canyon with vegetated surfaces; mid-severe global climate change scenario (+ 4.5 K) was taking into account; drawn point shown predicted *OHH* and $\Delta Q_{C,nd}$ in 2050 and *CHII* 8 K [26].

References

1. IPCC: Climate Change 2001: Synthesis Report
2. Vidrih, B., Dolinar, M., Medved, S.: The connection between the climate change model and a building's thermal response model: a case of Slovenia. J. Mech. Eng. **9**(52), 568–586 (2006)
3. Koppen climate classification, Wikipedia, en.wikipedia.org, access 1.1.2019.
4. Medved, S., Domjan, S., Arkar, C.: Passive and free cooling of buildings. In: Sarma, A., Shukla, A., Singh, R. (eds.) Low Carbon Energy Supply Technologies and Systems. CRC Press (2020). ISBN 9780367373405
5. Cartalis, C.: Modifications in Energy Demand in Urban Areas as a Result of Climate Changes: An Assessment for the Southeast Mediterranean Region
6. Akbari, H., Davis, S.: Cooling Our Communities—A Guidebook on Tree Planting and Light Color Surfaces. US Environmental protection Agency, Office of policy analysis, Climate Change Division (1992)
7. Santamouris, M.: Energy and Climate in the Urban Built Environment. University of Athens, GR (2001)
8. Oke, T.: Simulation of surface urban heat island under 'ideal' conditions at night. Boundary Layer Meteorol. **56**, 339–358 (1991)
9. IPCC Working Group II: Climate Change—The IPCC Impact Assessment, International Panel on Climate Change, pp. 3–5 (1990)
10. Vidrih, B.: Impact of Global Climate Changes and Urbanization on Micro-climate in Urban Environment; Ph.D. Thesis (supervisor assoc. prof. Sašo Medved); University of Ljubljana (2010). (in Slovenian language)
11. Taha, H.: Urban climates and heat islands: Albedo, evapotranspiration and anthropogenic heat. Energy Build. **25**, 99–103 (1997)
12. Santamouris, M.: Urban Canyon Experiments in Athens, International Report to the Polish Research Project. University of Athens, GR (1977)
13. Dimoudi, A.: Vegetation in the urban environment: microclimatic analysis and benefits. Energy Build. **35**, 69–76 (2003)
14. Craig, C., Lowry, W.: Reflections of the urban albedo. Am. Meteorol. Soc. 159–164 (1972)
15. Parker, D., Barkaszi, S.: Roof solar reflectance and cooling energy use: field research from Florida. Energy Build. **25**, 105–115 (1997)
16. Simpson, J.: The effects of roof albedo modification on cooling loads of scale model residences in Tucson, Arizona. Energy Build. **25**, 127–137 (1997)
17. Racissi, S.: Energy saving by proper tree plantation. Build. Environ. **34**, 565–570 (1999)
18. Takakura, T.: Cooling effect of greenery cover over a building. Energy Build. **31**, 1–6 (2000)
19. Rosenfeld, A.: Painting the Town White and Green. Lawrence Berkeley Laboratory
20. Arkar, C., et al.: Hydrological and thermal response of green roofs in different climate conditions. In: Sustainable Built Environment D-A-CH Conference 2019, Graz (2019)
21. ENVI-met: ENVI-MET GmbH, Germany
22. Vidrih, B., Medved, S.: The effects of changes in the climate on the energy demands of buildings. Int. J. Energy Res. **32**(11), 1016–1029 (2008)
23. Dolinar, M., Vidrih, B., Kajfež-Bogataj, L., Medved, S.: Predicted changes in energy demands for heating and cooling due to climate change. Phys. Chem. Earth, Parts A/B/C **35**(1–2), 100–106 (2010)
24. Akbari, H.: Calculating energy-saving potentials of heat-island reduction strategies. Energy Policy **33**, 721–756 (2005)
25. Fink, R.: Impact of Micro-Climate and Air Quality in Urban Environment on the Health and Well-Being of People with Cardiovascular Disease. Ph.D. Thesis (supervisor prof. Sašo Medved); University of Ljubljana (2013). (in Slovenian language)
26. Medved, S., Vidrih, B.: Numerical modelling and parametric analyse of outdoor thermal comfort in urban settlements with natural heat sinks. In: Urban Environment Pollution: Overcoming Obstacles to Sustainability and Quality of Life [conference] 2010. Elsevier, Boston, USA (2010)

The manufacturer's authorised representative in the EU is Springer Nature Customer Service Centre GmbH, Europaplatz 3, 69115 Heidelberg, Germany. If you have any concerns regarding our products, please contact ProductSafety@springernature.com

Printed and bound by CPI Group (UK) Ltd, Croydon, CR0 4YY
30/09/2026
02225294-0017